Sustainable Development and Biodiversity

Volume 44

Series Editor

Kishan Gopal Ramawat, Botany Department
Mohanlal Sukhadia University
Udaipur, India

Sustainable Development Goals are best achieved by mechanisms such as research, innovation, and knowledge sharing. This book series aims to help researchers by reporting recent progress and providing complete, comprehensive, and broad subject-based reviews about all aspects of sustainable development and ecological biodiversity.

The series explores linkages of biodiversity with delivery of various ecosystem services and it offers a discourse in understanding the biotic and abiotic interactions, ecosystem dynamics, biological invasion, ecological restoration and remediation, diversity of habitats and conservation strategies. It is a broad scoped collection of volumes, addressing relationship between ecosystem processes and biodiversity. It aims to support the global efforts towards achieving sustainable development goals by enriching the scientific literature. The books in the series brings out the latest reading material for botanists, environmentalists, marine biologists, conservationists, policy makers and NGOs working for environment protection.

We welcome volumes on the themes --

Agroecosystems, Agroforestry, Biodiversity, Biodiversity conservation, Conservation of ecosystem, Ecosystem, Endangered species, Forest conservation, Genetic diversity, Global climate change, Hotspots, Impact assessment, Invasive species, Livelihood of people, Plant biotechnology, Plant resource utilization, Sustainability of the environment, Sustainable management of forests, Sustainable use of terrestrial ecosystems and plants, Traditional methods, Urban horticulture.

Sumita Jha • Mihir Halder • Liliana Essi
Editors

Genetic Diversity in Medicinal Plants

Challenges, Threats, and Future Applications

Springer

Editors
Sumita Jha
Botany (Retired)
University of Calcutta
Kolkata, West Bengal, India

Mihir Halder
Botany
Barasat Government College
Kolkata, West Bengal, India

Liliana Essi
Biology
Universidade Federal de Santa Maria
Santa Maria, Rio Grande do Sul, Brazil

ISSN 2352-474X ISSN 2352-4758 (electronic)
Sustainable Development and Biodiversity
ISBN 978-981-95-7280-9 ISBN 978-981-95-7281-6 (eBook)
https://doi.org/10.1007/978-981-95-7281-6

This Springer imprint is published by the registered company Springer Nature Singapore Pte Ltd.
The registered company address is: 152 Beach Road, #21-01/04 Gateway East, Singapore 189721, Singapore

Preface

The genetic diversity (diverse gene pool) found in medicinal plants is crucial for their ability to adapt to changing environmental conditions and continue producing important phytochemicals. There is a global loss of genetic diversity due to human interference, ecological issues, and plant-specific challenges. Understanding these issues and conducting comprehensive assessments of genetic diversity, population structure, and phytochemical variations are necessary for effective conservation and management strategies to ensure the long-term survival and sustainable utilization of medicinal plants. It is significantly important for drug discovery, the quality control of herbal medicine, and the improvement of medicinal plants through classical and molecular plant breeding. The future of medicinal plant genetic diversity relies on the responsible and sustainable use of genetic diversity, with a focus on sustainable sourcing, effective conservation of wild populations through international cooperation, equitable benefit-sharing with indigenous communities, and technological advancements. The field has a bright future, as only a limited number of medicinal plants have undergone genetic evaluation. This book is an invaluable resource, offering comprehensive insights into all aspects of medicinal plant genetic diversity, including an overview of its current status, importance, associated threats, and conservation strategies, both globally and for specific medicinal plants from several countries and ecosystems, across its 17 chapters distributed into three parts. Chapters are supplemented with illustrations and figures to facilitate the reader's understanding.

The Part I comprises seven chapters covering broader aspects of the current state of genetic diversity in medicinal plants globally, the genetic mechanisms underlying phytochemical diversification, tracking genetic erosion and phytochemical shifts in medicinal plants, and recent applications and emerging trends in genetic diversity assessment techniques, including molecular techniques.

The Part II comprises eight chapters focused on various threats to medicinal plants. It explores possible conservation strategies to minimize genetic erosion and protect commercially valuable, endemic, threatened, or endangered medicinal plant species, in general and in specific biodiversity-rich regions of several countries and ecosystems. This part also examines the impact of cultivation, domestication, and climate change on the genetic and phytochemical diversity of medicinal plants.

The Part III presents two chapters that emphasize the application of medicinal plant genetic diversity across various fields, including conservation, drug discovery, and climate adaptation. Comprehensive overview of significant achievements in the production of secondary metabolites through traditional breeding and transgenic techniques, highlighting the importance of understanding and characterizing genetic diversity for targeted genome manipulation using conventional breeding, biotechnology, and molecular biology.

The primary audience comprises researchers, scientists, botanists, students, and academics interested in the genetic diversity of medicinal plants and its implications for pharmaceuticals, drug discovery, conservation, plant breeding, transgenic plant development, and sustainable use. Additionally, information on genetic diversity, assessment methods, and major threats to medicinal plants may benefit policymakers, healthcare professionals, and environmentalists in developing conservation strategies and promoting the sustainable use of medicinal plant resources.

Finally, we gratefully acknowledge all our contributors, who have made immense efforts to ensure the scientific quality of this book. We also thank all our colleagues at Springer for their excellent support.

We dedicate this book to our Series Editor, the Late Professor Dr. K. G. Ramawat, whose valuable guidance until his last days will always be cherished by the editors.

Kolkata, West Bengal, India — Sumita Jha
Kolkata, West Bengal, India — Mihir Halder
Santa Maria, Rio Grande do Sul, Brazil — Liliana Essi

Contents

Part II Threats and Conservation of Medicinal Plant Genetic Diversity

Editors and Contributors

About the Editors

Sumita Jha, FNASc., FWAST, served as a distinguished faculty member in the Botany Department at the University of Calcutta, Kolkata, India, where she taught from 1990 to 2017. She continued her research as an NASI Senior Scientist until 2022. She earned her M.Sc. in 1975 and Ph.D. in 1981 from the University of Calcutta. Throughout her career, she has made significant contributions to plant biology, molecular genetics, and biotechnology. Her research focuses on developing transgenic cultures in endangered medicinal plants to produce high-value pharmaceuticals. With over 160 research papers and 25 doctoral theses supervised, she has earned numerous accolades, including the "Science Academy Medal for Young Scientist" in 1983 and election as a fellow of the National Academy of Sciences, India, in 2008. She has also been a visiting scientist at renowned institutions in the UK and France.

Mihir Halder serves as an Assistant Professor in the postgraduate Department of Botany at Barasat Government College, West Bengal, India. He completed his B.Sc. with honors in 2007 and his M.Sc. in Botany in 2009, both from the University of Calcutta, Kolkata, India. He later earned his Ph.D. in Biochemistry in 2017 from the University of Calcutta, Kolkata, India. With over a decade of teaching experience, he specializes in cytogenetics, cell and molecular biology, and plant biotechnology. His research contributions to plant cytogenetics and biotechnology are well recognized, with more than 25 research papers and book chapters published, and he has supervised more than 20 M.Sc. projects. He showcases his expertise and dedication to advancing botanical sciences.

Liliana Essi, biologist, a researcher and Full Professor in Botany, earned her M.Sc. (2003) and Ph.D. (2007) in Botany from Universidade Federal do Rio Grande do Sul, Brazil. She completed postdoctoral research at Université Paris-Sud 11 in France (2008). She has been teaching plant systematics and genetics at Universidade Federal de Santa Maria since 2009. Her research contributions include several published papers and master's theses supervised, and is member of the Botanical Society of Brazil (SBB). Her work in plant genetics and systematics has earned her a reputation in the scientific community.

Contributors

A. M. K. L. Abeykoon Department of Plant Sciences, Faculty of Science, University of Colombo, Colombo, Sri Lanka

Samarathunga M. P. G. P. B. Abeyrathna Department of Botany, Faculty of Science, University of Jaffna, Jaffna, Sri Lanka

Pushpa Damayanthi Abeysinghe Department of Botany, University of Ruhuna, Matara, Sri Lanka

Anisa Anggraeni Department of Economic Plants and Biotechnology, Yunnan Key Laboratory for Wild Plant Resources, Kunming Institute of Botany, Chinese Academy of Sciences, Kunming, China

University of Chinese Academy of Science, Huairou, Beijing, China

Beauty Rani Basumatary Department of Botany, Gauhati University, Guwahati, Assam, India

Gülnur Ekşi Bona Department of Pharmaceutical Botany, Faculty of Pharmacy, İstanbul University-Cerrahpaşa, Istanbul, Turkey

Mehmet Bona Department of Botany, Faculty of Science, İstanbul University, Istanbul, Turkey

Kunal Boro Department of Botany, Gauhati University, Guwahati, Assam, India

Uddipta Borthakur Department of Botany, Gauhati University, Guwahati, Assam, India

Henrique da Rosa Ferreira Post Graduate Program in Agrobiology – Universidade Federal de Santa Maria, Santa Maria, Brazil

Milena de Melo Hettwer Post Graduate Program in Agrobiology – Universidade Federal de Santa Maria, Santa Maria, Brazil

Ali Devlet Department of Agronomy, Faculty of Agriculture and Natural Science, Bilecik Şeyh Edebali University, Bilecik, Turkey

Maurício Sedrez dos Reis Núcleo de Pesquisas em Florestas Tropicais, Universidade Federal de Santa Catarina (UFSC), Florianópolis, Santa Catarina, Brazil

Núcleo de Estudos em SocioAgroBiodiversidade, Pós-Graduação em Ecossistemas Agrícolas e Naturais, UFSC, Curitibanos, Santa Catarina, Brazil

Liliana Essi Department of Biology, Universidade Federal de Santa Maria, Santa Maria, Brazil

Nabela Fikkirya Aquatic Resource Management, Faculty of Fisheries and Marine Science, Universitas Jenderal Soedirman, Purwokerto, Indonesia

Sanghamitra Gogoi Department of Botany, Gauhati University, Guwahati, Assam, India

Mihir Halder Post-graduate Department of Botany, Barasat Government College, Kolkata, West Bengal, India

S. K. M. K. Herapathdeniya Department of Ayurveda Pharmacology, Pharmaceutics and Community Medicine, Faculty of Indigenous Medicine, University of Colombo, Colombo, Sri Lanka

Mesut Işık Department of Bioengineering, Faculty of Engineering, Bilecik Şeyh Edebali University, Bilecik, Turkey

H. D. Jayasinghe National Institute of Fundamental Studies, Kandy, Sri Lanka

Sumita Jha Department of Botany, University of Calcutta, Kolkata, West Bengal, India

Bikash Kalita Department of Botany, Assam Don Bosco University, Tapesia Garden, Kamarkuchi, Sonapur, Assam, India

H. S. Kathriarachchi Department of Plant Sciences, Faculty of Science, University of Colombo, Colombo, Sri Lanka

O. W. Ketakela Department of Plant Sciences, Faculty of Science, University of Colombo, Colombo, Sri Lanka

Bikash Kumar Kundu Department of Botany, Gauhati University, Guwahati, Assam, India

Nguyen Thi Le Thuy Department of Biotechnology, Ho Chi Minh City Open University, Ho Chi Minh City, Vietnam

Anrini Majumder Department of Botany, Harimohan Ghose College, Kolkata, West Bengal, India

Adelar Mantovani Núcleo de Pesquisas em Florestas Tropicais, Universidade Federal de Santa Catarina (UFSC), Florianópolis, Santa Catarina, Brazil

Laboratório de Engenharia Florestal, Pós-Graduação em Engenharia Florestal, Pós-Graduação em Produção Vegetal, Universidade do Estado de Santa Catarina, Lages, Santa Catarina, Brazil

Alexandre Mariot Núcleo de Pesquisas em Florestas Tropicais, Universidade Federal de Santa Catarina (UFSC), Florianópolis, Santa Catarina, Brazil

Drimys Agroambiental LTDA, Florianópolis, Santa Catarina, Brazil

D. M. R. G. Mayakaduwa Department of Plant Sciences, Faculty of Science, University of Colombo, Colombo, Sri Lanka

Yago Luís Nóbrega Mendes Post Graduate Program in Agrobiology – Universidade Federal de Santa Maria, Santa Maria, Brazil

Tiago Montagna Núcleo de Pesquisas em Florestas Tropicais, Universidade Federal de Santa Catarina (UFSC), Florianópolis, Santa Catarina, Brazil

Pós-Graduação em Recursos Genéticos Vegetais, UFSC, Florianópolis, Santa Catarina, Brazil

Rubar Hussein Muhammed-Salih Department of Biology, College of Education, University of Sulaimani, Sulaymaniyah, Kurdistan Region, Iraq

Nguyen Thi Hong Nhung Department of Biology, University of Medicine and Pharmacy at Ho Chi Minh City, Ho Chi Minh City, Vietnam

Sudip Pandey Wood Science Laboratory, Forest Biomaterials Science and Engineering Program, Institute of Applied Sciences, Madan Bhandari University of Science and Technology, Chitlang, Bagmati Province, Nepal

Poonam Pant Central Institute of Science and Technology, Kathmandu, Nepal

Sulagna Gautom Phukan Department of Botany, Gauhati University, Guwahati, Assam, India

R. J. Piumal Department of Plant Sciences, Faculty of Science, University of Colombo, Colombo, Sri Lanka

Endang C. Purba Institute of Ecology and Evolutionary Biology, National Taiwan University, Taipei, Taiwan

Henti H. Rachmat Research Center for Ecology, National Research and Innovation Agency (BRIN), Cibinong, Bogor, Indonesia

Gowri Rajkumar Department of Botany, Faculty of Science, University of Jaffna, Jaffna, Sri Lanka

Dipasree Roychowdhury Department of Botany, Surendranath College, Kolkata, West Bengal, India

Nibedita Sarma Department of Botany, Gauhati University, Guwahati, Assam, India

Núcleo de Pesquisas em Florestas Tropicais, Universidade Federal de Santa Catarina (UFSC), Florianópolis, Santa Catarina, Brazil

Núcleo de Estudos em SocioAgroBiodiversidade, Pós-Graduação em Ecossistemas Agrícolas e Naturais, UFSC, Curitibanos, Santa Catarina, Brazil

Hulia Juana Scherer Post Graduate Program in Agrobiology – Universidade Federal de Santa Maria, Santa Maria, Brazil

Marina Scheuer Post GraduateProgram in Agrobiology – Universidade Federal de Santa Maria, Santa Maria, Brazil

Táfinis Sturbelle Schiavon Graduate in Biological Sciences–Universidade Federal de Santa Maria, Santa Maria, Brazil

Nandini Sikdar Department of Botany, Gauhati University, Guwahati, Assam, India

Marina Silalahi Biology Education Department, Universitas Kristen Indonesia, Jakarta, Indonesia

Alexandre Siminski Núcleo de Pesquisas em Florestas Tropicais, Universidade Federal de Santa Catarina (UFSC), Florianópolis, Santa Catarina, Brazil

Centro Reforma de Restauração Ecológica, UFSC, Curitibanos, Santa Catarina, Brazil

Núcleo de Estudos em SocioAgroBiodiversidade, Pós-Graduação em Ecossistemas Agrícolas e Naturais, UFSC, Curitibanos, Santa Catarina, Brazil

Bhaben Tanti Department of Botany, Gauhati University, Guwahati, Assam, India

Nguyen Thanh Triet Department of Traditional Pharmacy, University of Medicine and Pharmacy at Ho Chi Minh City, Ho Chi Minh City, Vietnam

Didi Usmadi Research Center for Biota Systems, National Research and Innovation Agency (BRIN), Cibinong, Bogor, Indonesia

Kusumadewi S. Yulita Research Center for Ecology, National Research and Innovation Agency (BRIN), Cibinong, Bogor, Indonesia

Part I
Medicinal Plant Genetic Diversity and Assessment Tools

Chapter 1
General Perspective of Genetic Diversity, Threats, and Conservation of Medicinal Plants

Mihir Halder, Sumita Jha, and Liliana Essi

Abstract Genetic, species, and ecological diversity are the three distinct but hierarchically linked aspects of biodiversity. Genetic diversity represents the total intra- and inter-species variations in genetic material (DNA). It is the basis of species and ecosystem diversity, and they have bidirectional relationships. A comprehensive understanding and maintenance of the structure and distribution of genetic diversity within and among medicinal plants, as well as interconnected relationships, is essential for the continued access and long-term sustainable use of their phytochemical diversity (qualitative and quantitative) for existing and emerging remedies in traditional medicine and modern pharmacology. Continuous availability of this genotypic plasticity on hand ensures their ecological resilience, adaptability, and the development of effective conservation strategies and breeding programs. However, the existing genetic variation in medicinal plants is severely threatened by the combined impacts of anthropogenic and environmental factors. The erosion of genetic diversity compromises both the long-term survival of medicinal plant populations and their potential to provide future medicinal solutions. By differentiating genetic diversity from species and ecological diversity, this chapter offers the concept of genetic diversity, its significance, sources, and assessment methods. It also discusses multifaceted threats they face and presents current and emerging conservation strategies in the context of genetic diversity protection. This information is intended for researchers, students, conservation practitioners, and policymakers working in the fields of genetic diversity, ethnobotany, phytochemistry, and conser-

M. Halder (✉)
Post-graduate Department of Botany, Barasat Government College, Kolkata, West Bengal, India
e-mail: mihir.halder@bgc.ac.in

S. Jha
Department of Botany, University of Calcutta, Kolkata, West Bengal, India

L. Essi
Department of Biology, Universidade Federal de Santa Maria, Santa Maria, Brazil
e-mail: liliana.essi@ufsm.br

S. Jha et al. (eds.), *Genetic Diversity in Medicinal Plants*, Sustainable Development and Biodiversity 44, https://doi.org/10.1007/978-981-95-7281-6_1

vation biology. The chapte r emphasizes the urgent need for integrative comprehensive approaches that combine molecular genetics research with socio-economic sustainability, traditional ecological knowledge, sustainable harvesting, and community engagement to protect these precious biological assets for current and future generations.

Keywords Genetic variability · Phytochemical diversity · Climate change · Genetic diversity assessment · Conservation

1.1 Introduction

Genetic, species, and ecological diversity are the three distinct but hierarchically interconnected aspects of biodiversity. Genetic diversity of medicinal plants refers to the total number of genetic characteristics or variation in genetic material (DNA) within the unique gene pool of a species or its populations. It acts as the foundational platform for the other two hierarchical structures of biodiversity, that is, species diversity and ecosystem diversity. Genetic diversity contributes to phytochemical diversity and the adaptability of species (Terletskaya et al. 2023; Swamy and Kumar 2023). Species diversity significantly extends the horizon of therapeutic options, while ecosystem diversity encompasses habitats and processes that sustain all life and drive the evolutionary process. These three forms are interconnected and exhibit a bidirectional relationship with each other. For example, the loss of genetic diversity can have a significant adverse effect on both species and ecosystem diversity. Furthermore, the degradation of ecological diversity directly reduces habitat availability, which, in turn, may result in the extinction of medicinal plant species (loss of species diversity) and subsequently declines genetic diversity within the remaining populations.

Medicinal plants are a specific group of plants utilized as raw materials for therapeutic purposes or serving as the primary source of valuable therapeutic phytochemicals for both traditional healthcare systems, such as Ayurveda, traditional Chinese medicine, and Sowa-Rigpa, and multi-billion-dollar modern pharmaceutical industries (Qadir and Raja 2021; Najmi et al. 2022). A deep-rooted prehistoric relationship between medicinal plants and humanity can be traced back to a time long before the development of modern medical systems for treating various infections, chronic diseases, and health-related issues (Ijinu et al. 2022). Despite remarkable progress in new synthetic drug development, this relationship remains relevant in the era of modern medicine because the contribution of medicinal plants to promote human health remains equally important. A significant proportion of currently available drugs are derived from or inspired by natural products, predominantly from plant sources (Newman and Cragg 2020). A many-fold increase in the demand for herbal medicine has been observed in recent times, as more people have become aware of the potential negative side effects associated with synthetic or semi-synthetic medicines. According to Davis and Choisy (2024), medicinal plants and humans are partners engaged in a complex and poorly explored set of socio-ecological interactions that extend beyond domestication or cultivation but also contain commensalisms and mutualisms.

The remarkable ability of medicinal plants to produce a wide range of complex bioactive secondary metabolites (also referred to as specialized metabolites) with diverse therapeutic properties and efficacy is encoded in their genomic blueprint (Halder and Jha 2023a). The existing genetic variability in medicinal plant species is a result of their long evolutionary past, which is shaped by numerous adaptation and speciation events and influenced by environmental pressures, reproductive isolation, and genetic drift. Any variation in genetic characteristics within the genome can be regarded as genetic diversity. This variation in the genome can be evaluated at the species, population, or individual level. Genotypic and allelic variations are in the streamline because they are directly involved in the variations of the concentration and composition of secondary metabolites. Modern research has revealed a significant role for non-coding genetic materials, such as transcription factors and non-coding RNAs, in the phytochemical diversity as well as the growth and development of medicinal plants (Deng et al. 2024; Rabeh et al. 2025). Moreover, the interactions between the complex genetic networks and the specific environmental conditions in which they grow regulate and govern their phytochemical profiles (Raskar et al. 2022; Zhan et al. 2022). These profiles include the quality, quantity, and pharmacological potency of therapeutically relevant phytochemicals. The unique genetic makeup of chemotypes is responsible for producing different phytochemicals in varying quantities or entirely new phytochemicals (Xavier et al. 2022).

Despite their immense ecological, economic, cultural, spiritual, and pharmaceutical significance (Zahra et al. 2019), several medicinal plants are facing significant threats locally or globally during the ongoing global biodiversity crisis (Mir et al. 2021; Halder and Jha 2023b). Several intrinsic and extrinsic factors are threatening these invaluable resources at all three levels of biodiversity. Among the several potential threats, anthropogenic activities, such as destruction of natural habitats, rapid climate changes, overexploitation for commercial and medicinal purposes, unsustainable wild harvesting, cultivation, domestication, and monoculture farming, and the introduction of invasive species, are the major drivers of genetic diversity erosion in medicinal plants (Halder and Jha 2023b).

The maintenance and management of a high degree of genetic variability within and among populations is essential for their chemotypic stability, ecological resilience, adaptability, and long-term sustainability in response to changing climatic conditions and anthropogenic challenges (Srivastava et al. 2017; Monteiro et al. 2022). This genotypic plasticity ensures their survival and the continuous availability of existing remedies associated with these plants, which are prerequisites for targeted bioprospecting and the identification of potential drug leads. In contrast, genetic erosion from the ecosystem quickly reduces this vital genomic resource, compromising the long-term viability and bioactivity of critical medicinal species that directly impact their therapeutic efficacy. It creates a significant risk to both ecosystems and human health. A comprehensive understanding of the structure of genetic diversity, its distribution, and the mechanisms of its maintenance is crucial for developing effective conservation strategies, breeding programs, and ensuring the sustained resilience of the global pharmaceutical supply chain. Therefore, effective conservation and sustainable utilization demand an integrated approach that recognizes and addresses diversity at its genetic, species, and ecological scales. Each form of biodiversity has its own role and unique implications for research, conservation, and sustainable management of medicinal plant resources.

Interestingly, a majority of books and reviews available on the topic of medicinal plant biodiversity often only address the status, threats, and conservation prospects of species diversity. There are a very limited number of available publications that have specifically explored various aspects of genetic diversity in medicinal plants. This chapter concerns the general concept of genetic diversity, its different sources, and its significance. It will cover different methods of assessing genetic diversity, identify the leading threats causing genetic diversity loss, and effective conservation strategies being employed and proposed. Moreover, this chapter aims to provide a broad perspective on genetic diversity and highlight the need for urgent conservation efforts to preserve it. By differentiating genetic diversity from species and ecological diversity, this chapter aims to provide a foundational and accessible platform for researchers, conservation practitioners, and policymakers working in the fields of ethnobotany, phytochemistry, ecology, and conservation biology.

1.2 Sources of Genetic Diversity in Medicinal Plants

The genetic diversity of medicinal plant species or populations is highly dynamic. Various intrinsic and extrinsic forces modify the genetic material of medicinal plant species, contributing to their continuous evolution. Natural biological processes, such as spontaneous mutations and recombination, along with different evolutionary forces, such as gene flow, genetic drift, natural polyploidization, and natural hybridization, form the foundation for the development of natural genetic variation (Salgotra and Chauhan 2023). Additionally, various environmental factors (climate variables, soil conditions, and biotic interactions) and anthropogenic activities (habitat destruction, overexploitation, deliberate or accidental introduction of invasive species, and cultivation and domestication) also continuously influence and shape the genetic variability found within these plants and among their populations. Each of these factors acts on wild medicinal plant populations, either independently or in combination. Their impact and effectiveness are variable and are specific to each medicinal plant species.

Spontaneous mutation (random, spontaneous changes in genetic material at coding or non-coding regions of the genome) and recombination (the exchange of genetic segments between homologous chromosomes during meiotic cell division) are the two fundamental biological processes that serve as the primary sources of natural genetic variability. A single nucleotide change can evolve through spontaneous mutation and is the primary source of allelic variation. It can modify the biosynthesis or function of key enzymes involved in the secondary metabolism pathway, resulting in the qualitative or quantitative production of bioactive compounds. In contrast, genetic recombination reshuffles existing genetic variation to create new combinations of alleles during meiosis. Genetic recombination in combination with the independent assortment of chromosomes plays a significant role in the development of genetic diversity among sexually reproducing medicinal plants. These processes ensure that each offspring receives and inherits a unique genetic blueprint from its parents. Thus, these two molecular processes are key contributors to the raw material for evolutionary processes.

In addition to basic molecular processes, genetic diversity is also influenced by various evolutionary forces. These forces include gene flow (the transfer of genetic material between populations of the same species), genetic drift (random changes in allele frequencies from one generation to the next), and natural selection (the process by which individuals with advantageous traits are more likely to survive and reproduce than others). Populations of rare and endemic medicinal species frequently exist in fragmented, small, geographically isolated patches. In such populations, gene flow helps minimize the adverse effects of inbreeding depression (reduced fitness due to mating between close relatives) by introducing new alleles that may have emerged through mutations in other populations. The vector-mediated dispersal of pollen or seeds that carrying new alleles may enrich gene pools and alter allelic frequencies in recipient populations. In medicinal plant species with small populations, genetic drift can negatively impact genetic diversity by random changes in allele frequencies from one generation to the next. This process can maximize or minimize specific alleles, regardless of their adaptive significance. While genetic drift may sometimes be able to fix beneficial mutations, it can also cause the loss of rare alleles that often contribute to medicinal value or to population resilience. Overharvesting and habitat fragmentation frequently reduce population sizes, thereby increasing the susceptibility of medicinal plants to genetic drift.

Natural selection is another evolutionary force that can change genetic diversity within and among different medicinal plant populations. This process helps individuals with genes for specific adaptive traits related to defense mechanisms, stress tolerance, and attraction of pollinators that enhance their chances of survival and reproduction in their respective environments. Over time, these selective pressures can lead to the creation of unique populations that differ significantly in terms of their genetic diversity and adaptability. For example, medicinal plants with genes that enable the production of secondary metabolites, which serve as defense mechanisms against herbivores and pathogens, or attract pollinators, are more likely to survive and pass on their genes. Similarly, individuals with genes for stress tolerance or superior adaptations to environmental changes are more likely to be selected by natural selection. Both natural and induced polyploidization, which involves the duplication of entire sets of chromosomes, play a significant role in creating genetic diversity in numerous medicinal plants (Madani et al. 2021). This genetic alteration at the chromosomal level can lead to various morphological, anatomical, and biochemical changes, including alterations in metabolic profiles (Qi-Qing et al. 2018; Bhusare et al. 2021; Hamarashid et al. 2022). One notable effect is the production of new secondary metabolites or changes in the existing ones.

Environmental variables, including climate variables (such as temperature, rainfall, and sunlight intensity), soil conditions (such as soil type, pH, nutrient availability, and contamination by heavy metals or toxins), and biotic interactions, exert considerable directional selective pressures on many medicinal plants. These pressures ultimately affect genetic diversity, leading to a rapid shift in genetic frequencies within populations. The interactions of these environmental factors can promote local adaptations, where distinct populations of the same medicinal plant species evolve their unique genetic signatures and potentially different phytochemical profiles for functional specialization in their specific ecological role. This explains why the same species of plant can exhibit differences in chemical compounds and medicinal value, depending on its geographic origin.

Furthermore, several anthropogenic interventions, including habitat destruction, overexploitation, the deliberate or accidental introduction of invasive species, and the cultivation and domestication of medicinal plants, have significantly impacted genetic diversity. These intentional actions have altered the genetic diversity of landraces and cultivars with unique medicinal properties. With biotechnological advancements, humans have now achieved several technologies that can create genetic variations at specific or non-specific sites in a very short timescale. These techniques include induced polyploidization (Qi-Qing et al. 2018; Bhusare et al. 2021), intentional hybridization (Banadka et al. 2025), genetic modification (Wang et al. 2017; Khoshsokhan et al. 2022), gene editing (Shelake et al. 2023; Das et al. 2024), and induced mutation (Sawadogo et al. 2012; Magdy et al. 2020). These techniques have been applied to alter the genetic and phytochemical composition of various medicinal plants.

Hybridization in medicinal plants can occur naturally or through human intervention, either intentionally or unintentionally, between individuals of genetically distinct populations, species, or varieties. It may lead to hybrids containing new genetic combinations that produce new traits and phytochemical characteristics (chemotype). Hybridization can produce beneficial genetic combinations that may have considerable consequences on genetic diversity as it introduces new genes or alleles into the population. Hybridization, especially between different species, can cause genetic instability, leading to rearrangements, deletions, or duplications, and the loss of undesirable traits if not managed carefully. In some cases, it can also lead to the formation of allopolyploids, where chromosome sets from both parent species are retained and then duplicated. These events can drastically alter gene expression and the metabolic landscape of the plant.

1.3 Significance of Genetic Diversity

Genetic diversity within medicinal plant populations is of massive significance for their ecological resilience, adaptability to environmental changes, chemotypic stability, discovery and development of new pharmaceuticals, prioritization in biodiversity conservation, and medicinal plant improvement through breeding or genetic manipulation (Srivastava et al. 2017; Monteiro et al. 2022). Such DNA-level variation provides species with the adaptive capacity to withstand diverse environmental stressors, to ensure the long-term sustainability and availability of these invaluable natural resources. Furthermore, different genotypes can vary in their production of secondary metabolites, resulting in differences in production, availability, quality, and potential for new bioactive compounds. These bioactive compounds may have greater efficacy or fewer side effects, and can be used as novel candidates for new drug discovery. Moreover, maintaining genetic diversity supports the planning of breeding programs, which require genetic diversity for improving the quality of medicinal plants. Conservation and complete characterization of the genetic diversity of medicinal plants is also necessary for access to future medicine, new chemotypes of natural products, and safeguarding human health.

1.3.1 Adaptation and Resilience to Environmental Stress

The adaptive ability of a species to both existing and emerging biotic and abiotic stresses is contributed to by the genetic composition. The ability of a population of medicinal plants to deal with different environmental challenges is similarly regulated by their overall gene pool or genetic diversity, where different genotypes and allele combinations act as a genetic arsenal. More specifically, populations with rich genetic diversity are expected to have individuals harboring genotypes with beneficial traits for survival and reproduction, under the new selective pressures they may experience. It has been demonstrated that genetic diversity allows survival of this species under extreme selective pressures or climatic change. In contrast, less genetically diverse populations, or those with a monomorphic genetic structure, are highly vulnerable and may not be able to survive. Thus, genetic diversity supports the long-term survival of medicinal plant species in their natural habitats by ensuring their adaptive capacity and resilience to rapid climate change.

1.3.2 Relationship Among Genetic Diversity, Phytochemical Diversity, and Medicinal Value

The most direct and essential impact of genetic diversity in medicinal plants is its effect on biochemical and phytochemical diversity. The genetic diversity is directly linked with their metabolic plasticity, that is, their ability to produce a wide range of phytochemicals. Among these phytochemicals, secondary metabolites determine their medicinal values. These secondary metabolites offer a broader spectrum of pharmacological activities with different mechanisms of action that ensure ongoing human health benefits through traditional medicine systems and modern pharmacology (Mikail et al. 2022). Differences in genotype within a species, or ecotypic variants, can result in differences in the types, amounts, and ratios of active compounds, producing unique "chemotypes." The concept of chemotype is widespread in medicinal plants, and a single species may have multiple chemotypes. It refers to a chemically distinct population or individuals of a single species of medicinal plant that possess a different composition of secondary metabolites (Amin and Park 2025). This phytochemical variation (quantity and quality of secondary metabolites) in chemotypes is usually due to allelic variation in genes that encode for enzymes involved in metabolic pathways. Several studies have demonstrated that intra-specific genetic divergence is directly linked to chemodiversity in medicinal plants, which, in turn, influences the therapeutic efficacy and applications of plant-derived medicines (Patonay et al. 2021; Xavier et al. 2022; Tiwari et al. 2024). A rich genetic pool, therefore, provides a broader chemical space for drug screening and the identification of novel pharmacological agents.

Several research studies have demonstrated that plants sourced from one geographical region or population may have different therapeutic properties compared to those from a genetically distinct population of the same species from another location. The understanding and characterization of the complex relationship between genetic variation and phytochemical diversity is vital for standardizing herbal medicines to ensure consistent product quality and optimizing drug discovery efforts. It suggests that species-level identification of medicinal plants is not always effective; the chemotypic distinctions are essential for effective medicinal application.

1.3.3 Conservation and Cultivation Prioritization

Wild populations of high-demand medicinal plants are constantly facing intense collection pressures, habitat destruction, and climate change, which lead to loss of genetic diversity. This loss of genetic diversity, in turn, can affect the adaptive potential of medicinal plants, increase inbreeding depression, and ultimately lead to the extinction of species. The success and effectiveness of long-term conservation efforts are significantly linked to the genetic diversity of medicinal plant species, which determines the evolutionary potential of a species and enables its continuous adaptation in response to environmental challenges. The information on the genetic diversity of *Pilocarpus microphyllus* in the eastern Amazon (Monteiro et al. 2022) and Indian *Curcuma* species (Sahoo et al. 2021) is useful in the cultivation and conservation prioritization process. It helps them adopt more effective conservation strategies. Thus, incorporating these genotypic approaches into global conservation strategies is crucial to ensure the utility and sustain the ecological integrity of medicinal plants in the future.

1.3.4 Plant Breeding and Biotechnological Improvement Programs

Genetic diversity has been utilized in various classical and marker-assisted plant breeding programs for medicinal plants to develop improved varieties with desired traits, such as higher yields of bioactive compounds, disease resistance, faster growth rates, or adaptability to cultivation conditions (Dong et al. 2017; Wu et al. 2020). In breeding programs, the existing gene pool serves as the raw material from which breeders select individuals and then cross them to develop superior varieties. The success of plant breeding programs greatly depends on the successful identification and selection of suitable genotypes as parents, which, in turn, is entirely dependent upon the genetic heterogeneity available within the species. The rich genetic pool offers breeders more options to achieve their objectives in the experiment, whereas insufficient genetic diversity significantly reduces the potential for improvement.

Moreover, in the current period of genomics, specific genes or transcription factors can be used as transgenes in transgenic and metabolic engineering-mediated enhancement of secondary metabolites or to achieve any desired goals in the same or other medicinal species or any organism (Tian et al. 2024).

1.4 Significance of Assessing Genetic Diversity

The systematic assessment of genetic diversity found in medicinal plant species has a unique and vital perspective. The evaluation of this aspect of biological variation is significant because it is directly connected to their phytochemical diversity or variation (Chemotypes), which strongly influences the potency and consistency of their therapeutic compounds and ultimately affects the global healthcare system (Azizi et al. 2012; Alami et al. 2022). Additionally, genetic richness within a species or populations of medicinal plants provides the evolutionary tools. These varieties determine their internal adaptability and resilience against upcoming climatic change and the threat of newly emerging diseases, ultimately helping in their ongoing evolution. A precise and comprehensive understanding and mapping of the entire genome using next-generation sequencing (NGS) holds great potential to ensure highly targeted and effective long-term conservation management and sustainable use (Segelbacher et al. 2022; Salaria et al. 2025). This approach has the potential to extend beyond species conservation to the conservation of specific genetic lineages that are known to produce unique or potent phytochemical profiles. Several genome sequencing projects on medicinal plants, such as *Scutellaria baicalensis* (Pei et al. 2023) and *Ocimum tenuiflorum* (Upadhyay et al. 2015), have recently been completed, with a few more underway. The review of Sharma et al. (2022) provided ideas for applying NGS technologies to investigate the genomes of various endangered medicinal plants, offering valuable insights for the conservation of these vulnerable species. The authentication and quality control of commercialized products derived from medicinal plants is done by utilizing genetic markers (Kshirsagar et al. 2017; Dhabal et al. 2025; Kumar et al. 2025). Genetic markers provide genetic "fingerprints" to verify the species and origin of plant material. It ensures consumer safety and promotes sustainable use and conservation of endangered or endemic medicinal plants. Assessing genetic diversity in medicinal plants also enhances the success rate in medicinal plant breeding programs and makes their cultivation and domestication process more effective.

1.5 Methods of Assessing Genetic Diversity

The accurate evaluation of genetic diversity in medicinal plants can be done through several methods. Over the past few decades, assessment techniques have undergone rapid evolution, transitioning from simple traditional assessments to DNA-based molecular analysis (Sharma et al. 2023). Molecular analysis technologies that directly analyze the DNA of the genome have also evolved rapidly from random amplified polymorphic DNA (RAPD) and simple sequence repeats (SSRs) to sophisticated high-throughput NGS (Wang et al. 2024). Each method provides unique information about the genetic structure and evolutionary processes. This section provides an overview of the primary assessment methods, including their principles, advantages, limitations, and selected applications in medicinal plant research.

Traditional genetic diversity assessment methods primarily depend on visible morphological characteristics (phenotypic assessment), biochemical traits, and chemical composition for preliminary screening or verification. Morphological characterization is the most basic, simple, cost-effective, and indirect approach that focuses on the observation and measurement of specific qualitative and quantitative characteristics (growth and productivity) using low-tech equipment. A limited resolution power and phenotypic plasticity reduce its utility and reliability in differentiating between actual genetic variation and environmental variation in closely related populations.

Biochemical approaches are also relatively inexpensive, indirect methods that provide a more direct connection between genetics and variability in the biochemical output of medicinal plants. This output can be measured in either the form of an isozyme and phytochemical profile or a chemical fingerprint. In this method, genetic differences are predicted by evaluating visible differences in the banding patterns of isoenzymes (enzymes with the same charge and size) separated using gel electrophoresis. Chemical profiling is a highly relevant indirect approach to determining genetic variation, based on the concept that genetic variation has an immediate impact on metabolic pathways. This approach focuses on the qualitative and quantitative analysis of secondary metabolites. Low-resolution power and the influence of environmental factors, seasonal conditions, and developmental stage on gene expression variability are significant limitations of these biochemical markers.

These limitations associated with phenotypic and chemical analyses have led to a shift toward direct assessment of the plant's genome using molecular markers.

These molecular techniques allow scientists to read the DNA sequence directly, providing the most direct and unbiased measure of structural variation at the DNA level. Such techniques offer different levels of resolution and reproducibility and are largely independent of environmental factors, and developmental stage, which can detect subtle differences between closely related individuals or populations.

Many of the initial molecular techniques relied heavily on polymerase chain reaction (PCR) technologies to amplify random or semi-random genomic regions, followed by separation of DNA fragments by size on an agarose gel or polyacrylamide gel for the effective screening of large numbers of samples. This approach is cost-effective and does not require prior sequence information. RAPD, inter-simple sequence repeats (ISSRs), and amplified fragment length polymorphism (AFLP) are PCR-based dominant markers that can differentiate the presence or absence of an allele but are unable to distinguish between heterozygote and homozygote conditions. The RAPD technique utilizes short, random primers, typically 10 base pairs long, to amplify unknown, random segments of the genome. The ISSR technique utilizes arbitrary primers, 10–30 base pairs long, which anneal in microsatellite regions and amplify unknown regions flanked by those repeats. ISSR does not require previous knowledge of the target genome and usually presents high levels of polymorphism. ISSR usually results in a more reliable and higher number of bands, in comparison with RAPD products (Godwin et al. 1997; Nadeem et al. 2018). The AFLP combines restriction enzyme digestion and selective PCR amplification to generate AFLP profiles with many polymorphic fragments. RAPD often suffers from lower reproducibility across different laboratories or even different thermal cyclers. AFLP is a technically complex, highly sensitive, and reliable technology compared to RAPD, which remains valuable for assessing initial genetic variation where genomic information is limited.

Restriction fragment length polymorphism (RFLP) is a hybridization-based marker that generates a distinct banding pattern if genetic diversity exists at the restriction site. The resulting fragments of different lengths are then detected via Southern blotting. Due to its high cost, time-consuming nature, and requirement of large amounts of high-quality DNA, it is unsuitable for large-scale diversity screening. An adaptation of RFLP, the PCR-RFLP, is a technique that uses the same principle of digestion of DNA by restriction enzymes, however, to digest DNA target fragments amplified by PCR. It is a simple and less expensive technique (Nishida et al. 2025), which may be developed in low infrastructure laboratories, but is less informative than the original RFLP, due to limited ability to detect nucleotide variations (Hashim and Al-Shuhaib 2019).

Simple sequence repeats (SSRs) or microsatellites are highly trusted, sequence-based, co-dominant markers that differentiate between homozygous and heterozygous individuals. SSRs are short, tandem repeats of 2–6 base pairs, highly polymorphic, and abundant throughout plant genomes. Variability arises from differences in the number of motif repetitions, which indicates that they are excellent indicators of intraspecific variation. Their high level of polymorphism offers superior resolution and information, making this method suitable for quantifying allelic richness, heterozygosity, and genetic differentiation among populations of closely related individuals, cultivar identification, mapping, germplasm fingerprinting, and authentication purposes in medicinal plants. This marker is one of the most frequently used markers to access plant genetic diversity (Lima et al. 2021; Guler and Imamoglu 2023).

In recent times, genetic diversity assessment technology has entered a new era with the integration of molecular markers with NGS, which allows whole-genome or high-throughput genomic coverage. This technology advances beyond sequencing specific fragments to sequencing large parts of a genome with high efficiency and low cost. While SSRs reveal diversity at specific loci, the newest generation of methods harnesses high-throughput sequencing technology to analyze thousands, even millions, of loci simultaneously. Approaches, such as genotyping-by-sequencing and restriction site-associated DNA sequencing, efficiently and economically produce thousands of single-nucleotide polymorphism (SNP) markers across large quantities of germplasm. SNPs are highly stable and the most abundant form of genetic variation in plant genomes, characterized by variations at a single base pair in the DNA sequence within the genome. This offers extensive data with unparalleled resolution for comprehensive studies of genetic variation within and among populations, genotyping, gene discovery, tracing the origin of medicinal materials, authentication of plant materials, identifying cryptic hybridization events, identifying elite genotypes for conservation and cultivation, and supporting molecular breeding programs that aim to improve active compound production. Moreover, integrating RNA sequencing enables the concurrent evaluation of genetic polymorphisms and their transcriptional effects, which is crucial for linking specific genetic variants to metabolic pathways responsible for chemodiversity (Morris and Cheng 2023). However, the process is costly and generates large datasets that require advanced bioinformatics tools for analysis. These methods allow researchers to calculate metrics such as gene diversity (heterozygosity), allelic richness, and genetic distance, providing insights into population structure, gene flow, and the extent of genetic erosion.

Recent reports have demonstrated that the genetic diversity of different medicinal plants, such as *Withania somnifera* (Khabiya et al. 2024), *Tribulus terrestris* (Chaudhary and Sharma 2025), *Rauwolfia tetraphylla* (Rohela et al. 2019), *Hypericum* sp. (Ma et al. 2021), *Adhatoda vasica* (Varma and Shrivastava 2018), *Andrographis paniculata* (Varma and Shrivastava 2018), *Crepidiastrum denticulatum* (Do et al. 2019), and *Centella asiatica* (Mathavaraj and Sabu 2021), has been assessed using RAPD, start codon targeted (SCoT), ISSR, and SNP markers. Some studies used different combinations of the methods to counter their individual limitations.

1.6 Threats to Genetic Diversity of Medicinal Plants

Gene diversity of medicinal plants is a non-renewable natural resource with significant ecological, ethnobotanical, and pharmacological importance. This diversity is declining rapidly due to multiple threats, including both natural drivers and anthropogenic interventions (Jha and Halder 2023). Natural drivers are events caused by volcanic activity, cyclones, tsunamis, flooding, earthquakes, climate change, and so on. These events are beyond human control and not derived from anthropogenic activity. In contrast, anthropogenic interventions include the direct activities of humans, such as the destruction of wild habitats, the introduction of invasive alien species, the overexploitation of desired wild species, pollution activities, and human-induced climate change (Jha and Halder 2023).

Instead of describing all the threats, this section focuses on the most effective threats that are responsible for a significant loss of genetic diversity in medicinal plants. The combination of habitat destruction, unsustainable harvesting practices, climate change, and selective cultivation practices is significantly contributing to the erosion of genetic diversity. Many of these threats interact synergistically, compounding their impacts on genetic variation.

1.6.1 Loss and Fragmentation of Natural Habitat

Agricultural expansion, urbanization, infrastructure development, pollution, soil erosion, and mining are the leading causes of human-induced destruction of the natural habitats of various plants (Christenhusz and Govaerts 2025), including medicinal plants. Depending on the degree of habitat destruction, this can lead to partial or complete loss of the natural habitats of wild medicinal plant populations. Several studies on several medicinal plants demonstrated that habitat loss and habitat fragmentation cause declines in species and genetic diversity (Wang and Zhang 2022; Póvoa et al. 2023; Siqueira et al. 2025; Wariss et al. 2025), as they directly and rapidly transform the native ecosystems in which these plants thrive. The severity of such events has more pronounced impacts in biodiversity hotspot areas and for endemic medicinal plant species with specific and narrow ecological niches. The complete loss of the natural habitat may result in the total genetic erosion of the population if they are exclusively endemic to the concerned habitat. Partial degradation of the natural habitat may result in a single population or have more than one subpopulation. If the natural habitat of the population becomes significantly disturbed, the ecosystem becomes unstable and may reduce the success rate of recolonization by no longer supporting the growth and development of the original population.

Medicinal plants that experience very narrow ecological amplitudes or have a complete dependence on specific microclimates are more severely affected by this risk, resulting in the removal of unique ecotypes or localized alleles, thereby reducing genetic variation. In addition, partially degraded habitats may result in a significant decline in the effective population size of the surviving species, and consequently, the loss of rare alleles. This creates a demographic bottleneck, making the smaller population susceptible to random fluctuations in gene frequencies, such as genetic drift, which further triggers the loss of low-frequency alleles and an overall decline in heterozygosity within the population.

The fragmentation of large habitats results in smaller and isolated subpopulations. Fragmentation directly impacts genetic diversity in the remaining populations by imposing physical barriers that often disrupt essential ecological processes, such as gene flow. Thus, it promotes localized mating between related individuals within residual populations. The population has increased consanguinity, which can lead to a subsequent reduction in fitness for both reproduction and survival by inbreeding depression over time. This can deplete evolutionary capacity in a population before it even disappears.

1.6.2 Overexploitation and Unsustainable Harvesting

The global market for medicinal and aromatic plants is experiencing continuous growth at a significant rate, and it is projected to reach around $5 trillion by 2050 (Jan et al. 2020). Significant proportions (60–90%) of the 3000 medicinal plant species used in global trade are still harvested directly from their wild populations (Jenkins et al. 2018). High local or global demand in the pharmaceutical industry and the traditional healthcare system for various parts (roots, rhizomes, bark, or whole plants) of specific populations with desired, commercially valuable traits often drives selection-based, unsustainable, and destructive harvesting practices. This frequently reduces the survival and reproductive capacity of an individual plant. Moreover, unsustainable harvesting practices, such as uprooting entire mature plants, harvesting juvenile plants, or collecting during non-reproductive periods, or harvesting more than the actual requirements for economic advantages of collectors, exert intense directional selective pressure on targeted specific genetic diversity. For example, continuous over-exploitation of high-yielding genotypes or germplasm for the pharmaceutical industry also exerts intense selection pressure on superior genotypes with the desired characters. The consistent removal of these superior genotypes from populations faster than the remaining individuals can reproduce, causing a decline in effective population size and resulting in a significant reduction in allele frequencies or complete loss of the superior variety from the gene pool. This genetic "bottlenecking" reduces the range of genetic material available for future adaptations. Moreover, if plants with higher concentrations of a desired compound are preferentially harvested, the population may evolve over time to have lower concentrations, which reduces their medicinal value. However, several genetically determined biological abilities, such as growth rate, reproduction efficiency, and level of adaptability, can drive their capability to cope with this threat (Halder et al. 2021).

1.6.3 Cultivation, Domestication, and Monoculture

While the purposeful cultivation and domestication of wild medicinal plants are often proposed as a conservation solution, they have a significant anthropogenic impact on their genetic structure. They are considered a threat to their genetic diversity. The cumulative effect of cultivation, domestication, and monoculture practices ultimately results in an irreversible loss of genetic variation among medicinal plants, which are essential for unique chemotypes, secondary metabolite profiles, and inherent disease resistance capabilities. In the long run, the loss of genetic diversity will challenge the discovery of new therapeutic compounds, sustainable supply chains, and the efficacy and treatment viability of many plant-based medications. Additionally, it compromises adaptability of medicinal plant species to emerging biotic and abiotic threats.

Domestication and widespread cultivation of medicinal plants often focus on specific genotypes that exhibit desirable traits, such as maximum yield, consistent compound production, unique secondary metabolite profiles, disease resistance, or simplified harvesting. When wild medicinal plants are brought into cultivation, they undergo a domestication bottleneck. Continuous cultivation imposes selective pressures that differ from natural evolutionary forces, ultimately leading to a reduction in genetic diversity, whether intended or unintended. This intense selection pressure in these cultivated populations may result in genetic erosion and the rapid loss of valuable alternative chemotypes and adaptive traits found in wild or locally adapted populations. This purposeful selection bias towards the desired characteristics is responsible for further narrowing the genetic base of cultivated and domesticated accessions compared to their wild parental populations.

The risk of monoculture to diversity is well-documented in major staple crops, but it is increasingly relevant for heavily traded medicinal species. The extensive adoption of monoculture systems by farmers in medicinal plant cultivation has the potential to intensify further genetic erosion through the cultivation of genetically homogeneous or closely related plants across a broad agricultural landscape. This practice minimizes existing intraspecific diversity and overall allelic variations. This type of agriculturally homogeneous environment, which lacks genetic plasticity, is particularly vulnerable to widespread disease outbreaks, pest infections, and environmental fluctuations, thereby threatening the entire supply chain.

1.6.4 Climate Change

Global climatic change brings considerable environmental uncertainty and poses a threat to the integrity of global biodiversity, including the genetic diversity of medicinal plants. For example, a study examining the impact of climate change on medicinal plants in Brazilian Cerrado (a biodiversity hotspot) projected a trend of loss of 50–64% of suitable habitats in Brazil by 2060 (Guerra dos Santos et al. 2025). Also, depending on their intrinsic properties, these plants exhibit differential responses to this challenge. The genetic structure of individual plants controls their adaptive capacity to this threat.

Climate changes, such as altered precipitation patterns, prolonged drought, or thermal extremes, impose intense new selective pressures at the genetic level. Although selection is a natural process, the accelerated pace of anthropogenic climate change drastically shortens the time frame required for the adaptation of medicinal plants. This leads to a net reduction in overall intraspecific genetic diversity if the rate of change outpaces the rate of adaptive evolution. Individual genotypes that possess alleles suited to the 'new' climate thus survive. In contrast, other genotypes that are unable to cope with this climatic change face a high risk of localized extinction. It creates a genetic bottleneck and systematic loss of valuable, rare, or unique alleles that may be needed for future, unexpected environmental stresses. The adapted populations may also lose alleles that were previously advantageous in the "old" climate but are no longer suited to the "new" climate.

Moreover, climate change may induce habitat fragmentation, shift the biogeographical range of species, disrupt phenology (the timing of life events), increase phenotypic plasticity, and reduce chemodiversity. A decrease in genetic variance and an increase in environmental stress translate quantitatively to a change in the phytochemical profile. This may result in lower concentrations of essential compounds or alterations in the ratios of active versus inert metabolites. Consequently, the pharmaceutical reliability and traditional potency of wild-harvested medicinal resources are fundamentally compromised.

1.6.5 Invasive Alien Species

Drastic human population growth and human-driven intensive global trading remarkably promote intentional or unintentional transfer of invasive alien species beyond their natural geographical limits. The influence of competitive dominance, the introduction of new diseases, and allelopathic features of invasive alien species, such as *Parthenium hysterophorus, Mikania micrantha, Lantana camara, Acacia saligna, and Amorpha fruticosa*, drastically influence local selection pressures and resource dynamics (Poudel et al. 2019; Tozzi et al. 2021; Praleskouskaya and Venanzoni 2021). These disruptions caused by invasive species contribute to the fragmentation and limit the effective range of medicinal plant populations. This initial ecological disruption immediately initiates a process of genetic erosion by two interconnected mechanisms that place immense stress on endemic medicinal taxa and threaten the unique phytochemical profiles. Population fragmentation caused by invasive alien species that encroachment isolates gene pools and limits effective population sizes. This genetic bottleneck effect exacerbates genetic drift and increases the rate of inbreeding, which results in a measurable decrease in heterozygosity and the loss of rare, potentially adaptive alleles, whereas, in certain instances, genetically related invasive alien species may facilitate introgression hybridization with native medicinal germplasm. Such genetic pollution compromises the integrity of the native germplasm, potentially diluting specialized traits associated with secondary metabolite production and reducing overall fitness within the native ecological niche.

1.7 Consequences of Reduced Genetic Diversity

A progressive decline in genetic diversity in medicinal plants has been noticed over recent decades. The cumulative effect is a substantial reduction in the reservoir of unique genetic traits, which poses serious and complex consequences for biodiversity and its applications. It puts populations at a high risk of extinction, limits their capacity to adapt, threatens global health security, reduces biochemical efficacy, alters ecological impacts, narrows opportunities for novel therapeutic discovery, and results in adverse socio-economic outcomes.

1.7.1 Decline in Chemodiversity

Each genotype or allelic variation plays a significant role in the quality and quantity of secondary metabolite accumulation in medicinal plants by altering or modulating the expression or function of genes or regulatory elements that are involved in biosynthetic pathways (Jadaun et al. 2023). Distinct genetic variants within a single species may have unique profiles of secondary metabolites that are responsible for delivering therapeutics to both traditional healthcare systems and modern pharmacognosy. Medicinal plant species with a narrow gene pool can affect the range of chemical compounds (spectrum) and their concentrations that a species can synthesize. Low genetic variability can create genetically uniform plant populations, leading to the irreversible loss of specific genetic variants or unique chemotypes responsible for the biosynthesis of rare or highly potent compounds. This loss results in the permanent disappearance of one or more specific, effective compounds, or even entire classes of medically relevant molecules. It leads to a decrease in chemodiversity in medicinal plants. Such a reduction in chemical diversity puts the sustainability of traditional healing practices and herbal remedies at increasing risk. This change in chemical profiles and reduction in the production of specific active compounds may reduce the efficacy and effectiveness of existing medicines.

1.7.2 Decrease the Options for Pharmaceutical Development

The discovery of new herbal medicines or lead compounds for drug development often involves exploring the genetic and biochemical diversity within medicinal plant species to search for chemically and structurally novel phytomolecules (Najmi et al. 2022; Halder and Jha 2023a; Zhang et al. 2023). The loss of genetic diversity in these plants has significant negative impacts on the pharmaceutical research and drug development process (Gul et al. 2023). A genetically homogeneous population offers fewer novel pathways to explore, which significantly decreases future access to undiscovered, unique phytomolecules and innovative drugs with a unique mechanism of action, improved efficacy, or reduced side effects through bioprospecting. This genetic erosion restricts the development of treatments for current and future human disease, including antibiotic-resistant infections and emerging viral pathogens.

1.7.3 Reduce Effectiveness and Variability of Traditional Herbal Remedies

Traditional medicine systems totally depend on the rich spectrum of genetic and chemical variations (Rahman et al. 2022). The loss of diverse chemotypes within wild and cultivated medicinal plant species can destabilize traditional medicine systems by reducing the effectiveness and variability of existing traditional remedies. This situation negatively affects communities that rely on these plants for healthcare and threatens valuable indigenous knowledge associated with their use.

1.7.4 Increase Vulnerability to Pests, Diseases, and Environmental Stressors

A rich genetic diversity offers diverse genetic pools that are essential for natural defense against different diseases, invading pests, and abiotic stresses (Salgotra and Chauhan 2023). Medicinal plants with limited genetic diversity face a high risk of regional or global extinction due to a lack of broad genetic variability necessary for developing resistance to these challenges. Genetic uniformity makes them more vulnerable to such threats and significantly reduces their adaptive capacity. As a result, a single virulent pathogen or a significant adverse environmental event can destroy an entire population or species with little to no internal genetic resistance. It compromises the availability of the resource for local communities and international markets. This also negatively impacts the quality and safety of these plants. Moreover, the use of chemical pesticides to protect a susceptible population with low genetic diversity has its own negative impacts on the environment and human health. In contrast, genetically diverse populations comprise individual plants with a wide range of genes that confer resistance to specific pathogens, pests, or environmental conditions. On the arrival of an epidemic, they have a higher chance of surviving and ensuring the continuation of the species.

1.7.5 Compromise Loss of Adaptive Potential and Evolutionary Capacity

Genetic variation is the fundamental material for natural selection and evolution. When a medicinal plant species has a limited gene pool, it loses its "evolutionary toolkit," becoming genetically static. A key consequence of loss of genetic variability is that their adaptability and resilience are severely compromised. This makes them acutely vulnerable to rapidly changing environmental stressors, including anthropogenic climate change (higher or lower temperatures, altered precipitation patterns, and varied soil conditions), emerging pathogens, and habitat fragmentation, all of which significantly increase extinction probabilities. A genetically uniform population lacks the variation needed to adapt to these changes and faces increased risk from abrupt climatic shifts, pests, and diseases. These populations are particularly vulnerable to stress-induced displacement or extinction. A medicinal plant species with limited genetic diversity reduces its ability to evolve and respond to future challenges, such as climate change and emerging pathogens. This vulnerability increases the risk of extinction and the irreversible loss of unique chemotypes or entire species.

1.7.6 Threaten Long-Term Survival of Species

Genetic bottlenecks further significantly reduce the biological plasticity and evolutionary potential of these invaluable species. These affect their long-term survival. Additionally, such a decline also threatens the integrity of natural ecosystems by compromising ecological resilience and eroding valuable cultural heritage.

1.7.7 Effects on Indigenous Communities and Their Traditional Knowledge

The reduction of genetic diversity in medicinal plants has several implications for indigenous communities worldwide. These communities have relied significantly on cultivated or wild-harvested medicinal plants, as well as their ethnobotanical knowledge, to address health-related issues over generations (Ssenku et al. 2022; Ralte et al. 2024; Nugroho et al. 2025). The decline in genetic diversity can affect their cultural identity, healthcare systems, ecological stewardship, and economic performance, directly impacting their livelihoods. The loss of specific genotypes or species results in fewer effective treatment options for them. This compels reliance on costly medicines from outside their culture, which may not always be effective. For them, the medicinal plant species are living repositories of knowledge associated with these plants. Their extinction could lead to the loss of these knowledge in future. Furthermore, the genetic decline of medicinal plants can lead to cascading effects that ultimately impact the sustainability of environments vital to cultural traditions and the livelihoods of indigenous peoples.

1.7.8 Ecological Effects

Decreased genetic diversity of medicinal plants can lead to wider ecological disruption and minimize ecosystem values in addition to their medicinal values. Many of the plants also perform critical ecological functions in their native ecosystems through soil stabilization, nutrient cycling, and providing food or habitat to other organisms, such as pollinators and symbiotic microbes. The decline or extinction of a genetically uniform medicinal plant can then impact biodiversity at multiple trophic levels, leading to decreased ecosystem stability through cascading effects that may spread throughout the ecosystem.

1.8 Genetic Diversity Conservation Strategies

Biodiversity conservation for medicinal plants requires a comprehensive and multifaceted approach that integrates both in-situ and ex-situ conservation methods (Jha and Halder 2023). Conservation of the genetic diversity of medicinal plants requires a deeper level of understanding compared to species diversity conservation. Species diversity conservation emphasizes ecological surveys, habitat restoration, establishment of protected areas, and ex-situ collections. Genetic diversity conservation efforts begin with assessing genetic variation within and among populations using molecular techniques to identify genetic variations. The chemical composition of different plant populations is often evaluated using techniques, such as high-performance liquid chromatography (HPLC), gas chromatography-mass spectrometry (GC-MS), and nuclear magnetic resonance (NMR) to identify variations in bioactive compounds (Halder et al. 2024). A general overview of genetic diversity conservation strategies are briefly discussed here.

1.8.1 In Situ Conservation of Genetic Diversity and Limitations

In situ conservation is an effective approach to manage and protect wild populations of medicinal plants and their entire ecosystems in their natural habitats, where continuous evolutionary forces have shaped their unique genetic composition and biochemical profiles. Apart from maintaining total species or species richness, this conservation strategy enables the continuation of their evolutionary process through dynamic interactions with various drivers of genetic diversity. Knowledge of the population structure, demographic dynamics, and the spatial distribution of genetic variation plays a crucial role in the effective in-situ conservation of genetic diversity (Yang et al. 2024). Several research studies demonstrated that populations with high intraspecific variation and unique chemotypes often prioritize conservation (Singh et al. 2025; Chakraborty et al. 2024). Biodiversity hotspots generally have very high levels of genetic variation (Chi et al. 2017), which should be specifically targeted for in-situ conservation.

The establishment and management of protected areas, such as national parks, biosphere reserves, wildlife sanctuaries, sacred groves, and community-managed regions, enable the natural growth and protection of large, targeted populations of genetically diverse medicinal plants (Jha and Halder 2023; Pidigam et al. 2023). These areas function as living gene banks, where genetic resources are conserved in a manner that promotes the continuous spatial gene flow among different genetically diverse populations. It minimizes inbreeding depression and population extinction risk. Additionally, the implementation of effective scientific management strategies is crucial for long-term success, facilitating the control of invasive species, the development of fire management plans, and regular ecological monitoring.

On-farm agro-biodiversity approaches are applied to conserve local landrace varieties and wild relatives of medicinal and aromatic plant species that exist outside the protected network system or are part of agricultural landscapes. Indigenous communities have purposefully protected, domesticated, or semi-domesticated several medicinal plants and their adaptive genes. They frequently grow and manage these medicinally important plant species in diverse, small-scale plots within human-modified traditional agroforestry systems and their home gardens over generations. Such conservation practices adopted by rural and indigenous communities effectively conserve intraspecific genetic diversity and also meet their socio-economic and healthcare needs (Meghwal and Dwivedi 2017). It also reinforces the essential links between biological diversity and cultural traditions.

The involvement of indigenous and local communities, along with their traditional knowledge, has significant relevance to in-situ conservation, as these communities are the true custodians of profound Traditional Ecological Knowledge (Agize et al. 2022; Koricha and Jemal Adem 2024). It encompasses a comprehensive understanding of species distribution, phenology, sustainable harvesting practices, and the ethnobotanical uses of numerous plant varieties. Community-based in-situ conservation initiatives support their socio-economic well-being and more effective management and protection of medicinal plant resources by integrating science-based conservation practices with their traditional knowledge systems (Mbelebele et al. 2024). Fair-trade, equitable partnerships, benefit-sharing arrangements, and recognizing sovereignty over conventional lands encourage sustainable management and effective conservation of medicinal plants. Active participation of local communities makes conservation efforts culturally more appropriate and locally supported. It helps to ensure the minimization of destructive harvesting or habitat destruction by local communities. Together, this produces a sustainable and resilient means of protecting genetic diversity.

The success or failure of the in-situ conservation approach largely depends on the implementation of robust policy and legal frameworks that prevent harmful anthropogenic activities (Oluka et al. 2024). These frameworks can range from broad international agreements, such as the Convention on Biological Diversity (CBD) and the Convention on International Trade in Endangered Species of Wild Fauna and Flora (CITES), to specific national frameworks. For example, CBD is a comprehensive treaty designed to protect biodiversity, promote the sustainable use of its elements, and implement a fair and equitable sharing of benefits from the use of genetic resources. CITES is another example of an international framework that regulates international trade in endangered species to prevent overexploitation through trade and is an essential framework for protecting biodiversity. Similar to these international agreements, there are country- or issue-specific legal and policy frameworks that translate international commitments to domestic action. Many countries have frameworks that typically consist of various legislative acts, such as environmental protection legislation and wildlife protection legislation, for protecting and managing the environment and biodiversity to ensure ecological integrity and the sustainable use of resources (Arora 2022). These policies translate conservation objectives into directives to manage human behavior in a manner that does not compromise natural resources for future generations. In the absence of explicit policy direction, conservation initiatives often lack the enforceability and resources necessary to be effective in the face of anthropogenic pressures.

However, a significant gap often exists within policy and regulatory frameworks in establishing explicit protection for genetic diversity. CBD, CITES, and national legislation frequently focus on the safety and management of individual species, whereas establishing laws and policy instruments for maintaining and managing intraspecific genetic variation is insufficient (Hoban et al. 2021). This gap is critical, as genetic diversity is the essential basis of a species' ability to adapt to environmental change, support disease resistance, and provide long-term evolutionary potential.

1.8.2 Ex Situ Conservation of Genetic Diversity and Applications

Ex situ conservation approaches have been applied to protect the specific germplasms, genotypes, or chemotypes of various threatened medicinal plants outside their natural habitats in carefully controlled environments, such as seed banks, field gene banks, in vitro tissue cultures, and living collections in botanical gardens. While these strategies significantly compromise their natural evolution, they are highly effective in protecting endangered and rare medicinal plant populations that possess a broad spectrum of rare alleles or unique gene combinations and chemotypes, or those that face the risk of severe genetic erosion due to habitat destruction, overharvesting, or unexpected disasters. Effective conservation requires meticulous data collection, sampling, and data guides to incorporate a wide range of geographical landscapes and traditional landraces, maximizing the representation of the broad genetic pool. Exciting advances in molecular techniques are helping in the more reliable identification and prioritization of populations that are rich in genetic variation for conservation (Vaishnav et al. 2023; Rafiq et al. 2025). The rich genetic diversity of the planet can be conserved by focusing on these characteristics.

In seed gene banks, seeds with desiccation tolerance are dried to a low moisture content and stored in controlled, cool conditions (generally −18 °C to −20 °C) to conserve a vast genetic diversity over the long term in minimal space (Liu et al. 2023). In contrast, more advanced in vitro tissue culture techniques have been frequently applied to conserve genetically or phytochemically unique varieties or germplasm of medicinal plants. Plantlets or organ cultures are maintained in artificial culture media under slow-growth conditions and in an aseptic environment (Chandran et al. 2023; Nallakaruppan and Thiagarajan 2025). Cryopreservation also preserves explants, usually seeds, roots, apical meristems, or somatic embryos by storing them in cryoprotectants like liquid nitrogen (−196 °C) to halt all metabolic processes indefinitely and guarantee the long-term, genetically stable storage of valuable accessions (Yang et al. 2019; Sharma et al. 2021; Nits et al. 2024). A cryobank for medicinal plants (National Southern Medicinal Gene Bank) located in Hainan Province, China, has successfully cryopreserved 11,055 medicinal plant seeds. Zeng et al. (2024) reported successful cryopreservation of 154 medicinal plant seeds out of a total of 164 medicinal plant seeds that were attempted. Furthermore, field gene banks (Kumar 2025) and specialized botanic gardens (Chen and Sun 2018; O'Donnell and Sharrock 2018) maintain accessible collections of living plants in cultivation, ensuring that the comprehensive genetic variability underpinning medicinal potency remains available for future pharmacological use, research, phytochemical analysis, and species restoration.

1.9 Future Directions in Conservation of Genetic Diversity of Medicinal Plants

Proactive measures must be taken to conserve genetic diversity in medicinal plants, ensuring the preservation of existing genetic diversity for present and future generations, as the demand for plant-based medicines continues to increase. Involving scientists, policymakers, and local communities in integrating innovative science with indigenous knowledge and multidisciplinary approaches is crucial for creating a sustainable path to developing holistic conservation strategies that protect the genetic diversity of medicinal plants, their ecological services, and cultural heritage.

Effective and efficient conservation of medicinal plants' genetic diversity demands further synergy between scientific innovation and collaborative efforts from all stakeholders. Current conservation strategies mainly focus on species-level conservation, as these efforts do not distinguish between different genetic variants of the same species. However, the genetic structure of populations and their genetic differentiation should be taken into account in conservation strategies. The use of new, emerging ex-situ conservation approaches, such as cryopreservation and tissue cultures, is more effective and should be implemented more widely alongside traditional conservation methods, including in-situ protected areas and ex-situ seed banks, for the conservation of genetic diversity. Similarly, the integration and application of modern genomic technologies, such as DNA barcoding, NGS, and genome-wide association studies, as well as environmental DNA sampling, should be encouraged in medicinally important but less-exploited species. These technologies facilitate effective mapping and assessment of genetic diversity, tracking of adaptive traits, and real-time monitoring of genetic drift. The incorporation of artificial intelligence and machine learning methods holds excellent potential for conserving genetic diversity in medicinal plants over the next few years. They effectively predict vulnerability of the population through a large-scale, data-driven analytical process. Moreover, citizen science and local community-based monitoring should be encouraged for data collection and conservation, as well as empowering local stewardship of these medicinal plant species.

Cooperative efforts between governments and local communities need to be strengthened to reduce overexploitation and habitat degradation through sustainable harvests in protected areas. Furthermore, in-situ conservation must be enhanced through community-based initiatives or the creation of protected areas that are connected to local or traditional knowledge systems. This approach allows survival of natural populations in more effective manner. It cannot be denied that more legal policies and regulations are required in many medicinal plant-rich countries in the current situation to regulate trade, prevent biopiracy, and ensure equitable benefit-sharing from genetic resources. International agreements like the Nagoya Protocol must be reinforced in order to avoid unauthorized exploitation. Failure to implement effective ABS mechanisms can discourage local conservation efforts by indigenous communities. Receiving equitable benefits from commercial utilization of different medicinal plants or traditional knowledge systems is critical for incentivizing in-situ conservation and combating biopiracy. Government and local authorities should maintain a balance between broader access and mobilization of a genetic approach to support a more inclusive and dynamic platform of genetic biodiversity conservation. Moreover, an increase in funding allocation from the Government, private sources, and non-governmental organizations is required for conservation research and the enforcement of strict regulations in protected areas against deforestation and illegal harvesting to mitigate biodiversity loss. This strategy can help reduce the pressures of wild harvesting and preserve genetic heterogeneity.

Another approach for increasing genetic diversity could be to integrate conservation with sustainable livelihoods by involving indigenous people and farmers in medicinal plant cultivation. This would allow genetic diversity to be present not only in laboratories or protected areas but also in managed agroecosystems that incorporate traditional knowledge and practices. Participatory breeding programs and agroforestry models could facilitate a viable framework for an incentivized approach to the cultivation of medicinal plants while maintaining genetic diversity.

Education, capacity building, and public awareness will be essential in shaping the future of medicinal plant conservation. The exchange of knowledge and development of awareness will support grass-roots conservation. A multidisciplinary curriculum, workshops, online forums, and collaborative research networks can help create the next generation of conservation leaders and potentially bridge the gaps between scientists, policymakers, and local communities. Additionally, community outreach through documentaries, open-access publications, and museums can further raise public awareness about the ecological and economic importance of the genetic diversity of medicinal plants.

1.10 Conclusion

Recently, humanity has started to realize that conservation of genetic diversity is unique and different from simply protecting plant species. This realization has highlighted the need for targeted conservation of specific, inheritable internal variations within their gene pool. In medicinal plants, conservation of genetic diversity represents a previously overlooked area that has unique ecological, socio-economic, and health-related significance. Protecting existing species of medicinal value is not synonymous with conserving genetic diversity. Individual chemotypes within a single species of medicinal plant are the outcome of a long evolutionary process. The loss of these repositories of genetic information is directly linked to their medicinal properties. Thus, genetic erosion is equivalent to the deletion of existing or potential therapeutic molecules, which have significant negative impacts on human healthcare.

The impacts of different threats may vary from one ecotype or chemotype to another, which necessitate specialized conservation strategies for specific genotypes. Furthermore, it is essential to recognize that species diversity conservation does not always ensure the preservation of specific, valuable genetic diversity. A species may survive as a viable population. Yet, its potent or unique alleles may be lost through genetic drift, selective pressures favoring less medicinally relevant traits, or hybridization with less valuable varieties. In contrast, a species might be critically endangered, but the few remaining individuals could harbor a genetic legacy of immense medicinal worth. A deeper understanding and proactive measurement should be taken immediately to ensure the conservation of their invaluable genetic library. This library is a dynamic resource, essential for ensuring that these plants can be sustainably used for future generations and the planet.

Conservation of genetic diversity in medicinal plants is a multifaceted global challenge. An integrated approach that combines ecological knowledge, modern technology, and socio-political action is the future goal of effective conservation. Detailed characterization of genetic and chemical variation among these plants will help in adopting more effective in-situ and ex-situ conservation strategies and climate-adaptive breeding programs. The conservation of medicinal plant genetic diversity requires bold, visionary, and collaborative action from policymakers, researchers, governmental agencies, local communities, and educators to balance genetic and species-level interventions. This effort will minimize anthropogenic pressures and ultimately protect the foundation of medicine, cultural identity, and global well-being for future generations.

References

Agize M, Asfaw Z, Nemomissa S, Gebre T (2022) Ethnobotany of traditional medicinal plants and associated indigenous knowledge in Dawuro Zone of Southwestern Ethiopia. J Ethnobiol Ethnomed 18(1):48

Alami MM, Ouyang Z, Zhang Y, Shu S, Yang G, Mei Z, Wang X (2022) The current developments in medicinal plant genomics enabled the diversification of secondary metabolites' biosynthesis. Int J Mol Sci 23(24):15932

Amin A, Park S (2025) Chemotaxonomy, an efficient tool for medicinal plant identification: current trends and limitations. Plants 14(14):2234

Arora T (2022) Legal framework for conservation of biodiversity in India. In: Kaur S, Batish DR, Singh HP, Kohli R (eds) Biodiversity in India: status, issues and challenges. Springer Nature, Singapore, pp 589–610

Azizi A, Hadian J, Gholami M, Friedt W, Honermeier B (2012) Correlations between genetic, morphological, and chemical diversities in a germplasm collection of the medicinal plant *Origanum vulgare* L. Chem Biodivers 9(12):2784–2801

Banadka A, Toppo V, Kashyap K, Nandhini M, Akshita B, Sudheer WN, Nagella P (2025) Revolutionizing Ashwagandha breeding: from traditional to modern breeding techniques for Indian Ginseng, *Withania somnifera* (L.) Dunal. In: Al-Khayri JM, Jain SM, Penna S (eds) Biodiversity and genetic improvement of medicinal and aromatic plants II. Springer, Cham, pp 1–34

Bhusare BP, John CK, Bhatt VP, Nikam TD (2021) Colchicine induces tetraploids in in vitro cultures of *Digitalis lanata* Ehrh.: enhanced production of biomass and cardiac glycosides. Ind Crop Prod 174:114167

Chakraborty A, Mukherjee S, Santra I, Dey D, Mukherjee S, Ghosh B (2024) Secondary metabolite fingerprinting, anti-pathogenic activity, elite chemotype selection and conservation of *Curcuma caesia*-an ethnomedicinally underutilized species. 3 Biotech 14(6):155

Chandran S, Raghu AV, Mohanan KV (2023) In vitro conservation of rare, endangered, and threatened plants. In: Sukumaran ST, Keerthi TR (eds) Conservation and sustainable utilization of bioresources. Springer Nature, Singapore, pp 391–408

Chaudhary N, Sharma V (2025) Elucidation of genetic diversity in the medicinal herb *Tribulus terrestris* L. from North Indian plains using morphological and ISSR markers. Nucleus 25:1

Chen G, Sun W (2018) The role of botanical gardens in scientific research, conservation, and citizen science. Plant Divers 40(4):181–188

Chi X, Zhang Z, Xu X, Zhang X, Zhao Z, Liu Y, Wang Q, Wang H, Li Y, Yang G, Guo L (2017) Threatened medicinal plants in China: distributions and conservation priorities. Biol Conserv 210:89–95

Christenhusz MJM, Govaerts R (2025) Plant extinction in the Anthropocene. Bot J Linn Soc 207(3):183–196

Das S, Kwon M, Kim JY (2024) Enhancement of specialized metabolites using CRISPR/Cas gene editing technology in medicinal plants. Front Plant Sci 15:1279738

Davis CC, Choisy P (2024) Medicinal plants meet modern biodiversity science. Curr Biol 34(4):R158–R173

Deng K, Li Z, Huang T, Huang J (2024) Noncoding RNAs in regulation of plant secondary metabolism. Plant Physiol Biochem 211:108718

Dhabal S, Chakrabarty AK, Banerjee D, Katiyar CK, Rai RK, Dubey SK (2025) Internal transcribed spacer (ITS): the powerful DNA barcode and phylogenetic marker for successful authentication of *Withania somnifera*. Mol Biol Rep 52(1):77

Do HD, Jung J, Hyun J, Yoon SJ, Lim C, Park K, Kim JH (2019) The newly developed single nucleotide polymorphism (SNP) markers for a potentially medicinal plant, *Crepidiastrum denticulatum* (Asteraceae), inferred from complete chloroplast genome data. Mol Biol Rep 46(3):3287–3297

Dong LL, Chen ZJ, Wang Y, Wei FG, Zhang LJ, Xu J, Wei GF, Wang R, Yang J, Liu WL, Li XW (2017) DNA marker-assisted selection of medicinal plants (I). Breeding research of disease-resistant cultivars of *Panax notoginseng*. China J Chinese Materia Medica 42(1):56–62

Godwin ID, Aitken EA, Smith LW (1997) Application of inter simple sequence repeat (ISSR) markers to plant genetics. Electrophoresis 18(9):1524–1528

Guerra dos Santos LA, Vilela B, de Sousa FG, Ferreira Junior WS, de Paiva Silva D (2025) The potential effects of climate change on medicinal plants from the Brazilian Cerrado in South America. Biodivers Conserv 34(13): 4881

Gul MZ, Bhat MY, Ryan EP, Ghazi IA (2023) Unraveling medicinal plant chemical diversity for novel drug discovery through biotechnological interventions. In: Omics studies of medicinal plants. CRC Press, pp 45–74

Guler BA, Imamoglu E (2023) Molecular marker technologies in food plant genetic diversity studies: an overview. Foods Raw Mater 11(2):282–292

Halder M, Majumder A, Ray S, Jha S (2021) Medicinal plant research at crossroads: Biotechnological approaches for conservation, production and stability in tissue cultures and regenerated plants. In: Ekiert HM, Ramawat KG, Arora J (eds) Medicinal plants: Sustainable development and biodiversity, vol 28. Springer, Cham, pp 459-544

Halder M, Jha S (2023a) Medicinal plants and bioactive phytochemical diversity: a fountainhead of potential drugs against human diseases. In: Jha S, Halder M (eds) Medicinal plants: biodiversity, biotechnology and conservation. Springer Nature, Singapore, pp 39–93

Halder M, Jha S (2023b) The current status of population extinction and biodiversity crisis of medicinal plants. In: Jha S, Halder M (eds) Medicinal plants: biodiversity, biotechnology and conservation. Springer Nature, Singapore, pp 3–38

Halder M, Kundu A, Jha S (2024) Secondary metabolites identification techniques of the current era. In: Plant specialized metabolites, phytochemistry, ecology and biotechnology. Spinger, Cham, pp 1–41

Hamarashid SH, Khaledian Y, Soleimani F (2022) In vitro polyploidy-mediated enhancement of secondary metabolites content in *Stachys byzantina* L. Genet Resour Crop Evol 69(2):719–728

Hashim HO, Al-Shuhaib MBS (2019) Exploring the potential and limitations of PCR-RFLP and PCR-SSCP for SNP detection: a review. J Appl Biotechnol Rep 6(4):137–144

Hoban S, Campbell CD, da Silva JM, Ekblom R, Funk WC, Garner BA, Godoy JA, Kershaw F, MacDonald AJ, Mergeay J, Minter M (2021) Genetic diversity is considered important but interpreted narrowly in country reports to the Convention on Biological Diversity: current actions and indicators are insufficient. Biol Conserv 261:109233

Ijinu TP, George V, Pushpangadan P (2022) History of research on medicinal plants in India. In: Máthé Á, Khan IA (eds) Medicinal and aromatic plants of India, vol 1. Springer, Cham, pp 35–61

Jadaun JS, Yadav R, Yadav N, Bansal S, Sangwan NS (2023) Influence of genetics on the secondary metabolites of plants. In: Carocho M, Heleno SA, Barros L (eds) Natural secondary metabolites: from nature, through science, to industry. Springer, Cham, pp 403–433

Jan HA, Jan S, Bussmann RW, Wali S, Sisto F, Ahmad L (2020) Complementary and alternative medicine research, prospects and limitations in Pakistan: a literature review. Acta Ecol Sin 40(6):451–463

Jenkins M, Timoshyna A, Cornthwaite M (2018) Wild at home: exploring the global harvest, trade and use of wild plant ingredients. TRAFFIC, Cambridge. Accessed Oct 2022

Jha S, Halder M (2023) Medicinal plants: biodiversity, biotechnology and conservation. Springer

Khabiya R, Choudhary GP, Sairkar P, Silawat N, Jnanesha AC, Kumar A, Lal RK (2024) Unraveling genetic diversity analysis of Indian ginseng (*Withania somnifera* (Linn.) Dunal) insight from RAPD and ISSR markers and implications for crop improvement vital for pharmacological and industrial potential. Ind Crop Prod 210:118124

Khoshsokhan F, Babalar M, Salami SA, Sheikhakbari-Mehr R, Mirjalili MH (2022) Rosmarinic acid production in hairy root cultures of *Salvia nemorosa* L.(Lamiaceae). Biocat Agric Biotechnol 45:102494

Koricha HG, Jemal Adem M (2024) Investigated the role of community based approaches for biodiversity conservation and socio-economic development in Bale Mountains National Park, Southeast Ethiopia. Sci Rep 14(1):12241

Kshirsagar P, Umdale S, Chavan J, Gaikwad N (2017) Molecular authentication of medicinal plant, *Swertia chirayita* and its adulterant species. Proc Natl Acad Sci, India Sect B Biol Sci 87(1):101–107

Kumar K (2025) Field Genebank: an efficient method for conservation and sustainable utilization of seeds. In: Roy B, Shukla G, Dunna V, Sharma P, Shukla SP (eds) Advances in seed quality evaluation and improvement. Springer Nature, Singapore, pp 465–490

Kumar J, Choudhary K, Singh H, Kumar A, Bagga P (2025) Molecular approaches for authentication and identification of medicinal plants. Plant Mol Biol Report 25:1–21

Lima MFRC, Kraetzig LC, Praça FB, Burtet LW, Stefanello R, Pereira TA, Benhard EL, Essi L (2021) Marcadoresmolecularesutilizados para estudo da diversidadegenética de plantasameaçadas de extinção no Brasil. In: Zuffo AM, Aguilera JG (eds) Pesquisas Agrárias e Ambientais, vol 5. Pantanal Editora, Nova Xavantina, pp 133–141

Liu U, Gianella M, Dávila Aranda P, Diazgranados M, Flores Ortíz CM, Lira-Saade R, Bacci S, Mattana E, Milliken W, Mitrovits O, Pritchard HW (2023) Conserving useful plants for a sustainable future: species coverage, spatial distribution, and conservation status within the Millennium Seed Bank collection. Biodivers Conserv 32(8):2791–2839

Ma S, Khayatnezhad M, Minaeifar AA (2021) Genetic diversity and relationships among *Hypericum* L. species by ISSR Markers: a high value medicinal plant from Northern of Iran. Caryologia 74(1):97–107

Madani H, Escrich A, Hosseini B, Sanchez-Muñoz R, Khojasteh A, Palazon J (2021) Effect of polyploidy induction on natural metabolite production in medicinal plants. Biomolecules 11(6):899

Magdy AM, Fahmy EM, Abd EL-Rahman MA, Awad G (2020) Improvement of 6-gingerol production in ginger rhizomes (*Zingiber officinale* Roscoe) plants by mutation breeding using gamma irradiation. Appl Radiat Isot 162:109193

Mathavaraj S, Sabu KK (2021) Genetic diversity and structure revealed by genomic microsatellite markers in *Centella asiatica* (L.) Urb., a plant with medicinal potential. Mol Biol Rep 48(11):7387–7396

Mbelebele Z, Mdoda L, Ntlanga SS, Nontu Y, Gidi LS (2024) Harmonizing traditional knowledge with environmental preservation: sustainable strategies for the conservation of indigenous medicinal plants (IMPs) and their implications for economic well-being. Sustainability 16(14):5841

Meghwal RR, Dwivedi NK (2017) Agro-biodiversity collection and conservation from arid and semi-arid regions of India. In: Goyal MR (ed) Emerging technologies in agricultural engineering. Apple Academic Press, pp 209–229

Mikail HG, Mohammed M, Umar HD, Suleiman MM (2022) Secondary metabolites: the natural remedies. In: Secondary metabolites-trends and reviews. IntechOpen

Mir TA, Jan M, Khare RK, Bhat MH (2021) Medicinal plant resources: threat to its biodiversity and conservation strategies. In: Aftab T, Hakeem KR (eds) Medicinal and aromatic plants: healthcare and industrial applications. Springer, Cham, pp 717–739

Monteiro WP, Dalapicolla J, Carvalho CS, Costa Veiga J, Vasconcelos S, Ramos SJ, Gastauer M, Jaffé R, Caldeira CF (2022) Genetic diversity and structure of an endangered medicinal plant species (*Pilocarpus microphyllus*) in eastern Amazon: implications for conservation. Conserv Genet 23(4):745–758

Morris R, Cheng F (2023) Herbal medicine and RNA-seq. In: Cheng F, Morris R (eds) RNA-Seq in drug discovery and development. CRC Press, pp 205–232

Nadeem MA, Nawaz MA, Shahid MQ, Doğan Y, Comertpay G, Yildiz M, Hatipoğlu R, Ahmad F, Alsaleh A, Ladhane N, Özkan H, Chung G, Baloch FS (2018) DNA molecular markers in plant breeding: current status and recent advancements in genomic selection and genome editing. Biotechnol Biotechnol Equip 32:261–285

Najmi A, Javed SA, Al Bratty M, Alhazmi HA (2022) Modern approaches in the discovery and development of plant-based natural products and their analogues as potential therapeutic agents. Molecules 27(2):349

Nallakaruppan N, Thiagarajan K (2025) Efficient in vitro propagation and conservation of the endangered medicinal plant *Hedyotis sithiravaraiensis*. Plant Cell Tissue Organ Cult 161(2):43

Newman DJ, Cragg GM (2020) Natural products as sources of new drugs over the nearly four decades from 01/1981 to 09/2019. J Nat Prod 83(3):770–803

Nishida T, Sawa S, Sugimura K (2025) Development of a low-cost, simple, and rapid identification method for *Glycyrrhiza uralensis* using PCR–RFLP and evaluation of seeds distributed on the market. J Nat Med 76:1419–1432

Nits OS, Sementsova MV, Osipova ES, Tereshonok DV, Gladkov EA (2024) IPPRAS cryobank for the conservation of orthodox seeds of rare, endangered, medicinal, and ornamental plant species—current research. Plants 13(10):1354

Nugroho Y, Soendjoto MA, Suyanto S, Matatula J, Alam S, Wirabuana PY (2025) Traditional medicinal plants and their utilization by local communities around Lambung Mangkurat Education Forests, South Kalimantan, Indonesia. https://doi.org/10.13057/biodiv/d230137

O'Donnell K, Sharrock S (2018) Botanic gardens complement agricultural gene bank in collecting and conserving plant genetic diversity. Biopreserv Biobank 16(5):384–390

Oluka OB, Sindhu RK, Dagur P, Kumar S (2024) Protecting heritage: the role of legal frameworks in safeguarding African traditional medicines. Biochem Cellular Arch 24:3317

Patonay K, Szalontai H, Radácsi P, Zámboriné-Németh É (2021) Chemotypes and their stability in *Mentha longifolia* (L.) L.—a comprehensive study of five accessions. Plants 10(11):2478

Pei T, Zhu S, Liao W, Fang Y, Liu J, Kong Y, Yan M, Cui M, Zhao Q (2023) Gap-free genome assembly and CYP450 gene family analysis reveal the biosynthesis of anthocyanins in *Scutellaria baicalensis*. Hortic Res 10(12):uhad235

Pidigam S, Geetha A, Pandravada SR, Khan S, Thuraga V, Sivaraj N, Nagaraju K (2023) Conservation strategies for Indian medicinal plants. In: Máthé Á, Khan IA (eds) Medicinal and aromatic plants of India, vol 2, pp 279–302

Poudel M, Adhikari P, Thapa K (2019) Biology and control methods of the alien invasive weed *Mikania micrantha*: a review. Environ Contam Rev 2(2):06–12

Póvoa O, Lopes V, Barata AM, Farinha N (2023) Monitoring genetic erosion of aromatic and medicinal plant species in Alentejo (South Portugal). Plants 12(14):2588

Praleskouskaya S, Venanzoni R (2021) Effects of invasive alien species on riparian vegetation over a 20-year time-lapse: a case study from the Tiber river in central Italy. Biodiversitas 22(1–2):67–81

Qadir SU, Raja V (2021) Herbal medicine: old practice and modern perspectives. In: Bhat RA, Hakeem KR, Dervash MA (eds) Phytomedicine. Academic Press, pp 149–180

Qi-Qing LI, Zhang J, Ji-Hua LI, Bo-Yang YU (2018) Morphological and chemical studies of artificial *Andrographis paniculata* polyploids. Chin J Nat Med 16(2):81–89

Rabeh K, Hnini M, Oubohssaine M (2025) A comprehensive review of transcription factor-mediated regulation of secondary metabolites in plants under environmental stress. Stress Biol 5(1):15

Rafiq S, Malla BA, Wagay NA, Kaloo ZA, Mahajan R, Hadi A (2025) Genetic diversity assessment of *Aconitum chasmanthum* Stapf ex Holmes, a critically endangered and endemic medicinal plant of Kashmir Himalaya, using inter simple sequence repeat markers. Genet Resour Crop Evol 72(1):255–269

Rahman MH, Roy B, Chowdhury GM, Hasan A, Saimun MS (2022) Medicinal plant sources and traditional healthcare practices of forest-dependent communities in and around Chunati Wildlife Sanctuary in southeastern Bangladesh. Environ Sustain 5(2):207–241

Ralte L, Sailo H, Singh YT (2024) Ethnobotanical study of medicinal plants used by the indigenous community of the western region of Mizoram, India. J Ethnobiol Ethnomed 20(1):2

Raskar S, Purkar V, Sardesai M, Mitra S (2022) Assessing the impact of geographical distribution and genetic diversity on metabolic profiles of a medicinal plant, *Embelia ribes* Burm. fd. Plants 11(21):2861

Rohela GK, Jogam P, Bylla P, Reuben C (2019) Indirect regeneration and assessment of genetic fidelity of acclimated plantlets by SCoT, ISSR, and RAPD markers in *Rauwolfia tetraphylla* L.: an endangered medicinal plant. BioMed Res Intl 2019(1):3698742

Sahoo A, Behura S, Singh S, Jena S, Ray A, Dash B, Kar B, Panda PC, Nayak S (2021) EST-SSR marker-based genetic diversity and population structure analysis of Indian *Curcuma* species: significance for conservation. Braz J Bot 44(2):411–428

Salaria N, Kumari I, Kumar R (2025) Utilization and future prospective of next generation sequencing (NGS) in biodiversity conservation of endangered medicinal plants. In: Biotechnological innovations for sustainable biodiversity and development. CRC Press, pp 12–26

Salgotra RK, Chauhan BS (2023) Genetic diversity, conservation, and utilization of plant genetic resources. Genes 14(1):174

Sawadogo WR, Maciuk A, Banzouzi JT, Champy P, Figadere B, Guissou IP, Nacoulma OG (2012) Mutagenic effect, antioxidant and anticancer activities of six medicinal plants from Burkina Faso. Nat Prod Res 26(6):575–579

Segelbacher G, Bosse M, Burger P, Galbusera P, Godoy JA, Helsen P, Hvilsom C, Iacolina L, Kahric A, Manfrin C, Nonic M (2022) New developments in the field of genomic technologies and their relevance to conservation management. Conserv Genet 23(2):217–242

Sharma N, Gowthami R, Devi SV, Malhotra EV, Pandey R, Agrawal A (2021) Cryopreservation of shoot tips of *Gentiana kurroo* Royle–a critically endangered medicinal plant of India. Plant Cell Tissue Organ Cult 144(1):67–72

Sharma R, Patil C, Majeed J, Kumar S, Aggarwal G (2022) Next-generation sequencing in the biodiversity conservation of endangered medicinal plants. Environ Sci Pollut Res 29(49):73795–73808

Sharma RS, Vaidya N, Maloo SR, Kumar A, Sharma S, Ramkrishnan RS, Kumari V (2023) Application of molecular markers in assessment of genetic diversity of medicinal plants. In: Kumar N (ed) Molecular marker techniques: a potential approach of crop improvement. Springer Nature, Singapore, pp 103–116

Shelake RM, Jadhav AM, Bhosale PB, Kim JY (2023) Unlocking secrets of nature's chemists: potential of CRISPR/Cas-based tools in plant metabolic engineering for customized nutraceutical and medicinal profiles. Plant Physiol Biochem 203:108070

Singh V, Pandey Y, Sharma N, Singh N, Gupta AK, Tiwari G (2025) A novel species-specific SCAR marker-based unambiguous molecular identification and conservation of *Mentha spicata* L. Genet Resour Crop Evol 31:1–5

Siqueira MVBM, Carlos JS, Orcini W, Bajay MM, Martins K, Melo ATDO, Veasey EA, Tambarussi EV, Konzen ER (2025) Genomic diversity and structure of *Copaifera langsdorffii* populations from a transition zone between the Atlantic Forest and the Brazilian savanna. Plants 14(18):2858

Srivastava S, Misra A, Mishra P, Shukla P, Kumar M, Sundaresan V, Negi KS, Agrawal PK, Rawat AK (2017) Molecular and chemotypic variability of forskolin in *Coleus forskohlii* Briq., a high value industrial crop collected from Western Himalayas (India). RSC Adv 7(15):8843–8851

Ssenku JE, Okurut SA, Namuli A, Kudamba A, Tugume P, Matovu P, Wasige G, Kafeero HM, Walusansa A (2022) Medicinal plant use, conservation, and the associated traditional knowledge in rural communities in Eastern Uganda. Trop Med Health 50(1):39

Swamy MK, Kumar A (2023) Phytochemical genomics: plant metabolomics and medicinal plant genomics. Springer Nature

Terletskaya NV, Turzhanova AS, Khapilina ON, Zhumagul MZ, Meduntseva ND, Kudrina NO, Korbozova NK, Kubentayev SA, Kalendar R (2023) Genetic diversity in natural populations of *Rhodiola* species of different adaptation strategies. Genes 14(4):794

Tian Q, Han W, Wang D, Wang Z (2024) Heterologous expression of MYB gene (Rosea1) or bHLH Gene (Delila) from *Antirrhinum* increases the phenolics pools in *Salvia miltiorrhiza*. Int J Mol Sci 25(22):11917

Tiwari S, Shukla PK, Chauhan D, Khatoon S (2024) Variability study of active metabolites and identification of elite chemotype of *Withania somnifera* L. Dunal collected from Central India. J Planar Chromat–Mod TLC 37(2):189–198

Tozzi FP, Carranza ML, Frate L, Stanisci A (2021) The impact of *Acacia saligna* on the composition and structure of the Mediterranean maquis. Biodiversity 22(1–2):53–66

Upadhyay AK, Chacko AR, Gandhimathi A, Ghosh P, Harini K, Joseph AP, Joshi AG, Karpe SD, Kaushik S, Kuravadi N, Lingu CS (2015) Genome sequencing of herb Tulsi (*Ocimum tenuiflorum*) unravels key genes behind its strong medicinal properties. BMC Plant Biol 15(1):212

Vaishnav K, Tiwari V, Durgapal A, Meena B, Rana TS (2023) Estimation of genetic diversity and population genetic structure in *Gymnema sylvestre* (Retz.) R. Br. ex Schult. populations using DAMD and ISSR markers. J Genet Eng Biotechnol 21(1):42

Varma A, Shrivastava N (2018) Genetic structuring in wild populations of two important medicinal plant species as inferred from AFLP markers. Plant Biosyst 152(5):1088–1100

Wang Q, Zhang HX (2022) Population genetic structure and biodiversity conservation of a relict and medicinal subshrub *Capparis spinosa* in arid Central Asia. Diversity 14(2):146

Wang HM, Jeng ST, To KY (2017) In vitro regeneration, *Agrobacterium*-mediated transformation, and genetic assay of chalcone synthase in the medicinal plant *Echinacea pallida*. Plant Cell Tissue Organ Cult 130(1):117–130

Wang M, Lin H, Lin H, Du P, Zhang S (2024) From species to varieties: how modern sequencing technologies are shaping medicinal plant identification. Genes 16(1):16

Wariss HM, Liu T, Zhang H, Wu J, Yang Z, Li W (2025) Genetic diversity and population structure of the endangered medicinal plant *Ferula sinkiangensis*. Glob Ecol Conserv 58:e03437

Wu WG, Dong LL, Chen SL (2020) Development direction of molecular breeding of medicinal plants. China J Chinese Materia Medica 45(11):2714–2719

Xavier JK, Baia TG, Alegria OV, Figueiredo PL, Carneiro AR, Moreira EC, Maia JG, Setzer WN, da Silva JK (2022) Essential oil chemotypes and genetic variability of *Cinnamomum verum* leaf samples commercialized and cultivated in the Amazon. Molecules 27(21):7337

Yang X, Popova E, Shukla MR, Saxena PK (2019) Root cryopreservation to biobank medicinal plants: a case study for *Hypericum perforatum* L. In Vitro Cell Dev Biol Plant 55(4):392–402

Yang Y, Ke X, Ji Q, Lang T, Lai Z, Guan Y (2024) Conservation implications of population structure and dynamics in medicinal arbor *Albizia odoratissima* on Hainan Island, China. Forests 15(12):2227

Zahra W, Rai SN, Birla H, Singh SS, Rathore AS, Dilnashin H, Keswani C, Singh SP (2019) Economic importance of medicinal plants in Asian countries. In: Keswani C (ed) Bioeconomy for sustainable development. Springer, Singapore, pp 359–377

Zeng L, Sun Z, Fu L, Gu Y, Li R, He M, Wei J (2024) Cryopreservation of medicinal plant seeds: strategies for genetic diversity conservation and sustainability. Plants 13(18):2577

Zhan X, Chen Z, Chen R, Shen C (2022) Environmental and genetic factors involved in plant protection-associated secondary metabolite biosynthesis pathways. Front Plant Sci 13:877304

Zhang W, Zeng Y, Jiao M, Ye C, Li Y, Liu C, Wang J (2023) Integration of high-throughput omics technologies in medicinal plant research: the new era of natural drug discovery. Front Plant Sci 14:1073848

Chapter 2
Genetic Mechanisms Underlying the Diversification of Secondary Metabolism

Nguyen Thi Hong Nhung, Nguyen Thi Le Thuy, and Nguyen Thanh Triet

Abstract Plant genetic diversity underpins the biosynthesis of a vast array of bioactive compounds that play essential roles in ecological adaptation, human health, and drug discovery. Secondary metabolites such as alkaloids, terpenoids, flavonoids, and phenolics are highly diverse, shaped by both genetic and environmental factors, and often occur in specific cultivars, populations, or species. Advances in evolutionary theory, genomics, and pan-genomics have revealed that processes including gene duplication, whole-genome duplication, horizontal gene transfer, and epigenetic regulation drive chemical diversification and the emergence of novel metabolites. Biosynthetic gene clusters and transcriptional regulators further coordinate specialized metabolism, reflecting a complex interplay between genome architecture and ecological selection pressures. Case studies from *Centella asiatica*, *Panax ginseng*, *Panax vietnamensis*, and *Thymus vulgaris* highlight the close association between genetic variation and chemical phenotypes, demonstrating how genetic diversity determines chemotype formation, metabolite profiles, and therapeutic potential. These insights emphasize biodiversity as an invaluable resource for developing novel pharmaceuticals, nutraceuticals, and value-added products while underscoring the urgent need for conservation and sustainable use of plant genetic resources in the face of global biodiversity loss.

N. T. H. Nhung
Department of Biology, University of Medicine and Pharmacy at Ho Chi Minh City, Ho Chi Minh City, Vietnam
e-mail: nguyenthihongnhung@ump.edu.vn

N. T. Le Thuy
Department of Biotechnology, Ho Chi Minh City Open University, Ho Chi Minh City, Vietnam
e-mail: thuy.ntl@ou.edu.vn

N. T. Triet (✉)
Department of Traditional Pharmacy, University of Medicine and Pharmacy at Ho Chi Minh City, Ho Chi Minh City, Vietnam
e-mail: nguyenthanhtriet1702@ump.edu.vn

S. Jha et al. (eds.), *Genetic Diversity in Medicinal Plants*, Sustainable Development and Biodiversity 44, https://doi.org/10.1007/978-981-95-7281-6_2

Keywords Plant genetic diversity · Secondary metabolites · Chemical diversity · Biosynthetic gene clusters · Genomics and pan-genomics · Medicinal plants

2.1 Introduction

Biodiversity, particularly plant genetic diversity, serves as the foundation for maintaining ecosystem stability and provides essential raw materials for various industries, ranging from food and pharmaceuticals to cosmetics and biomaterials. Today, the majority of pharmaceutical drugs are derived from natural products of plant origin. Natural products are composed of numerous bioactive compounds, which exhibit biological activities against certain pathogens. To date, a wide array of secondary metabolites with diverse structures and pharmacological properties has been identified from plants. The medicinal properties of plants often result from different combinations of these natural compounds, collectively known as phytochemicals. Generally, phytochemicals are categorized into primary and secondary compounds. Primary compounds include chlorophyll, proteins, and carbohydrates, whereas secondary compounds encompass terpenoids, alkaloids, flavonoids, and phenolics (Naism et al. 2022).

Many fruits, vegetables, and herbs are rich in phytochemicals, such as phenolic compounds (phenolic acids, flavonoids, quinones, coumarins, lignans, stilbenes, and tannins), nitrogen-containing compounds (alkaloids, amines, and betalains), vitamins (A, C, D, and E), and terpenes (including carotenoids). Different phytochemicals exert distinct pharmacological effects—for instance, terpenoids demonstrate antibacterial, anti-inflammatory, anticancer, antimalarial, and antiviral activities; alkaloids are primarily known for their anesthetic properties; phenolic compounds play a crucial role in neutralizing free radicals; and flavonoids, one of the largest phenolic groups, possess multiple clinical properties such as anti-atherosclerotic, anti-inflammatory, antitumor, and antiviral effects. Consequently, plant-derived compounds are considered a rich source of natural antioxidants and contribute significantly to the development of modern drugs for treating diseases such as tumors, liver disorders, and arthritis (Saboon et al. 2019).

Biodiversity exists at three levels: species, habitat, and genetic diversity. Natural products, particularly secondary metabolites (alkaloids, flavonoids, terpenoids, phenolics, etc.), extracted from plants embody these three levels of biodiversity (Pushpangadan et al. 2018). Importantly, bioactive compounds often occur only in specific lines, cultivars, or species due to genetic variation, which constitutes valuable genetic resources. Approximately 70,000 medicinal plants have been used in therapeutic applications, primarily within Asian traditional medicine (Naism et al. 2022). In Vietnam, approximately 5100–5117 medicinal plant species are utilized in traditional medicine as well as in the development of modern pharmaceutical products (Ministry of Agriculture and Rural development, 2021). Some typical Vietnamese medicinal plants are illustrated in Figs. 2.1 and 2.2.

Fig. 2.1 Some typical Vietnamese medicinal plants. (**a**) *Camellia yokdonensis* Dung & Hakoda. (Photos by Le Viet Hung), (**b**) *Camellia dormoyana* (Pierre ex Laness.) Sealy. (Photos by Doan Thanh Luan), (**c**) *Camellia cattienensis* Orel. (Photo by Le Viet Hung), (**d**) *Distichochlamys rubrostriata* W.J. Kress & Rehse. (Photos by Le Viet Hung), (**e**) *Stahlianthus involucratus* (King ex Baker) Craib ex Loes. (Photo by Nguyen Thanh Triet), and (**f**) *Distichochlamys orlowii* K. Larsen & M. F. Newman. (Photos by Le Viet Hung)

Medicinal plants possess an inexhaustible repository of ethnobotanical compounds currently exploited in diverse drug discovery and development programs. Some potential bioactive substances in Vietnamese medicinal plants are presented in Table 2.1. The inherent potential of plant biodiversity as a key resource for developing novel value-added products in food, medicine, cosmetics, and other natural products is increasingly recognized. Furthermore, plant genetic resources determine species adaptability to climate change, diseases, and environmental fluctuations. Biodiversity is thus an invaluable natural asset that must be conserved and sustainably used for the benefit of both present and future generations. Yet, humanity has

Fig. 2.2 Some typical Vietnamese medicinal plants. (**a**) *Goniothalamus takhtajanii* Ban. (Photo by Phan Le Y), (**b**) Panax vietnamensis Ha et Grushv. (Photo by Vo Ngoc Linh Giang), (**c**) *Aeginetia indica* L. (Photo by Doan Thanh Luan), and (**d**) *Cyclacanthus coccineus* S. Moore. (Photo by Nguyen Thanh Triet)

only tapped into a fraction of nature's vast genetic library. Therefore, the sustainable utilization of biodiversity is of paramount importance. However, biodiversity worldwide is currently under serious threat of depletion due to overexploitation.

A newly emerging biotechnology application concerning biodiversity is known as bioprospecting, defined as "the systematic search for genes, natural compounds, designs, and whole organisms in the wild with potential product development value, using biological observations along with physiological, biochemical, and genetic methods without causing harm to nature." Within this field, genetic prospecting has become a promising domain. With advances in DNA fingerprinting technologies, it is now possible to identify and characterize genetic variations within a species or population at the DNA level, describe and evaluate genetic resources (particularly endemic species), and provide a foundation for discovering new drugs and drug targets, elucidating mechanisms of drug action, and maintaining records of developed pharmaceuticals and their therapeutic efficacy (Pushpangadan et al. 2018).

Table 2.1 Content of bioactive compounds in the potential of Vietnamese medicinal plants

Sl. No.	Medicinal plants (Vietnamese name)	Family	Parts of the plant used	Active constituents	Biological activities	References
1.	*Scutellaria barbata* (Bán chi liên)	Lamiaceae	Leaves	Scutellarein (flavonoid)	Anticancer, anti-inflammatory, anti-obesity, hepatoprotective activities	Nguyen and Ninh (2025)
2.	*Artemisia annua* (Thanh hao hoa vàng)	Asteraceae	Stems, leaves	Artemisinin	Anti-breast cancer activities	Lai et al. (2006)
3.	*Scutellaria baicalensis* (Hoàng cầm)	Lamiaceae	Roots	Baicalein	Anti-proliferative and apoptotic effects	Sato et al. (2012)
4.	*Garcinia blongifolia* (Bứa)	Clusiaceae	Branch	Benzophenones	Anticancer, antioxidant, antifungal, apoptotic, and anti-pathogenic activities	Li et al. (2016)
5.	*Lumnitzera racemosa* (Cóc trắng)	Combretaceae	Leaves, twigs	Steroids, triterpenoids, flavonoids, lignans, phenolic compounds	Anticancer activity against MCF-7 and HeLa cell lines	Eswaraiah et al. (2020)
6.	*Bruguiera sexangula* (Lour.) Poir. (Vẹt đen)	Rhizophoraceae	Barks, stems	Alkaloids, dithiobenzoquinone, diterpenes, phenolic glycosides, flavonoids, triterpenoids	A-549 human lung cancer and HL-60 human leukemic cell lines	Li et al. (2010)

(continued)

Table 2.1 (continued)

Sl. No.	Medicinal plants (Vietnamese name)	Family	Parts of the plant used	Active constituents	Biological activities	References
7.	*Hedyotis diffusa* (Bạch hoa xà thiệt thảo)	Rubiaceae		Polysaccharides, triterpenes, anthraquinones	Cancers of the liver, colon, lungs, brain, and pancreas	Chen et al. (2016)
8.	*Morus alba* (Dâu tằm trắng)	Moraceae	*Leaves, roots*	Albanol A	HL-60 (human leukemia) and Crl1579 (human melanoma) cell lines	Kikuchi et al. (2010)
9.	*Paris polyphylla* (Thất diệp nhất chi hoa)	Melanthiaceae	Rhizomes	Steroidal saponins	A549 cell lines	Li et al. (2013)
10.	*Perilla frutescens* (Tía tô xanh)	Lamiaceae	Leaves	Isoegomaketone	Huh-7 hepatoma cell carcinoma	Ying et al. (2012)

2.2 Relationship Between Genetic Diversity and Secondary Metabolite Biosynthesis

Plants produce a wide range of secondary metabolites—compounds not directly involved in primary growth processes but strongly influencing flavor, human and animal health, and serving as an important source of pharmaceutical discovery. These metabolites play key roles in ecological interactions such as pollinator attraction, seed dispersal, defense against herbivores and pathogens, and allelopathy. To date, tens of thousands of compounds have been described (e.g., ~21,000 alkaloids, 22,000 terpenoids, and 5000 flavonoids and tannins) (Kroymann 2011), though the actual number is likely much higher. Chemical diversity in plants is expressed through variation in both the quality and quantity of secondary metabolites within and across species, reflecting the richness of gene families involved. Recent advances in understanding the diversity and adaptability of secondary metabolism in plants have been applied to address questions of evolutionary origins, mechanisms of formation, adaptive roles, and the maintenance of chemical diversity in plant lineages.

2.2.1 Genetic Diversity Drives Chemical Diversification

Plants are widely recognized as natural chemical factories, with each species capable of producing dozens to hundreds of unique secondary metabolites (Afendi et al. 2012). These metabolites are classified into major chemical groups such as terpenoids, alkaloids, phenylpropanoids, and steroids, and their biosynthesis is regulated by multiple factors, including physiology, genotype, and environment. They are synthesized via distinct biochemical pathways, each characteristic of a particular class of chemical entities produced in plants (Vleminckx et al. 2018; Forrister et al. 2019). The abundance and relative composition of these metabolites constitute a chemical phenotype referred to as chemical diversity. Studies have demonstrated that intraspecific genetic diversity can lead to substantial variation in chemical diversity across individuals or populations. Genetics plays a pivotal role in determining the types and quantities of metabolites produced, while environmental factors modulate their expression and accumulation. Essentially, genes provide the blueprint for metabolite biosynthesis, whereas the environment shapes their manifestation (Forrister et al. 2022). Plants generate chemical novelty ranging from new metabolites to unique molecular scaffolds. The evolution of novel compounds or distinctive combinations of metabolites is referred to as chemical traits, which enhance adaptability, promote health, and facilitate species coexistence (Forrister et al. 2022). Understanding the origin and maintenance of chemical diversity thus lies at the core of plant evolutionary biology and ecology. This raises a fundamental question: how do plants generate chemical diversity, and what evolutionary processes drive the emergence of novel metabolites and distinctive chemical traits?

The "Escape and Radiate" model, proposed by Ehrlich and Raven (1964), provides a theoretical framework in evolutionary ecology to explain co-evolutionary dynamics between plants and herbivores as a mechanism driving species diversification. According to this model, when a plant lineage evolves a novel defense trait—such as toxic secondary metabolites, specialized morphological structures, or physiological mechanisms that reduce herbivory—it can effectively "escape" selective pressure imposed by herbivores. Freed from these constraints, the lineage may expand into new ecological niches and undergo adaptive radiation, giving rise to multiple descendant lineages adapted to different environments. Over time, herbivores may evolve counter-adaptations that overcome plant defenses, reestablishing interactions and initiating a new evolutionary cycle in which plants once again develop novel strategies. This iterative process is considered a major driver of rapid diversification in both plants and herbivores. Random mutations in biosynthetic genes generate new defensive metabolites, often through progressive elaboration of core chemical scaffolds into increasingly complex derivatives. If these novel compounds provide stronger deterrence or broader protection against diverse enemies, natural selection favors the corresponding genotypes.

A study of 100 tropical tree species in the genus *Inga* (Fabaceae) tested the "Escape and Radiate" hypothesis by examining the stepwise chemical evolution of secondary metabolites (Forrister et al. 2022). Inga, comprising ~300 species across Neotropical rainforests, provides a valuable model to investigate secondary metabolic evolution over short phylogenetic distances. The study integrated untargeted metabolomics with comparative phylogenetics to characterize chemical traits across species. Inga species invest heavily in soluble secondary metabolites, particularly phenolics and saponins, which are closely associated with defense against herbivores and pathogens. Remarkably, ~10,000 unique compounds were identified across the genus, the majority belonging to phenolics and saponins. Phenolic chemistry in *Inga* derives from the flavonoid pathway, with heavily modified flavone and flavan monomer/polymer backbones, while saponins are glycosylated triterpenoids synthesized via the mevalonic acid pathway, thus representing distinct biosynthetic origins. Each Inga species harbors structurally unrelated compounds, reflecting high levels of phytochemical diversification. Closely related species exhibit strikingly different chemical profiles, with unique sets of compounds and compound classes showing little or no phylogenetic signal. For example, each species expresses distinct classes of defensive metabolites, including quinic acid gallates, tyrosine derivatives and related depsides, flavones, flavonoids, flavan-3-ols, and saponins. Phylogenetic signal was detected for certain traits such as quinic acid gallates, tyrosine-related depsides, and saponin glycosides, but was absent for flavonoid subclasses. Chemical similarity was highest within species but declined sharply between species, and even within species, similarity decreased between individuals from different geographic locations. Closely related species in sympatry displayed highly divergent chemical profiles, often more distinct than those of sister species in nearby regions, suggesting a selective advantage for chemical differentiation within ecological communities. Such divergence indicates rapid chemical evolution consistent with adaptive divergence, further supported by the fact that most compounds are highly labile and many compound classes exhibit weak or absent phylogenetic signal.

Ultimately, these results challenge the "Escape and Radiate" model of gradual scaffold elaboration, as each *Inga* species maximizes phytochemical diversification by producing structurally unrelated compounds (Forrister et al. 2022). This suggests that chemical diversification in Inga is not constrained to incremental modification of ancestral metabolites but instead involves rapid, extensive, and independent evolution of novel metabolic traits.

2.2.2 Genetic Mechanisms Underlying the Diversification of Secondary Metabolism

Genetics plays a central role in determining the types and quantities of secondary metabolites produced by plants. Although these compounds are not essential for plant growth and development, they confer a wide range of ecological and adaptive advantages, being influenced by both genetic factors and environmental conditions. Fundamentally, genes provide the blueprint for metabolite biosynthesis, while the environment modulates their expression and accumulation. Secondary metabolites are synthesized through distinct pathways characteristic of specific chemical classes in plants. These biosynthetic pathways often follow linear or branched architectures, in which any genetic or regulatory change along the pathway—whether in a single gene or gene cluster—can affect metabolite production and accumulation (Smith et al. 2019).

Recent studies of metabolic diversity in plants have been strengthened by integrative approaches, including comparative genomics, transcriptomics, and gene expression profiling, leading to more precise insights into metabolic divergence among related species (Aboud 2024). For example, genome comparisons between *Arabidopsis thaliana* and its sister species, *Arabidopsis lyrata,* revealed that genes in the thalianin (T10) biosynthetic pathway are present in both genomes. However, *A. lyrata* lacks two additional reductase genes, resulting in the production of a novel triterpenoid, epi-thalianin (T17), an epimer of thalianin, in its roots (Aboud 2024).

A crucial first step in tracing the evolution of metabolic diversity is the identification of structural genes (encoding enzymes) that underlie biosynthetic pathways (Smith et al. 2019). Primary metabolism provides the essential metabolites required for growth, development, and reproduction, while secondary (or specialized) metabolism is typically described as serving ecological functions, enabling plants to survive under complex biotic and abiotic conditions. Genes underlying primary metabolism are subject to strong purifying selection and are highly conserved across lineages, whereas genes associated with specialized metabolites are often shaped by relaxed or diversifying selection. Even minor genetic variations can give rise to novel metabolites with previously unknown bioactivities, forming chemical traits that represent a distinct dimension of biodiversity crucial for both natural and managed ecosystems. Recent evidence also suggests that primary and secondary metabolic diversification can be integrated, with specialized downstream pathways branching from primary metabolism. With the advancement of molecular biology tools, genome sequencing has revealed clear links between metabolic diversity and genetic variation (Smith et al. 2019).

Gene and Genome Duplication

Since the sequencing of the *A. thaliana* genome in 2000 (The Arabidopsis Genome Initiative 2000), hundreds of plant genomes of varying depth and quality have been published. High-quality genomes, in combination with transcriptomic and metabolomic data, have propelled evolutionary studies of chemical innovation and metabolite distribution. Gene duplication and whole-genome duplication (WGD) are widely recognized as fundamental drivers of metabolic expansion in plants (Zhou and Liu 2022). Gene duplication underlies the diversification of many enzyme families such as cytochrome P450 oxidoreductases, hydrolases, acyltransferases, glycosyltransferases, methyltransferases, and lipases/esterases, all of which play central roles in specialized metabolism (Barakat et al. 2011; Leong and Last 2017; Louveau and Osbourn 2019). Duplicated genes can diversify in multiple ways to acquire new enzymatic functions, generate novel metabolic classes, or alter the diversity within existing metabolite classes (Zhou and Liu 2022).

The complexity of metabolic pathways, particularly those involved in the formation of highly elaborate chemical structures or exhibiting cross-interactions with other pathways, poses substantial challenges to investigating metabolic diversification driven by WGD. To date, only a few examples of WGD-driven metabolic diversification have been reported in plants (Zhou and Liu 2022).

Metabolic diversification associated with WGD has been identified across various classes of plant compounds. The expansion of the *CYP79* gene family, originating from the second meso-polyploid WGD event in the Brassicaceae, laid the foundation for glucosinolate diversification in this family (e.g., *A. thaliana*) (Zhou and Liu 2022). The differential accumulation of oleic and linoleic acids produced in olive (*Olea europaea*) compared to its closely related oilseed species in the sesame family is partly attributable to the divergence of the *FAD2* gene, encoding a fatty acid desaturase, which arose from WGD (Zhou and Liu 2022). *N-methylputrescine oxidase* (*MPO1*), required for nicotine biosynthesis in *Nicotiana attenuata*, is thought to have originated from a Solanaceae-specific whole-genome triplication (Zhou and Liu 2022). In *Papaver somniferum*, three genes (*T6ODM, CODM,* and *COR*) were likely duplicated through WGD, and their encoded enzymes catalyze the final steps of the morphine biosynthetic pathway (Zhou and Liu 2022). Notably, WGD can result in the duplication of entire biosynthetic pathways, as observed in *Eriobotrya japonica* (loquat), where WGD-derived gene copies have been retained and coordinated to yield high levels of ursane-type triterpenes. Nevertheless, assessing the precise impact of WGD remains challenging due to the complexity of metabolic pathways and frequent post-duplication gene loss. Current evidence indicates that WGD not only fosters chemical diversification but also shapes unique evolutionary trajectories of plant metabolism (Zhou and Liu 2022).

Another instance, repeated WGD events in the Brassicales expanded gene families responsible for glucosinolate biosynthesis, thereby promoting the diversification of these defensive compounds through subsequent functional evolution of both regulatory and structural homologs. A similar pattern is observed in the evolution of acylsugars in the Solanaceae (Fig. 2.3; Smith et al. 2019). Hundreds of defensive acylsugars can be found on a single leaf of some species, arising from the functional diversification of the first three acylsugar acyltransferases (ASATs) in the biosynthetic pathway through complex molecular processes, including lineage-specific gene duplication, neofunctionalization, allelic divergence, and gene loss within a single enzyme family (Smith et al. 2019). Likewise, duplication of B: Benzyl alcohol-O-acetyltransferase (BEAT); A: Anthocyanin-Ohydroxycinnamoyltransferase (AHCT); H: N-Hydroxycinnamoyl/benzoyltransferase (HCBT); D: Deacetylvindoline-4-O-acetyltransferase (DAT) (BAHD) acyltransferase genes in *Ipomoea batatas* (Convolvulaceae) gave rise to ASAT lineages (ASAT1, ASAT2, and ASAT3) present

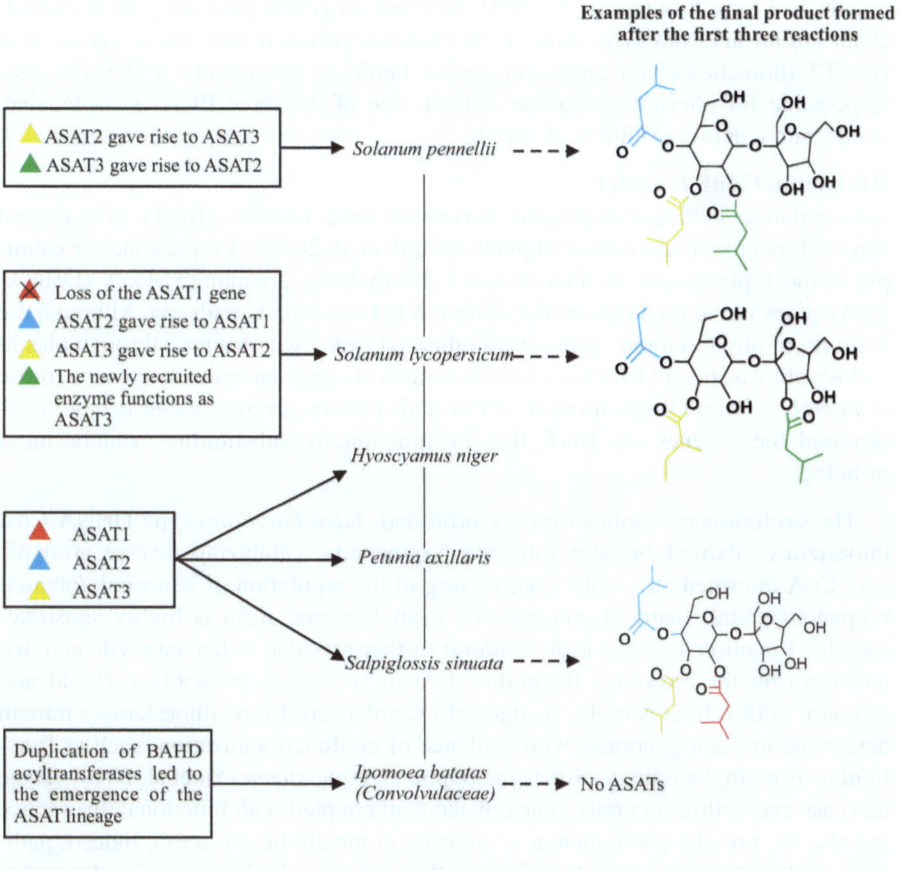

Fig. 2.3 The evolution of the first three acylsugar acyltransferases (ASATs) in the acylsugar biosynthetic pathway and their impact on acylsugar phenotypes in the Solanaceae family (Smith et al. 2019)

in Solanaceae, with species-specific expansions enabling the synthesis of diverse acylsugar products bearing four to seven acyl chains.

Some other studies have also identified gene family expansions in non-model plants as well. Examples include *CYP719* in *Coptis chinensis* (protoberberine alkaloids), *CHS* in *Mangifera indica* (urushiols) (Liu et al. 2021), and *CYP71D* in *Salvia miltiorrhiza* (tanshinones) (Ma et al. 2021). RNA-seq has also been instrumental in identifying biosynthetic genes, such as CHS-like enzymes associated with styrylpyrone biosynthesis of kavalactones in *Piper methysticum* (Pluskal et al. 2019).

Mutations in the genome, including single-nucleotide polymorphisms (SNPs), indels, domain losses, and gene fusions, can alter coding sequences, expression patterns, or translational efficiency of metabolic genes, thereby affecting downstream pathways. Genome-wide association studies (GWASs) have identified thousands of metabolic trait loci. For instance, advanced genetic mapping approaches have successfully identified quantitative trait loci (QTLs) that regulate variation in multiple metabolite classes, including amino acids, organic acids, glucosinolates, carotenoids, and tocochromanols. A GWAS and haploid genotyping study on branched-chain amino acid (BCAA) traits in Arabidopsis revealed that allelic variation at BCAT2 (branched-chain aminotransferase family)—specifically SNP5373—was responsible for altering isoleucine content, one of the three BCAAs (isoleucine, valine, and leucine) (Smith et al. 2019).

Horizontal Gene Transfer

Accumulating evidence highlights horizontal gene transfer (HGT) as a pivotal driver of metabolic innovation in plants (Smith et al. 2019). A representative example is the replacement of thioesterase 1,4-dihydroxy-2-naphthoyl-CoA (DHNA-CoA) genes in the phylloquinone (vitamin K1) biosynthetic pathway. Although the majority of phylloquinone genes were inherited from cyanobacteria through plastid endosymbiosis, plant DHNA-CoA thioesterases are phylogenetically closer to those of *Lactobacillales* (Widhalm et al. 2009). This pattern strongly suggests that plants acquired these genes via HGT, thereby functionally substituting cyanobacterial orthologs.

The evolutionary implications are profound. *Lactobacillales*-type DHNA-CoA thioesterases exhibit broader substrate promiscuity, catalyzing diverse aromatic acyl-CoA intermediates while contributing to the regulation of benzenoid/phenylpropanoid metabolism. In contrast, the cyanobacterial form is highly substrate-specific. Functional assays in *A. thaliana* further revealed a dual catalytic activity, underscoring the enzyme's integrative role in secondary metabolism (Field and Osbourn 2008). Interestingly, vestiges of cyanobacterial-type thioesterases remain detectable in plant genomes, with evidence of neofunctionalization, such as functioning as methylketone synthases in *Solanum lycopersicum* (Yu et al. 2010). Thus, this case exemplifies not only gene replacement coupled with functional divergence but also the broader phenomenon of convergent metabolic evolution. Indeed, pathways such as those for naphthoquinones, flavonoids, and other specialized metabolites repeatedly demonstrate independent origins yet yield analogous biochemical outcomes—for instance, in betalain, caffeine, and flavonoid biosynthesis. Collectively, these findings underscore gene duplication and HGT as complementary evolutionary forces that foster metabolic diversification and enable the emergence of novel plant traits (Smith et al. 2019).

Biosynthetic Gene Clusters

Biosynthetic gene clusters (BGCs)—defined as physically co-localized sets of genes encoding enzymes of a single metabolic pathway—are well established in bacteria and fungi but have only recently been recognized in plants (Osbourn 2010; Rokas et al. 2018). The maize DIMBOA cluster, responsible for benzoxazinone defense metabolites, was the first reported example (Field and Osbourn 2008). Since then, over 20 additional BGCs have been identified across diverse plant lineages, suggesting that clustering represents a recurring yet independently evolved genomic strategy (Nutzmann et al. 2016). Strikingly, both oats and *A. thaliana* developed distinct clusters for triterpenoid biosynthesis (avenacin and thalianol, respectively), converging on similar gene complements, including oxidosqualene cyclases, cytochrome P450 monooxygenases, and acyltransferases (Field and Osbourn 2008). Gene duplication and specialization are repeatedly implicated as major sources of functional novelty within these clusters, paralleling innovations in non-clustered pathways such as flavonoid biosynthesis (Chu et al. 2011; Boycheva et al. 2014).

The organization of metabolic pathways into BGCs is hypothesized to provide evolutionary advantages through co-regulation and co-inheritance of functionally related genes, thereby reducing the risk of toxic intermediate accumulation (Yu et al. 2008; McGary et al. 2013). Nevertheless, compared to fungi and bacteria, plants exhibit far fewer BGCs despite their extraordinary metabolic diversity. Several genomic and population-level constraints may underlie this discrepancy: the relatively conserved gene order in plant genomes (contrasting with rapid genome rearrangements in fungi), smaller effective population sizes, and the abundance of intergenic DNA, all of which reduce opportunities for functionally linked genes to cluster (Lynch and Conery 2003).

Taken together, the distribution of BGCs in plants reflects a balance between ecological selection pressures and genomic constraints. Rather than mirroring microbial strategies, plant BGCs illustrate a distinctive evolutionary solution for expanding specialized metabolism—one that relies heavily on the interplay between gene duplication, genomic architecture, and selective ecological advantages.

Transcriptional Regulators

The expression and induction of secondary metabolite biosynthesis are strongly governed by transcription factors (Yang et al. 2017). TFs are DNA-binding proteins that attach to the promoter regions of target genes to regulate the rate of transcription initiation mediated by RNA polymerase. They act as integrators of intracellular and extracellular signals to control the expression of enzyme-encoding genes, thereby modulating the accumulation of SMs (Aboud 2024).

In recent decades, numerous TFs have been identified, and their regulatory mechanisms in SM biosynthetic pathways have been increasingly elucidated (Naing and Kim 2018). Different TF families may act simultaneously, independently, or

Table 2.2 Transcription factors and their main functions in regulating the production of secondary metabolites in plants

Transcription factor family	Primary regulatory function	Species	Type of transcription factor	Validated target genes/ proteins	References
R2R3-MYB	Involved in the phenylpropanoid pathway and regulation of anthocyanin biosynthesis	Lotus	MnMYB5	GST	Sun et al. (2016)
		Rose	RcMYB1	ABG	He et al. (2023)
bHLH	Regulates diverse biological processes, including growth, development, and environmental adaptation	*Arabidopsis*	bHLH122	CYP707A3	Chen et al. (2013)
		Paeonia suffruticosa Andrews	PsbHLH1	PsDFR	Guo et al. (2021)
WD40	Participates in the formation of the MYB-bHLH-WD (MBW) complex associated with anthocyanin biosynthesis	Wheat	TaWD40-4B.1	TaCAT1 and TaCAT3	Tian et al. (2023)
		Actinida chinensis	AcWD40.063 (AcTTG1)	AcMYB75	Yang et al. (2025)

synergistically in the regulation of SM production in plants. A detailed catalog of TFs and their primary functions is presented in Table 2.2.

Epigenetic Modifications Facilitating Divergence

Changes in the regulation of gene expression involving biosynthetic enzyme-encoding genes (structural genes) enable distantly related species to produce the same compound, while closely related species may synthesize different metabolites (Moore et al. 2014; Courtois et al. 2016). For example, one sister species may produce saponins, whereas its close relative produces phenolics, thereby imposing distinct detoxification challenges on herbivores and pathogens (Mith et al. 2019). Similarly, the production of S-linalool, a monoterpene alcohol that contributes to the floral scent of *Clarkia breweri*, has been attributed to the divergent expression of the biosynthetic enzyme S-linalool synthase (Dudareva et al. 1996). The mRNA of this enzyme accumulates at high levels in the petals, stigmas, styles, and stamens of *C. breweri* flowers but is only weakly detectable in the stigma of its scentless close relative, *C. concinna*, leading to extremely low production of S-linalool and its oxide derivatives, insufficient to attract pollinators (Raguso and Pichersky 1995).

Regulation as a model of chemical evolution implies that species maintain a complete repertoire of biosynthetic enzymes within their genomes, differentially upregulated or downregulated among taxa, with "unused" genes having to retain their functionality over evolutionary time scales. Preliminary genomic analyses in two *Inga* species indicate that core biosynthetic genes related to flavonoid and saponin production are present across species, even in those that do not accumulate these compound classes. The persistence of such apparently "unused" enzymes may be essential due to the profound conservation and pleiotropy of core biosynthetic enzymes. Several hypotheses explain how these "unused" genes are maintained. First, many intermediates do not accumulate to physiologically significant levels, yet their enzymes must still be transcribed and functional because they are required for downstream products. This is exemplified by phenylpropanoid intermediates linking the shikimate pathway with the flavonoid pathway. Second, certain metabolites absent in leaves may instead accumulate in other tissues (Smith et al. 2019). Key epigenetic mechanisms underlying such regulation include RNA interference (RNAi), DNA methylation, and histone modifications.

RNA Interference

Secondary metabolism is regulated by non-coding RNAs (ncRNAs). In plants, ncRNAs act as important transcriptional regulators of secondary metabolites and participate in the biosynthesis of bioactive compounds. Four major ncRNA classes have been identified to date: circular RNAs, long ncRNAs, small interfering RNAs (siRNAs), and microRNAs. Genes associated with secondary metabolite biosynthesis in plants are modulated by these ncRNAs. For instance, under heat stress, species such as *Petunia hybrida*, *Glycine max*, and *A. thaliana* activate siRNA-mediated biosynthetic mutations; siRNAs regulate *CHS* genes and flavonoid production (Aboud 2024).

DNA Methylation

DNA methylation, an epigenetic modification present in both prokaryotes and eukaryotes, primarily occurs at cytosine bases. When localized to promoter regions, methylation can suppress transcription by hindering transcription factor binding or recruiting repressor proteins. In plants, two main forms of methylation exist: de novo and maintenance methylation, with DRM methyltransferases playing a central role in de novo initiation (Aboud 2024). The stability of DNA methylation allows it to be inherited across generations, shaping development, physiology, and environmental adaptation (Kooke et al. 2019; Zhang et al. 2019).

Beyond regulating gene expression, DNA methylation also affects transposon activity, chromosomal stability, and metabolite accumulation (Aboud 2024). Studies in *A. thaliana* suggest a mediating role of small RNAs in transcriptional regulation via methylation (Přibylová et al. 2019). Experimental evidence directly links DNA methylation with secondary metabolite production: anthocyanins in apple, ginsenosides in ginseng, phenolics in *S. miltiorrhiza* hairy root cultures, and benzylisoquinoline alkaloids in opium poppy (Leung and Gaudin 2020). Thus, DNA methylation is regarded as a critical genetic–epigenetic regulatory mechanism contributing to the biosynthesis and accumulation of plant secondary metabolites.

Histone Modifications

Histone modifications such as acetylation, methylation, phosphorylation, and ubiquitination play central roles in the epigenetic regulation of secondary metabolism in plants. Histone acetylation promotes transcription by loosening chromatin structure, thereby facilitating transcription factor binding at biosynthetic gene regulatory regions, whereas deacetylation results in chromatin compaction and transcriptional repression. Other histone modifications also modulate chromatin dynamics and gene regulation, profoundly influencing secondary metabolism. Understanding these epigenetic mechanisms provides crucial insights into plant development, flowering, and longevity and offers applications in crop improvement and agricultural resilience (Aboud 2024).

2.2.3 Explaining Chemical Diversification Through Evolutionary Models

The diversity of natural products in living organisms can be interpreted through the lens of various evolutionary theories. Evolutionary models such as the *Fluctuating Selection Hypothesis*, the *Dominance Reversal Hypothesis*, the *Interaction Diversity Hypothesis (IDH)*, the *Synergy Hypothesis*, and the *Screening Hypothesis* have been proposed as potential mechanisms maintaining or promoting chemodiversity via natural selection and environmental interactions (Johnson et al. 2018).

The *Fluctuating Selection Hypothesis* posits that selective pressures vary over time due to environmental factors such as climate, seasonality, pathogens, or biotic competition. These fluctuations cause the fitness advantage of individual variants to shift dynamically, with certain variants being advantageous at one point and disadvantageous at another, thereby promoting the coexistence of multiple variants within a population, especially under "storage effects" or rare-allele advantages. For instance, in *Trifolium repens* (white clover), cyanogenic phenotypes that produce hydrogen cyanide (HCN) deter herbivores but are frost-sensitive. Consequently, the frequency of HCN-producing individuals increases in warmer seasons and decreases in colder ones, thus sustaining chemical diversity within the population (Johnson et al. 2018).

The *Dominance Reversal Hypothesis* explains why heterozygotes may maintain a selective advantage and contribute to the preservation of genetic diversity. According to this hypothesis, the dominance level of an allele is not fixed but varies depending on the environment or trait in question. In one context, allele A may be dominant over allele a, whereas in another, allele a becomes dominant. As a result, heterozygotes (Aa) can achieve higher fitness compared to homozygotes (AA or aa), since one allele will always provide an adaptive advantage depending on environmental conditions. A classic example is pesticide resistance in insects: allele R confers resistance but reduces reproductive fitness in pesticide-free environments, while allele S enhances reproduction but is susceptible to pesticides. When pesticides are applied intermittently, R is dominant under pesticide pressure and S in its absence, granting heterozygotes (RS) a balanced advantage. This mechanism represents a form of balancing selection that sustains genetic variants in natural populations (Connallon and Chenoweth 2019).

The *IDH* emphasizes that the stability and functionality of ecosystems depend not only on species richness but also on the diversity of interactions among species. Ecosystems with simple or limited interaction networks are less resilient to environmental perturbations compared to those with complex, functionally complementary, and redundant interactions. This perspective holds particular significance in natural product research, as the biosynthesis and accumulation of secondary metabolites (e.g., alkaloids, flavonoids, and terpenoids) are often shaped by biotic interactions. For example, in the genus *Capsicum*, selective pressures from both frugivorous animals (birds and rodents) and insect herbivores promoted the evolution of capsaicinoids, compounds that simultaneously deter pathogens and insects while attracting birds for seed dispersal. Similarly, in *Curcuma longa* (turmeric), diverse interactions with soil microbes, rhizome-boring insects, and pollinators exerted selective pressures that drove the biosynthesis of curcuminoids—polyphenolic compounds with antimicrobial, antioxidant, and tissue-protective properties. Hence, the richness of biotic interaction networks contributes fundamentally to the unique chemodiversity observed in nature (Gaüzère et al. 2022).

The *Synergy Hypothesis* suggests that the biological efficacy of compound mixtures is often greater or qualitatively different compared to that of individual isolated molecules. Such synergistic effects may manifest as enhanced pharmacological activity, reduced toxicity, or improved bioavailability. This hypothesis provides a rationale for why crude extracts frequently exhibit stronger biological activity than purified compounds. A notable example is curcumin from *Curcuma longa*, which demonstrates poor bioavailability and weak anti-inflammatory effects in isolation, but its co-administration with piperine from black pepper enhances curcumin's bioavailability by nearly 20-fold, thereby amplifying its therapeutic efficacy. Likewise, in *Camellia sinensis* (green tea), epigallocatechin gallate synergizes with caffeine to enhance energy metabolism. Furthermore, supportive compounds in natural extracts may inhibit microbial resistance mechanisms, such as flavonoids, reducing efflux pump activity when combined with antibiotics like erythromycin. These findings underscore the importance of the Synergy Hypothesis in natural product research and highlight its translational potential in developing multi-component pharmaceuticals and harnessing traditional medicinal resources (Wagner and Ulrich-Merzenich 2009).

Finally, the *Screening Hypothesis* (Firn and Jones 2003) proposes that the extraordinary diversity of secondary metabolites results from an evolutionary strategy to increase the probability of generating biologically active molecules. While the likelihood of any single compound having potent bioactivity is low, the production of large chemical repertoires increases the probability of discovering a molecule with significant biological function. For example, in the genus *Curcuma*, beyond the well-studied anti-inflammatory and antioxidant curcumin, plants produce numerous curcuminoids, terpenoids, and essential oils. Although most of these compounds exhibit modest or negligible activity, a few display notable pharmacological potential. Similarly, *Streptomyces* species generate hundreds of secondary metabolites, some of which have become foundational antibiotics in modern medicine. Thus, the Screening Hypothesis highlights chemical diversification as a

Table 2.3 Comparison of five evolutionary hypotheses on the chemodiversity of bioactive compounds

Hypothesis	Advantages	Limitations	Potential Applications
Fluctuating Selection Hypothesis	Explains how environmental variability maintains genetic and metabolic diversity; highlights the adaptive value of natural products under changing conditions.	Difficult to test experimentally due to the need for long-term ecological data; may not account for the stability of certain metabolites across lineages.	Understanding ecological drivers of metabolite diversity; guiding conservation strategies for medicinal plants in variable environments.
Dominance Reversal Hypothesis	Clarifies how genetic variation is maintained through fitness trade-offs between alleles in different environments.	Evidence mainly from population genetics; direct links to metabolite diversity are still limited.	Breeding programs exploiting heterozygote advantage; predicting plant chemical responses to climate change.
IDH	Emphasizes the role of multi-species interactions (e.g., plant–herbivore–microbe) in shaping metabolite diversity.	Complex to disentangle specific drivers; requires integrative ecological and metabolomic approaches.	Designing polyculture systems that enhance plant chemical defenses; exploring co-evolutionary dynamics for drug discovery.
Synergy Hypothesis	Explains why mixtures of compounds can be more effective than isolated molecules; accounts for functional complementarity.	Quantifying synergistic effects is challenging; disentangling true synergy from additive effects requires robust experimental design.	Development of multi-component drugs and phytomedicines; valorization of whole extracts in pharmaceutical and nutraceutical industries.
Screening Hypothesis	Highlights the stochastic nature of metabolite evolution through gene duplication, mutation, and ecological filtering.	May overemphasize randomness and underplay adaptive mechanisms; difficult to predict specific outcomes.	Discovery of novel metabolites through genome mining; guiding synthetic biology approaches by mimicking natural "trial-and-error" evolution.

probabilistic evolutionary strategy for producing bioactive natural products (Firn and Jones 2003). To further clarify the applicability of these hypotheses, their respective strengths and limitations are summarized in Table 2.3.

The analysis of the five evolutionary hypotheses discussed above reveals that each theoretical framework elucidates a distinct aspect of the origin and maintenance of natural product diversity. The Fluctuating Selection Hypothesis emphasizes the role of environmental variability; the dominance reversal hypothesis explains the preservation of genetic variants at the allelic level; while the *IDH* extends the perspective to multi-species ecological networks. The Synergy Hypothesis provides a compelling rationale for the persistence of multi-component chemical mixtures, whereas the screening hypothesis highlights the stochastic and exploratory nature of metabolic evolution. Although each of these hypotheses has inherent limitations, together they constitute a multidimensional evolutionary framework to account for the persistence and expansion of chemodiversity. Ultimately, the diversification of natural compounds emerges as the outcome of complex interactions among selection, genetics, ecology, and molecular mechanisms. This integrative perspective not only advances our understanding of the evolutionary mechanisms underlying secondary metabolism but also holds significant applied potential, ranging from novel strategies in crop breeding and medicinal plant utilization to the design of sustainable agricultural systems.

2.3 Applications of Genomics and Pan-Genomics in the Study of Secondary Metabolism in Plants

A current trend is the use of genomics and pan-genomics tools to gain deeper insights into the genetic basis controlling the biosynthesis of secondary metabolites and to identify key variants that influence their chemical repertoire. Secondary metabolism in plants gives rise to numerous biologically active compounds, including alkaloids, flavonoids, terpenoids, and glucosinolates (Zhou et al. 2022). These compounds not only play essential ecological roles—such as providing defense against herbivores and pathogens or attracting pollinators—but also possess considerable medicinal and nutritional value. However, elucidating the biosynthetic pathways of secondary metabolites remains challenging due to the complexity of plant genomes, species-specific traits, and the prevalence of large structural variations (SVs). In this context, genomics and pan-genomics have emerged as powerful approaches to decode and exploit natural chemical diversity (Zhou et al. 2022).

Applications of genomics have facilitated the identification of BGCs in plant genomes (Bayer et al. 2020). With the aid of tools such as *plantiSMASH* (Kautsar et al. 2017), numerous clusters responsible for the biosynthesis of alkaloids, terpenoids, and flavonoids have been characterized. The integration of transcriptomic and metabolomic datasets enables co-expression analyses, which can predict missing genes within metabolic pathways (Ho et al. 2020). Moreover, approaches based on mGWAS and eQTL mapping link genetic variants to metabolite accumulation, thereby pinpointing candidate loci underlying metabolic diversity. For example, in *P. somniferum* (Liu et al. 2020), genome sequencing has led to the discovery of gene clusters controlling benzylisoquinoline alkaloid biosynthesis, including the key gene *STORR*. In *Camellia sinensis* (Wei et al. 2018), WGD events have expanded gene families involved in the phenylpropanoid pathway, contributing to the enhanced accumulation of catechins and the unique biochemical properties of tea.

While single-reference genomics provides a foundational view, pan-genomics offers the capacity to explain chemical diversity across cultivars, accessions, and closely related species (Zhou et al. 2022). A pan-genome, which consists of both a core genome and an accessory genome, reveals presence–absence variation (PAV) and SV in biosynthetic genes or clusters—variations that are often missed when relying on a single reference genome (Zhou et al. 2022). A prominent example is the *Solanum lycopersicum* pan-genome, where the identification of the rare allele

TomLoxC (Gao et al. 2019) has been directly associated with improved flavor mediated by apocarotenoid volatiles. Similarly, in the Brassica genus, PAVs of glucosinolate-related genes explain differences in flavor and defensive capacity among morphotypes. Notably, graph-based pan-genome representations are becoming a new paradigm, enabling the precise mapping of complex SVs and their direct integration with metabolomic data (Zhou et al. 2022).

Overall, the combination of genomics and pan-genomics not only advances our understanding of the evolution of secondary metabolic pathways but also provides actionable tools for molecular breeding and synthetic biology. Looking forward, the integration of next-generation sequencing technologies with multi-omics and machine learning algorithms is expected to accelerate the discovery of novel genes and metabolites, thereby promoting the efficient utilization of the natural chemical repertoire in plants.

2.4 Case Study

Example 1: Centella asiatica—Impact of Genetic Diversity on the Content of Four Centellosides (Pentacyclic Triterpenoids)

Centella asiatica (L.) Urban (family Apiaceae) (Gray et al. 2017; Kunjumon et al. 2022) is an important medicinal herb, renowned for its neuroprotective, memory-enhancing, cardioprotective, antioxidant, anti-inflammatory, anticancer, and wound-healing properties. To date, more than 130 secondary metabolites have been isolated from *C. asiatica.*

A study on 50 accessions collected from South India and cultivated under uniform conditions ensured that chemical variation was genetically determined rather than influenced by environmental factors. The results revealed pronounced differences in the content of the four major centellosides—asiatic acid, madecassic acid, asiaticoside, and madecassoside—ranging from 0.00% to 0.29%. The biotransformation of these centellosides involves two key triterpenoid modifications: C6 hydroxylation by CYP450-dependent monooxygenases and C28 glycosylation by UDP-Glc glucosyltransferases. The hydroxylated forms at C6 (madecassic acid: asiatic acid ratio 4.00; madecassoside: asiaticoside ratio 3.61) and the glycosylated forms at C28 (asiaticoside: asiatic acid ratio 14.67; madecassoside: madecassic acid ratio 13.25) predominated over their corresponding non-derivatives. This highlights these two reactions as the principal biotransformation pathways in *C. asiatica.*

Chemical clustering combined with inter-simple sequence repeat (ISSR) markers across accessions from India and China further revealed four distinct chemotypes of *C. asiatica*, corresponding to the four main centellosides.

Conclusion: C. asiatica exemplifies how genetic origin plays a pivotal role in regulating the levels of four secondary saponins, the centellosides.

Example 2: Panax ginseng—Genetic Variation and Ginsenoside Content
Ginsenosides, the principal bioactive compounds in ginseng, are widely employed in the treatment of various diseases. Studies have shown that transcriptional regulation, including alternative splicing, contributes to differences in ginsenoside content between cultivated and wild ginseng. Transcription factors such as MYB, AP2/ERF, NAC, and bZIP have been implicated in modulating the expression of ginsenoside biosynthetic genes (Schlag and Mclntosh 2013; Liu et al. 2023; Fang et al. 2024).

Additional research demonstrated that ginseng germ cells harbor over 1.5 million SNPs, associated with variability in the accumulation of 16 distinct ginsenosides, suggesting that genetic variation underlies differential biosynthesis. A slight reduction in genetic diversity was observed between wild and cultivated ginseng, with approximately 330 genes identified as being related to ginsenoside composition. Furthermore, genetic distance was strongly correlated with divergence in ginsenoside profiles, indicating that different populations exhibit distinct chemical signatures aligned with genetic differentiation.

Conclusion: Genetic variation in *Panax ginseng* directly influences the ginsenoside biosynthetic network, leading to variation in composition and abundance of bioactive compounds across cultivars and ecological populations.

Example 3: Panax vietnamensis—Genetic Diversity and Chemotype in Ginsenoside Composition
Research on *Panax vietnamensis* (Ngoc Linh ginseng) (Tien et al. 2021), an endemic species of Vietnam, has provided compelling evidence for the relationship between genetic diversity and chemical variability in ginsenoside composition. High-performance liquid chromatography analysis revealed significant differences in the content and profiles of ginsenosides—particularly characteristic compounds such as majonoside-R2, ginsenoside-Rb1, Rg1, and Rh1—among natural populations and cultivated lines across different geographical locations. In parallel, molecular marker analyses (Simple Sequence Repeat (SSR) and Amplified Fragment Length Polymorphism (AFLP)) demon In Viet Nam, approximately strated substantial levels of genetic variation among populations, reflecting both evolutionary history and ecological adaptation. By integrating these datasets, the study demonstrated that variations in ginsenoside content and structure are closely linked to genetic differentiation at the population level, thereby explaining the formation and maintenance of chemotypes in this species. These findings not only elucidate the biological basis underlying ginsenoside diversity in *P. vietnamensis* but also hold practical implications for genetic resource conservation, elite line selection, and the development of high-quality medicinal products.

Example 4: Thymus vulgaris—Chemotypes Associated with Genetic Variation
The term "chemotype" refers to an intraspecific subgroup distinguished by the production of specific secondary metabolites not typically present in the standard form of the species. A classic example of multiple chemotypes is *Thymus vulgaris* (Woldetsadik et al. 2016; Pluhár et al. 2024). Although morphologically indistinguishable, individuals of *T. vulgaris* can be classified into seven chemotypes based on essential oil composition, dominated by thymol, carvacrol, linalool, geraniol,

sabinene hydrate (thuyanol), α-terpineol, or eucalyptol. Such designations (e.g., *T. vulgaris* ct. thymol or *T. vulgaris* ct. geraniol) are not taxonomically valid but reflect close links between genetic variation and secondary metabolite composition. Population genetic studies combined with chemical analyses demonstrated that chemotype differentiation has ecological value—facilitating environmental adaptation—as well as practical importance for medicinal and essential oil industries. ISSR markers have been employed to discriminate chemotypes (e.g., thymol and carvacrol), with each appearing as genetically distinct clusters. According to Thompson et al. (2003), essential oil polymorphism represents an adaptive strategy to specific environments. Long-term adaptation has resulted in genetically regulated pathways of essential oil biosynthesis, driven by variation in monoterpene composition. This is further supported by the wide chromosomal variation observed within the genus *Thymus* ($2n = 24–90$), encompassing diploid, tetraploid, and hexaploid forms, likely derived from the basic numbers $x = 14$ and $x = 15$ (Kalvandi et al. 2012).

Conclusion: Thymus vulgaris provides clear evidence of the link between genetic variation and chemical phenotypes (chemotypes), with significant applications in breeding and essential oil production.

The representative examples of *C. asiatica, P. ginseng, T. vulgaris,* and *Panax vietnamensis* illustrate that genetic diversity not only underpins differences in productivity and ecological adaptation but also plays a central role in determining the content, composition, and applied value of secondary metabolites. This knowledge provides a critical foundation for the exploitation, conservation, and breeding of medicinal plants.

2.5 Conclusion

Biodiversity, particularly plant genetic diversity, represents a cornerstone of natural chemical innovation. The interplay between genetic variation and secondary metabolism not only underpins the ecological adaptability of plants but also serves as the molecular foundation for the discovery of novel bioactive compounds. Gene duplication, HGT, biosynthetic gene clustering, and epigenetic regulation collectively drive the diversification of metabolic pathways, generating immense chemical diversity across species. Advances in genomics, pan-genomics, and integrative multi-omics approaches have further elucidated the genetic architecture controlling secondary metabolite biosynthesis, enabling the identification of key genes, variants, and regulatory networks underlying chemical phenotypes.

Case studies across *Centella asiatica*, *Panax ginseng*, *Panax vietnamensis*, and *Thymus vulgaris* clearly demonstrate that intraspecific genetic diversity translates directly into variation in secondary metabolite composition, resulting in distinct chemotypes and pharmacological profiles. These findings highlight that genetic diversity is not merely an evolutionary relic but a dynamic engine of biochemical innovation with profound ecological and biotechnological implications.

Ultimately, the sustainable conservation and utilization of biodiversity are essential not only for maintaining ecosystem resilience but also for unlocking the full potential of nature's chemical repertoire in developing next-generation pharmaceuticals, nutraceuticals, and other bio-based products.

References

Aboud NMAI (2024) Unlocking the genetic potential: strategies for enhancing secondary metabolite biosynthesis in plants. J Saudi Soc Agric Sci 23(8):542–554

Afendi FM, Okada T, Yamazaki M, Morita AH, Nakamura Y, Nakamura K, Ikeda S, Takahashi H, Amin MAU, Darusman LK, Saito K, Kanaya S (2012) KNApSAcK family databases: integrated metabolite–plant species databases for multifaceted plant research. Plant Cell Physiol 2(53):1–12

Barakat A, Choi A, Yassin NBM, Park JS, Sun Z, Carlson JE (2011) Comparative genomics and evolutionary analyses of the *O*-methyltransferase gene family in Populus. Gene 479(1–2):37–46

Bayer PE, Golicz AA, Scheben A, Batley J, Edwards D (2020) Plant pan-genomes are the new reference. Nat Plants 6:914–920

Boycheva S, Daviet L, Wolfender JL, Fitzpatrick TB (2014) The rise of operon-like geneclusters in plants. Trends Plant Sci 19:447–459

Chen Y, Li F, Ma Y, Chong K, Xu Y (2013) Overexpression of *OrbHLH001*, a putative helix-loop-helix transcription factor, causes increased expression of AKT1 and maintains ionic balance under salt stress in rice. J Plant Physiol 170(1):93–100

Chen R, He J, Tong X, Tang L, Liu M (2016) The *Hedyotis diffusa* Willd. (Rubiaceae): a review on phytochemistry, pharmacology, quality control and pharmacokinetics. Molecules 21(6):710–740

Chu HY, Wegel E, Osbourn A (2011) From hormones to secondary metabolism: the emergence of metabolic gene clusters in plants. Plant J 66:66–79

Connallon T, Chenoweth SF (2019) Dominance reversals and the maintenance of genetic variation for fitness. PLoS Biol 17:e3000118

Courtois EA, Dexter KG, Paine C, Timothy E, Stien D, Engel J, Baraloto C, Chave J (2016) Evolutionary patterns of volatile terpene emissions across 202 tropical tree species. Ecol Evol 6:2854–2864

Dudareva N, Cseke L, Blanc VM, Pichersky E (1996) Evolution of floral scent in *Clarkia*: novel patterns of *S*-linalool synthase gene expression in the *C. breweri* flower. Plant Cell 8:1137–1148

Ehrlich PR, Raven PH (1964) Butterflies and plants: a study in coevolution. Evolution 18:586–608

Eswaraiah G, Peele KA, Krupanidhi S, Indira M, Kumar RB, Venkateswarulu TC (2020) GC–MS analysis for compound identification in leaf extract of *Lumnitzera racemosa* and evaluation of its *in vitro* anticancer effect against MCF7 and HeLa cell lines. J King Saud Univ Sci 32(1):780–783

Fang X, Zhou X, Wang Y, Zhang W, Wu Hao XL, Sun M, Xiao H (2024) Determining the genetic basis of ginsenosides variation during ginseng domestication by evolutionary transcriptomics. Ind Crop Prod 212:118369

Field B, Osbourn AE (2008) Metabolic diversification: independent assembly of operon-like gene clusters in different plants. Science 320:543–547

Firn RD, Jones CG (2003) A simple model to explain chemical diversity. Nat Prod Rep 20(4):382–391

Forrister DL, Endara MJ, Younkin GC, Coley PD, Kursar TA (2019) Herbivores as drivers of negative density dependence in tropical forest saplings. Science 363:1213–1216

Forrister DL, Endara JM, Soule IA, Younkin CG, Mills GA, Lokvam J, Dexter GK, Pennington RT, Kidner CA, Nicholls AJ, Loiseau O, Kursar TA, Colley PD (2022) Diversity and divergence: evolution of secondary metabolism in the tropical tree genus *Inga*. New Phytol 237(2):631–642

Gao L, Gonda I, Sun H, Ma Q, Bao K, Tienman DM, Chang EAB, Fish TL, Stromberg KA, Sacks GL, Thannhauser TW, Foolad DJM, Blanca J, Canizares J, Xu Y, Knaap EVD, Huang S, Klee HJ, Giovannoni JJ, Fei Z (2019) The tomato pan-genome uncovers new genes and a rare allele regulating fruit flavor. Nat Genet 51:1044–1051

Gaüzère P, O'Connor L, Botella C, Poggiato G, Münkemüller T, Pollock LJ, Brose U, Maiorano L, Harfoot M, Thuiller W (2022) The diversity of biotic interactions complements functional and phylogenetic facets of biodiversity. Curr Biol 32(9):2093–2100.e3

Gray NE, Alcazar MA, Lak P, Wright KM, Quinn J, Stevens JF, Maier CS, Soumyanath A (2017) *Centella asiatica* - Phytochemistry and mechanisms of neuroprotection and cognitive enhancement. Proc Phytochem Soc Eur 17(1):161–194

Guo J, Sun B, He H, Zhang Y, Tian H, Wang B (2021) Current understanding of bHLH transcription factors in plant abiotic stress tolerance. Intern J Mol Sci 22(9):4921

He G, Zhang R, Jiang S, Wang H, Ming F (2023) The *MYB* transcription factor *RcMYB1* plays a central role in rose anthocyanin biosynthesis. Horticult Res 10(6):uhad080

Ho SS, Urban AE, Mills RE (2020) Structural variation in the sequencing era. Nat Rev Genet 21:171–189

Johnson MTJ, Prashad CM, Lavoignat M, Saini HS (2018) Contrasting the effects of natural selection, genetic drift and gene flow on urban evolution in white clover (*Trifolium repens*). Proc Royal Soc B 285:20181019

Kalvandi R, Hejazi SMH, Atri M, Mirza M, Jamzad Z, Safikhani K (2012) Karyotpe analysis among 10 populations of *Thymus eriocalyx* (Ronniger) Jalas species in Iran. Ann Biol Res 3(8):3916–3925

Kautsar SA, Suarez Duran HG, Blin K, Osbourn A, Medema MH (2017) PlantiSMASH: automated identification, annotation and expression analysis of plant biosynthetic gene clusters. Nucleic Acids Res 45:W55–W63

Kikuchi T, Nihei M, Nagai H, Fukushi H, Tabata K, Suzuki T, Akihisa T (2010) Albanol A from the root bark of *Morus alba* L. induces apoptotic cell death in HL60 human leukemia cell line. Chem Pharm Bull 58(4):568–571

Kooke R, Morgado L, Becker F, Eekelen HV, Hazarika R, Zheng Q, De Vos RCH, Johannes F, Keurentjes JJB (2019) Epigenetic mapping of the arabidopsis metabolome reveals mediators of the epigenotype-phenotype map. Genome Res 29(1):96–106

Kroymann J (2011) Natural diversity and adaption in plant secondary metabolism. Curr Opin Plant Biol, 14(3): 246–251

Kunjumon R, Johnson AJ, Remadevi RKSR, Baby S (2022) Assessment of major centelloside ratios in *Centella asiatica* accessions grown under identical ecological conditions, bioconversion clues and identification of elite lines. Sci Rep 12:8177

Lai H, Singh NP (2006) Oral artemisinin prevents and delays the development of 7: 12-dimethylbenz[a]anthracene (DMBA)-induced breast cancer in the rat. Cancer Lett 231(1):43–48

Leong BJ, Last RL (2017) Promiscuity, impersonation and accommodation: evolution of plant specialized metabolism. Curr Opin Struct Biol 47:105–112

Leung J, Gaudin V (2020) Who rules the cell? An epi-tale of histone, DNA, RNA, and the metabolic deep state. Plant Cell Biol 11:181

Li L, Huang CG, Wang CY, Guo YW (2010) Sexangulic acid, a new cytotoxic triterpenoid from the Chinese mangrove *Bruguiera sexangula*. Nat Prod Res 24(11):1044–1049

Li Y, Gu JF, Zou X, Wu J, Zhang MH, Jiang J, Qin D, Zhou JY, Liu BXZ, Zhu YT, Jia XB, Feng L, Wang RP (2013) The anti-lung cancer activities of steroidal saponins of P. polyphylla Smith var. chinensis (Franch.) Hara through enhanced immunostimulation in experimental Lewis tumor-bearing C57BL/6 mice and induction of apoptosis in the A549 cell line. Molecules 18(10):12916–12936

Li P, Senthilkumar HA, Wu S, Bo L, Guo Z, Fata JE, Kennelly EJ, Long C (2016) Comparative UPLC-QTOF-MS-based metabolomics and bioactivities analyses of *Garcinia oblongifolia*. J Chromatogr B Analyt Technol Biomed Life Sci 1011:179–195

Liu Z, Duran HGS, Harnvanichvech Y, Stephenson MJ, Schranz ME, Nelson D, Medema MH, Osbourn A (2020) Drivers of metabolic diversification: how dynamic genomic neighbourhoods generate new biosynthetic pathways in the Brassicaceae. New Phytol 227:1109–1123

Liu Y, Wang B, Shu S, Li Z, Song C, Liu D, Niu Y, Liu J, Zhang J, Liu H, Hu Z, Huang B, Liu X, Liu W, Jiang L, Alami MM, Zhou Y, Ma Y, He X, Yang Y, Zhang T, Hui H, Barker MS, Chen S, Wang X, Nie J (2021) Analysis of the *Coptis chinensis* genome reveals the diversification of protoberberine-type alkaloids. Nat Commun 12:3276

Liu S, Jiang Y, Wang Y, Huo H, Cilkiz M, Chen P, Han Y, Li L, Wang K, Zhao M, Zhu L, Leo J, Wang Y, Zhang M (2023) Genetic and molecular dissection of ginseng (*Panax ginseng* Mey.) germplasm using high-density genic SNP markers, secondary metabolites, and gene expressions. Front. Plant Sci 14:1165349

Louveau T, Osbourn A (2019) The sweet side of plant-specialized metabolism. Cold Spring Harb Perspect Biol 11(12):a034744

Lynch M, Conery JS (2003) The origins of genome complexity. Science 302:1401–1404

Ma Y, Cui G, Chen T, Ma X, Wang R, Jin B, Yang J, Kang L, Tang J, Lai C, Wang Y, Zhao Y, Shen Y, Zeng W, Peters RJ, Qi X, Guo J, Huang L (2021) Expansion within the CYP71D subfamily drives the heterocyclization of tanshinones synthesis in *Salvia miltiorrhiza*. Nat Commun 12:685

McGary KL, Slot CJ, Rokas A (2013) Physical linkage of metabolic genes in fungi is an adaptation against the accumulation of toxic intermediate compounds. Proc Nat Acad Sci 110:11481–11486

Ministry of Agriculture and Rural Development (2021) Untapped potential of medicinal plants in Vietnam. *Vietnam Agriculture.*. Retrieved from https://vietnamagriculture.nongnghiep.vn/untapped-potential-of-medicinal-plants-in-vietnam-d284901.html. Accessed 30 Oct 2025

Moore BD, Andrew RL, Külheim C, Foley WJ (2014) Explaining intraspecific diversity in plant secondary metabolites in an ecological context. New Phytol 201:733–750

Naing AH, Kim CK (2018) Roles of *R2R3-MYB* transcription factors in transcriptional regulation of anthocyanin biosynthesis in horticultural plants. Plant Mol Biol 98(1–2):1–18

Nasim N, Sandeep SS, Mohanty S (2022) Plant-derived natural products for drug discovery: current approaches and prospects. Springer Nature 65(3):399–411

Nguyen TT, Ninh TS (2025) Scutellarein: a review of chemistry and pharmacology. J Pharm Pharmacol 77(3):352–370

Nutzmann HW, Huang AC, Osbourn A (2016) Plant metabolic clusters—from genetics to genomics. New Phytol 211:771–789

Osbourn A (2010) Secondary metabolic gene clusters: evolutionary toolkits for chemical innovation. Trends Genet 26:449–457

Pluhár Z, Gosztola B, Kun R, Cservenka J, Neumayer É, Sárosi TS, Radácsi P (2024) Variations in essential oil composition and chemotype patterns of wild Thyme (*Thymus*) species in the natural habitats of Hungary. Horticulturae 10(2):150

Pluskal T, Torrens-Spence MP, Fallon TR, Abreu AD, Shi CH, Weng JK (2019) The biosynthetic origin of psychoactive kavalactones in kava. Nat Plants 5:867–878

Přibylová A, Čermák V, Tyč D, Fischer L (2019) Detailed insight into the dynamics of the initial phases of de Novo RNA-directed DNA methylation in plant cells. Epigenetics Chromatin 12:54

Pushpangadan P, George V, Ijinu TP, Chithra MA (2018) Biodiversity, bioprospecting, traditional knowledge, sustainable development and value added products: a review. J Tradit Med Clin Naturopathy 7(1):1000256

Raguso RA, Pichersky E (1995) Floral volatiles from *Clarkia breweri* and *C. concinna* (Onagraceae): recent evolution of floral scent and moth pollination. Plant Syst Evol 194:55–67

Rokas A, Wisecaver JH, Lind AL (2018) The birth, evolution and death of metabolic gene clusters in fungi. Nat Rev Microbiol 16:731–744

Saboon, Chaudhari SK, Arshad S, Amjad MS, Akhtar MS (2019) Natural compounds extracted from medicinal plants and their applications. In: Natural bio-active compounds. Springer, Singapore, pp 193–207

Sato D, Kondo S, Yazawa K, Mukudai Y, Li C, Kamatani T, Katsuta H, Yoshihama Y, Shirota T, Shintani S (2012) The potential anticancer activity of extracts derived from the roots of *Scutellaria baicalensis* on human oral squamous cell carcinoma cells. Mol Clin Onco 11(1):105–111

Schlag EM, McIntosh MS (2013) The relationship between genetic and chemotypic diversity: American ginseng (*Panax quinquefolius* L.). Phytochemistry 93:96–104

Smith SD, Angelovici R, Karolina H, Maeda HA, Moghe GD, Pires JC, Joshua RW, Wisecaver HJ (2019) The renaissance of comparative biochemistry. Am J Bot 106(1):3–13

Sun SS, Gugger PF, Wang QF, Chen JM (2016) Identification of a *R2R3-MYB* gene regulating anthocyanin biosynthesis and relationships between its variation and flower color difference in lotus (*Nelumbo* Adans.). Peer J 1(4):e2369

The Arabidopsis Genome Initiative (2000) Analysis of the genome sequence of the flowering plant. *Arabidopsis thaliana*. Nature 408:796–815

Thompson JD, Chalchat JC, Michet A, Linhart YB, Ehlers B (2003) Qualitative and quantitative variation in monoterpene co-occurrence and composition in the essential oil of *Thymus vulgaris* chemotypes. J Chem Ecol 29:859–880

Tian G, Wang S, Wu J, Wang Y, Wang X, Liu S, Han D, Xia G, Wang M (2023) Allelic variation of *TaWD40-4B.1* contributes to drought tolerance by modulating catalase activity in wheat. Nat Comun 14:1200

Tien NQĐ, Ma X, Man LQ, Chi DTK, Huy NX, Nhut DT, Rombauts S, Ut T, Loc NH (2021) *De novo* whole-genome assembly and discovery of genes involved in triterpenoid saponin biosynthesis of Vietnamese ginseng (*Panax vietnamensis* Ha et Grushv.). Physiol Mol Biol Plants (27)10:2215–2229

Vleminckx J, Salazar D, Fortunel C, Mesones I, Dávila N, Lokvam J, Beckley K, Baraloto C, Fine PVA (2018) Divergent secondary metabolites and habitat filtering both contribute to tree species coexistence in the Peruvian Amazon. Front Plant Sci 9:836

Wagner H, Ulrich-Merzenich G (2009) Synergy research: approaching a new generation of phytopharmaceuticals. Phytomedicine 16(2–3):97–110

Wei C, Yang H, Wang S, Jian Z, Liu C, Gao L, Xia E, Lu Y, Tai Y, She G, Sun J, Cao H, Tong W, Gao Q, Li Y, Deng W, Jiang X, Wang W, Chen Q, Zhang S, Li H, Wu J, Wang P, Li P, Shi C, Zheng F, Jian J, Huang B, Shan D, Shi M, Fang F, Jian J, Huang B, Shan D, Shi M, Fang C, Yue Y, Li F, Li D, Wei S, Han B, Jiang C, Yin Y, Xia T, Zhang Z, Bennetzen JL, Zhao S, Wan X (2018) Draft genome sequence of *Camellia sinensis* var. *sinensis* provides insights into the evolution of the tea genome and tea quality. Proc Natl Acad Sci USA 115(18):E4151–E4158

Widhalm JR, Oostende VC, Furt F, Basset JG (2009) A dedicated thioesterase of the Hotdog-fold family is required for the biosynthesis of the naphthoquinone ring of vitamin K-1. Proc Natl Acad Sci 106(14):5599–5603

Woldetsadik TF, Tesfaye K, Feye TA, Shiferaw E (2016) Assessing genetic diversity and population structure of Thyme (*Thymus Schimperi* Ronniger) in Eastern, Central and Northern Highlands of Ethiopia. JAD 6(1):1

Yang N, Zhou W, Jiao S, Wang X, Li L, Wang L, Cao X, Wang Z (2017) Overexpression of SmMYC2 increases the production of phenolic acids in *Salvia miltiorrhiza*. Front Plant Sci 8:1804

Yang S, Wang Y, Wang W, Wang N, Yan R, Li S, Zhang T, Liu J, Zeng X, Zhao S, Zhang X, Dong Q, Luan H, Guo S, Qi G, Jia P (2025) Analysis of *WD40* genes in kiwifruit reveals the key role of the light-induced *AcTTG1-AcMYB75-AcbHLH2* complex in anthocyanin accumulation. Int J Biol Macromol 297:139758

Ying W, Xiaoxu H, Jianming H, Wenling Z & Wenli M (2012) Extract of Perilla frutescens inhibits tumor proliferation of HCC via PI3K/AKT signal pathway. Afr J Tradit Complement Altern Med 10(2): 251–257.

Yu G, Nguyen TTH, Guo Y, Abatedaga I, Auldridge ME, Bhuiyan N, Israel IB, Iijima Y, Fridman E, Noel JP, Pichersky E (2010) Enzymatic functions of wild tomato methylketone synthases1 and 2. Plant Physiol 154:67–77

Yu L, Lopez A, Anaflous A, Bali BE, Hamal A, Ericson E, Heisler LE, McQuibban A, Giaever G, Nislow C, Boone C, Brown GW, Bellaoui M (2008) Chemical -genetic profiling of imidazo [1,2-a]pyridines and – pyrimidines reveals target pathways conserved between yeast and human cells. Plos Genet, 4(11).

Zhang Y, Ji A, Zhichao X, Luo H, Song J (2019) The AP2/ERF transcription factor SmERF128 positively regulates diterpenoid biosynthesis in *Salvia Miltiorrhiza*. Plant Mol Biol 100(1–2):83–93

Zhou X, Liu Z (2022) Unlocking plant metabolic diversity: a (pan)-genomic view. Plant Commun 3(2):100300

Chapter 3
Climate Change and the Future of Healing: Tracking Genetic Erosion and Phytochemical Shifts in Medicinal Plants

Sudip Pandey and Poonam Pant

Abstract Global climate change is increasingly affecting the distribution, growth, and chemical composition of medicinal plants, posing severe threats to conservation and sustainable use. In Nepal, which is both a country with rich biodiversity and rich traditional knowledge, medicinal plants make significant contributions to local healthcare systems and national economies. However, rising temperatures, altered precipitation patterns, and elevated rates of extreme weather occurrences are disrupting essential natural habitats and threatening the genetic integrity and phytochemical contents of many significant medicinal plants, which undermines the efficacy and quality of herbal drugs for traditional applications and new phytopharmaceutical uses. This review examines the impact of climate change on phytochemical and genetic variation of medicinal plants in Nepal, identifying vulnerable species, ecological transformation, and adaptation mechanisms. Emphasis is placed on integrating conservation efforts that harmonize traditional practices, scientific research, and climate-resilient cultivation techniques in an effort to protect Nepal's valuable medicinal plant diversity against climate change.

Keywords Climate change · Medicinal plants · Genetic diversity · Phytochemical composition · Nepal

S. Pandey (✉)
Wood Science Laboratory, Forest Biomaterials Science and Engineering Program, Institute of Applied Sciences, Madan Bhandari University of Science and Technology, Chitlang, Bagmati Province, Nepal
e-mail: sudip.pandey@mbust.edu.np

P. Pant
Central Institute of Science and Technology, Kathmandu, Nepal

S. Jha et al. (eds.), *Genetic Diversity in Medicinal Plants*, Sustainable Development and Biodiversity 44, https://doi.org/10.1007/978-981-95-7281-6_3

3.1 Introduction

Medicinal plants have served as a cornerstone of human health and well-being for millennia (WHO 2023; Chaachouay and Zidane 2024). From Ayurvedic and Traditional Chinese Medicine (TCM) to indigenous knowledge systems in Africa, the Americas, and Oceania, botanical remedies constitute an essential part of primary healthcare for nearly 80% of the world's population (Bussmann and Sharon 2006; WHO 2009; De Boer and Cotingting 2014; Oyebode et al. 2016; Tinitana et al. 2016; Turpin et al. 2022; Sánchez-Ramos et al. 2024; Meñiza et al. 2024). These plants are not only valued for their therapeutic properties but also form the basis for the discovery of many modern drugs, including aspirin (*Salix* spp.), artemisinin (*Artemisia annua*), and morphine (*Papaver somniferum*) (Balunas and Kinghorn 2005; Newman and Cragg 2020; Nasim et al. 2022; Pandey and Pant 2023; Chaachouay and Zidane 2024). However, the rapidly intensifying global climate crisis poses a significant threat to the sustainability, accessibility, and pharmacological reliability of these invaluable resources (Pant et al. 2021; El Gendy et al. 2023; Alum 2024). Key stressors include rising temperatures, changing rainfall patterns, prolonged droughts, increased atmospheric CO_2 levels, and the growing frequency of extreme weather events (Parmesan and Yohe 2003; IPCC 2023). These factors directly influence the distribution, phenology, and reproductive success of medicinal plant species (Saslis-Lagoudakis et al. 2014; Woldearegay 2020). In some cases, climate-induced habitat shifts are leading to altitudinal and latitudinal migrations, which may push certain species into ecological bottlenecks or result in local extinctions (Pandey et al. 2018; Baumbach et al. 2021; Maharjan et al. 2023). Such changes disrupt the intricate ecological networks that support plant growth and the synthesis of bioactive compounds.

Two of the most concerning results of climate stress on medicinal plants are genetic erosion and phytochemical shifts. Genetic erosion refers to the loss of genetic variation within and among plant populations, a trend exacerbated by habitat fragmentation, overharvesting, and rapid environmental change (Khoury et al. 2022; Póvoa et al. 2023). Loss of genetic diversity can reduce the adaptive capacity of species, making them more susceptible to pests, diseases, and future climatic fluctuations (Jump and Peñuelas 2005; Priyadarshan 2016). This is particularly critical for medicinal plants with narrow ecological ranges or those that are endemic to climate-sensitive regions such as alpine zones, tropical rainforests, and arid savannas. Ecological surveys have reported a dramatic shrinkage in the natural habitat of *Nardostachys jatamansi*, especially in Uttarakhand and Himachal Pradesh, due to a combination of over extraction, habitat degradation, and the impacts of climate change (Kala 2005; Kaur et al. 2020). Similarly, *Picrorhiza kurroa* (kutki) has experienced 50% genetic diversity loss in heavily harvested populations, particularly affecting genes responsible for producing medicinal Picrosides (Yadav and Singh 2024).

Phytochemical shifts, on the other hand, involve alterations in the production and concentration of secondary metabolites, such as alkaloids, flavonoids, terpenoids, and glycosides, which are responsible for the therapeutic actions of medicinal plants (Zidorn 2010; Qaderi et al. 2023). These compounds are highly sensitive to environmental conditions. For example, elevated temperatures and drought stress have been shown to increase the concentration of certain phenolic compounds in *Hypericum perforatum* (St. John's Wort) but simultaneously reduce overall biomass and flowering (Zobayed et al. 2005). Similarly, *A. annua* grown under elevated CO_2 conditions exhibited increased artemisinin yield in some studies, but variability in concentration and bioavailability remains a concern (Oladipo and Ojumu 2025). *Centella asiatica* (Gotu Kola), widely used in Ayurveda and traditional Southeast Asian medicine, exhibits lower triterpenoid content under prolonged heat stress, diminishing its neuroprotective and wound-healing properties (Sun et al. 2020). Likewise, *Morinda citrifolia* (Noni), valued for its anti-inflammatory and antioxidant effects, suffers from fruit abortion and reduced phytochemical yields during droughts (Hou et al. 2025). These climate-induced physiological changes weaken plant populations, making them more susceptible to diseases and habitat loss. Furthermore, these chemical changes have downstream implications for safety, efficacy, and dosage consistency issues that are already challenging in the regulation of herbal medicine (Ekor 2014). Shifts in phytochemistry can render a traditionally effective remedy less potent or, in some cases, toxic. This threatens not only healthcare access for vulnerable populations but also the pharmacological pipeline for future drug discovery.

To address these emerging challenges, this chapter explores the interplay between climate change, genetic erosion, and phytochemical variability in medicinal plants. We examine the mechanisms behind these changes, present illustrative case studies, and evaluate the tools and methodologies used to monitor them, including molecular markers, metabolomic profiling, and ecological modeling. We also discuss adaptive conservation strategies, such as in-situ and ex-situ preservation, climate-resilient cultivation techniques, and the integration of indigenous knowledge systems. In doing so, we aim to provide a comprehensive understanding of how climate change is reshaping the pharmacological landscape and outline urgent priorities for research, policy, and practice.

3.2 Medicinal Plant Diversity and Climate Sensitivity

Climate change significantly impacts medicinal plant biodiversity, affecting species distribution, growth, ecosystem dynamics, and bioactive compounds (Shrestha et al. 2022; Kunwar et al. 2023; Alum 2024). Rising temperatures, altered precipitation patterns, and increased atmospheric CO_2 levels are key drivers of these changes (Rahman 2019; Harrison 2020). Various studies have highlighted that medicinal plants exhibit species-dependent responses to climate change, leading to varying gains and losses in their potential habitats across different geographical regions, including areas such as Crete, Greece, and the Tibetan Plateau (Bariotakis et al. 2023; Yang et al. 2024) (Table 3.1). In particular, the Himalayas is identified as a

Table 3.1 Climate change effects on secondary metabolites in medicinal plants by global region (examples)

Region	Key species at risk	Climate threat	Impacts	References
Himalayas	*Nardostachys jatamansi*	Warming temperatures	Reduced essential oil yield	Li et al. (2019), Bose et al. (2022), Wen et al. (2022)
	Withania somnifera	Rising temperatures	Withanolide content	Mirjalili et al. (2009), Rehman et al. (2010)
	Panax ginseng	Erratic rainfall	Decrease ginsenoside content	Wu et al. (2024)
	Picrorhiza kurroa	Glacier melt, erratic snowmelt	Decrease in Picrosides	Kumari et al. (2020), Rawat et al. (2022)
Southeast Asia	*Aquilaria malaccensis*	Increased fungal attacks	Reduced resin production	Chhipa and Kaushik (2017), Nor Mazila Ramli et al. (2022)
	Centella asiatica	Heat waves	Decrease triterpenoids content	Sun et al. (2020), Oyenihi et al. (2023)
	Curcuma longa	Erratic rainfall, flooding	Decrease curcuminoids in waterlogged soil	Sasikumar (2005), Chintakovid et al. (2022)
Africa	*Balanites aegyptiaca*	Desertification	Decline in anti-diabetic compounds	Murthy et al. (2020)
	Artemisia annua	Drought	Decrease artemisinin content	Ogwang et al. (2012), Carrà et al. (2014)
	Prunus africana	Soil degradation	Reduce lignans and sterols in bark	Nyamai et al. (2015), Komakech and Kang (2019)
South America	*Tabernaemontana undulata*	Deforestation and droughts	Loss of alkaloid-rich variants	Brito-da-costa et al. (2020)
	Aniba rosaeodora	Increasing droughts and fires	Decrease linalool	Chantraine et al. (2009)
	Lepidium meyenii	Frost reduction	Decrease glucosinolates	Tafuri et al. (2019)
	Cinchona officinalis	UV radiation	Increase quinine alkaloids	Shi and Liu (2021)
Mediterranean	*Salvia officinalis*	Heat	Increase essential oils' short-term drought	Nowak et al. (2010)
	Lavandula angustifolia	Water scarcity	Decrease rosmarinic acid in chronic stress	Saunier et al. (2022)

critically sensitive region that experiences remarkable shifts in the distributions of medicinal plant species, pushing them to higher altitudes or threatening their existence altogether (Maikhuri et al. 2018; Manish 2022; Shrestha et al. 2022). Similarly, predictions show that climate scenarios could potentially eliminate suitable habitats for many species if global temperatures rise unchecked, emphasizing the precarious future of medicinal flora in regions such as Nepal, India, and Bangladesh (Ghosh et al. 2022; Kunwar et al. 2023; Ahmed et al. 2024). Climate change (temperature fluctuation, drought stress, and elevated CO_2) is also impacting the production of secondary metabolites in medicinal plants that are essential for therapeutic properties. This underscores the importance of understanding how climate dynamics influence plant chemistry to ensure the continued efficacy and availability of plant-based medicines (Pant et al. 2021; Hounsou et al. 2024; Ngezahayo et al. 2025).

Climate change also impacts pollination and seed dissemination systems by altering the timing of plant phenological events, such as blooming and fruiting (Fitter and Fitter 2002; Gaira and Dhar 2020). Plants and their pollinators may become mismatched as a result of these alterations, which could lower long-term survival and reproductive success. Vassvik et al. (2024) investigated the quantity and quality of seed production on perennial *Ranunculus acris,* indicating the importance of interactive effects of change in direct abiotic and indirect biotic factors for responses to snowmelt timing. Similarly, Kudo and Cooper (2019) showed the regional difference in temperature seasonality affecting flowering phenology and pollinator interaction in herbaceous plants. Medicinal species, such as *C. asiatica* (Gotu Kola) and *Andrographis paniculata* (King of Bitters), are experiencing earlier flowering but shorter reproductive cycles, leading to mismatches with pollinator activity and reduced seed viability (Gaira and Dhar 2020). Scartezzini et al. (2006) showed that delayed monsoon rain disrupts fruiting in *Emblica officinalis,* lowering vitamin C content. Drought stress forces *Alpinia galanga* into early dormancy, shrinking rhizome size and curcuminoid production (Yang and Eilerman 1999; Gaira and Dhar 2020). Phenological disruptions threaten both biodiversity and human health in Southeast Asia. Urgent action—combining traditional knowledge with climate-smart agriculture—is needed to safeguard medicinal plant efficacy.

3.3 Genetic Erosion Under Climate Stress

Genetic erosion refers to the loss of genetic diversity within and among populations of a species, often due to environmental change, habitat degradation, and unsustainable exploitation (Khoury et al. 2022; van Oosterhout et al. 2025). Under climate stress, genetic erosion is accelerating, particularly in plant species that are ecologically specialized, geographically restricted, or already under anthropogenic pressure. This process undermines species' resilience, adaptability, and long-term survival (Bijlsma and Loeschcke 2012; Póvoa et al. 2023). In medicinal and endemic plants, climate-induced genetic erosion is particularly alarming. For example, studies on Himalayan medicinal herbs, such as *N. jatamansi* and *P. kurroa,* indicate reduced gene flow and increasing population fragmentation due to upward habitat shifts and reduced habitat connectivity (Kaur et al. 2020; Kunwar et al. 2023). Similarly, *Scutellaria baicalensis*, a medicinal plant widely used in TCM, has experienced a significant decline in wild populations over recent decades. To meet growing demand, large-scale cultivation was introduced; however, this shift has led to a notable reduction in genetic diversity. Cultivated populations lost 10 of the 25 identified alleles and showed increased levels of homozygosity, indicating a genetic bottleneck (Yuan et al. 2010). *Panax quinquefolius* (American ginseng) is widely harvested in the United States for its medicinal value, with wild roots considered more potent than cultivated ones. Heavy harvesting has reduced wild populations below viable levels, pushing the species toward extinction (Cruse-Sanders and Hamrick 2004). Cruse-Sanders et al. (2005) revealed that random harvest of *P. quinquefolius* resulted in significant genetic erosion, especially its allelic wealth relative to the initial level, using computer simulation. In the market, the demand and price are high for wild-gathered ginseng roots compared to cultivated ones.

Panax vietnamensis (Vietnamese ginseng) has experienced population declines due to its inability to shift its range rapidly in response to warming (Le et al. 2022; Vu et al. 2024). Climate change alters the flowering and fruiting cycles of *Rauvolfia serpentina* (Indian snakeroot), disrupting pollination synchrony with insect partners (Anoosha et al. 2020). Shil and Dewanjee (2022) showed that prolonged drought reduces genetic diversity in *Withania somnifera* by killing drought-sensitive variants. *Curcuma longa* (turmeric) plantations in India now suffer from rampant rhizome rot due to loss of disease-resistant wild relatives (Sasikumar 2005). A study in China on *A. annua* showed the decrease in artemisinin content by 15–20% with a rise in temperature, eroding heat-sensitive genotypes (Marchese and Rehder 2001). Monoculture plantation of *Catharanthus roseus* for vincristine production replaced wild populations, eliminating drought-resistant landraces (Birat et al. 2022). All research calls for immediate action for safeguarding living pharmacopeias combining genomics, traditional knowledge, and policy reform.

3.4 Phytochemical Shifts and Therapeutic Efficacy

Climate change is profoundly reshaping the chemical profiles of medicinal plants, with significant implications for their therapeutic efficacy, safety, and consistency (Fig. 3.1). Medicinal properties of plants largely depend on secondary metabolites, such as alkaloids, phenolics, terpenoids, glycosides, and essential oils, whose biosynthesis is highly sensitive to environmental conditions, including temperature, light, soil moisture, and atmospheric CO_2 concentrations (Isah 2019; Pant et al. 2021). Climate-induced abiotic stresses can either enhance or suppress the production of these bioactive compounds. For instance, moderate stress conditions may induce the accumulation of certain metabolites as a defense mechanism, but prolonged or extreme stress often leads to metabolite degradation, shifts in biosynthetic

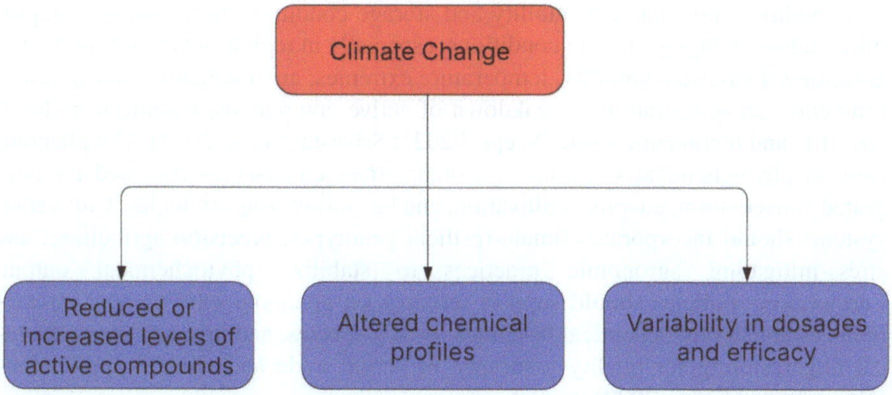

Fig. 3.1 Impacts of climate change on medicinal potency, safety, and standardization

pathways, or reduced yields, directly impacting the pharmacological value of the plant (Ashraf et al. 2018; Emami Bistgani et al. 2024).

Numerous studies illustrate how climate stress alters phytochemistry in key medicinal species. In *A. annua*, one of the most important antimalarial plants, elevated temperatures and water stress have been shown to reduce artemisinin content, the compound responsible for its antimalarial properties (Ferreira et al. 2005; Oladipo and Ojumu 2025). Similarly, *W. somnifera* (Ashwagandha), widely used in Ayurvedic medicine, shows variable levels of withanolides under drought and heat stress, influencing its adaptogenic effects (Singh et al. 2018). In *C. roseus*, a critical source of the anticancer alkaloids vincristine and vinblastine, increased temperatures and salinity stress affect the expression of genes involved in terpenoid indole alkaloid biosynthesis, altering the concentration of therapeutic compounds (Liu et al. 2021). Furthermore, climate change may alter not only quantitative yields but also qualitative profiles of phytochemicals. This can result in changes to the synergistic interactions among compounds that are crucial for medicinal efficacy. For example, essential oil-producing species such as *Mentha* spp., *Thymus vulgaris*, and *Ocimum sanctum* have shown climate-driven changes in both oil yield and chemical composition, thereby influencing their antimicrobial, anti-inflammatory, or anxiolytic properties (Soković et al. 2009; Vassiliou et al. 2023). Such alterations can cause variability in drug potency, complicate standardization of herbal products, and potentially lead to reduced effectiveness or unintended side effects. These shifts are particularly concerning in the context of traditional medicine systems, such as Ayurveda, TCM, and indigenous pharmacopoeias, where timing, location, and method of harvest are historically aligned with phytochemical peak periods. As climate change disrupts these environmental cues, traditional knowledge systems may face challenges in ensuring the consistent efficacy and safety of herbal remedies (Leonti and Casu 2013; WHO 2014). Moreover, pharmacological research and pharmaceutical industries that depend on wild-harvested or semi-domesticated plant resources face increased uncertainty regarding raw material quality.

In addition, post-harvest stability and storage conditions may also be compromised under changing climate conditions, especially in regions where infrastructure is lacking. Increased humidity, temperature extremes, and fluctuating storage environments can accelerate the breakdown of active compounds, leading to reduced shelf life and therapeutic value (Szepesi 2021; Srivastava et al. 2023). The phenomenon of phytochemical shifts under climate stress underscores the need for integrated conservation, adaptive cultivation, and bio-monitoring strategies. Cultivation systems should incorporate climate-resilient genotypes, precision agriculture, and stress-mitigating agronomic practices to stabilize phytochemical output. Furthermore, policies should support research on genotype–environment interactions, establish phytochemical baselines for key species, and promote phytochemical fingerprinting for quality assurance in herbal trade and clinical applications (Newman and Cragg 2020).

3.5 Methods for Tracking Genetic and Phytochemical Changes

Recent advances in molecular biology and analytical chemistry have enabled comprehensive monitoring of climate-induced genetic and phytochemical changes in medicinal plants. The application of DNA-based markers has revolutionized our ability to quantify genetic erosion in medicinal plant populations. Simple sequence repeats have been particularly valuable, as demonstrated by Tnah et al. (2011) in their assessment of *Eurycoma longifolia* populations across Peninsular Malaysia. For broader genome coverage, single-nucleotide polymorphism (SNP) analysis through genotyping-by-sequencing has proven effective. Chen et al. (2014) identified 12 drought-responsive SNPs in *A. paniculata* populations from Thailand and Myanmar, with three SNPs strongly correlated with andrographolide production. Next-generation sequencing technologies now enable comprehensive genomic studies. A landmark study by the Malaysian Agricultural Research Institute employed whole-genome resequencing of 120 *C. asiatica* accessions, identifying significant copy number variations in genes involved in triterpenoid biosynthesis under heat stress conditions (Gbolahan et al. 2016). Similarly, Pootakham et al. (2022) published the first chromosome-level assembly of *Mitragyna speciosa*, revealing structural variations in alkaloid biosynthesis genes between drought-stressed and control populations.

Advanced phytochemical methods like high-performance liquid chromatography (HPLC) remain the gold standard for quantifying medicinal compounds. Pan et al. (2016) developed a validated High-Performance Liquid Chromatography with Diode-Array Detection (HPLC-DAD) method that documented a 35–60% reduction in vincristine and vinblastine content in drought-stressed *C. roseus*. For volatile compounds, two-dimensional gas chromatography Tandem mass spectrometry (MS/MS) has shown superior resolution. A study on *C. longa* demonstrated that waterlogging stress reduced curcuminoid content by 28% while increasing undesirable sesquiterpenes by 15% (Li 2011). Recent developments in NMR spectroscopy have enabled real-time metabolic tracking. A study on *C. asiatica* cultivars grown in a greenhouse using a clustering technique helps to identify cultivars with desirable phytochemical traits for medicinal use (Alam et al. 2025; Hein et al. 2025).

Combining transcriptomics and metabolomics provides a powerful strategy to understand how gene expression changes influence metabolic pathways. Transcriptomics reveals which genes are active under specific conditions, while metabolomics identifies the small molecules produced. Integrated transcriptomic and metabolomic analyses uncover the key pathways of *Limonium bicolorin* response to low and high salt stress (Zhu et al. 2024). Yumeng et al. (2025) unveil the spatial dynamics of flavonoid metabolism in *Panax quinquefolius* and establish a Mediated Regulation of UDP-glycosyltransferase (MYB-UGT) regulatory axis, providing critical insights for metabolic engineering and bioactive compound optimization in medicinal plants. Transcriptomics techniques help to understand the relationship between the effects of grafting on changes in special metabolites and gene expression patterns in *Acanthopanax senticosus* leaves, revealing the transcriptional changes in secondary metabolites of medicinal plants via heterozygous grafting, helping to improve the production and breeding of plants (Wang et al. 2023). Also, the integrated metabolomic and transcriptomic analysis clarifies the drought stress-induced changes in specialized metabolites and regulating mechanisms in *Sophora alopecuroides,* helping biotechnologists to manipulate the levels of therapeutic specialized metabolites (Huang et al. 2023). Likewise, Zhang uses the technique to reveal key antioxidant mechanisms in *Dendrobium huoshanense* under combined salt and heat stress, providing insights into molecular mechanisms for improving stress resilience in medicinal orchids (Zhang et al. 2025). Advanced computational methods are transforming climate impact prediction for tracking genetic and phytochemical changes in medicinal plants, providing insights for crop improvement and other applications (Yoo et al. 2020; Varghese et al. 2025). Iqbal et al. (2023) develop supervised machine learning approach for predicting Ginger yield using Bangladesh climatic data, providing smart decisions about handling crops, such as where to put resources and how to reduce risks. A review has been conducted to investigate the application of deep learning in the classification of medicinal plants, focusing on delineating the specific plant components and providing potential improvement in the classification and recognition of medicinal plant species (Mulugeta et al. 2023).

Another important method for tracking genetic erosion and phytochemical changes in medicinal plants is through field monitoring and traditional knowledge integration. Nowadays, hyperspectral imaging has emerged as a powerful tool for large-scale monitoring of medicinal plants' diversity and origin (Meng et al. 2023; Gano et al. 2024). Satellite-derived Normalized Difference Vegetation Index (NDVI) data provide a quick, non-destructive means to measure the biomass yield of *Moringa oleifera* (Tshabalala et al. 2021). The study shows that mountain greening due to climate and land-use changes has led to significant genomic erosion in Ironwort, a medicinal grassland plant. Satellite NDVI data accurately predicted population declines, highlighting remote sensing's role in biodiversity monitoring (Theodoridis et al. 2025). Ethnobotanical documentation with integration of scientific knowledge has proven invaluable for understanding the phytochemical profile in medicinal plants. A study on the Tamang people in Nepal identifies many high-value and high-priority medicinal plant species, indicating high potential for economic development through sustainable collection and trade (Uprety et al. 2010). Cordero et al. (2022) documented 131 medicinal plant species used by the Panay Bukidnon in the Philippines to treat 91 ailments. Leaves and decoctions were most commonly used, with *C. longa*, *Artemisia vulgaris*, and *Chromolaena odorata* standing out for their cultural and medicinal significance.

3.6 Conservation and Adaptation Strategies

The accelerating threats of climate change and habitat destruction necessitate robust conservation and adaptation strategies to safeguard Southeast Asia's medicinal plant biodiversity. A combination of in-situ and ex-situ conservation, climate-smart cultivation, and community-based initiatives is critical to mitigating genetic erosion and ensuring sustainable utilization of these valuable species. In-situ conservation preserves plants within their natural habitats, such as through ecosystems-oriented protected areas or wild nurseries maintaining ecological relationships and maximizing genetic diversity (Chen et al. 2016). Similarly, ex-situ conservation complements these efforts by preserving plants outside their natural setting (botanical garden, seed banks, or cryopreservation) (Dronamraju 2019). Advanced methods such as in vitro storage, field gene banks, and cryobanks allow for long-term germplasm preservation, even for species with seeds that resist drying or conventional storage (Zeng et al. 2024). *Nothapodytes nimmoniana*, known for its anticancer compound camptothecin, is conserved in the Western Ghats' sacred groves, which are traditionally protected forest patches managed by local communities (Kala 2005). The indigenous community not only preserves biodiversity but also reflects the deep cultural and spiritual connection between people and nature. *R. serpentina*, a plant used for treating hypertension, was conserved within the Cauvery Wildlife Sanctuary in India to support natural regeneration and protection of wild populations (Ved and Goraya 2007). In Uganda's Bwindi Impenetrable Forest, local communities play a key role in conserving medicinal plants like *Prunus africana*, widely used in traditional medicine for treating prostate disorders. Sustainable use is promoted through collaborative forest management strategies (Hamilton 2004). For species with recalcitrant seeds (e.g., *Myristica fragrans* (Nutmeg)) with very limited geographical distribution, micropropagation and tissue culture techniques are being developed to conserve genetic material (Iyer 2007).

Another important practice for conservation and sustainable use of medicinal plants under changing environmental conditions is to cultivate climate-resilient varieties. In India, drought-resistant varieties of *Zingiber officinale* (Ginger) are selectively bred to withstand water scarcity (Praveen Kumar et al. 2025). Retana-Cordero et al. (2022) found that shade and mulching reduce heat stress in *Z.*

officinale and turmeric (*C. longa*), producing a yield up to 87% higher in ginger and 47% higher in turmeric in an open field. Furthermore, soil moisture conservation through organic farming and agroecological practices enhances the resilience of root-based medicinal plants, such as *Angelica sinensis* (Ang), in saline-affected regions (Guo et al. 2025). These sustainable agricultural approaches not only support ecosystem resilience but also offer economic opportunities to local communities by diversifying income sources, preserving soil health, and promoting biodiversity.

Engaging local communities in conservation efforts ensures the preservation of both biological and cultural knowledge. *Trichopus zeylanicus* led to the development of the herbal drug "Jeevani." The benefits from this drug were shared with the community under a benefit-sharing model, highlighting how community engagement can lead to both conservation and socioeconomic gains (Pushpangadan et al. 1988). The Ifugao people of the Philippines use selective harvesting for *Blumea balsamifera* (Sambong), a diuretic herb. They avoid uprooting entire plants and collect leaves only during specific lunar phases, which traditional beliefs suggest enhances potency (Baddu and Ouano 2018). In Malaysia, the Temuan Indigenous People document uses of *Eurycoma longifolia* (Tongkat Ali) through storytelling and herbarium collaborations with universities (Nordin 2014). Similarly, the Hmong ethnic group in Vietnam cultivates *Amomum villosum* (Black Cardamom) in the forest under stories, maintaining wild population buffers through cultivation. Such a system demonstrates how traditional ecological knowledge can guide climate-resilient cultivation methods that protect both plants and livelihoods (Khuat et al. 2022). From the Khasi tribe's protection of *Taxus wallichiana* (Himalayan Yew) in sacred groves (Gajurel et al. 2015) to Bali temple gardens preserving *Plectranthus barbatus* (Caneva et al. 2017), these culturally protected areas have inadvertently created refuges for medicinal species. The local knowledge system and experience of local environment, socio-economy and culture, and technological innovations have been effectively used for sustainable conservation of biodiversity in the area in Nepal (Bajracharya et al. 2007). The conservation of Southeast Asia's medicinal plants requires a multidisciplinary approach that integrates ecological, biotechnological, and socio-cultural strategies. By combining in-situ habitat protection, ex-situ genetic preservation, climate-adaptive farming, and community engagement, stakeholders can mitigate genetic erosion and secure these resources for future generations.

3.7 Endemic and Threatened Medicinal Plants of Nepal

Nepal, a landlocked Himalayan nation spanning approximately 147,181 km^2, exhibits its extraordinary biogeographical diversity due to its dramatic altitudinal gradient from the lowland Terai plains at <200 m above sea level to the towering peaks of the Himalayas exceeding 8000 m (CBS 2021). This topographic heterogeneity, coupled with varied climatic zones ranging from tropical to alpine, fosters one of the world's richest repositories of vascular plant species, with over 6500 angiosperms documented, including 229 endemic taxa (Kunwar et al. 2013; Tiwari et al. 2019; Rokaya and Sigdel 2024). Among these, Nepal harbors approximately 1515–2331 medicinal and aromatic plant species, representing 25–35% of its flora, which have been integral to indigenous healthcare systems for millennia (Fig. 3.2) (Kunwar et al. 2021; Rokaya and Sigdel 2024). Within this floral diversity, only a small fraction is regarded as strictly endemic (i.e., confined to specific ecological zones or districts) (Kunwar et al. 2022; DPR 2015). These plants support traditional Ayurvedic, Amchi (Tibetan), and folk medicinal practices among Nepal's 125 ethnic groups, addressing ailments from respiratory disorders to gastrointestinal issues,

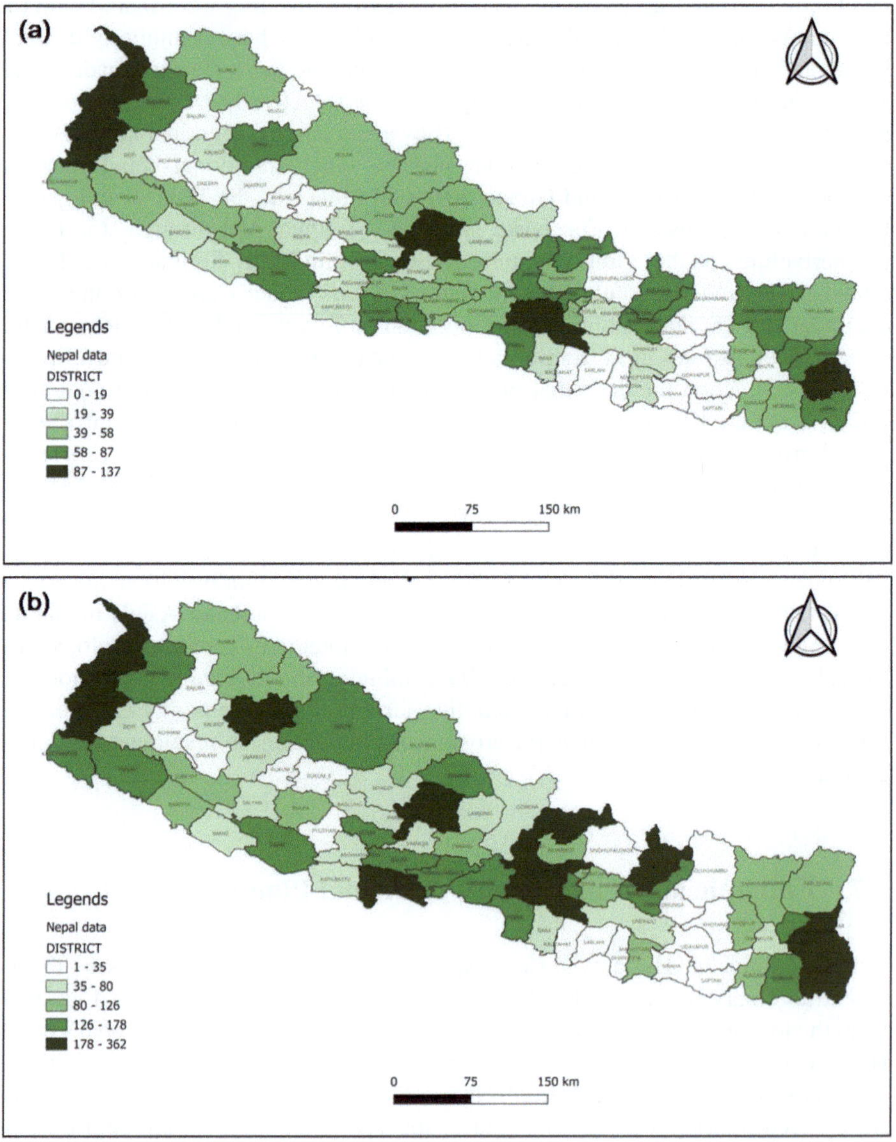

Fig. 3.2 Spatial distribution of (**a**) traded and (**b**) used medicinal and aromatic plants across Nepal

while also supporting livelihoods through trade valued at millions annually (Pyakurel et al. 2018; Adhikari et al. 2019; Kunwar et al. 2022).

Example includes *Swertia chirayita* var. *nepalensis*, a bitter tonic used for fever and liver disorders (Kumar and Van Staden 2016), and *Aconitum spicatum* Stapf., a potent analgesic with cardiotonic properties derived from its diterpenoid alkaloids (Shyaula et al. 2016) (Table 3.2). These endemic taxa occupy narrow altitudinal bands, making them particularly vulnerable to habitat loss, overharvesting, and climate induced range shifts. The levels of endemism may vary in a predictable way

Table 3.2 Selected endemic and threatened medicinal plants of Nepal

Botanical name	Local name(s)	Traditional medicinal uses	Conservation status and notes
Allium hypsistum Stearn	Ban Lasun, Jimbu	A key culinary and medicinal herb in the Himalayas, used as a spice and to treat colds, coughs, and stomach ailments.	Not fully assessed, but habitat-specific. Overharvesting is a concern (Rokaya et al. 2012)
Berginia ciliate (Haw.) Sternb.	Pakhanbed	Rhizome is used to treat fevers, diarrhea, wounds, and urinary problems. Considered a vital "Jadi Buti" (medicinal herb).	Vulnerable due to high demand and unsustainable harvesting (Kunwar et al. 2022)
Dactylorhiza hatagirea (D. Don) Soo.	Panch Aule	One of the most valuable and threatened species. Tuber is used as a tonic, aphrodisiac, and to treat dysentery, diarrhea, and chronic fever.	Endangered (International Union for Conservation of Nature (IUCN), Convention on International Trade in Endangered Species of Wild Fauna and Flora (CITES) listed). Critically threatened by illegal trade (Shrestha et al. 2021; Dorjey et al. 2022; Thakar and Dave 2024)
Nardostachys grandiflora (D. Don) DC.	Jatamansi	Rhizome and roots are used for mental disorders, insomnia, epilepsy, and as a sedative. Highly valued for its essential oil.	Critically Endangered (IUCN). Severely over-exploited for national and international trade (Kaur et al. 2020)
Neopicrorhiza scrophulariiflora (Pennell) Hong	Kutki	Rhizome is a potent bitter tonic used for fevers, liver disorders, and digestive issues. Often used as a substitute for *Picrorhiza kurroa*.	Vulnerable. High demand leads to unsustainable harvesting from the wild (Kafle et al. 2018; Rokaya et al. 2020)
Rheum australe D. Don.	Padamchal	Rhizome is used as a purgative, for treating wounds, burns, and indigestion.	Data deficient on IUCN, but considered threatened locally due to harvest pressure (Pandith et al. 2018; Paudel et al. 2025)
Swertia chirayita (Roxb. ex Fleming) Karsten	Chirayito	The whole plant is a renowned bitter tonic for malaria, fever, and diabetes. It is one of the most widely traded medicinal herbs.	Critically endangered due to premature harvesting and habitat destruction (Kumar and Van Staden 2016)

along gradients of primary environmental variables such as rainfall, temperature, productivity, and microtopography. In order to conserve this endemic species, the Government of Nepal is strengthening in-situ conservation with the community and protected forests, coupled with ex-situ propagation programs.

3.8 Future Directions and Research Gaps

The accelerating pace of climate change presents unprecedented challenges to medicinal plant conservation, necessitating urgent research to address knowledge gaps in genetic erosion and phytochemical stability. While existing studies have documented species vulnerability and traditional conservation practices (Corlett 2016), critical uncertainties remain regarding long-term adaptation strategies. A key priority lies in developing predictive models that integrate climate projections with species-specific responses (Table 3.3).

For instance, Gajurel et al. (2014) emphasize the need for advanced species distribution modeling to forecast range shifts of high-value medicinal plants like *Taxus wallichiana* (Himalayan Yew) under various climate scenarios. Such models must incorporate genomic data to identify populations with adaptive traits, as demonstrated by Hoban et al. (2022) in their work on climate-resilient genotypes of *Glycyrrhiza glabra* (Licorice). However, significant gaps persist in linking these genetic adaptations to therapeutic compound production, particularly for species like *W. somnifera* (Ashwagandha), where drought-induced phenotypic plasticity may alter withanolide concentrations (Sangwan et al. 2017; Khabiya et al. 2024). Understanding the biochemical pathways underlying climate-phytochemical relationships represents another critical research frontier. Recent metabolomic studies by Liu et al. (2021) reveal that elevated CO_2 levels differentially affect alkaloid biosynthesis in *C. roseus* (Madagascar periwinkle), suggesting complex stress-response mechanisms. Marchese and Rehder (2001) further highlight the value of transcriptomic analyses in identifying heat-responsive genes regulating artemisinin production in *A. annua*. Yet, few studies examine how interacting stressors—such as

Table 3.3 Summary of future direction

Research direction	Core objective
Long-term genetic monitoring	Detect and prevent genetic erosion in medicinal plant populations
Landscape genomics	Identify adaptive genetic variants linked to climate resilience
Mechanistic ecological modeling	Predict how climate variables influence phytochemical profiles
eDNA surveillance	Monitor species presence and genetic health non-invasively
Genotype–phytochemical integration	Link genetic architecture to chemical resilience under stress

combined heat and soil salinity—impact medicinal species like *Aloe vera*, creating a knowledge gap with direct implications for cultivation practices. Equally understudied are rhizosphere microbiome interactions; Kolařík et al. (2025) propose that symbiotic microbes may buffer climate impacts on secondary metabolism, offering potential bioaugmentation strategies for threatened species.

Community-based conservation approaches, while increasingly recognized, require robust empirical validation. Some studies integrate traditional shade of cultivation; however, there is a lack of comparative studies quantifying efficacy against conventional methods. Also, more research can be conducted on the economic analysis of community-managed medicinal plant reserve, which could bridge the gap by systematically evaluating species based on ecological vulnerability and cultural significance. Finally, there should be transdisciplinary collaboration prioritizing long-term monitoring, development of climate resilient cultivation protocol with funding for participatory research. These efforts will be essential to safeguard both medicinal plant biodiversity and the traditional healthcare systems that depend on them.

3.9 Conclusion

Climate change is having a serious impact on medicinal plants around the world. Rising temperatures, unpredictable rainfall, and changing habitats are causing many plants to lose their natural genetic diversity. This genetic erosion reduces their ability to adapt and survive in new environmental conditions. As a result, some important species may disappear from certain areas or become less effective in the future. At the same time, climate stress also affects the way plants produce their healing compounds, called phytochemicals. These compounds are what give medicinal plants their power to treat diseases. Climate change can alter the quality and quantity of these phytochemicals even if the plant survives. To protect the future of natural medicine, we need to take strong and smart actions. This includes closely monitoring plant populations over time, studying how climate affects their growth and chemistry, and using new technologies such as genetic analysis and climate models. It is also important to work together with local communities and traditional healers, who have deep knowledge of these plants and how they are used. In conclusion, conserving medicinal plants is not just about saving nature, but it is also about protecting human health, cultural traditions, and the ability to treat future illnesses. Climate change makes this work more urgent than ever.

Acknowledgments SP is thankful to the Madan Bhandari University of Science and Technology, Chitlang, Nepal, and PP gratefully acknowledges the CiST College affiliated to Pokhara University, Nepal, for continuous support and encouragement in research activities.

References

Adhikari M, Thapa R, Kunwar RM, Devkota HP, Poudel P (2019) Ethnomedicinal uses of plant resources in the Machhapuchchhre Rural Municipality of Kaski District, Nepal. Medicine 6(2):69

Ahmed MS, Oishe SJ, Rahman F, Basri R, Haque M, Monalisa MN, Rayhana N, Ahsan S, Dev S, Rahman MA (2024) Ethnomedicinal plant diversity, traditional knowledge, and comparative analysis across regions: a study on medicinal plant utilization. J Angiother 8(11):1–8

Alam MN, Marney L, Yang L, Choi J, Cerruti N, Techen N, Bassett S, Smith J, Brown K, Cabey K et al (2025) Clustering of chemical profiles of *Centella asiatica* cultivars, grown in greenhouses, allows grouping of metabolites with similar production trends. Ind Crop Prod 231:121159

Alum EU (2024) Climate change and its impact on the bioactive compound profile of medicinal plants: implications for global health. Plant Signal Behav 19(1):2419683

Anoosha V, Saini S, Kaushik HD (2020) Efficient pollinators of threatened taxa, Sarpagandha (*Rauvolfia serpentina*) under North Indian conditions. Indian J Anim Res 54(4):482–487

Ashraf MA, Iqbal M, Rasheed R, Hussain I, Riaz M, Arif MS (2018) Environmental stress and secondary metabolites in plants. In: Plant metabolites and regulation under environmental stress, pp 153–167

Baddu VD, Ouano NB (2018) Ethnobotanical survey of medicinal plants used by the Y'Apayaos of Sta. Praxedes in the Province of Cagayan, Philippines. Mindanao J Sci Technol 16(1):128–153

Bajracharya S, Gurung GB, Basnet K (2007) Learning from community participation in conservation area management. J For Livelihood 6(2):54–66

Balunas MJ, Kinghorn AD (2005) Drug discovery from medicinal plants. Life Sci 78(5):431–441. https://doi.org/10.1016/J.LFS.2005.09.012

Bariotakis M, Georgescu L, Laina D, Koufaki M, Souma M, Douklias S, Giannakakis KA, Chouli KN, Paoli L, Loppi S et al (2023) Climate change dependence in ex situ conservation of wild medicinal plants in Crete, Greece. Biology (Basel) 12(10):1327. https://doi.org/10.3390/BIOLOGY12101327/S1

Baumbach L, Warren DL, Yousefpour R, Hanewinkel M (2021) Climate change may induce connectivity loss and mountaintop extinction in Central American forests. Commun Biol 4(1):1–12. https://doi.org/10.1038/s42003-021-02359-9

Bijlsma R, Loeschcke V (2012) Genetic erosion impedes adaptive responses to stressful environments. Evol Appl 5(2):117–129. https://doi.org/10.1111/J.1752-4571.2011.00214.X

Birat K, Siddiqi TO, Mir SR, Aslan J, Bansal R, Khan W, Dewangan RP, Panda BP (2022) Enhancement of vincristine under in vitro culture of *Catharanthus roseus* supplemented with *Alternaria sesami* endophytic fungal extract as a biotic elicitor. Int Microbiol 25(2):275–284. https://doi.org/10.1007/s10123-021-00213-w

Bose B, Kumaria S, Tandon P (2022) Physiological insights into the role of temperature and light conditions on in vitro growth, membrane thermostability and antioxidative activity of *Nardostachys jatamansi*, an IUCN Red-listed critically endangered therapeutic plant. S Afr J Bot 146:365–374. https://doi.org/10.1016/J.SAJB.2021.11.001

Brito-da-costa AM, Dias-da-silva D, Gomes NGM, Dinis-oliveira RJ, Madureira-carvalho Á (2020) Toxicokinetics and Toxicodynamics of Ayahuasca alkaloids N,N-Dimethyltryptamine (DMT), Harmine, Harmaline and Tetrahydroharmine: clinical and forensic impact. Pharmaceuticals 13(11):334. https://doi.org/10.3390/PH13110334

Bussmann RW, Sharon D (2006) Traditional medicinal plant use in Northern Peru: tracking two thousand years of healing culture. J Ethnobiol Ethnomed 2(1):1–18. https://doi.org/10.1186/1746-4269-2-47

Caneva G, Traversetti L, Sujarwo W, Zuccarello V (2017) Sharing ethnobotanical knowledge in traditional villages: evidence of food and nutraceutical "Core Groups" in Bali, Indonesia. Econ Bot 71(4):303–313. https://doi.org/10.1007/S12231-017-9395-X

Carrà A, Bagnati R, Fanelli R, Bonati M (2014) Fast and reliable artemisinin determination from different *Artemisia annua* leaves based alimentary products by high performance liquid chromatography- tandem mass spectrometry. Food Chem 142:114–120. https://doi.org/10.1016/j.foodchem.2013.07.052

CBS (2021) National Population and Housing Census 2021 - National Report. Accessed 2025 Oct 13. www.censusnepal.cbs.gov.np

Chaachouay N, Zidane L (2024) Plant-derived natural products: a source for drug discovery and development. Drugs Drug Candidates 3(1):184–207. https://doi.org/10.3390/ddc3010011

Chantraine JM, Dhénin JM, Moretti C (2009) Chemical variability of Rosewood (*Aniba rosaeodora* Ducke) essential oil in French Guiana. J Essent Oil Res 21(6):486–495. https://doi.org/10.1080/10412905.2009.9700225

Chen R, Wang XY, Song YN, Feng ZY, Liang WP, Li M, Zhong GY (2014) Genetic diversity analysis of *Andrographis paniculata* in China based on SRAP and SNP. J Chinese Mater Medica 39(23):4559–4565. https://doi.org/10.4268/cjcmm20142316

Chen SL, Yu H, Luo HM, Wu Q, Li CF, Steinmetz A (2016) Conservation and sustainable use of medicinal plants: problems, progress, and prospects. Chin Med 11(1):1–10. https://doi.org/10.1186/s13020-016-0108-7

Chhipa H, Kaushik N (2017) Fungal and bacterial diversity isolated from *Aquilaria malaccensis* tree and soil, induces agarospirol formation within 3 months after artificial infection. Front Microbiol 8:1286. https://doi.org/10.3389/fmicb.2017.01286

Chintakovid N, Tisarum R, Samphumphuang T, Sotesaritkul T, Cha-um S (2022) Evaluation of curcuminoids, physiological adaptation, and growth of *Curcuma longa* under water deficit and controlled temperature. Protoplasma 259(2):301–315. https://doi.org/10.1007/s00709-021-01670-w

Cordero CS, Meve U, Alejandro GJD (2022) Ethnobotanical documentation of medicinal plants used by the indigenous Panay Bukidnon in Lambunao, Iloilo, Philippines. Front Pharmacol 12. https://doi.org/10.3389/FPHAR.2021.790567/PDF

Corlett RT (2016) Restoration, reintroduction, and rewilding in a changing world. Trends Ecol Evol 31(6):453–462. https://doi.org/10.1016/j.tree.2016.02.017

Cruse-Sanders JM, Hamrick JL (2004) Genetic diversity in harvested and protected populations of wild American ginseng, *Panax quinquefolius* L. (Araliaceae). Am J Bot 91(4):540–548. https://doi.org/10.3732/AJB.91.4.540

Cruse-Sanders JM, Hamrick JL, Ahumada JA (2005) Consequences of harvesting for genetic diversity in American ginseng (*Panax quinquefolius* L.): a simulation study. Biodivers Conserv 14(2):493–504. https://doi.org/10.1007/S10531-004-7308-7/METRICS

De Boer HJ, Cotingting C (2014) Medicinal plants for women's healthcare in Southeast Asia: a meta-analysis of their traditional use, chemical constituents, and pharmacology. J Ethnopharmacol 151(2):747–767. https://doi.org/10.1016/J.JEP.2013.11.030

DFRS, 2015. State of Nepal's Forests. Forest Resource Assessment (FRA) Nepal, Department of Forest Research and Survey (DFRS). Kathmandu, Nepal

Dorjey K, Maurya AK, Sinha D (2022) *Dactylorhiza hatagirea* (D. Doon) Soo, an important medicinal herb of the Himalaya and urgent need for its conservation-a review. Indian J Nat Prod Resour 13(3):265–273. https://doi.org/10.56042/ijnpr.v13i3.58263

Dronamraju KR (2019) Conservation of medicinal plants. In: Biol Soc Issues Biotechnol Shar, pp 79–100. https://doi.org/10.4324/9780429460883-9

Ekor M (2014) The growing use of herbal medicines: issues relating to adverse reactions and challenges in monitoring safety. Front Neurol 4(JAN):177. https://doi.org/10.3389/fphar.2013.00177

El Gendy ANG, Fouad R, Omer EA, Cock IE (2023) Effects of climate change on medicinal plants and their active constituents. In: Hasanuzzaman M (ed) Climate-Resilient Agriculture, Vol. 1 Crop Responses and Agroecological Perspectives. Springer, Cham, pp 125–156. https://doi.org/10.1007/978-3-031-37424-1_6

Emami Bistgani Z, Barker AV, Hashemi M (2024) Physiology of medicinal and aromatic plants under drought stress. Crop J 12(2):330–339. https://doi.org/10.1016/j.cj.2023.12.003

Ferreira JFS, Laughlin JC, Delabays N, de Magalhães PM (2005) Cultivation and genetics of *Artemisia annua* L. for increased production of the antimalarial artemisinin. Plant Genet Resour 3(2):206–229. https://doi.org/10.1079/pgr200585

Fitter AH, Fitter RSR (2002) Rapid changes in flowering time in British plants. Science 296(5573):1689–1691. https://doi.org/10.1126/science.1071617

Gaira KS, Dhar U (2020) Phenological change modelling for selected Himalayan medicinal herbs using herbarium records: a case study. Ecol Inform 60:101177. https://doi.org/10.1016/j.ecoinf.2020.101177

Gajurel JP, Werth S, Shrestha KK, Scheidegger C (2014) Species distribution modeling of *Taxus wallichiana* (Himalayan Yew) in Nepal Himalaya. Asian J Conserv Biol 3(2):43

Gajurel JP, Shrestha KK, Werth S, Scheidegger C (2015) *Taxus wallichiana* (Himalayan Yew) for the livelihood of local people in some protected areas of Nepal. J Nat Hist Mus 28:1–8. https://doi.org/10.3126/JNHM.V28I0.14162

Gano B, Bhadra S, Vilbig JM, Ahmed N, Sagan V, Shakoor N (2024) Drone-based imaging sensors, techniques, and applications in plant phenotyping for crop breeding: a comprehensive review. Plant Phenome J 7(1). https://doi.org/10.1002/PPJ2.20100

Gbolahan BW, Abiola AI, Kamaldin J, Ahmad MA, Atanassova MS (2016) Accession in *Centella asiatica*; current understanding and future knowledge. J Pure Appl Microbiol 10(4):2485–2494. https://doi.org/10.22207/JPAM.10.4.02

Ghosh S, Roy N, Behera S, Ansar W (2022) Climate change and medicinal plants in India: an overview. In: Biodiversity, Conserv Sustain Asia Vol 2 prospect challenges South Middle Asia, vol 2. Springer, Cham, pp 887–909. https://doi.org/10.1007/978-3-030-73943-0_49

Guo X, Shengli Z, Yin R, Zhu S, Ji B, Peng Y, Shuhan Z, Xiao X, Peng C, Wang Y (2025) Soil nutrients, rhizosphere bacteria, and endophytes interact to influence the accumulation of amino acid-short peptide components in *Angelica sinensis* roots, each playing distinct roles. Rhizosphere 33:101027. https://doi.org/10.1016/J.RHISPH.2025.101027

Hamilton AC (2004) Medicinal plants, conservation and livelihoods. Biodivers Conserv 13(8):1477–1517. https://doi.org/10.1023/B:BIOC.0000021333.23413.42

Harrison S (2020) Plant community diversity will decline more than increase under climatic warming. Philos Trans R Soc B 375(1794):20190106. https://doi.org/10.1098/RSTB.2019.0106

Hein ZM, Gopalakrishna PK, Kanuri AK, Thomas W, Hussan F, Naik VR, Shantakumari N, Che Ramli MD, Mohd Moklas MA, Che Mohd Nassir CMN, Vishnumukkala T (2025) *Centella asiatica*: advances in extraction technologies, phytochemistry, and therapeutic applications. Life 15(7):1081. https://doi.org/10.3390/life15071081

Hoban S, Archer FI, Bertola LD, Bragg JG, Breed MF, Bruford MW, Coleman MA, Ekblom R, Funk WC, Grueber CE, Hand BK (2022) Global genetic diversity status and trends: towards a suite of Essential Biodiversity Variables (EBVs) for genetic composition. Biol Rev 97(4):1511–1538. https://doi.org/10.1111/brv.12852

Hou S, Ma D, Wu S, Hui Q, Hao Z (2025) *Morinda citrifolia* L.: a comprehensive review on phytochemistry, pharmacological effects, and antioxidant potential. Antioxidants 14(3):295. https://doi.org/10.3390/antiox14030295

Hounsou KE, Sonibare MA, Elufioye TO (2024) Climate change and the future of medicinal plants research. Bioact Compd Heal Dis 7(3):152–169. https://doi.org/10.31989/BCHD.V7I3.1310

Huang X, Chu G, Wang J, Luo H, Yang Z, Sun L, Rong W, Wang M (2023) Integrated metabolomic and transcriptomic analysis of specialized metabolites and isoflavonoid biosynthesis in *Sophora alopecuroides* L. under different degrees of drought stress. Ind Crop Prod 197:116595. https://doi.org/10.1016/J.INDCROP.2023.116595

IPCC (2023) Summary for policymakers. In: Climate change 2023: synthesis report. Contribution of Working Groups I, II and III to the Sixth Assessment Report of the Intergovernmental Panel on Climate Change, Geneva, Switzerland.; Accessed 2025 Jun 24. https://doi.org/10.59327/IPCC/AR6-9789291691647.001

Iqbal LB, Rukhsara L, Sarker MMR, Mamun S (2023) GingerBangla: a supervised machine learning approach for predicting ginger yield using Bangladeshi climate data. In: 2023

14th Int Conf Comput Commun Netw Technol ICCCNT 2023. https://doi.org/10.1109/ICCCNT56998.2023.10307178

Isah T (2019) Stress and defense responses in plant secondary metabolites production. Biol Res 52(1):39. https://doi.org/10.1186/s40659-019-0246-3

Iyer RI (2007) In vitro propagation of Nutmeg, Myristica Fragrans Houtt. In: Protoc Micropropag Woody trees fruits. Springer, Netherlands, pp 335–344. https://doi.org/10.1007/978-1-4020-6352-7_31

Jump AS, Peñuelas J (2005) Running to stand still: adaptation and the response of plants to rapid climate change. Ecol Lett 8(9):1010–1020. https://doi.org/10.1111/j.1461-0248.2005.00796.x

Kafle G, Bhattarai I, Siwakoti M, Shrestha AK (2018) Demand, end-uses, and conservation of alpine medicinal plant *Neopicrorhiza scrophulariiflora* (Pennell) D. Y. Hong in central Himalaya. Evid based Complem Altern Med 2018:6024263. https://doi.org/10.1155/2018/6024263

Kala CP (2005) Indigenous uses, population density, and conservation of threatened medicinal plants in protected areas of the Indian Himalayas. Conserv Biol 19(2):368–378. https://doi.org/10.1111/J.1523-1739.2005.00602.X;PAGE:STRING:ARTICLE/CHAPTER

Kaur H, Lekhak MM, Chahal S, Goutam U, Jha P, Naidoo D, Ochatt SJ, Kumar V (2020) *Nardostachys jatamansi* (D.Don) DC.: an invaluable and constantly dwindling resource of the Himalayas. S Afr J Bot 135:252–267. https://doi.org/10.1016/j.sajb.2020.08.010

Khabiya R, Choudhary GP, Sairkar P, Silawat N, Jnanesha AC, Kumar A, Lal RK (2024) Unraveling genetic diversity analysis of Indian ginseng (*Withania somnifera* (Linn.) Dunal) insight from RAPD and ISSR markers and implications for crop improvement vital for pharmacological and industrial potential. Ind Crop Prod 210:118124. https://doi.org/10.1016/J.INDCROP.2024.118124

Khoury CK, Brush S, Costich DE, Curry HA, de Haan S, Engels JMM, Guarino L, Hoban S, Mercer KL, Miller AJ, Nabhan GP (2022) Crop genetic erosion: understanding and responding to loss of crop diversity. New Phytol 233(1):84–118. https://doi.org/10.1111/nph.17733

Khuat QV, Kalashnikova EA, Kirakosyan RN, Nguyen HT, Baranova EN, Khaliluev MR (2022) Improvement of in vitro seed germination and micropropagation of *Amomum tsao-ko* (Zingiberaceae Lindl.). Horticulturae 8(7):640. https://doi.org/10.3390/horticulturae8070640

Kolařík M, Stodůlková E, Kajzrová S, Semerád J, Hubert J, Kuzma M, Šulc M, Císařová I, Jašica A, Wennrich JP, Hulcr J (2025) Secondary metabolites and their impact on symbiotic interactions in the ambrosia fungus *Geosmithia eupagioceri*. Fungal Ecol 74:101414. https://doi.org/10.1016/J.FUNECO.2025.101414

Komakech R, Kang Y (2019) Ethnopharmacological potential of African cherry [*Prunus africana*]. J Herb Med 17–18. https://doi.org/10.1016/j.hermed.2019.100283

Kudo G, Cooper EJ (2019) When spring ephemerals fail to meet pollinators: mechanism of phenological mismatch and its impact on plant reproduction. Proc R Soc B 286(1904). https://doi.org/10.1098/RSPB.2019.0573

Kumar V, Van Staden J (2016) A review of *Swertia chirayita* (Gentianaceae) as a traditional medicinal plant. Front Pharmacol 6(JAN):308. https://doi.org/10.3389/FPHAR.2015.00308

Kumari M, Joshi R, Kumar R (2020) Metabolic signatures provide novel insights to *Picrorhiza kurroa* adaptation along the altitude in Himalayan region. Metabolomics 16(7):1–17. https://doi.org/10.1007/S11306-020-01698-8/METRICS

Kunwar RM, Mahat L, Acharya RP, Bussmann RW (2013) Medicinal plants, traditional medicine, markets and management in far-west Nepal. J Ethnobiol Ethnomed 9(1):1–10. https://doi.org/10.1186/1746-4269-9-24

Kunwar RM, Sher H, Bussmann RW (eds) (2021) Ethnobotany of the Himalayas. Springer International Publishing, Cham. https://doi.org/10.1007/978-3-030-57408-6

Kunwar RM, Baral B, Luintel S, Uprety Y, Poudel RC, Adhikari B, Adhikari YP, Subedi SC, Subedi CK, Poudel P, Paudel HR (2022) Ethnomedicinal landscape: distribution of used medicinal plant species in Nepal. J Ethnobiol Ethnomed 18(1):1–11. https://doi.org/10.1186/s13002-022-00531-x

Kunwar RM, Thapa-Magar KB, Subedi SC, Kutal DH, Baral B, Joshi NR, Adhikari B, Upadhyaya KS, Thapa-Magar S, Ansari AS, Thapa GJ (2023) Distribution of important medicinal plant species in Nepal under past, present, and future climatic conditions. Ecol Indic 146:109879. https://doi.org/10.1016/J.ECOLIND.2023.109879

Le HTT, Nguyen LN, Pham HLB, Le HTM, Luong TD, Huynh HTT, Nguyen VT, Van NH, Teixidor-Toneu I, De Boer HJ, Manzanilla V (2022) Target capture reveals the complex origin of Vietnamese Ginseng. Front Plant Sci 13:814178. https://doi.org/10.3389/fpls.2022.814178

Leonti M, Casu L (2013) Traditional medicines and globalization: current and future perspectives in ethnopharmacology. Front Pharmacol 4:92. https://doi.org/10.3389/FPHAR.2013.00092

Li S (2011) Chemical composition and product quality control of turmeric (*Curcuma longa* L.). Pharm Crop 5(1):28–54. https://doi.org/10.2174/2210290601102010028

Li J, Wu J, Peng K, Fan G, Yu H, Wang W, He Y (2019) Simulating the effects of climate change across the geographical distribution of two medicinal plants in the genus *Nardostachys*. PeerJ 7(4):e6730. https://doi.org/10.7717/PEERJ.6730

Liu Y, Patra B, Singh SK, Paul P, Zhou Y, Li Y, Wang Y, Pattanaik S, Yuan L (2021) Terpenoid indole alkaloid biosynthesis in *Catharanthus roseus*: effects and prospects of environmental factors in metabolic engineering. Biotechnol Lett 43(11):2085–2103. https://doi.org/10.1007/s10529-021-03179-x

Maharjan SK, Sterck FJ, Raes N, Zhao Y, Poorter L (2023) Climate change induced elevational range shifts of Himalayan tree species. Biotropica 55(1):53–69. https://doi.org/10.1111/BTP.13159

Maikhuri RK, Phondani PC, Dhyani D, Rawat LS, Jha NK, Kandari LS (2018) Assessment of climate change impacts and its implications on medicinal plants-based traditional healthcare system in Central Himalaya, India. Iran J Sci Technol Trans A Sci 42(4):1827–1835. https://doi.org/10.1007/S40995-017-0354-2/METRICS

Manish K (2022) Medicinal plants in peril due to climate change in the Himalaya. Ecol Inform 68:101546. https://doi.org/10.1016/J.ECOINF.2021.101546

Marchese JA, Rehder VLG (2001) The influence of temperature in the production of artemisinin in *Artemisia annua* L. Rev Bras Plantas Med 4(1):89–93

Meng J, You X, Zhang X, Shi T, Zhang L, Chen X, Zhao H, Xu M (2023) Remote sensing application in Chinese medicinal plant identification and acreage estimation—a review. Remote Sens 15(23):5580. https://doi.org/10.3390/RS15235580

Meñiza JF, Pasco MM, Alimbon JA (2024) A review of ethnobotanical studies reveals over 500 medicinal plants in Mindanao, Philippines. Plant Divers 46(5):551–564. https://doi.org/10.1016/J.PLD.2024.05.001

Mirjalili MH, Moyano E, Bonfill M, Cusido RM, Palazón J (2009) Steroidal lactones from *Withania somnifera*, an ancient plant for novel medicine. Molecules 14(7):2373–2393. https://doi.org/10.3390/MOLECULES14072373

Mulugeta AK, Sharma DP, Mesfin AH (2023) Deep learning for medicinal plant species classification and recognition: a systematic review. Front Plant Sci 14:1286088. https://doi.org/10.3389/FPLS.2023.1286088

Murthy HN, Yadav GG, Dewir YH, Ibrahim A (2020) Phytochemicals and biological activity of desert date (*Balanites aegyptiaca* (L.) Delile). Plants 10(1):32. https://doi.org/10.3390/PLANTS10010032

Nasim N, Sandeep IS, Mohanty S (2022) Plant-derived natural products for drug discovery: current approaches and prospects. Nucleus 65(3):399. https://doi.org/10.1007/S13237-022-00405-3

Newman DJ, Cragg GM (2020) Natural products as sources of new drugs over the nearly four decades from 01/1981 to 09/2019. J Nat Prod 83(3):770–803

Ngezahayo E, Komariah K, Masykuri M (2025) The climate change influence on Indonesia medicinal plants: a review. E3S Web Conf 605:03062. https://doi.org/10.1051/E3SCONF/202560503062

Nor Mazila Ramli A, Yusof S, Bhuyar P, Aminan AW, Tajuddin SN (2022) Fungi mediated agarwood (*A. malaccensis*) production and their pharmaceutical applications: a systematic review. Int J Plant Based Pharm 2(2):261–270. https://doi.org/10.29228/ijpbp.8

Nordin MS (2014) Distribution of the population of Tongkat Ali (*Eurycoma* Spp.) in Malaysia based on data taken from herbarium records. Med Aromat Plants 03(02):155. https://doi.org/10.4172/2167-0412.1000155

Nowak M, Kleinwächter M, Manderscheid R, Weigel HJ, Selmar D (2010) Drought stress increases the accumulation of monoterpenes in sage (*Salvia officinalis*), an effect that is compensated by elevated carbon dioxide concentration. J Appl Bot Food Qual 83(2):133–136

Nyamai DW, Mawia AM, Wambua FK, Njoroge A, Matheri F, Lagat R, Burugu M (2015) Phytochemical profile of *Prunus africana* stem bark from Kenya. J Pharmacogn Nat Prod 1(1):8. https://doi.org/10.4172/jpnp.1000110

Ogwang PE, Ogwal JO, Kasasa S, Olila D, Ejobi F, Kabasa D, Obua C (2012) *Artemisia Annua* L. infusion consumed once a week reduces risk of multiple episodes of malaria: a randomised trial in a Ugandan community. Trop. J Pharm Res 11(3):445–453. https://doi.org/10.4314/tjpr.v11i3.14

Oladipo B, Ojumu TV (2025) Supercritical CO_2 extraction of artemisinin from *Artemisia annua* plant and the biotechnological production of artemisinin precursors: a dual-focus review. Biotechnol Notes 6:143–156. https://doi.org/10.1016/j.biotno.2025.05.003

Oyebode O, Kandala NB, Chilton PJ, Lilford RJ (2016) Use of traditional medicine in middle-income countries: a WHO-SAGE study. Health Policy Plan 31(8):984–991. https://doi.org/10.1093/heapol/czw022

Oyenihi AB, George TT, Oyenihi OR, Obilana AO, Opperman M (2023) Three decades of research on *Centella asiatica*: insights and future trends from bibliometric analysis. J Herb Med 39:100662. https://doi.org/10.1016/J.HERMED.2023.100662

Pan Q, Saiman MZ, Mustafa NR, Verpoorte R, Tang K (2016) A simple and rapid HPLC-DAD method for simultaneously monitoring the accumulation of alkaloids and precursors in different parts and different developmental stages of *Catharanthus roseus* plants. J Chromatogr B Anal Technol Biomed Life Sci 1014:10–16. https://doi.org/10.1016/j.jchromb.2016.01.034

Pandey S, Pant P (2023) Possibilities and challenges for harnessing tree bark extracts for wood adhesives and green chemicals and its prospects in Nepal. Forest Sci Technol 19(1):68–77. https://doi.org/10.1080/21580103.2023.2175729

Pandey S, Carrer M, Castagneri D, Petit G (2018) Xylem anatomical responses to climate variability in Himalayan birch trees at one of the world's highest forest limit. Perspect Plant Ecol Evol Syst 33. https://doi.org/10.1016/j.ppees.2018.05.004

Pandith SA, Dar RA, Lattoo SK, Shah MA, Reshi ZA (2018) *Rheum australe*, an endangered high-value medicinal herb of North Western Himalayas: a review of its botany, ethnomedical uses, phytochemistry and pharmacology. Phytochem Rev 17(3):573–609. https://doi.org/10.1007/s11101-018-9551-7

Pant P, Pandey S, Dall'Acqua S (2021) The influence of environmental conditions on secondary metabolites in medicinal plants: a literature review. Chem Biodivers 18(11):e2100345. https://doi.org/10.1002/cbdv.202100345

Parmesan C, Yohe G (2003) A globally coherent fingerprint of climate change impacts across natural systems. Nature 421(6918):37–42. https://doi.org/10.1038/nature01286

Paudel BR, Subedi CK, Ghimire SK, Pyakurel D, Rajbhandari M, Chaudhary RP (2025) Impacts of climate change on the distribution pattern of Himalayan Rhubarb (*Rheum australe* D. Don) in Nepal Himalaya. PLoS One 20:e0323755. https://doi.org/10.1371/journal.pone.0323755

Pootakham W, Yoocha T, Jomchai N, Kongkachana W, Naktang C, Sonthirod C, Chowpongpang S, Aumpuchin P, Tangphatsornruang S (2022) A chromosome-scale genome assembly of *Mitragyna speciosa* (Kratom) and the assessment of its genetic diversity in Thailand. Biology (Basel) 11(10):1492. https://doi.org/10.3390/biology11101492

Póvoa O, Lopes V, Barata AM, Farinha N (2023) Monitoring genetic erosion of aromatic and medicinal plant species in Alentejo (South Portugal). Plants 12(14):2588. https://doi.org/10.3390/plants12142588

Praveen Kumar R, Ramesh Kumar S, Mohanraj K, Anbarasan P (2025) Ginger cultivation in India: impact of climate change and sustainability strategies–a potential review. Cogent Food Agric 11(1):2446653. https://doi.org/10.1080/23311932.2024.2446653;WGROUP:STRING:PUBLICATION

Priyadarshan PM (2016) Genetic diversity and erosion in *Hevea* Rubber. In: Ahuja MR, Mohan Jain S (eds) Genetic diversity and erosion in plants: case histories. Springer International Publishing, Cham, pp 233–267. https://doi.org/10.1007/978-3-319-25954-3_6

Pushpangadan P, Rajasekharan S, Ratheshkumar PK, Jawahar CR, Nair VV, Lakshmi N, Amma LS (1988) "Arogyappacha" (*Trichopus zeylanicus* gaerin), the "ginseng" of Kani Tribes of Agashyar Hills (Kerala) for ever green health and vitality. Anc Sci Life 8(1):13–16. https://pmc.ncbi.nlm.nih.gov/articles/PMC3331343/

Pyakurel D, Bhattarai Sharma I, Smith-Hall C (2018) Patterns of change: the dynamics of medicinal plant trade in far-western Nepal. J Ethnopharmacol 224:323–334. https://doi.org/10.1016/j.jep.2018.06.004

Qaderi MM, Martel AB, Strugnell CA (2023) Environmental factors regulate plant secondary metabolites. Plants 12(3):447. https://doi.org/10.3390/PLANTS12030447

Rahman MA (2019) Assessing the impact of climate change on plant life: understanding adaptations, stress responses, and ecosystem alterations in a warming world. Adv Herb Res 2(1):1–5. https://doi.org/10.25163/AHI.219914

Rawat N, Purohit S, Painuly V, Negi GS, Bisht MPS (2022) Habitat distribution modeling of endangered medicinal plant *Picrorhiza kurroa* (Royle ex Benth) under climate change scenarios in Uttarakhand Himalaya, India. Ecol Inform 68:101550. https://doi.org/10.1016/J.ECOINF.2021.101550

Rehman S, Keefover-Ring K, Haq IU, Dilshad E, Khan MI, Akhtar N, Mirza B (2010) Drier climatic conditions increase withanolide content of *Withania coagulans* enhancing its inhibitory potential against human prostate cancer cells. Appl Biochem Biotechnol 188(2):460–480. https://doi.org/10.1007/s12010-018-02933-8

Retana-Cordero M, Flores S, Freyre R, Gómez C (2022) Strategies to reduce radiation stress in open-field ginger and turmeric production. Agronomy 12(8):1910. https://doi.org/10.3390/AGRONOMY12081910

Rokaya MB, Sigdel SR (eds) (2024) Flora and vegetation of Nepal, 1st edn. Springer Cham, Cham. https://doi.org/10.1007/978-3-031-50702-1

Rokaya MB, Münzbergová Z, Shrestha MR, Timsina B (2012) Distribution patterns of medicinal plants along an elevational gradient in central Himalaya, Nepal. J Mt Sci 9(2):201–213. https://doi.org/10.1007/s11629-012-2144-9

Rokaya MB, Parajuli B, Bhatta KP, Timsina B (2020) *Neopicrorhiza scrophulariiflora* (Pennell) Hong: a comprehensive review of its traditional uses, phytochemistry, pharmacology and safety. J Ethnopharmacol 247:112250. https://doi.org/10.1016/j.jep.2019.112250

Sánchez-Ramos M, Ruiz-Betancourt A, Tadeo-Cuenca SA, Román-Guerrero A, Columba-Palomares MC, Guerrero-Alonso A, Bernabé-Antonio A, Ojeda-Ramírez D, Cruz-Sosa F (2024) The role of Latin America medicinal plants in wound healing. Front Chem Eng 6:1514962. https://doi.org/10.3389/FCENG.2024.1514962/XML/NLM

Sangwan NS, Tripathi S, Srivastava Y, Mishra B, Pandey N (2017) Phytochemical genomics of ashwagandha. In: Kaul SC, Wadhwa R (eds) Science of Ashwagandha: preventive and therapeutic potentials. Springer International Publishing, Cham, pp 3–36. https://doi.org/10.1007/978-3-319-59192-6_1

Sasikumar B (2005) Genetic resources of *Curcuma* : diversity, characterization and utilization. Plant Genet Resour 3(2):230–251. https://doi.org/10.1079/pgr200574

Saslis-Lagoudakis CH, Hawkins JA, Greenhill SJ, Pendry CA, Watson MF, Tuladhar-Douglas W, Baral SR, Savolainen V (2014) The evolution of traditional knowledge: environment shapes medicinal plant use in Nepal. Proc R Soc B Biol Sci 281(1780). https://doi.org/10.1098/rspb.2013.2768

Saunier A, Ormeño E, Moja S, Fernandez C, Robert E, Dupouyet S, Despinasse Y, Baudino S, Nicolè F, Bousquet-Mélou A (2022) Lavender sensitivity to water stress: comparison between eleven varieties across two phenological stages. Ind Crop Prod 177:114531. https://doi.org/10.1016/j.indcrop.2022.114531

Scartezzini P, Antognoni F, Raggi MA, Poli F, Sabbioni C (2006) Vitamin C content and antioxidant activity of the fruit and of the Ayurvedic preparation of *Emblica officinalis* Gaertn. J Ethnopharmacol 104(1–2):113–118. https://doi.org/10.1016/J.JEP.2005.08.065

Shi C, Liu H (2021) How plants protect themselves from ultraviolet-B radiation stress. Plant Physiol 187(3):1096. https://doi.org/10.1093/PLPHYS/KIAB245

Shil S, Dewanjee S (2022) Impact of drought stress signals on growth and secondary metabolites (SMs) in medicinal plants. J Phytopharm 11(5):371–376. https://doi.org/10.31254/phyto.2022.11511

Shrestha B, Tsiftsis S, Chapagain DJ, Khadka C, Bhattarai P, Shrestha NK, Kolanowska MA, Kindlmann P (2021) Suitability of habitats in Nepal for *Dactylorhiza hatagirea* now and under predicted future changes in climate. Plants 10(3):1–21. https://doi.org/10.3390/plants10030467

Shrestha UB, Lamsal P, Ghimire SK, Shrestha BB, Dhakal S, Shrestha S, Atreya K (2022) Climate change-induced distributional change of medicinal and aromatic plants in the Nepal Himalaya. Ecol Evol 12(8):e9204. https://doi.org/10.1002/ECE3.9204

Shyaula SL, Tamang T, Ghouri N, Adhikari A, Marasini S, Bajracharya GB, Manandhar MD, Choudhary MI (2016) Antileishmanial diterpenoid alkaloids from *Aconitum spicatum* (Bruhl) Stapf. Nat Prod Res 30(22):2590–2593. https://doi.org/10.1080/14786419.2015.1114941

Singh R, Gupta P, Khan F, Singh SK, Sanchita MT, Kumar A, Dhawan SS, Shirke PA (2018) Modulations in primary and secondary metabolic pathways and adjustment in physiological behaviour of *Withania somnifera* under drought stress. Plant Sci 272:42–54. https://doi.org/10.1016/j.plantsci.2018.03.029

Soković MD, Vukojević J, Marin PD, Brkić DD, Vajs V, Van Griensven LJLD (2009) Chemical composition of essential oils of *Thymus* and *Mentha* species and their antifungal activities. Molecules 14(1):238–249. https://doi.org/10.3390/molecules14010238

Srivastava K, Singh S, Singh A, Jain T, Datta R, Kohli A (2023) Effect of temperature (cold and hot) stress on medicinal plants. In: Med plants their response to abiotic stress, pp 153–168. https://doi.org/10.1007/978-981-19-5611-9_5

Sun B, Wu L, Wu Y, Zhang C, Qin L, Hayashi M, Kudo M, Gao M, Liu T (2020) Therapeutic potential of *Centella asiatica* and its triterpenes: a review. Front Pharmacol 11:568032. https://doi.org/10.3389/fphar.2020.568032

Szepesi Á (2021) Plant metabolites and regulation under environmental stress. Plants 10(10). https://doi.org/10.3390/PLANTS10102013

Tafuri S, Cocchia N, Carotenuto D, Vassetti A, Staropoli A, Mastellone V, Peretti V, Ciotola F, Albarella S, Del Prete C et al (2019) Chemical analysis of *Lepidium meyenii* (Maca) and its effects on redox status and on reproductive biology in Stallions. Molecules 24(10):1981. https://doi.org/10.3390/MOLECULES24101981

Thakar A, Dave D (2024) Population assessment, distributional pattern and ethnomedicinal significance of *Dactylorhiza hatagirea* (D. Don) Soo: an endangered species of the Himalayan region. Front For Glob Chang 7:1475620. https://doi.org/10.3389/ffgc.2024.1475620

Theodoridis S, Hickler T, Nogues-Bravo D, Ploch S, Mishra B, Thines M (2025) Satellite-observed mountain greening predicts genomic erosion in a grassland medicinal herb over half a century. Curr Biol 35(12):2761–2770.e5. https://doi.org/10.1016/J.CUB.2025.04.007

Tinitana F, Rios M, Romero-Benavides JC, de la Cruz RM, Pardo-de-Santayana M (2016) Medicinal plants sold at traditional markets in southern Ecuador. J Ethnobiol Ethnomed 12(1):1–18. https://doi.org/10.1186/S13002-016-0100-4/TABLES/4

Tiwari A, Uprety Y, Rana SK (2019) Plant endemism in the Nepal Himalayas and phytogeographical implications. Plant Divers 41(3):174–182. https://doi.org/10.1016/j.pld.2019.04.004

Tnah LH, Lee CT, Lee SL, Ng KKS, Ng CH, Hwang SS (2011) Microsatellite markers of an important medicinal plant, *Eurycoma longifolia* (Simaroubaceae), for DNA profiling. Am J Bot 98(5):e130–e132. https://doi.org/10.3732/ajb.1000469

Tshabalala T, Abdel-Rahman EM, Ncube B, Ndhlala AR, Mutanga O (2021) Leveraging of hyperspectral remote sensing on estimating biomass yield of *Moringa oleifera* Lam. medicinal plant. S Afr J Bot 140:37–49. https://doi.org/10.1016/J.SAJB.2021.03.035

Turpin G, Ritmejerytė E, Jamie J, Crayn D, Wangchuk P (2022) Aboriginal medicinal plants of Queensland: ethnopharmacological uses, species diversity, and biodiscovery pathways. J Ethnobiol Ethnomed 18(1):1–15

Uprety Y, Asselin H, Boon EK, Yadav S, Shrestha KK (2010) Indigenous use and bio-efficacy of medicinal plants in the Rasuwa District, Central Nepal. J Ethnobiol Ethnomed 6(1):1–10

van Oosterhout C, Speak SA, Birley T, Hitchings LW, Bortoluzzi C, Percival-Alwyn L, Urban L, Groombridge JJ, Segelbacher G, Morales HE (2025) Genomic erosion in the assessment of species' extinction risk and recovery potential. BioRxiv 2022-09. https://doi.org/10.1101/2022.09.13.507768

Varghese R, Shringi H, Efferth T, Ramamoorthy S (2025) Artificial intelligence driven approaches in phytochemical research: trends and prospects. Phytochem Rev 27:1–16

Vassiliou E, Awoleye O, Davis A, Mishra S (2023) Anti-inflammatory and antimicrobial properties of *Thyme* oil and its main constituents. Int J Mol Sci 24(8):6936. https://doi.org/10.3390/IJMS24086936

Vassvik L, Vandvik V, Östman SAH, Nielsen A, Halbritter AH (2024) Temporal and spatial variation in the direct and indirect effects of climate on reproduction in alpine populations of *Ranunculus acris* L. Alp Bot 134(2):123–139. https://doi.org/10.1007/S00035-024-00317-9/FIGURES/7

Ved DK, Goraya GS (2007) Demand and supply of medicinal plants in India, vol 18(85). NMPB, New Delhi & FRLHT, Bangalore, India, pp 210–252

Vu DD, Nguyen MT, Nguyen MD, Nguyen PLH, Bui TTX, Phan KL, Vu DG, Pham QT, Nguyen TPT (2024) Genetic population structure of the Vietnamese ginseng (*Panax vietnamensis* Ha et Grushv.) detected by microsatellite analysis. Braz J Biol 84:e264369. https://doi.org/10.1590/1519-6984.264369

Wang Q, Deng K, Ai J, Yingping W, Yougui W, Ren Y, Zhang N (2023) Integrated transcriptomic and metabolomic analyses reveal the effects of grafting on special metabolites of *Acanthopanax senticosus* leaves. Molecules 28(12):4877. https://doi.org/10.3390/MOLECULES28124877/S1

Wen J, Zhou L, Liu L, He Y (2022) Analysis of the impact of climate change on the distribution and active compound content of the plateau medicinal plant *Nardostachys jatamansi* (D. Don) DC. Ind Crop Prod 187. https://doi.org/10.1016/J.INDCROP.2022.115438

WHO (2009) The Use of Herbal Medicines in Primary Health Care. Accessed 2025 Jun 24. https://iris.who.int/handle/10665/206476

WHO (2014) WHO Traditional Medicine Strategy 2014–2023. Accessed 2025 Aug 12. https://www.who.int/publications/i/item/9789241506096

WHO (2023) Traditional medicine has a long history of contributing to conventional medicine and continues to hold promise. Accessed 2025 Jun 24. https://www.who.int/news-room/feature-stories/detail/traditional-medicine-has-a-long-history-of-contributing-to-conventional-medicine-and-continues-to-hold-promise

Woldearegay M (2020) Climate change impacts on the distribution and phenology of plants: a review. Trop Plant Res 7(1):196–204. https://doi.org/10.22271/tpr.2020.v7.i1.025

Wu D, Xiong F, Wang H, Liu S, Zhu J, Zhao D, Yang J, Ma W, Guo L, Kang C (2024) Temperature seasonality and soil phosphorus availability shape ginseng quality via regulating ginsenoside contents. BMC Plant Biol 24(1):824. https://doi.org/10.1186/S12870-024-05518-2

Yadav N, Singh S (2024) Genetic resources and variations in *Picrorhiza kurroa* Royle ex Bentham. In: Nandave M, Joshi R, Upadhyay J (eds) Ethnopharmacology and OMICS advances in medicinal plants Vol 2: reveal revealing the secrets of medicinal plants. Springer Nature, Singapore, pp 55–74. https://doi.org/10.1007/978-981-97-4292-9_4

Yang X, Eilerman RG (1999) Pungent principal of *Alpinia galangal* (L.) Swartz and its applications. J Agric Food Chem 47(4):1657–1662. https://doi.org/10.1021/jf9808224

Yang L, Zhu X, Song W, Shi X, Huang X (2024) Predicting the potential distribution of 12 threatened medicinal plants on the Qinghai-Tibet Plateau, with a maximum entropy model. Ecol Evol 14(2):e11042

Yoo S, Yang HC, Lee S, Shin J, Min S, Lee E, Song M, Lee D (2020) A deep learning-based approach for identifying the medicinal uses of plant-derived natural compounds. Front Pharmacol 11:584875. https://doi.org/10.3389/FPHAR.2020.584875/FULL

Yuan QJ, Zhang ZY, Hu J, Guo LP, Shao AJ, Huang LQ (2010) Impacts of recent cultivation on genetic diversity pattern of a medicinal plant, *Scutellaria baicalensis* (Lamiaceae). BMC Genet 11(1):1–13

Yumeng W, Jiaxin Z, Wang P, Li Y, Yihan W, Yan Y, Chi J, Chen J, Lian J, Piao X et al (2025) Integrated transcriptomic and metabolomic analysis reveals tissue-specific flavonoid biosynthesis and MYB-mediated regulation of UGT71A1 in *Panax quinquefolius*. Int J Mol Sci 26(6):2669. https://doi.org/10.3390/IJMS26062669/S1

Zeng L, Sun Z, Fu L, Gu Y, Li R, He M, Wei J (2024) Cryopreservation of medicinal plant seeds: strategies for genetic diversity conservation and sustainability. Plants 13(18):2577. https://doi.org/10.3390/PLANTS13182577

Zhang X, Li G, Dai J, Wei P, Du B, Li F, Yulu W, Yujuan W (2025) Integrated transcriptomic and metabolomic analyses reveal key antioxidant mechanisms in *Dendrobium huoshanense* under combined salt and heat stress. Plants 14(15):2303. https://doi.org/10.3390/PLANTS14152303

Zhu Z, Zhou Y, Liu X, Meng F, Xu C, Chen M (2024) Integrated transcriptomic and metabolomic analyses uncover the key pathways of *Limonium bicolor* in response to salt stress. Plant Biotechnol J 23(3):715. https://doi.org/10.1111/PBI.14534

Zidorn C (2010) Altitudinal variation of secondary metabolites in flowering heads of the Asteraceae: trends and causes. Phytochem Rev 9(2):197–203. https://doi.org/10.1007/S11101-009-9143-7

Zobayed SMA, Afreen F, Kozai T (2005) Temperature stress can alter the photosynthetic efficiency and secondary metabolite concentrations in St. John's wort. Plant Physiol Biochem 43(10–11):977–984. https://doi.org/10.1016/j.plaphy.2005.07.013

Chapter 4
Applications and Emerging Trends in Techniques for Assessing Genetic Diversity in Medicinal Plants

Dipasree Roychowdhury and Mihir Halder

Abstract The vast genetic diversity of medicinal plants is the basic foundation that dictates their phytochemical diversity, adaptability, and resilience against changing environments. The therapeutic efficacy of medicinal plants is directly correlated with their phytochemical diversity, which acts as a major source of modern drug discovery, traditional medicine, and sustainable healthcare systems. Various anthropogenic activities and climate change are imposing tremendous pressure on the genetic diversity of medicinal plants. The assessment and conservation of genetic diversity have been largely overlooked and less well exploited than species diversity. The application of traditional morphology- and phytochemical-based genetic diversity assessment is challenging due to low resolution, environmental influences, and cryptic speciation. In contrast, molecular approaches, such as random amplified polymorphic DNA, amplified fragment length polymorphism, inter-simple sequence repeat, simple sequence repeat marker, DNA barcoding, and single-nucleotide polymorphism (SNP), are more reliable and effective in unlocking the genetic architecture of medicinal plants. During their application, their advantages and limitations should be considered. Moreover, next-generation sequencing-based approaches, such as SNPs and whole-genome sequencing, are adding further resolution, precision, and consistency to genetic profiling and are frequently applied to understand intra- and interspecific genetic variation, identify unique, rare, or threatened genotypes, and understand population structure by genetic characterization, thereby supporting the adoption of effective ex-situ and in-situ conservation strategies. The integration of multi-omics techniques, bioinformatics, artificial intelligence, and machine learning models is a current trend in this field that may help reveal complex biosynthetic pathways of secondary metabolites (SMs), identify genes responsible

D. Roychowdhury
Department of Botany, Surendranath College, Kolkata, West Bengal, India

M. Halder (✉)
Post-graduate Department of Botany, Barasat Government College, Kolkata, West Bengal, India
e-mail: mihir.halder@bgc.ac.in

S. Jha et al. (eds.), *Genetic Diversity in Medicinal Plants*, Sustainable Development and Biodiversity 44, https://doi.org/10.1007/978-981-95-7281-6_4

for desired SMs, and accelerate the development of superior varieties using genomics-assisted breeding and transgenic approaches. Moreover, comprehensive genetic assessment helps to achieve sustainable utilization, informed conservation, and continued discovery of novel natural products essential for global health and well-being in the twenty-first century.

Keywords Genetic diversity assessment · Single-nucleotide polymorphism · Simple sequence repeat · Cytological diversity · DNA barcoding

4.1 Introduction

Genetic diversity refers to the variety of different genetic characteristics present within the genome of a population or species. This concept encompasses variations in individual nucleotide sequences in both coding and non-coding regions, as well as structural rearrangements in chromosomes and variations in the number of chromosomes. Genetic diversity can exist on multiple levels, such as within individuals, among individuals in the same population, and among different populations or species. Genetic diversity can be measured by estimating the degree of polymorphism (variation) at specific genetic loci in individual samples, whereas at population level, genetic diversity can be estimated based on allele frequency (the proportion of a specific allele in a population), heterozygosity (the proportion of observed heterozygous individuals in a population), polymorphic loci (the percentage of loci that have more than one allele), and genetic distance/similarity (a measure of genetic divergence or relatedness between populations or individuals). The variation in gene frequencies and allelic richness of individual populations of a single species (Intraspecific genetic diversity) of medicinal plants is the foundation for their evolutionary capacity, as it directly contributes to their adaptability, resilience, and survival in a constantly changing environment (Salgotra and Chauhan 2023). It serves as the raw material that is closely linked to chemo-diversity within and between populations in medicinal plants, which are responsible for their therapeutic potential. Genetic variations can lead to various metabolic pathways, resulting in a wider array of compounds or varying concentrations of existing ones, thereby creating distinct "chemotypes" within the same species (Xavier et al. 2022). This extensive genetic diversity is essential for a long-term, steady supply of these plants as a natural resource. It ensures a reliable supply of traditional and modern medicine with therapeutic benefits.

Accurate and comprehensive assessments, coupled with a deep understanding of their genetic diversity, enable researchers to uncover unique chemical profiles and discover entirely new compounds with previously unknown therapeutic activities. It is an essential step in both effective conservation efforts and ensuring their sustainability for future generations. For example, intraspecific variation provides the necessary detail for localized management, directly guiding population-level conservation strategies to ensure genetic robustness, local adaptation, and ecological resilience within a particular species, while interspecific genetic diversity analysis informs broader biodiversity policy and germplasm collection priorities. This macro-level perspective ensures that we are not simply saving isolated populations, but preserving the underlying evolutionary potential of entire ecosystems. By integrating these two perspectives, we establish a robust management system that secures both the immediate utility of critical plant resources and their sustained, long-term survival against future environmental pressures. Understanding their evolutionary relationship through genetic diversity studies provides the foundational material necessary for improving desirable traits in these plants through breeding programs. The methodologies employed to assess genetic diversity in medicinal plants have undergone significant advancements over time. The history of genetic diversity assessment techniques represents a progression from traditional methods of observing morphological differences to precise analysis of protein variants and modern molecular techniques (Rahayu et al. 2025).

Historically, classical techniques used for genetic variability assessment primarily relied on morphological characterization and basic ethnobotanical descriptions, which provided limited insights into genetic architecture and population dynamics. Observable differences in morphological (phenotypic) traits in their natural habitats served as the first markers of genetic diversity assessment that involves a systematic observation, documentation, and comparative analysis of qualitative or quantitative characteristics, which includes both macro- and microscopic morphological and anatomical characteristics, usually collected from field-grown plants, traditional ethnobotanical knowledge, and herbarium collections. This method has very limited application due to its low resolution compared to molecular approaches, as all existing genetic variations within the genome do not directly affect observable morphological or anatomical changes. Moreover, the variability in many morphological traits has no genetic basis, and these variations may be influenced by environmental conditions, developmental phases, and phenotypic plasticity (Stotz et al. 2021; Kanta et al. 2024). These factors can induce morphological changes regardless of genetic background, which ultimately complicates the process of determining genetic diversity using morphological markers.

The innovation and introduction of DNA-based molecular approaches, such as restriction fragment length polymorphism (RFLP), random amplified polymorphic DNA (RAPD), amplified fragment length polymorphism (AFLP), simple sequence repeats (SSRs), inter-simple sequence repeat (ISSR), and single-nucleotide polymorphisms (SNPs), have enhanced the ability to assess genetic diversity and conduct population genetics studies with more precision (Hussain and Nisar 2020). Currently, the integration of next-generation sequencing (NGS) technologies, transcriptomics, proteomics, metabolomics, epigenomics, and advanced bioinformatics tools holds great potential to uncover new insights into genetic diversity, proteins, and metabolites of medicinal plant species (Kumar et al. 2019; Pandita et al. 2021). These advancements enable whole-genome assessments and the identification of functional genes within medicinal plants, which was previously not possible, marking the beginning of a new era in assessing genetic diversity.

Despite these remarkable advancements in genetic and genomic technologies, significant challenges persist in achieving comprehensive, scalable, and universally applicable evaluations of genetic diversity across diverse ecosystems and species. The deep-rooted methodological limitations related to inconsistencies in data standardization, lack of inter-platform comparability among genomic tools, cost-effectiveness for large-scale field applications, and the challenge of effective translation of complex genomic information into actionable conservation strategies and practical breeding programs. A thorough understanding of the evolutionary perspective on these assessment methodologies, ranging from basic phenotypic observations to sophisticated genomic investigations, is essential, as each method holds its strengths and limitations. This knowledge can help identify persistent limitations and guide a strategic path for future innovations that maintain a balance of precision, practicality, and affordability. An integrative approach that adopts a multifaceted research framework combining morphological, biochemical, and molecular techniques can yield a more holistic understanding of genetic diversity within medicinal plants.

Researchers and conservationists can prioritize at-risk genetic lineages for conservation, enhance breeding programs to develop improved cultivars, and foster sustainable utilization of plant genetic resources by combining these methods. Ultimately, these efforts will effectively safeguard the genetic resources of these invaluable natural pharmacies and unlock their full potential for drug discovery, phytomedicine development, and improved healthcare outcomes. Strategic collaborations among geneticists, conservation biologists, bioinformaticians, and policy-makers will be pivotal in translating scientific insights into scalable, real-world solutions.

4.2 Biochemical and Chemotaxonomic Approach for Genetic Diversity Assessment

Molecular markers can generally be classified as biochemical markers and DNA-based markers. The biochemical markers are considered the final downstream product of the genome, and they can be represented as specific (particularly isoenzymes) or general enzymes, proteins, and metabolites (particularly secondary metabolites (SMs)). They often indirectly reflect information about genetic variation in the genetic code that would be changed in the gene products by alterations in their expression and function. Isozymes are considered the preferred biochemical marker, but evaluating overall protein profiles gives a broader perspective of expressed genes compared to focusing solely on isozymes. Isozymes are allelic variants of the same enzyme that catalyze the same biochemical reactions. They exhibit differences in their electrophoretic mobility patterns in an electric field due to variations in net charge or conformation. This outcome is the result of nucleotide substitutions directly at the specific enzyme-encoding gene, located at a single locus, or at different loci that change the amino acid sequence. Thus, banding patterns of isozymes on starch or polyacrylamide gels serve as a direct reflection of genetic variation at a specific gene encoding an enzyme. The majority of these studies primarily used esterase, peroxidase, acid phosphatase, alkaline phosphatase, alcohol dehydrogenase, malate dehydrogenase, polyphenol oxidase, superoxide dismutase, cytochrome oxidase, and glucose-6-phosphate dehydrogenase. Various studies have demonstrated that the use of these biochemical markers effectively reveals moderate to low levels of polymorphism at the species, genus, population, or germplasm level in numerous medicinal plants (Habiba et al. 2019; Hake et al. 2025; Setyawan et al. 2014). Although biochemical markers offer a direct snapshot of the physiological state and functional variations, they are highly sensitive to genetic and environmental interactions, thus making it difficult to solely determine purely genetic effects. It can be difficult to directly link specific changes in metabolite concentration to underlying genes as well.

The underlying principle of using a chemotaxonomic (chemosystematics or chemical profiling) approach for assessing genetic diversity involves qualitative and quantitative analysis of SMs. This analysis is performed both within and between populations. In contrast to highly conserved primary metabolites, SMs include a diverse range of compounds that have various ecological roles and therapeutic properties (Halder and Jha 2023). The biosynthesis of these SMs relies on complex SM-specific linear or often branched biosynthetic pathways that originate from primary metabolic pathways and ultimately diversify.

Variations in these chemical profiles directly reflect underlying genetic variations in enzyme-encoding single genes or gene clusters, as well as their regulatory elements, like transcription factors. Genetic diversity is a heritable influencer of these biosynthetic pathways. They directly alter/modulate the metabolic flux of the biosynthetic pathway, leading to variations in the types and concentrations of specific metabolites (Jadaun et al. 2023). These distinctions often manifest as distinct chemotypes—genetically distinct varieties of a species that often produce unique chemical profiles or varying concentrations of specific compounds. Medicinal plants with several chemotypes have been identified in several medicinal plants such as *Artemisia annua*, *Mentha* species, *Ocimum* species, *W. somnifera*, and several others (Jadaun et al. 2023). These distinctions in chemical profile are often relevant for medicinal plants because the resulting chemical composition is directly correlated with the efficacy and therapeutic properties.

The process of assessing phytochemical diversity includes the identification and collection of medicinal plants, the extraction of chemical compounds using a solvent system appropriate for the polarity of the target compounds, separation, qualitative and quantitative analysis, identification of the extracted compounds, determination, and identification of the chemical profile using various technologies. These techniques include thin-layer chromatography, high-performance liquid chromatography, gas chromatography (GC), supercritical fluid chromatography, mass spectrometry (MS), ultraviolet–visible spectroscopy, nuclear magnetic resonance spectroscopy, GC-MS, and liquid chromatography-MS (Halder et al. 2024). The chemical profiles include the presence, absence, or relative concentrations of specific compounds, which are used to assess diversity. These data are converted into numerical data, and statistical methods such as cluster analysis, principal component analysis (PCA), and discriminant analysis are then applied to group samples based on their chemical similarities and dissimilarities, thereby revealing patterns of genetic diversity.

It is now well-documented that the synthesis and accumulation of SMs are also affected by up- or down-regulation of concerned genes. These changes are driven by natural or artificial factors, including internal developmental genetic circuits (regulated gene, enzyme), cell, tissue, organ, and developmental stage, as well as external environmental factors such as light, temperature, water, and salinity (Kong et al. 2017; Cheng et al. 2018). The influence of external environmental and physiological conditions on the chemical profile is the major drawback of this approach.

4.3 Assessing Genetic Diversity Using Classical and Molecular Cytogenetics Techniques

Various classical and molecular plant cytogenetic techniques are still considered one of the unique and fundamental tools for assessing genetic diversity. Molecular techniques detect diversity at the DNA sequence level, while cytogenetic techniques assess genetic variations through the direct visualization and analysis of the physical organization and characteristics of the genome at the chromosome level. These methods offer a unique perspective on the structural and numerical variations that shape the genetic landscape of plant populations. Cytogenetic analysis evaluates features, such as chromosome number, size, morphology, meiotic behavior, and the physical distribution of specific DNA sequences in somatic ($2n$) or gametic (n) cells. Additionally, it considers meiotic abnormalities, chromosome rearrangements, variation in ploidy levels, and the presence of B-chromosomes to differentiate medicinal plants at both inter- and intra-species levels. These techniques have great potential for providing additional complementary information on genetic diversity and evolutionary relationships of medicinal plants at the generic, species, or below-species level. These parameters are often an excellent source for determining genetic diversity, identifying cytotypes, monitoring genetic stability, elucidating evolutionary relationships, understanding reproductive strategies, and guiding targeted breeding programs that have the potential for genetic improvement of medicinal plants. Cytogenetics also enables researchers to identify unique chromosomal patterns and behavior that may relate to specific traits, adaptive potential, or even phytochemical profiles. This section presents a few selected examples of potential contributions of different cytogenetic methods to the characterization of genetic diversity in medicinal plants.

In Karyotype analysis, the unique chromosomal architecture of the complete set of somatic metaphase chromosomes is examined in detail, including their number, size, centromere position, and morphological features of secondary constriction-bearing chromosomes. Each medicinal plant species has its unique karyotype characteristics that serve as its own special "fingerprint" at the chromosome level. This method helps to understand genome organization and stability. Several studies demonstrated that detailed classical cytological studies have the ability to reveal the existence of inter- and intra-species genetic variation within wild populations, or species of *Thymus* (Tohidi et al. 2021), *Cinnamomum* (Firdausi et al. 2022), and *Phyllanthus* (Rahman et al. 2024). Karyotype analysis detected different chromosomal races or cytotypes within wild populations of *Rumex* collected from the Northwest Himalayas (Jeelani et al. 2017), *Gentiana kurroo* grown at various altitudes in the Kashmir Himalayas (Jeelani et al. 2021), *Syzygium cumini* from North India (Sharma et al. 2020), and *Inula racemosa* from the Kashmir Himalayas (Mudassir Jeelani et al. 2022).

Singh et al. (2018) also studied chromosomal behavior during male meiosis in 29 species from 54 populations across 18 genera of Lamiaceae in the Sirmaur district of Himachal Pradesh. They found B-chromosomes in cytotypes of five species, namely, *A. bracteosa* ($n = 32+ 0–2B$), *Leucas lanata* ($n = 11+ 0–1B$), *Lycopus europeus* ($n = 11+ 0–1B$), *Phlomis bracteosa* ($n = 11+ 0–1B$), and *Vitex negundo* ($n = 16+ 0–1B$).

In addition to basic morphology, differential staining of chromosomes using DNA base-specific fluorochromes such as 4′,6-diamidino-2-phenylindole (DAPI) and Chromomycin A3 (CMA) produces different and specific fluorescent chromosome banding patterns along the chromosome length based on the number, location, and percentages of GC- and AT-rich repetitive sequences in the genome. Detailed karyotype characterization and differential chromosome banding with CMA and DAPI studies demonstrated higher resolution of this technique as it detect genetic variations in the distribution and composition of DNA and chromatin condensation in various medicinal plants, including Indian species of *Drimia* (Nath et al. 2015), three *Rauvolfia* species (Rashid et al. 2019), three *Cinnamomum* species from Bangladesh (Firdausi et al. 2022), five Indian species of *Dipcadi* (Samanta et al. 2023), in two *Solanum* species (Biswas et al. 2024), three Indian germplasms of *Gloriosa superba* (Jha et al. 2024), and five *Phyllanthus* species (Rahman et al. 2024). All these studies highlight the effectiveness of cytogenetic techniques for differentiating the genetic diversity of different wild populations of medicinal plants.

Fluorescence in-situ hybridization (FISH) and genomic in-situ hybridization (GISH) are advanced, highly sensitive molecular cytogenetic techniques that enable the visualization of specific DNA sequences, genes, repetitive sequences on chromosomes, or whole genomes using specific probes. Zhou et al. (2019) highlighted the effectiveness of pre-labeled oligoprobe-FISH in better understanding the genome structure of four ginseng-related species of *Aralia elata*, *Dendropanax morbiferus*, *Eleutherococcus sessiliflorus*, and *Kalopanax septemlobus* over the conventional FISH method. Chromosome identification, karyotypic variation analysis, and understanding evolutionary dynamics through a genetic diversity study have been reported in five varieties and cultivars of *Zanthoxylum armatum* using Oligo-FISH (He et al. 2023), in four related species of *Dracaena* (Zhao et al. 2021), and in 11 *Senna* species using comparative triple-color FISH mapping (Nguyen et al. 2021). Chen et al. (2024) reported an effective method of non-denaturing FISH in *Cichorium intybus* using new oligoprobes. Genetic diversity and phylogenetic relationships among 40 *Aloe* species have been studied using FISH and GISH in combination with chloroplast DNA markers, revealing the origin of an allodiploid hybrid species from two parental species, *A. vera* and *A. arborescens* (Lee et al. 2016). Similarly, FISH and GISH technology were applied in five *Chlorophytum* species to differentiate them based on cytological parameters (Lekhak et al. 2024).

Several studies have clearly demonstrated that genome size (the total amount of DNA in a cell) and ploidy level estimation are reliable strategies for producing additional information about the genetic diversity among medicinal plants. The C-value index is a species-specific characteristic that is highly useful in systematics, genome size estimation, and many other biological fields. Flow cytometric analysis is usually effectively used for determining the genome size and ploidy level by measuring the fluorescence intensity of a DNA-binding dye (e.g., propidium iodide and DAPI) in a suspension of intact nuclei (Nath et al. 2015; Esfahani et al. 2016; Than et al. 2017; Morovati et al. 2024). Nath et al. (2015) reported successful detection of genetic variation at the species level in five Indian species of *Drimia* (*Drimia indica, Drimia polyantha, Drimia razii, Drimia wightii*, and *Drimia coromandeliana*) using fluorescent chromosome banding patterns and nuclear DNA content analysis. In a similar study, 12 ecotypes of *Drimia indica* have been identified from populations of *D. indica* located in different geographical locations of Odisha, further illustrating the utility of integrating karyotype analysis with 2C DNA content for the identification of genome-specific cytotypes (Patel et al. 2024). They found a correlation between the 2C DNA values and ploidy variation at the ecotype level, which varies from 16.54 pg ($2n = 16$) in ecotype "Nayagarh" to 36.26 pg ($2n = 40$) in ecotype "Nuagaon." Esfahani et al. (2016) reported that 2C DNA value estimation by flow cytometry analysis is effectively able to distinguish unidentified Persian *Papaver bracteatum* in Northern Iran from other identified or unidentified *Papaver* species with similar morphological traits. Genetic variation at the karyotype, ploidy, and genome size levels was detected in 15 populations of both *Datura stramonium* and *D. innoxia* (Morovati et al. 2024). In another study conducted on the critically endangered medicinal plant *Swertia chirayita* and its allied species, *S. nervosa* and *S. bimaculata,* the existing genetic diversity was estimated by fluorescence chromosome banding patterns and nuclear genome size estimation (Than et al. 2017). Chromosome number and fluorescent banding patterns revealed the distinct differences in genetic composition among these species. *S. chirayita* has $2n = 26$ chromosomes, among which CMA^{+ve} signals were detected in six chromosomes, while all $2n = 26$ chromosomes of *S. nervosa* displayed CMA^{+ve} signals. CMA^{+ve} signals were detected on only two chromosomes in *S. bimaculata* ($2n = 26$). Moreover, nuclear genome size estimation by flow cytometry revealed that *S. bimaculata* (9.82 pg/2C) has approximately nine- and fivefold higher 2C DNA content than *S. chirayita* (1.09 pg/2C) and *S. nervosa* (1.95 pg/2C), respectively (Than et al. 2017). The lowest amount of GC-rich CMA sites, along with the highest DNA content, distinguished *S. bimaculata* from the other two species.

The field of plant cytogenetics continues to adapt and grow through its unique adaptability and integration with different new technologies, such as molecular biology, NGS, automated image analysis systems, artificial intelligence (AI), and machine learning (ML). The classical cytogenetics already incorporates sophisticated molecular techniques that extend its ability from simple visualization of chromosome morphology to providing sequence-level details of DNA within chromosomes. Furthermore, NGS data facilitate high-resolution cytological mapping that bridges the gap between chromosomal observation and DNA sequence analysis. The integration of automated image analysis systems equipped with sophisticated cameras and software now rapidly captures high-resolution microscopic images and processes them to quantify various karyomorphological parameters and detect chromosomal aberrations with impressive speed and accuracy (Kang et al. 2024). AI algorithms and ML models are also stepping down in this field, expanding their application in more precise identification of chromosomal aberrations and chromosomal banding patterns that are challenging to detect by human analysis (Zhou et al. 2024; Rosenblum et al. 2025). However, this approach only reflects genetic diversity at the chromosomal level. Although it is stable and not influenced by environmental or developmental conditions, the resolution power of this technique is limited.

4.4 Molecular Markers

4.4.1 Restriction Fragment Length Polymorphism

RFLP is one of the classic non-PCR-based DNA marker technologies that is more accurate and reliable than biochemical markers. It is a co-dominant marker that can distinguish between homozygous (two identical copies of the gene or locus) and heterozygous plants (two different copies of the gene or locus). In this method, isolated genomic DNA is treated with specific restriction endonucleases that recognize and cut DNA at specific, short recognition sites. Then, the mixture of DNA fragments is separated by size using gel electrophoresis, transferred onto a membrane (Southern Blot), and visualized using labeled DNA probes. The number of DNA fragments produced and their length directly correlated with the recognition sites. Any change in the DNA sequence will either create a new restriction site or remove an existing restriction site, which leads to a unique molecular fingerprint (banding pattern). Large amounts of high-quality DNA and the process being labor-intensive are major drawbacks to this method. Despite these challenges, RFLP has been used in genetic diversity assessment in different plants, including *Glycyrrhiza glabra, G. uralensis, G. echinata, G. pallidiflora* (Yamazaki et al. 1994), *Hypericum perforatum* (Halušková and Čellárová 1997), *Brassica juncea* (Pradhan et al. 2003), *Elaeis oleifera,* and *E. guineensis* (Barcelos et al. 2002). Current application of RFLP in resolving genetic diversity has been shifted to PCR-RFLP, which involves a simple modification of traditional RFLP by inclusion of PCR amplification as the first step, followed by restriction digestion of the amplified region instead of cutting the whole genome. The modified PCR-RFLP method has been extensively used in the study of medicinal plants, including *Salacia* species (Zhu et al. 2021), *Adenostemma lavenia* (Wu et al. 2022), *Foeniculum vulgare* (Križman and Jakše 2022), *Glycyrrhiza uralensis* (Nishida et al. 2025), and so on. PCR-RFLP has been successfully utilized in the study of genetic diversity in *F. vulgare* populations from the Istria region, indicating high variation at inter-population as well as within the populations studied (Križman and Jakše 2022). PCR-RFLP has been effectively utilized in *Salacia* species as a convenient molecular marker for species identification (Zhu et al. 2021). Wu et al. (2022) similarly developed an authentication method for *A. lavenia* using PCR-RFLP to accurately distinguish the species from its adulterants, ensuring safe consumption. Recently, Nishida et al. (2025) developed a simple and cost-effective PCR-RFLP-based method for the identification of *G. uralensis,* distinguishing it from its related species.

4.4.2 PCR-Based DNA Markers

To overcome the limitations of traditional non-PCR-based methods that either relied on morphological traits or biochemical profiling, the PCR-based molecular marker techniques have emerged. These techniques have significantly changed the previous conceptualization of genetic diversity assessment in medicinal plants. They provide a direct and meaningful, informative insight into the genetic code of the plant and present results with higher accuracy and efficiency. Their specificity and sensitivity allow reliable detection of small variations in the genome with confidence. PCR-based DNA markers include RAPD, AFLP, and SSRs. Together, these markers allow greater insight into population structure, facilitate species authentication, improve germplasm management, and accelerate breeding efforts. The selection of an appropriate marker system depends on the specific research objectives, available resources, and the specific organism under investigation.

4.4.2.1 Random Amplified Polymorphic DNA

RAPD is a simple, quick, and relatively inexpensive yet powerful PCR-based method classified as a dominant molecular marker. This technique is ideal for non-model plants because it does not require prior DNA sequence information and requires small amounts of template DNA. The technique uses short, randomly chosen primers of usually ten nucleotides in length that bind to numerous complementary sites scattered throughout the genome. When two such binding sites are sufficiently close and oriented inversely on opposite strands, it amplifies the DNA segment between them during PCR. A large number of polymorphic markers can be generated without prior sequence information throughout the genome, which can be used for screening of broader genetic differences. Polymorphisms arise from sequence variations at the primer binding sites or insertions/deletions between them that will prevent or enable amplification. These variations are then detected as the presence or absence of specific DNA fragments on an agarose gel. Several research studies have applied this marker for analysis of genetic diversity in medicinal plants, such as *Solanum trilobatum* (Shilpha et al. 2013), *Aloe* species (Ezzat et al. 2016), *Ceropegia juncea, Gymnema sylvestre, Oxystelma esculentum, Pentatropis capensis, Wattakaka volubilis* (Boomibalagan et al. 2021), *Salvia* species (Zheng et al. 2021), and so on. However, RAPD has significant limitations. Due to its dominant nature, it cannot distinguish heterozygotes from homozygotes. Its reproducibility is low compared to other markers, mainly due to their high sensitivity to PCR conditions, and non-specific amplification can occur.

4.4.2.2 Amplified Fragment Length Polymorphism

AFLP is a PCR-based technique used to make numerous copies of specific DNA sequences to detect genetic differences with high reproducibility. AFLP combines DNA cutting by restriction enzymes with a highly selective PCR step. Similar to RAPD and ISSR, AFLP is a dominant marker technique. This means that it indicates the presence or absence of DNA fragments, but cannot distinguish between two gene copies. Importantly, no prior DNA sequence information is needed. The process starts with the fragmentation of genomic DNA using two restriction enzymes, followed by the attachment of short, double-stranded DNA pieces called adapters to the fragment ends. Then, selective PCR is performed using specific primers designed based on the adapter and restriction site regions, with a few extra bases at the 3′ end. Only fragments with a complementary sequence to the primers are amplified. This generates unique DNA banding patterns (fingerprints) on polyacrylamide gels, and bands are detected using radioactive or fluorescent labels. This method can produce numerous DNA fragments from each digestion and provides a detailed overview of genetic variation. Several research studies have shown that AFLP is effective in resolving complex family relationships, detecting genetic diversity within populations, and revealing rare genetic types. For example, Kumar et al. (2022) studied the genetic diversity and population structure of 60 different *A. vera* genotypes with the help of AFLP markers. The analyses divided the 60 genotypes into three major clusters based on genetic polymorphism, which coincided with the groups according to geographical regions of collection. The analysis showed more than 4% genetic variation within clusters, while, among all the genotypes, more than 95% genetic variation was noted. *Glebionis discolor* (d'Urv.) Cano was earlier considered as *G. coronaria* var. *discolor*. Genetic diversity between these two plants was investigated using AFLP analysis with samples collected from Italy, Spain, and Portugal, in which the five primer combinations studied generated 1347 bands with more than 99% polymorphism (Gallucci et al. 2023). Similarly, AFLP analysis has been successfully used in many other medicinal plants, including *Papaver bracteatum* (Hadipour et al. 2020), *Solanum nigrum* (AL-Kiyyam et al.

2024), *Hibiscus sabdariffa* (Mulyaningsih et al. 2025), and so on. However, it can be labor-intensive, requires very pure DNA, has a more complex protocol, and usually costs more than simpler methods such as RAPD and ISSR.

4.4.2.3 Inter-Simple Sequence Repeat

ISSRs are PCR-based markers that use a single, short, usually repetitive primer of $(GA)_8C$ or $(CA)_8T$ designed from SSRs (microsatellites). Unlike SSRs, which need prior sequence information to design species-specific primers, ISSR primers are universal and randomly anneal to complementary sequences located in microsatellite regions. The methodology includes the isolation of genomic DNA, followed by the binding of primers to the complementary microsatellite regions. The DNA segments flanked by these closely spaced, oppositely oriented microsatellite sites are then amplified using PCR. The polymorphism in DNA fragments that appear on the gel after electrophoresis may be due to size variability caused by insertions, deletions, or mutations within the inter-repeat region. The presence or absence of a band corresponding to a specific molecular weight represents a specific allele, thus creating a unique "fingerprint" for each plant.

Despite its low resolution due to its limitation in distinguishing between homozygous and heterozygous genotypes, no requirement for prior sequence information makes ISSR an ideal choice for genetic analysis in understudied medicinal plants. This method has been applied in *Hypericum* species (Ma et al. 2021), *Gymnema sylvestre* (Vaishnav et al. 2023), *Justicia adhatoda* (Chaudhary and Sharma 2025), *Aloe littoralis* (Farahani et al. 2025), *Costus pictus* (Ragul et al. 2025), and many more for the evaluation of genetic diversity, studying population structure, development of ISSR-based molecular markers for the identification of species, assessment of genetic fidelity among in vitro multiplied plants, and so on.

4.4.2.4 SSR Markers or Microsatellites

SSRs or microsatellites are short tandem repeats of one to six base pairs found throughout eukaryotic genomes that are highly polymorphic in length due to variations in the number of repeats. SSR markers are co-dominant markers that have high reproducibility and reliability. SSR markers are developed by designing locus-specific primers based on existing sequencing information or database mining that flank the repeat regions. Length polymorphisms are revealed through PCR amplification and detected using gel electrophoresis or automated capillary electrophoresis with fluorescently labeled primers. Despite its several advantages, the use of SSR markers is restricted because it requires prior sequence information to develop specific primers, which is time-consuming and expensive for non-model species. It also requires high-resolution gel electrophoresis or automated sequencers for precise determination of length. They are exceptionally useful for detailed analyses of population genetics, parentage testing, and the identification of unique genetic resources.

These markers have been used in medicinal plants such as *Docynia delavayi* (Peng et al. 2021), *Polygonatum cyrtonema* (Liu et al. 2023), *Meconopsis integrifolia* (Wu et al. 2024), *Tribulus terrestris* (Chaudhary et al. 2025), *Arctium lappa* (Su et al. 2025), and so on for preliminary genetic diversity screening, species discrimination, population structure analysis, hybrid identification, and genetic purity testing.

4.4.2.5 Start Codon Targeted Markers

Start codon targeted (SCoT) markers are dominant, PCR-based markers that use short, single primers designed to anneal to "start codon" (ATG sequences) and their reverse complement sequences (CAT). Thus, primer design is very simple and requires no prior sequence information. Since start codons are found in gene-rich regions of the genome, SCoT primers predominantly focus on functional regions of the genome. Moreover, the combination of simplicity, efficiency, relevance, cost-effectiveness, and higher polymorphism makes SCoT accessible even to laboratories with limited resources. Plant researchers have successfully applied SCoT markers in a diverse range of medicinal plants, unraveling their genetic secrets. The application of SCoT markers in medicinal plant research includes study of genetic diversity, population structure, phylogenetic relationships, germplasm characterization, variety or cultivar identification, as well as evaluation of genetic fidelity in micropropagated plants, for example, in *Anacamptis coriophora*, *Dactylorhiza umbrosa*, *Himantoglossum affine*, *Orchis collina*, *Orchis mascula*, *Orchis simia*, *Ophrys schulzei*, and *Ophrys straussii* (Gholami et al. 2021), *Melissa officinalis* (Koohdar and Sheidai 2022), *Plumbago zeylanica* (Roy et al. 2022), *Annamocarya sinensis* (Pan et al. 2024), *Artemisia fragrans*, *Artemisia absinthium*, *Artemisia scoparia*, *Artemisia biennis* and *A. annua* (Jamshidi et al. 2025), *Bacopa monnieri* (Roy Choudhury et al. 2025), and so on.

4.5 Single-Nucleotide Polymorphisms

SNPs are the most abundant genetic variation that exists on average once per 100–1000 base pairs in the genome of all individuals. Any single-base pair change (point mutation) in coding and non-coding regions of genomic DNA develops a new SNP. Their high prevalence in the genome, biallelic nature, compatibility with high-throughput analysis, and stability make SNPs invaluable and a superior genetic marker system for assessing genetic diversity across a wide range of species or populations of medicinal plants. With the significant progress of NGC technologies, the detection of millions of SNPs has become very feasible and cost-effective. In the last few decades, SNPs have emerged as exceptionally powerful tools, and their utilization has increased as genetic markers. The high-throughput sequencing technologies have deeply facilitated the creation of organelle-based and species-specific molecular markers in plant sciences.

Several studies performed on medicinal plants highlighted the effective usefulness of SNPs for investigating genetic structure, gene flow, population structure, genetic divergence, and genetic relatedness within and among medicinal plant populations, species, or cultivars (Osman et al. 2003; Jo et al. 2015; Yang et al. 2017; Kim et al. 2018; Xu et al. 2021, 2025). For example, SNPs facilitate the identification of different natural populations of *Eurycoma longifolia* (Osman et al. 2003). In *Zingiber officinale*, 64,026 SNP sites along with 7034 indel polymorphisms were identified after analyzing extensive EST libraries (Chandrasekar et al. 2009). Interestingly, this study revealed that the frequency of molecular markers significantly varied in a tissue-specific manner in rhizomes (1.08/100 bp), roots (0.82/100 bp), and leaves (0.63/100 bp). De novo transcriptome assembly from mature roots of four cultivars of *Panax ginseng* and *Panax quinquefolius,* followed by mapping analyses, effectively identified 34 reliable HRM–SNP primer pairs that showed a high degree of polymorphism (Jo et al. 2015). Further screening identified four HRM–SNP primers that were useful in distinguishing ginseng cultivars. The genetic differentiation of a newly developed G-1 cultivar from other native Korean cultivars of *P. quinquefolius* has also been possible using SNP markers targeting genetic diversity at the 45S ribosomal DNA region (Yang et al. 2017). Kim et al. (2018) also revealed the effective differentiation of closely related species of *Allium ochotense* and *Allium microdictyon* using species-specific primers designed based on the information from SNP sequences in ITS and by using the Amplification-Refractory Mutation System (ARMS)-PCR technique. In *Sarcandra glabra* and *S. glabra* ssp. *brachystachys* 25,620 SSRs, 726,476 SNPs, and 42,939 insertion-deletions were identified, along with assembly of 141,954 unigenes with an average length of 646.63 bp from 57.756 million pair-end reads (Xu et al. 2021). Analysis of SNP-based markers demonstrated that high genetic homogeneity and a relatively large overall genetic differentiation in populations of *Horsfieldia hainanensis*, although genetic differentiation varies among populations depending on their geographical isolation (Xu et al. 2025).

Besides SNPs in the nuclear genome, some studies developed and use SNP markers derived from the chloroplast genome (cpDNA) because it has a more stable structure, a high copy number, and a maternal inheritance (Do et al. 2019; Makki et al. 2019; Giang et al. 2020; Lee et al. 2021). The complete characterization of the *Crepidiastrum denticulatum* chloroplast genome was reported by Do et al. (2019) using NGS via the MiSeq platform. They also employed the ARMS technique to develop highly-specific, chloroplast genome-based SNP markers located within the rbcL and matK genes, which are capable of distinguishing this species from related species in an effective and simple manner using a PCR protocol. In a similar study performed by Makki et al. (2019), unique chloroplast SNPs were identified by performing a complete molecular characterization using the Illumina HiSeq 2000 platform, which were able to distinguish among eight varieties of *Catharanthus roseus* and provided experimental support to the idea of heteroplasmy in the chloroplast genome in this medicinally important species. According to Ying et al. (2022), discrimination of plant species, cultivars, and landraces is challenging because plants have high phenotypic and genotypic resemblance. However, a comprehensive comparative analysis of their complete chloroplast genomes identified 1128 cpDNA-derived SNPs in 72 protein-coding genes within the seven *Panax* species (Giang et al. 2020) that simultaneously represent their genetic diversity and clarify evolutionary relationships. Moreover, they developed two or three species-specific markers for each species, which are able to differentiate among these species. The sequencing of the plastid genome revealed considerable variations in the genome structure and identified two highly variable regions (*trnC-petN* and *rpl32-trnL*) in four cultivars of *Polygonatum odoratum* (Lee et al. 2021). They found variability in the total genome size, the size of the large single-copy region, the size of the small single-copy region, and the size of two inverted repeats. In contrast, genome comparisons revealed a slight variation in the sequence across the five accessions.

The significance of geographic influences on genetic differentiation is also evident in the study by Qiao et al. (2021) on *Swertia mussotii*, a Tibetan medicinal plant. Genotyping-by-sequencing (GBS) revealed 4095 SNPs with a minor-allele frequency >0.25 and an average read depth <200 among 68 samples of *S. mussotii* collected from five geographic locations in China and four samples from related species, from different geographical locations. Based on analyses of all SNPs, substantial genetic differentiation in genes known or hypothesized to be involved in various metabolic processes was determined by the population differentiation survey, which was further supported by structural and phylogenetic topological analyses. The relationship between genetic differentiation and geographic distance is very strong, and local adaptive pressures are a possible cause of this phenomenon.

These diverse applications collectively illustrate the growing importance of SNP discovery and utilization for the molecular mapping of genes, precise plant identification, authentication, and understanding population genetics in the study of valuable medicinal flora.

4.6 Expressed Sequence Tag-SSRs

Expressed sequence tag-SSRs (EST-SSRs) are a distinct class of SSR markers, located within EST sequences. These markers are short, single-pass sequences that are derived from cDNA libraries. Genetic polymorphism in an EST-SSR indicates a genetic difference in the actively transcribed, protein-coding regions of the genome. Because this variation links directly to genes of known biological function, these markers are often designated as functional genetic markers. EST-SSRs have become effective, reproducible, and cost-effective tools for assessing genetic diversity at the intra-specific level in medicinal plants. Compared to genomic SSRs, EST-SSRs are more effective for comparative studies and for developing broadly useful markers to evaluate genetic variation in medicinal plants. Using EST-SSRs in a strategic way helps researchers and conservationists support long-term survival, breeding, and sustainable, responsible use of these vital botanical assets. This practice secures their continued availability for healthcare and human well-being. However, applying EST-SSRs depends on having pre-existing EST databases for the target species.

These markers also cannot detect polymorphism found in highly variable non-coding regions. Below are a few selected recent applications of EST-SSR markers for genetic diversity assessment in medicinal plants, such as *Paris polyphylla* var. *yunnanensis* (Gao et al. 2022), *Monochasma savatieri* (Yang et al. 2022), *Pseudostellaria heterophylla* (Xu et al. 2023), *Pinellia ternate* (Wang et al. 2025), and *Euphorbia resinifera* (Abd-dada et al. 2023).

Gao et al. (2022) analyzed the genetic diversity of 136 individuals from 10 cultivated populations of *Paris polyphylla* var. *yunnanensis* using 10 EST-SSR markers associated with polyphyllin backbone biosynthesis. The results confirmed high genetic polymorphism across the studied loci. The interpretation of the Shannon index demonstrated that the majority of genetic diversity is within populations rather than among them. This data suggests low overall population differentiation despite high genetic diversity among individuals. This pattern implies extensive gene flow and pervasive hybridization across the cultivated groups. In a similar study on 46 individuals of *Monochasma savatieri* collected from four provinces (Fujian, Hunan, Jiangxi, and Zhejiang), 208 alleles with an average of 6.303 per locus and high overall polymorphism were detected using 33 newly developed EST-SSR markers (Yang et al. 2022). They detected a relatively low degree of genetic diversity among the assessed four populations, with the highest genetic diversity ($I = 0.535$ and $H_e = 0.342$) in the Zhejiang population. Despite this, a significant genetic differentiation ($F_{st} = 0.497$) was observed, which indicates restricted gene flow ($N_m = 0.253$). A significant positive correlation between geographical and genetic distances was detected, which indicates that genetic differentiation was spatially correlated. Xu et al. (2023) efficiently developed 20 effective, high-performing transcriptome-derived EST-SSR markers for *Pseudostellaria heterophylla*, which detected genetic diversity within 10 selected cultivars, characterized by a mean observed heterozygosity (H_o) of 0.330, a Shannon's Information Index (I) of 0.560, and a mean Polymorphic Information Content (PIC) of 0.505. Recently, a comprehensive genetic diversity analysis in 255 germplasm resources of *Pinellia ternata* (117 wild, 138 cultivated) from China has been conducted using 17 highly polymorphic primer pairs, which revealed 99.21% polymorphism, an average of 1.98 observed alleles, and a mean PIC of 0.6482 (Wang et al. 2025). They reported wild germplasm contained greater genetic diversity than cultivated resources, though both types exhibited low overall genetic differentiation. According to this study, 91% of the total genetic variation was confined within populations, and only 9% distributed among them.

4.7 DNA Barcoding

Several studies have demonstrated that different amounts of genetic (DNA sequence) variation in specific, short, and standardized DNA regions (DNA barcoding loci) within chloroplast or nuclear genomes can be detected using DNA barcoding techniques in various medicinal plants. This molecular-based technique is remarkable for its accuracy, repeatability, and rapidity, which are not affected by morphological diversity, developmental stages, and environmental conditions. Both the coding and the noncoding regions of the chloroplast and nuclear genomes of the medicinal plant can be used as barcodes. The ribulose-1,5-bisphosphate carboxylase/oxygenase large subunit (*rbcL*) gene, maturase K (*matK*), *trnL*, and *trnH-psbA* are the plastid DNA regions commonly used for barcoding in medicinal plants. In addition, a few nuclear DNA regions, including the internal transcribed spacer (*ITS*), have also been used for this purpose (Bhamra et al. 2015). Unfortunately, in spite of advances in DNA barcoding, there is not a single universal DNA barcode that can be used to identify all plant species. The degree of effectiveness of individual DNA barcodes for differentiating medicinal plants at the species, genus, and family level can vary (Zhang et al. 2015; Hashim et al. 2021). DNA barcodes that are highly effective at distinguishing species may not be suitable at the genus or family levels (Zhang et al. 2015).

The presence of significant genetic variability in the matK barcode region among medicinal plants of Rauvolfioideae has been used for correct species identification by Mahadani et al. (2013). A comparative DNA barcoding study based on four loci (*ITS*, *trnH-psbA*, *matK*, and *rbcL*) has revealed a significant genetic variation in 135 individuals representing 33 species of Schisandraceae (Zhang et al. 2015). They reported that differentiation of all these species at both family and genus levels is feasible through the use of single and multi-locus combinations of genetic variations. They found more sequence variability in the *ITS* and *trnH-psbA* regions than in the two coding regions, *matK* and *rbcL*, among these species (Zhang et al. 2015). Similar genetic variation was reported at the *ITS* and *trnH-psbA* regions among the different species of tulsi plants (Bhamra et al. 2015). Hashim et al. (2021) deciphered the genetic relationships among nine endangered endemic plant species in Saint Katherine Protectorate by determining genetic variability using *rbcL*, *ITS*, and *ycf1* DNA barcodes (Hashim et al. 2021).

In another study, 212 DNA barcoding sequences were detected at four DNA barcoding regions (*ITS2, matK, rbcL,* and *trnL*) using specific primers for each region in 61 Indonesian medicinal plant species belonging to 30 families (Cahyaningsih et al. 2022). A study of 20 medicinal plant species of Caryophyllales using the nuclear barcode *ITS2* and the plastid DNA barcodes (*rbcL* and *matK*) found that the *ITS2* region has more genetic variability than *rbcL* and *matK* and was better in differentiating the tested species (Jamdade et al. 2022). Using a combination of more than one method can help distinguish morphologically similar or closely related taxa (Jamdade et al. 2022). Recent research on three *Plumbago* species, namely, *P. indica, P. zeylanica,* and *P. auriculata,* highlighted that the *ITS2* regions have more genetic variability among six DNA barcoding loci (*matK, ITS, ITS2, rbcL, trnH-psbA,* and *trnL-F*) tested (Intharuksa et al. 2024).

Insufficient genetic variability within traditional barcoding regions, an inability to effectively differentiate closely related species, and the reliance on limited markers may overlook significant genetic diversity present in other parts of the genome are some challenges of standard DNA barcoding. Advancements in genomics and sequencing technologies have further enhanced the efficiency and feasibility of large-scale genetic assessments, including complete sequencing of plastomes and nuclear genomes. To better understand genetic diversity, future research should integrate DNA barcoding with complementary "omics" approaches, including whole-genome sequencing (WGS), transcriptomics, and metabolomics. Additionally, the development of DNA databases specifically dedicated to medicinal plant genome sequences is also important. These genetic markers can help understand biodiversity and the development of effective conservation strategies of medicinal plants by allowing rapid species identification and identification of sources of medicinal compounds, and can contribute to studying genetic evolutionary relationships by accurate evaluation of intra- and inter-population variability in medicinal plant populations. This information is essential for understanding gene flow, population dynamics, and the specific effects of habitat fragmentation, as well as helping track and detect the illegal trade of endangered species.

4.8 Diversity Arrays Technology

Diversity Arrays Technology (DArT) is a sequence-independent, hybridization-based molecular marker technique that can detect and score hundreds of allelic diversity in a sample (Bidyananda et al. 2024). The fundamental methodology for this technology involves the isolation of genomic DNA from plant samples, their fragmentation, hybridization on a microarray platform, and PCR amplification of the signal of hybridized fragments. The microarray imaging system is then used to scan and detect thousands of DNA fragments simultaneously, and the collected data are further analyzed using specialized software. It focuses on SNPs and DArT seq-based genomic profiling. It can be applied to a wide variety of plant species, even those with limited genomic information. The presence or absence of specific DNA fragments generates a binary data set of thousands of markers that indicate genomic variation.

The ability to produce genome-wide information in a high-throughput and cost-effective manner, the use of a dominant marker system, its wider applicability, and high reproducibility among plant species make it an emerging popular tool. The technology is relatively inexpensive compared to most of the other high-throughput genotyping methods, such as NGS. It has been applied for a better understanding of genetic structure, population differentiation, whole-genome profiling, high-density genetic mapping, and phylogenetic relationships within and between populations of many crop and fruit species (Schouten et al. 2012; Deres and Feyissa 2023). It is expected that DArT will be employed more frequently in medicinal plant research in the near future for genome-wide genetic variability determination. It can assist in strengthening our critical understanding of genetic and population diversity. In future research, there is a need to perform further optimization of DArT protocols, explore its applications in various ecological contexts, and encourage interdisciplinary collaboration to fully utilize its potential for conservation of the genetic diversity of medicinal plants.

4.9 Application of Combined Marker

All the different types of markers discussed above come with their own set of disadvantages. Outcomes of solitary marker systems in the assessment of plant genetic diversity can be effectively substantiated and validated by the study of one or more additional markers, giving rise to the use of a multiple-marker approach. Employing several molecular marker types together improves the analytical accuracy and reliability of the study, contributing toward effective evaluation of genetic diversity, population structure, or conservation strategies. Several examples of such a combined marker approach have been observed in recent studies, including in medicinally important plant species. Baruah et al. (2017) employed 20 RAPD and 15 ISSR primers for studying 100 genotypes of *Cymbopogon flexuosus* and 27 genotypes of *Cymbopogon winterianus* from north-east India. Notably, 13 RAPD primers and 13 ISSR primers effectively revealed genetic variation among the genotypes investigated. Interestingly, UPGMA analyses with both the marker types were found to generate similar dendrograms. In addition, cluster analyses were in agreement with PCA based on individual markers as well as combined marker data, producing similar clusters with few exceptions in the *Cymbopogon* species (Baruah et al. 2017). A similar combination of ISSR and RAPD markers has been utilized in *Nilgirianthus ciliatus* by Rameshkumar et al. (2019) with 35 ISSR and 30 RAPD primers demonstrating 77% and 89.9% polymorphism, respectively. Two separate cluster analyses and PCA divided the populations into distinct clusters based on geographical locations of the population, indicating the effectiveness of both the markers in the identification of genetic variation and phylogenetic relationship in the species (Rameshkumar et al. 2019). Hadipour et al. (2020) employed 7 ISSR and 10 AFLP markers to study 67 genotypes representing 9 wild populations of *Papaver bracteatum* and found both markers to divide the genotypes into three major clusters and one minor one; however, they suggested AFLP to be a more efficient genetic marker compared to ISSR in the species, supported by high mean marker resolution. Some studies even used multiple molecular markers together in the assessment of medicinal plants, for example, 12 accessions of *Peganum harmala* studied with ISSR, ITS-RFLP, ITS-SSCP (ITS-single strand conformation polymorphism), and SSR revealed that the highest level of polymorphism could be detected with SSR, followed by ISSR and ITS-RFLP, and the least variation was identified with ITS-SSCP (Ranya and Ahmed 2018). The study further suggested partial out-crossing reproductive system in the species indicated by nearly similar inbreeding coefficient as assessed by the three co-dominant markers. Further, the accessions were clearly divided into distinct clusters, based on geographical locations coinciding with isolation by distance (Ranya and Ahmed 2018). A combination of DNA barcoding, ISSR, RAPD, and RAPD-derived SCAR (sequence characterized amplified region) markers was used by Mahgoub et al. (2025) for the study of four *Artemisia* species, successfully identifying ecotype-specific SCAR markers for the rapid identification of the species. Besides applying multiple molecular markers together in assessment of the genetic diversity, several studies combine the morphological, biochemical, and molecular markers in the evaluation process; for example, in *Reynoutria* species, Stafiniak et al. (2025) proposed usefulness of combined morphological, biochemical, and molecular markers (SCoT and sequence-related amplified polymorphisms) in the evaluation of identity of the species and can be further utilized in genetic diversity assessment.

4.10 Emerging Trends and Future Perspectives

The genetic diversity of medicinal plants is the outcome of an ongoing evolutionary process driven by various biological phenomena, such as mutations, genetic recombination, gene flow, hybridization, and polyploidization, natural and anthropogenic selection pressure, and so on. Unique genetic blueprints (genetic diversity) are strongly associated with phytochemical diversity in medicinal plants. Unfortunately, medicinal plants are experiencing a crisis of rapid decline in their diversity at species, genetic, and ecosystem levels due to increased pressures imposed primarily by several anthropogenic activities. The assessment and conservation of genetic diversity in medicinal plants is essential to ensure the sustainable use of their chemical diversity for therapeutic purposes.

NGS has provided a powerful tool for assessing genetic diversity and conducting population genetic studies. Barcoding genes and genomes is now possible with advanced NGS-based genomic resources. The unlocked genetic information by NGS can be used for the identification of different types of novel molecular markers, including SNPs, insertions/deletions (indels), SSRs, and structural variants that, in turn, can be used to evaluate intra- and inter-specific genetic diversity, a deeper understanding of population genetics, and their evolutionary relationships. In addition, the genetic information generated by this technology offers incredible opportunities. It can be used to develop more effective conservation strategies, optimize germplasm management, facilitate marker-assisted breeding of superior chemotypes, authenticate species, and bioprospect medicinal plants. The development of new advanced NGS-based technologies, such as WGS, GBS, restriction-site associated DNA sequencing, and RNA sequencing, is very effective and relevant to genetic diversity assessment because it allows simultaneous sequencing of entire genomes, transcriptomes, or a large number of targeted genomic regions at an unprecedented scale in a more rapid, accurate, and cost-effective manner. The technological advances in NGS approaches have not only reduced research costs and time but also significantly advanced medicinal plant research in various directions.

Future research on the genetic diversity of medicinal plant species should adopt a new approach, gathering holistic information rather than in a fragmented, isolated form. The synergistic integration of different "omics" technologies, that is, a "multi-omics" approach, provides deep and clear insights into a holistic framework for understanding the complex dynamic between the genetic makeup of plant species, gene expression, protein function, and the production of therapeutic compounds by gathering biological information from each technology. This comprehensive framework offers a resolution previously unattainable, elucidating not just the presence of genetic diversity but also its functional consequences and its relationship with environmental factors. Ultimately, multi-omics tools allow researchers to identify and prioritize populations with unique and valuable genetic resources for targeted conservation, develop strategies for sustainable harvesting, and accelerate the development of superior medicinal plant varieties with improved yield and quality of active compounds.

The development and integration of sophisticated and automated bioinformatics tools, AI, and ML algorithms have great potential for efficient data management, pre-processing, and quality control, analysis, functional annotation to genes, proteins, and metabolites, interpretation, and prediction. The successful implementation of multi-omics approaches to gain a comprehensive understanding of the genetic basis of their therapeutic properties heavily relies on advanced bioinformatics and computational capabilities. AI/ML techniques can be employed to identify and classify medicinal plants based on their genetic makeup, analyze large-scale genomic data to identify genetic markers associated with specific therapeutic properties, predict the structure and function of proteins involved in the biosynthesis of bioactive compounds, and develop predictive models to assess the impact of environmental factors on the genetic makeup of medicinal plants.

In conclusion, assessing, understanding, and maintaining the genetic diversity of medicinal plants is no longer an option. While there has been significant advancement in genetic diversity assessment methods over the last three decades, from morphological markers to PCR-based markers and more advanced NGS-based SNPs, there is further scope for improvements and applicability for integrating technological advancement and a multi-omics approach with existing methods to holistically assess genetic diversity, population structure, evolutionary relationships, and their connection to protein or SMs. For the academic community, prioritizing such assessments aligns with both scientific responsibility and the broader goal of maintaining a resilient, diverse, and accessible medicinal plant heritage for present and future generations.

Acknowledgments MH gratefully acknowledges the Principal of Barasat Government College, Kolkata, for the continuous support and encouragement in research activities. DRC acknowledges with gratitude the Principal of Surendranath College for invaluable support and motivation toward research pursuits.

References

Abd-Dada H, Bouda S, Khachtib Y, Ait Bella Y, Haddioui A (2023) Use of ISSR markers to assess the genetic diversity of an endemic plant of Morocco (*Euphorbia resinifera* O. Berg). J Genet Eng Biotechnol 21(1):91

AL-Kiyyam MA, Shibli RA, Zatimeh A, Tahtamouni RW, AlQudah TS, Hassan S (2024) Evaluation of genetic diversity in Jordanian *Solanum nigrum* plants and genetic stability of in vitro grown plant using (AFLP) technique. Jordan J Biol Sci 17(1):123–128

Barcelos E, Amblard P, Berthaud J, Seguin M (2002) Genetic diversity and relationship in American and African oil palm as revealed by RFLP and AFLP molecular markers. Pesq Agrop Brasileira 37:1105–1114

Baruah J, Gogoi B, Das K, Ahmed NM, Sarmah DK, Lal M, Bhau BS (2017) Genetic diversity study amongst *Cymbopogon* species from NE-India using RAPD and ISSR markers. Ind Crop Prod 95:235–243

Bhamra S, Heinrich M, Howard C, Johnson M, Slater A (2015) DNA authentication of tulsi (*Ocimum tenuiflorum*) using the nuclear ribosomal internal transcribed spacer (ITS) and the chloroplast intergenic spacer trnH-psbA. Planta Med 81(16):PW_20

Bidyananda N, Jamir I, Nowakowska K, Varte V, Vendrame WA, Devi RS, Nongdam P (2024) Plant genetic diversity studies: insights from DNA marker analyses. Int J Plant Biol 15(3):607–640

Biswas D, Santra I, Ghosh B (2024) EMA-based chromosome characterization and karyotype analysis in two wild spiny medicinally and nutritionally important Solanaceaous species–*Solanum sisymbriifolium* Lam. and *Solanum virginianum* L. Cytologia 89(1):65–70

Boomibalagan P, Subramanian SR, Rajasekharan PE, Karpakal S, Veeranan U, Saminathan E, Narayanan V, Kathiresan D (2021) Genetic relationship and polymorphism of selected medicinal plants of Asclepiadaceae using RAPD molecular analysis method. Ecol Genet Genom 21:100101

Cahyaningsih R, Compton LJ, Rahayu S, Magos Brehm J, Maxted N (2022) DNA barcoding medicinal plant species from Indonesia. Plants 11(10):1375

Chandrasekar A, Riju A, Sithara K, Anoop S, Eapen SJ (2009) Identification of single nucleotide polymorphism in ginger using expressed sequence tags. Bioinformation 4(3):119

Chaudhary D, Sharma V (2025) Evaluation of genetic diversity and population structure of the medicinal plant *Justicia adhatoda* L. from North India using ISSR markers. Ecol Genet Genom 34:100326

Chaudhary N, Chaudary D, Kumar R, Aggarwal K, Sharma H, Roy J, Gupta RC, Sharma V (2025) SSR based genetic diversity and population structure of important Herb *Tribulus terrestris* L. from North India. J Appl Res Med Aromat Plants 49:100675

Chen M, Jiang C, Huang D, Zheng Z, Yang W, Li G, Fu C, Liao H, Long W, Yang Z, Yang Y (2024) ND-FISH with new oligo probes for chromosome identification of *Cichorium intybus* revealing karyotypic variation and divergence of Asteraceae species. Plants 13(22):3135

Cheng L, Han M, Yang LM, Li Y, Sun Z, Zhang T (2018) Changes in the physiological characteristics and baicalin biosynthesis metabolism of *Scutellaria baicalensis* Georgi under drought stress. Ind Crop Prod 122:473–482

Deres D, Feyissa T (2023) Concepts and applications of diversity array technology (DArT) markers for crop improvement. J Crop Impr 37(6):913–933

Do HD, Jung J, Hyun J, Yoon SJ, Lim C, Park K, Kim JH (2019) The newly developed single nucleotide polymorphism (SNP) markers for a potentially medicinal plant, *Crepidiastrum denticulatum* (Asteraceae), inferred from complete chloroplast genome data. Mol Biol Rep 46(3):3287–3297

Esfahani ST, Karimzadeh G, Naghavi MR (2016) 2C DNA value of Persian poppy (*Papaver bracteatum* Lindl.) medicinal plant as revealed by flow cytometry analysis; a quick, effective criteria for distinguishing unidentified papaver species. Int J Adv Biotechnol Res 7:573–578

Ezzat SM, El Sayed AM, Salama MM (2016) Use of random amplified polymorphic DNA (RAPD) technique to study the genetic diversity of eight *Aloe* species. Planta Med 82(15):1381–1386

Farahani F, Shaker M, Noormohammadi Z (2025) Impact gamma rays on somaclonal features of in vitro regenerated *Aloe littoralis*: Morphological, physiological and molecular analysis. J Radiat Res Appl Sci 18(4):101952

Firdausi J, Dash CK, Rashid MH, Sultana SS (2022) Karyotype characterization of three important aromatic *Cinnamomum* L. species with special emphasis on reversible chromosome banding. J Appl Res Med Aromat Plants 31:100430

Gallucci A, Musarella CM, Ana CO, Fuentes JC, Canas RQ, Villano C (2023) Preliminary assessment of genetic diversity between *Glebionis coronaria* and *G. discolor* (Asteraceae) by AFLP markers. Acta Agric Slov 119(1):1–7

Gao X, Su Q, Yao B, Yang W, Ma W, Yang B, Liu C (2022) Development of EST-SSR markers related to polyphyllin biosynthesis reveals genetic diversity and population structure in *Paris polyphylla*. Diversity 14(8):589

Gholami S, Vafaee Y, Nazari F, Ghorbani A (2021) Exploring genetic variations in threatened medicinal orchids using start codon targeted (SCoT) polymorphism and marker-association with seed morphometric traits. Physiol Mol Biol Plants 27(4):769–785

Giang VN, Waminal NE, Park HS, Kim NH, Jang W, Lee J, Yang TJ (2020) Comprehensive comparative analysis of chloroplast genomes from seven *Panax* species and development of an authentication system based on species-unique single nucleotide polymorphism markers. J Ginseng Res 44(1):135–144

Habiba SD, Haque ME, Islam MA, Mukherjee A, Sikdar B (2019) Assessment of genetic diversity and phenetic relationships among some Bangladeshi cultivars of cucumber (*Cucumis sativus* L.) using isozyme and protein profiling. Vegetos 32(4):494–507

Hadipour M, Kazemitabar SK, Yaghini H, Dayani S (2020) Genetic diversity and species differentiation of medicinal plant Persian Poppy (*Papaver bracteatum* L.) using AFLP and ISSR markers. Ecol Genet Genom 16:100058

Hake AA, Jha SM, Jha SK, Mahatma MK (2025) Utility of isozyme markers for understanding genetic relatedness in Karanja (*Derris indica* L.,). Indian J Plant Genet Resour 38(02):168–174

Halder M, Jha S (2023) Medicinal plants and bioactive phytochemical diversity: a fountainhead of potential drugs against human diseases. In: Halder M, Jha S (eds) Medicinal plants: biodiversity, biotechnology and conservation. Springer Nature, Singapore, pp 39–93. https://doi.org/10.1007/978-981-19-9936-9_2

Halder M, Kundu A, Jha S (2024) Secondary metabolites identification techniques of the current era. In: Mérillon JM, Ramawat KG (eds) Plant specialized metabolites: phytochemistry, ecology and biotechnology. Springer, Berlin/Heidelberg, Germany, pp 1–41. https://doi.org/10.1007/978-3-031-30037-0_31-1

Halušková J, Čellárová E (1997) RFLP analysis of Hypericum perforatum L. somaclones and their progenies. Euphytica 95(2):229–235

Hashim AM, Alatawi A, Altaf FM, Qari SH, Elhady ME, Osman GH, Abouseadaa HH (2021) Phylogenetic relationships and DNA barcoding of nine endangered medicinal plant species endemic to Saint Katherine protectorate. Saudi J Biol Sci 28(3):1919–1930

He Z, Lei Y, Gong W, Ye M, Luo X (2023) Karyotype and phylogenetic relationship analysis of five varieties and cultivars of *Zanthoxylum armatum* based on oligo-FISH. Genes 14(7):1459

Hussain H, Nisar M (2020) Assessment of plant genetic variations using molecular markers: a review. J Appl Biol Biotechnol 8(5):99–109

Intharuksa A, Phrutivorapongkul A, Thongkhao K (2024) Integrating DNA barcoding, microscopic, and chemical analyses for precise identification of *Plumbago indica* L., a prominent medicinal plant. Microchem J 199:110038

Jadaun JS, Yadav R, Yadav N, Bansal S, Sangwan NS (2023) Influence of genetics on the secondary metabolites of plants. In: Natural secondary metabolites: from nature, through science, to industry. Springer International Publishing, Cham, pp 403–433

Jamdade R, Mosa KA, El-Keblawy A, Al Shaer K, Al Harthi E, Al Sallani M, Al Jasmi M, Gairola S, Shabana H, Mahmoud T (2022) DNA barcodes for accurate identification of selected medicinal plants (Caryophyllales): toward barcoding flowering plants of The United Arab Emirates. Diversity 14(4):262

Jamshidi B, Etminan A, Mehrabi AM, Pour-Aboughadareh A, Shooshtari L, Ghorbanpour M (2025) Genetic diversity of Artemisia species based on CAAT-box derived polymorphism (CBDP) and start codon targeted (SCoT) markers. Genet Resour Crop Evol 71(7):3803–3813

Jeelani SM, Farooq U, Gupta AP, Lattoo SK (2017) Phytochemical evaluation of major bioactive compounds in different cytotypes of five species of *Rumex* L. Ind Crop Prod 109:897–904

Jeelani SM, Singh J, Sharma A, Rather GA, Ali SA, Gupta AP, Singh S, Lattoo SK (2021) In-vitro cytotoxicity in relation to chemotypic diversity in diploid and tetraploid populations of *Gentiana kurroo* Royle. J Ethnopharmacol 274:113966

Jha TB, Chakroborty P, Halder M (2024) Karyotype analysis on *Gloriosa superba* using enzymatic maceration and air-drying-based Giemsa, DAPI, and CMA staining techniques. Cytologia 89(3):203–209. https://doi.org/10.1508/cytologia.89.203

Jo IH, Lee SH, Kim YC, Kim DH, Kim HS, Kim KH, Chung JW, Bang KH (2015) De novo transcriptome assembly and the identification of gene-associated single-nucleotide polymorphism markers in Asian and American ginseng roots. Mol Gen Genomics 290(3):1055–1065

Kang S, Han J, Lee I, Joo H, Chung Y, Yang S (2024) Chromosome analysis method based on deep learning: counting chromosomes and detecting abnormal chromosomes. Biomed Signal Proc Control 91:105891

Kanta C, Kumar A, Chauhan A, Singh H, Sharma IP (2024) The interplay between plant functional traits and climate change. In: Kumar N, Singh H (eds) Plant functional traits for improving productivity. Springer Nature, Singapore, pp 41–58

Kim YB, Ramekar RV, Choi SJ, Choi BG, Kim SW, Moon YK, Noh HS, Lee JK, Hong JS, Park NI, Choi IY (2018) Molecular identification of *Allium ochotense* and *Allium microdictyon* using multiplex-PCR based on single nucleotide polymorphisms. Hortic Environ Biotechnol 59(6):865–873

Kong DX, Li YQ, Bai M, He HJ, Liang GX, Wu H (2017) Correlation between the dynamic accumulation of the main effective components and their associated regulatory enzyme activities at different growth stages in *Lonicera japonica* Thunb. Ind Crop Prod 96:16–22

Koohdar F, Sheidai M (2022) Genetic diversity and population structure in medicinal plant *Melissa officinalis* L. (Lamiaceae). Genet Resour Crop Evol 69(5):1753–1758

Križman M, Jakše J (2022) Chemical and genetic variability of Istrian *Foeniculum vulgare* wild populations. Plants 11(17):2239

Kumar A, Kumar S, Thomas TD, Ramchiary N, Swamy MK, Ahmad I (2019) Linking omics approaches to medicinal plants and human health. In: Akhtar M, Swamy M (eds) Natural bio-active compounds: volume 3: biotechnology, bioengineering, and molecular approaches. Springer, Singapore, pp 31–57

Kumar R, Salar RK, Naik PK, Yadav M, Kumar A, Kumar A, Yogi R, Kumar M, Chhokar V (2022) Elucidation of genetic diversity and population structure of sixty genotypes of *Aloe vera* using AFLP markers. S Afr J Bot 147:1146–1155

Lee YS, Park HM, Kim NH, Waminal NE, Kim YJ, Lim KB, Baek JH, Kim HH, Yang TJ (2016) Phylogenetic relationship of 40 species of genus *Aloe* L. and the origin of an allodiploid species revealed by nucleotide sequence variation in chloroplast intergenic space and cytogenetic in situ hybridization. Genet Resour Crop Evol 63(2):235–242

Lee SY, Chen Z, Chen Z, Chen J, Zhang X, Pan J, Fan Q, Liao W (2021) Plastid genome sequencing, identification of nuclear SNP markers, and quality assessment of medicinal rhizomatous herb *Polygonatum odoratum* (Asparagaceae) cultivars. Ecol Evol 11(12):7660–7676

Lekhak MM, Basu S, Chougule RN, Suzuki G, Yamamoto M, Mukai Y (2024) Cellular and molecular cytogenetics of Safed Musli (*Chlorophytum*): gaps and the way forward. Nucleus 67(3):531–545

Liu H, Cheng H, Xu J, Hu J, Zhao C, Xing L, Wang M, Wu Z, Peng D, Yu N, Liu J (2023) Genetic diversity and population structure of *Polygonatum cyrtonema* Hua in China using SSR markers. PLoS One 18(8):0290605

Ma S, Khayatnezhad M, Minaeifar AA (2021) Genetic diversity and relationships among *Hypericum* L. species by ISSR Markers: a high value medicinal plant from Northern of Iran. Caryologia 74(1):97–107

Mahadani P, Sharma GD, Ghosh SK (2013) Identification of ethnomedicinal plants (Rauvolfioideae: Apocynaceae) through DNA barcoding from Northeast India. Pharmacogn Mag 9(35):255

Mahgoub YA, Mahmoud HA, El-Sebakhy NA, Abdallah II (2025) Unravelling *Artemisia* species genetic variation via DNA barcoding, ISSR and RAPD with the development of eco-specific SCAR markers. BMC Plant Biol 25(1):1034

Makki RM, Saeedi AA, Khan TK, Ali HM, Ramadan AM (2019) Single nucleotide polymorphism analysis in plastomes of eight *Catharanthus roseus* cultivars. Biotechnol Biotechnol Equip 33(1):419–428

Morovati Z, Karimzadeh G, Nagahavi MR, Monfared SR (2024) Chromosome, ploidy analysis, and flow cytometric genome size estimation of *Datura stramonium* and *D. innoxia* medicinal plant. Caryologia 77(3):53–61

Mulyaningsih ES, Hartati N, Dyan Anggraheni YG, Harmoko R, Indrayani S, Rahman N, Fitriani H, Nuro F, Hapsari Y, Andika NR (2025) Morpho genetic variability and Anthocyanine (Cyanidin-3-OGlucoside) Concent of Indonesia Roselle (*Hibiscus sabdariffa* L.). Int J Adv Sci Eng Inf Technol 15(1):240

Mudassir Jeelani S, Shahnawaz M, Prakash Gupta A, Lattoo SK (2022) Phytochemical diversity in relation to cytogenetic variability in Inula racemosa Hook. f., an endangered medicinal plant of Himalayas. Chem Biodivers 19(12):e202200486

Nath S, Jha TB, Mallick SK, Jha S (2015) Karyological relationships in Indian species of *Drimia* based on fluorescent chromosome banding and nuclear DNA amount. Protoplasma 252(1):283–299. https://doi.org/10.1007/s00709-014-0679-z

Nguyen TH, Waminal NE, Lee DS, Pellerin RJ, Ta TD, Campomayor NB, Kang BY, Kim HH (2021) Comparative triple-color FISH mapping in eleven Senna species using rDNA and telomeric repeat probes. Hortic Environ Biotechnol 62(6):927–935

Nishida T, Sawa S, Sugimura K (2025) Development of a low-cost, simple, and rapid identification method for Glycyrrhiza uralensis using PCR–RFLP and evaluation of seeds distributed on the market. J Nat Med 79: 1419–1432

Osman A, Jordan B, Lessard PA, Muhammad N, Haron MR, Riffin NM, Sinskey AJ, Rha C, Housman DE (2003) Genetic diversity of *Eurycoma longifolia* inferred from single nucleotide polymorphisms. Plant Physiol 131(3):1294–1301

Pan H, Deng L, Zhu K, Shi D, Wang F, Cui G (2024) Evaluation of genetic diversity and population structure of *Annamocarya sinensis* using SCoT markers. PLoS One 19(9):e0309283

Pandita D, Pandita A, Wani SH, Abdelmohsen SA, Alyousef HA, Abdelbacki AM, Al-Yafrasi MA, Al-Mana FA, Elansary HO (2021) Crosstalk of multi-omics platforms with plants of therapeutic importance. Cells 10(6):1296

Patel P, Dehery SK, Jena SN, Pradhan C, Das AB (2024) Chromosomal variations in twelve ecotypes of a medicinal plant *Drimia indica* (Roxb.) Jessop: karyotypes and 2C DNA content analysis. Genet Resour Crop Evol 71(2):621–634

Peng J, Shi C, Wang D, Li S, Zhao X, Duan A, Cai N, He C (2021) Genetic diversity and population structure of the medicinal plant *Docynia delavayi* (Franch.) Schneid revealed by transcriptome-based SSR markers. J Appl Res Med Arom Plants 21:100294

Pradhan A, Gupta V, Mukhopadhyay A, Arumugam N, Sodhi Y, Pental D (2003) A high-density linkage map in *Brassica juncea* (Indian mustard) using AFLP and RFLP markers. Theoret Appl Genet 106(4):607–614

Qiao Y, Guo F, Huo N, Zhan L, Sun J, Zuo X, Guo Z, Gu YQ, Wang Y, Liu Y (2021) Genotyping-by-sequencing to determine the genetic structure of a Tibetan medicinal plant *Swertia mussotii* Franch. Genet Resour Crop Evol 68(2):469–484

Ragul P, Rajagopal C, Loudursamy DK, Thanga Hemavathy A, Vellaikumar S, Rajagopal B, Karthikeyan P (2025) Integrated analysis of genetic diversity in *Costus pictus* through phenotypic characterization, molecular markers and UHPLC-MS-based diosgenin profiling. Genet Resour Crop Evol 72:9069–9082

Rahayu S, Hariri MR, Martiansyah I, Subositi D, Windarsih G, Cahyaningsih R (2025) Molecular markers in medicinal plant development: a systematic review. In: Molecular markers and crop improvement: principle and application, vol 28, pp 113–123

Rahman MS, Dash CK, Sultana SS (2024) Chromosomal characterization of five medicinally significant *Phyllanthus* species in Bangladesh by DNA base-specific fluorochrome banding technique. Int J Geogr Inf Syst 30(4):442–455

Rameshkumar R, Pandian S, Rathinapriya P, Selvi CT, Satish L, Gowrishankar S, Leung DW, Ramesh M (2019) Genetic diversity and phylogenetic relationship of *Nilgirianthus ciliatus* populations using ISSR and RAPD markers: implications for conservation of an endemic and vulnerable medicinal plant. Biocatal Agric Biotechnol 18:101072

Ranya EB, Ahmed DA (2018) Evaluation of genetic diversity in wild populations of *Peganum harmala* L., a medicinal plant. J Genet Eng Biotechnol 16(1):143–151

Rashid MH, Dash CK, Alam SS, Sultana SS (2019) Differential chromosome banding of three *Rauvolfia* species. Cytologia 84(2):143–146

Rosenblum LS, Holmes J, Taghiyev AF (2025) The emergence of artificial intelligence-guided Karyotyping: a review and reflection. Genes 16(6):685

Roy Choudhury D, Sharma L, Suma A, Singh GP, Singh R (2025) Exploring the genetic diversity and population structure of *Bacopa monnieri* (L.) using random amplified (RAPD and ISSR) and gene-targeted (SCoT and CBDP) markers. Genet Resour Crop Evol 72(1):1043–1063

Roy A, Sharma N, Bharadvaja N (2022) Assessment of phytochemical and genetic diversity analysis of *Plumbago zeylanica* L. accessions. Genet Resour Crop Evol 69(1):209–219

Salgotra RK, Chauhan BS (2023) Genetic diversity, conservation, and utilization of plant genetic resources. Genes 14(1):174

Samanta T, Jha TB, Ray S, Jha S (2023) Comparative cytogenetics and fluorescent chromosome banding in five Indian species of *Dipcadi* Medik. Plants 12(13):2534

Schouten HJ, Van De Weg WE, Carling J, Khan SA, McKay SJ, van Kaauwen MP, Wittenberg AH, Koehorst-van Putten HJ, Noordijk Y, Gao Z, Rees DJ (2012) Diversity arrays technology (DArT) markers in apple for genetic linkage maps. Mol Breed 29(3):645–660

Setyawan AD, Wiryanto W, Suranto S, Bermawie N (2014) Variation in isozymic pattern of germplasm from three ginger (*Zingiber officinale*) varieties. Nusantara Biosci 6(1):86–93

Sharma N, Singh B, Wani MS, Gupta RC, Habeeb TH (2020) Morphological, cytological, and chemotypic variation of essential oils in *Syzygium cumini* (L.) skeels. Anal Chem Lett 10(5):609–619

Shilpha J, Silambarasan T, Karutha Pandian S, Ramesh (2013) Assessment of genetic diversity in *Solanum trilobatum* L., an important medicinal plant from South India using RAPD and ISSR markers. Genet Resour Crop Evol 60(3):807–818

Singh V, Gupta RC, Sharma K, Sharma V, Sharma M, Kaur K (2018) Male meiotic studies in 29 species of Lamiaceae from Sirmaur District of Himachal Pradesh, India. Cytologia 83(3):235–243

Stafiniak M, Bielecka M, Kujawa K, Jezierska-Domaradzka A, Pencakowski B, Basiak A, Matkowski A, Nawrot-Hadzik I (2025) Integrative morphological, phytochemical, and molecular identification of three invasive and medicinal *Reynoutria* species. Sci Rep 15(1):6001

Stotz GC, Salgado-Luarte C, Escobedo VM, Valladares F, Gianoli E (2021) Global trends in phenotypic plasticity of plants. Ecol Lett 24(10):2267–2281

Su Y, Fu J, Xie H, Huang Z, Li Y, Luo Y, Zhou X, Li Y, Li J, Sun Y, Liu Y (2025) SSR markers development and their application in genetic diversity of burdock (*Arctium lappa* L.) germplasm. BMC Plant Biol 25(1):196

Than MM, Samaddar T, Bhowmick BK, Jha S (2017) Fluorescent chromosome banding and genome size estimation in three species of *Swertia*. Cytologia 82(5):513–520. https://doi.org/10.1508/cytologia.82.513

Tohidi B, Rahimmalek M, Arzani A (2021) Variation in phytochemical, morphological, and ploidy levels of Iranian *Thymus* species. Chem Biodivers 18(2):e2000911

Vaishnav K, Tiwari V, Durgapal A, Meena B, Rana TS (2023) Estimation of genetic diversity and population genetic structure in *Gymnema sylvestre* (Retz.) R.. Br. Ex Schult. populations using DAMD and ISSR markers. J Genet Eng Biotechnol 21(1):42

Wang M, Xu R, Luo M, Li X, Wei L, Lei M, Liu D, Miao Y (2025) Analysis of genetic diversity and population structure of *Pinellia ternata* using EST-SSR markers. J Appl Res Med Aromat Plants 44:100611

Wu K, Liu Y, Yang B, Kung Y, Chang K, Lee M (2022) Rapid discrimination of the native medicinal plant *Adenostemma lavenia* from its adulterants using PCR-RFLP. Peer J 10:13924

Wu J, Yang Q, Zhao W, Miao X, Qin Y, Qu Y, Zheng P (2024) Assessment of population genetic diversity of medicinal *Meconopsis integrifolia* (Maxim.) Franch. using newly developed SSR markers. Plants 13(18):2561

Xavier JK, Baia TG, Alegria OV, Figueiredo PL, Carneiro AR, Moreira EC, Maia JG, Setzer WN, da Silva JK (2022) Essential oil chemotypes and genetic variability of *Cinnamomum verum* leaf samples commercialized and cultivated in the Amazon. Molecules 27(21):7337

Xu Y, Tian S, Li R, Huang X, Li F, Ge F, Huang W, Zhou Y (2021) Transcriptome characterization and identification of molecular markers (SNP, SSR, and Indels) in the medicinal plant *Sarcandra glabra* spp. Biomed Res Int 1:9990910

Xu L, Li P, Su J, Wang D, Kuang Y, Ye Z, Chen M (2023) EST-SSR development and genetic diversity in the medicinal plant *Pseudostellaria heterophylla* (Miq.) pax. J Appl Res Med Aromat Plants 33:100450

Xu J, Zou R, Chai S, Jiang H, Zhong W, Tang J, Wei X (2025) Single nucleotide polymorphism markers based on hyper-seq revealed the genetic diversity and genetic relationship of *Horsfieldia hainanensis* in China. Forests 16(2):217

Yamazaki M, Sato A, Shimomura K, Saito K, Murakoshi I (1994) Genetic relationships among *Glycyrrhiza* plants determined by RAPD and RFLP analyses. Biol Pharm Bull 17:1529–1531

Yang DU, Kim MK, Mohanan P, Mathiyalagan R, Seo KH, Kwon WS, Yang DC (2017) Development of a single-nucleotide-polymorphism marker for specific authentication of Korean ginseng (*Panax ginseng* Meyer) new cultivar "G-1". J Ginseng Res 41(1):31–35

Yang W, Bai Z, Wang F, Zou M, Wang X, Xie J, Zhang F (2022) Analysis of the genetic diversity and population structure of *Monochasma savatieri* Franch. ex Maxim using novel EST-SSR markers. BMC Genomics 23(1):597

Ying Z, Awais M, Akter R, Xu F, Baik S, Jung D, Yang DC, Kwak GY, Wenying Y (2022) Discrimination of *Panax ginseng* from counterfeits using single nucleotide polymorphism: a focused review. Front Plant Sci 13:903306

Zhang J, Chen M, Dong X, Lin R, Fan J, Chen Z (2015) Evaluation of four commonly used DNA barcoding loci for Chinese medicinal plants of the family Schisandraceae. PLoS One 10(5):e0125574

Zhao H, Li S, Yang C, Li G, Wang Y, Peng J, Yan Z, Li R, Wang Y, Zhang L (2021) FISH-based karyotype analyses of four *Dracaena* species. Cytogenet Genome Res 161(5):272–277

Zheng R, Zhao S, Khayyatnezhad M, Shah SA (2021) Comparative study and genetic diversity in *salvia* (Lamiaceae) using RAPD Molecular Markers. Caryologia 74(2):45–56

Zhou HC, Pellerin RJ, Waminal NE, Yang TJ, Kim HH (2019) Pre-labelled oligo probe-FISH karyotype analyses of four Araliaceae species using rDNA and telomeric repeat. Genes Genom 41(7):839–847

Zhou Y, Xu L, Zhang L, Shi D, Wu C, Wei R, Song N, Wu S, Chen C, Li H (2024) Enhancing chromosomal analysis efficiency through deep learning-based artificial intelligence graphic analysis. Discov Appl Sci 6(6):299

Zhu S, Liu Q, He J, Nakajima N, Samarakoon SP, Swe S, Zaw K, Komatsu K (2021) Genetic identification of medicinally used *Salacia* species by nrDNA ITS sequences and a PCR-RFLP assay for authentication of Salacia-related health foods. J Ethnopharmacol 274:113909

Chapter 5
Anatomy and Chloroplast Barcoding for Genetic Diversity of Medicinally Important *Cinnamomum* Species Found in Sri Lanka

Pushpa Damayanthi Abeysinghe

Abstract The genus *Cinnamomum* belongs to the family Lauraceae, which is found and diverse in tropical and subtropical regions, comprising approximately 55 genera with 3000 species. Among the economically important Laurales, the genus *Cinnamomum* is significant not only for its economic value but also for its health benefits. This chapter discusses the use of anatomical characteristics (macro- and micro-morphology, petiole, leaf, stem, and root) and molecular techniques (developed before and after the advent of polymerase chain reaction (PCR) technology) for genetic diversity studies in *Cinnamomum* species from Sri Lanka. Although there are eight *Cinnamomum* species in Sri Lanka, *C. zeylanicum* is the most economically important species and is used for multiple purposes. In the farmer's field, a variation is found based on morphological characters. Therefore, accurate species identification and complete and comprehensive studies of the genus are essential for conservation and breeding purposes. Although very few anatomical studies have been conducted on the *Cinnamomum* species found in Sri Lanka, this chapter explains the anatomical variation using light microscopy and scanning electron microscopy among the *Cinnamomum* species. This chapter aims to fill the data gap on the anatomical characters of the leaf, petiole, stem, and root and molecular markers of *Cinnamomum* species. Cuticular sculptures and cuticular features could serve as excellent diagnostic criteria in the delimitation of *Cinnamomum* species. The distinguishing characters of taxonomic value include the variation in shape of the petioles, arrangement and shape of petiole vascular bundles, presence/absence of hairs, hair types, distribution of hairs on the petiole, type and distribution of crystals, pigments and sclerides, wood anatomical features, including the presence and types of trichomes, crystals, sclereids, and vessel arrangements, which are valuable in delimiting and identifying *Cinnamomum* species. In all species, the barks were characterized by different colored (dark golden brown, dark red, or dark brown)

P. D. Abeysinghe (✉)
Department of Botany, University of Ruhuna, Matara, Sri Lanka
e-mail: pushpa@bot.ruh.ac.lk

S. Jha et al. (eds.), *Genetic Diversity in Medicinal Plants*, Sustainable Development and Biodiversity 44, https://doi.org/10.1007/978-981-95-7281-6_5

pigments, the presence of islands of sclerenchyma in the pericycle, mucilage or oil droplets in the cortex, radially arranged vascular system consisted of diffuse-porous or semi-ring porous in wood, growth ring boundaries distinct or intermediate between distinct and indistinct, and solitary vessels in radial or diagonal multiples of 2–3 cell. Vessels of most species were evenly distributed, round to oval, with simple or scalariform perforation, bordered and alternate inter-vessel pitting, with tyloses, vasicentric axial parenchyma, uniseriate rays, and multiseriate rays. Moreover, allozyme variation and DNA-based markers, including randomly amplified polymorphic DNA, SRAP (sequencerelated amplified polymorphism), PCR-AFLP (PCR-restriction fragment length polymorphism), ISSR (inter-simple sequence repeat), and chloroplast barcoding of the *matK*, *rbcL*, *rpoB*, *rpoC*1, ITS of *trnH-psbA*, *trnL-F*, *atpF-atpH*, and *psbK-psbI* have been employed in the identification and diversity studies of *Cinnamomum* species. Moreover, the DNA variation of the *Cinnamomum* species found in Sri Lanka and other Laurales found in the world has been discussed.

Keywords Anatomy · Chloroplast DNA · *Cinnamomum* species · Diagnostic characters · DNA sequencing · Genetic diversity · Identification

5.1 Introduction

Lauraceae Juss., an angiosperm family within the order Laurales, comprises roughly 3000 species distributed across approximately 30 neotropical genera in India, Sri Lanka, Japan, China, and Australia and extends to temperate zones of both hemispheres (Dao et al. 2013; Liu et al. 2013; Alves et al. 2025; Andianto et al. 2015; Wu et al. 2025). One notable genus within the Lauraceae family is *Cinnamomum*, which comprises approximately 250 species, primarily found in the tropical and subtropical regions of Asia and Oceania/Australasia (Tao 2001). Although there are approximately 250 species of the genus *Cinnamomum* within the family Lauraceae worldwide, in Sri Lanka, this family is represented by 41 species across 9 genera (Senaratna 2001; Bandusekara et al. 2025). Eight species are present in Sri Lanka, in the genus *Cinnamomum* (*C. capparu-coronde* Blume, *C. citriodorum* Thw., *C. dubium* Nees, *C. litseaefolium* Thw., *C. ovalifolium* Weight, *C. rivulorum* Kosterm, and *C. sinharajaense* Kosterm (Liyanage and Senanayake 2010; Kumarathilake 2012; Ariyarathne et al. 2018; Bandusekara et al. 2020), and *Cinnamomum zeylanicum* Blume (syn. *Cinnamomum verum* Berchthold & Presl) distributed in Upcountry (Nuwara Eliya, Badulla), South (Galle, Matara, and Hambantota districts), South Western (Colombo, Gampaha, and Kalutara districts), Rathnapura, Kandy, Monaragala, and Matale districts (Abeysinghe et al. 2020). Moreover, the conservation status of the *Cinnamomum* species has been identified; *C. capparu-coronde*, *C. litseaefolium*, *C. ovalifolium*, *C. sinharajaense,* and *C. zeylanicum* were considered vulnerable species, *C. citriodorum, C. rivulorum* were endangered, and *C. dubium* was a near-threatened species (The National Red List

2020 2020). *Cinnamomum* has evergreen aromatic trees and shrubs, notable for the aromatic oils in their leaves and bark. They include many economically significant trees that are utilized in the production of cinnamon spice.

Cinnamomum species have great importance due to their economic value. Notably, Sri Lanka accounts for approximately 70% of the world's cinnamon bark production, highlighting its significant economic importance. Cinnamon is widely utilized in the food industry, enhancing the flavor of various products such as pickles, cakes, biscuits, sweets, and liquors (Lungarini et al. 2008; Wang et al. 2022). In addition to its flavoring properties, it acts as a food preservative and finds applications in the soap industry. Beyond the bark, cinnamon produces essential oils from its leaves, stem bark, and root bark, yielding unique oil types (Guo et al. 2024; Huang et al. 2025). Historically, the fat extracted from cinnamon fruits was also employed in producing church candles (Wijesekara et al. 1974; Purseglove 1977; Bavappa et al. 1981). Cinnamon species are one of the economically and medicinally important cash crops in Sri Lanka and worldwide. In the farmer fields, it is possible to observe morphologically heterogeneous plants of the same species, indicating that morphology cannot be used as a reliable marker for the correct identification, diversity studies, conservation, and breeding purposes of these species. Therefore, the use of other marker systems, such as anatomy and other molecular markers, is discussed in this chapter as an objective of this chapter.

5.2 Phytochemistry and Medicinal Properties of *Cinnamomum* Species (Pharmacology)

Cinnamon and other spices such as pepper and vanilla are widely recognized and used globally (Spence 2024). Several spices, including cinnamon, cloves, bay leaves, and turmeric, have demonstrated the ability to enhance insulin activity in vitro (Spence 2024). Many fragrant species from the Lauraceae family, especially *C. zeylanicum*, are highly valued worldwide for their medicinal properties (Broadhurst et al. 2000; Khan et al. 2003; Shan et al. 2007; Singh et al. 2007; Elumalai et al. 2010; Ranasinghe et al. 2013; Spence 2024). Traditionally, numerous Lauraceae species known for their pleasant aromas have been used in folk medicine. However, other *Cinnamomum* species are less economically significant than *C. zeylanicum*. Due to their lower cost, species such as *Cassia* are often sold as cinnamon in markets globally, especially in Canada, the United States, and European countries (Pathirana and Senaratne 2020). For instance, "Indonesian cinnamon" can refer to *C. burmanii*, and some other *Cassia* species, like *C. aromaticum*, are also sold as cinnamon.

Pathirana and Senaratne (2020) have reported the importance of the genus *Cinnamomum* in terms of economic and health benefits, as well as chemical composition. Chemical composition and its variability of *C. zeylanicum*, the inter- and intra-species phytochemical diversity, volatile oil profiles, and environmental influences have been examined using leaf and bark extracts using HPLC (High-performance liquid chromatography), the leaf volatile oil profiles through GC-MS (Gas Chromatography–Mass Spectrometry), and the leaf oil yield of wild *Cinnamomum* species found in Sri Lanka (Azad et al. 2024; Bandusekara et al. 2025).

These results show that different species have different chemical compositions, indicating that these species have unutilized chemicals, and such results can be used for breeding programs. According to the previous research, they identified key compounds such as cinnamaldehyde in bark, eugenol in leaves and bark, and cinnamyl acetate in bark. Chemical analyses of *C. zeylanicum* leaves have yielded a diverse array of bioactive substances, including polyphenols such as condensed tannins, terpenoids, and flavonoids, as well as a variety of other compounds, including coumarins, alkaloids, and steroids (Senevirathne et al. 2022; Gonawala et al. 2024).

The effect of the use of cinnamon during pregnancy and gynecological health (Jaafarpour et al. 2015; John and Shantakumari 2015; Biggs and Banerjee 2025), managing diabetes (Senevirathne et al. 2022), cardioprotective, anti-inflammatory effects (Pagliari et al. 2023), and antimicrobial and antifungal activities (Didehdar et al. 2022; Biggs and Banerjee 2025) have also been investigated. The potential applications of the bioactive compounds present in *Cinnamomum* species have been utilized to develop various formulations for mouth rinses, toothpastes, and root canal irrigants used in dentistry (Yanakiev 2020).

5.3 Diversity Studies of *Cinnamomum* Species

5.3.1 Morphological Diversity

Accurate species identification and a comprehensive study of the genus are necessary for conservation and breeding purposes (Bandusekara et al. 2020). The data for delimiting the species can be gathered from different sources, including morphology, anatomy, palaeobotany, cytology, and molecular biology for diversity studies. The *Cinnamomum* species, such as *C. zeylanicum* and its wild relatives, exhibit considerable morphological (Azad et al. 2015; Geekiyanage et al. 2024) and flower morphological diversity (Azad et al. 2018; Geekiyanage et al. 2024). Morphological and phytochemical variabilities are often influenced by their adaptation to isolated habitats (Geekiyanage et al. 2024; Bandusekara et al. 2025). In Australia, the family Lauraceae has been revised using only morphological characters (Weber 1981; Hyland 1986). Furthermore, Abeywikrama (1973) conducted a taxonomic revision of *Cinnamomum* species found in Sri Lanka. Bandusekara et al. (2020) studied leaf morphology in detail and developed a key for field identification of these species. Significant intraspecific diversity has been observed due to cross-pollination in wild species, such as *C. capparu-coronde, C. dubium, C. litseifolium, C. ovalifolium, and C. sinharajaense*, with the highest variation found in *C. zeylanicum*, followed by *C. dubium* (Bandusekara et al. 2025).

5.3.2 Identification and Delimitation of Flora Using Anatomical Diversity (Phytotaxonomy)

Environmental conditions affect morphological characters (Wang et al. 2022; Xie et al. 2022). Different anatomical characters are extensively used in plant diversity studies: infra-generic classification of Hydrangeaceae (Yang et al. 2024), *Hordeum* (Ro et al. 2025), delineation at the species level for genera like *Flemingia* (Thacker et al. 2021) and cultivar identification (Korir et al. 2012), especially when the floral parts or fruits are unavailable, authentication of medicinal plants (Anurag et al. 2023), and identification of species of the genus *Ficus* (Binnoubah et al. 2023). However, anatomical research is scarce, specifically focused on the diversity within the genus *Cinnamomum*. Few researchers have undertaken anatomical studies to investigate the diversity of Sri Lankan *Cinnamomum* species (Abeysinghe and Scharaschkin 2019; Abeysinghe et al. 2020; Abeysinghe 2023, 2024), and anatomical descriptions of leaf, root, and stem material of *Cinnamomum* species are scarce.

5.3.2.1 Petiole Anatomy

Petiole anatomical data has been effectively utilized in delimiting various taxa: *Clematis* L.–Ranunculaceae (Park et al. 2021), *Curcuma* species (Anu and Dan 2020), and *Acer* (Mohtashamian et al. 2022), and petiole anatomy is recognized as an important biomarker (Ingole and Kaikade 2015). In a comprehensive investigation of the petiole anatomy of the *Cinnamomum* species, variations have been investigated and found to be of taxonomic significance. The petiole anatomical structure of *Cinnamomum* has allowed for the species delimitations. The petiole anatomical structure of *Cinnamomum* enables distinguishing among species, as each exhibits unique structural properties and clear interspecific differences. Key diagnostic characters include variations in petiole shape, the number, arrangement, and shape of vascular bundles, presence or absence of hairs, hair types, distribution patterns of hairs, types and distributions of crystals, pigments, and sclereids within the petioles (Abeysinghe and Scharaschkin 2019). Each petiole has its structural properties and shows distinct differences among the examined types. For example, *C. capparucorondo* displays winged extensions on the adaxial side. Similar winged features are observed in Lauraceae members such as *C. pauciflorum* Nees and *Ocotea duckei* (Baruah and Nath 2006; Coutinho et al. 2006). This chapter also explains the utilization of the anatomy of leaf, petiole, stem, and root to study the diversity of economically and medicinally significant species of this genus and how anatomy has been used in diversity studies of *Cinnamomum* species.

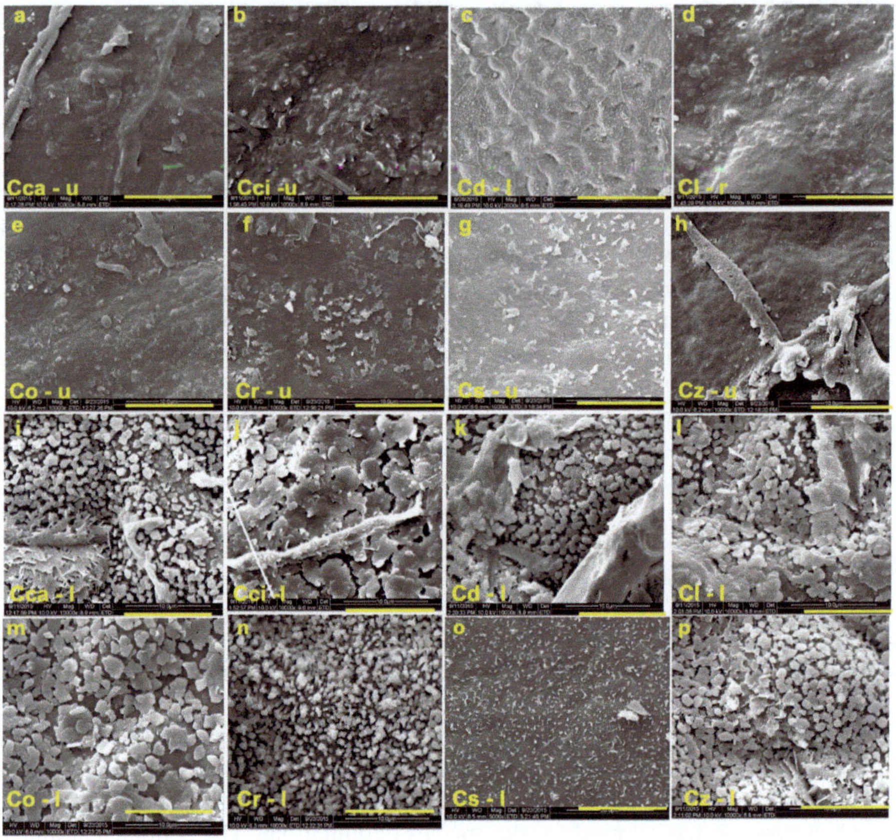

Fig. 5.1 Closer view of the SEM of wax deposition on adaxial surfaces (**a-h**) and abaxial surfaces (**i-p**) of the *Cinnamomum* species. (**a**) *C. capparu-coronde*—undulating; (**b**) *C. citriodorum*—irregularly deposited wax; (**c**) *C. dubium*—wavy; (**d**) *C. litseaefolium*—undulating; (**e**) *C. ovalifolium*—small wax crystals unevenly distributed; (**f**) *C. rivulorum*—granules; (**g**) *C. sinharajaense*—sands; (**h**) *C. zeylanicum*—large clusters; (**i**) *C. capparu-coronde*—has single, irregular shaped crystals; (**j**) *C. citriodorum*—larger single platelets/flake platelets; (**k**) *C. dubium*—transversely ridged rodlets; (**l**) *C. litseaefolium*—crusts or platelets; (**m**) *C. ovalifolium*—platelets; (**n**) *C. rivulorum*—clusters of tubules; (**o**) *C. sinharajaense*—irregular, undulating pattern; and (**p**) *C. zeylanicum*—crusts. (Abeysinghe 2024, with permission to reuse the figure published in Cey J Sci 53(3):413–425 DOI:10.4038/cjs.v53i3.8235)

5.3.2.2 Cuticle, Micromorphology, and Trichomes, Vascular Bundle of the Leaf

Scanning electron microscopy (SEM) revealed that all species possess a thick, waxy cuticle on both leaf surfaces (upper and lower). These cuticles show continuous or discontinuous epicuticular wax layers with uniseriate epidermal cells, consistent across species, but exhibit species-specific ultrastructural features (Fig. 5.1).

Variations in cuticle striation patterns further serve as diagnostic features. Prior studies (Bakker et al. 1992; Nishida and van der Werff 2007; Yang et al. 2012; Nishida et al. 2016) support the utility of cuticular features in taxonomic differentiation within Lauraceae, including fossil specimens like *Machilus maomingensis* from the Eocene (Tang et al. 2016) and Palaeocene fossils of leaves (Tang et al. 2016). Several researchers have paid attention to the utilization of cuticular features in the taxonomy of the Lauraceae (Christophel and Rowett 1996; Li and Christophel 2000; Loutfy 2001; Yang et al. 2012; Nishida et al. 2016)—not only in living plants, but also in extinct plants such as *Machilus maomingensis* (Tang et al. 2016), foliar fossils from the late Palaeocene (Tang et al. 2016), and *Cinnamomum*, from the Oligocene period (Shi et al. 2014).

Cinnamomum species were characterized by simple, unicellular, or falcate, non-glandular trichomes present on both surfaces of their leaves, predominantly on the lower surface. For example, *C. dubium* featured unicellular, falcate trichomes, while *C. capparu-coronde* has short and thick trichomes (Abeysinghe 2024). In contrast, *C. rivulorum* displayed unicellular trichomes that can be either straight or falcate, varying in length. *C. capparu-coronde*, *C. dubium,* and *C. ovalifolium* exhibit densely packed simple, unicellular/falcate, long, and thin trichomes on their adaxial surfaces (Abeysinghe 2024). Notably, some species lack trichomes on the adaxial surface, aligning with previous studies on *Cinnamomum* and other members of the Lauraceae family (Bakker et al. 1992; Gonçalves et al. 2018). The types of trichomes varied among species, exhibiting shapes that range from simple to falcate, straight, or undulated and vary in length, depending on the specific species (Abeysinghe 2024). These variations in trichomes in different species may be useful in breeding programs. TS of leaves along the midrib revealed various outline shapes, allowing classification of leaves into six categories based on the outline/shape of sections along the midrib. In all species, a transverse section along the blade, on both upper and lower epidermal layers, shows a cuticle layer. The pavement cells of the epidermis displayed a range of shapes, including isodiametric, rectangular, tabloid forms, tetragonal, polygonal with varying sizes, and some exhibiting sinuous walls (Abeysinghe 2024).

Between the epidermal layers, there was a single layer of densely packed palisade parenchyma cells. Below the palisade layer and above the lower epidermis, several layers of loosely arranged mesophyll cells were observed, characteristic of spongy parenchyma with intercellular spaces, indicative of a dorsiventral leaf structure. The tissue arrangement in the transverse section of the lamina along the vascular bundle, which was an open arc with no lateral bundles, was similar across most species. In the single central vascular bundle, the xylem faced the adaxial side of the leaf, and the phloem was positioned abaxially. The midrib contained a single central vascular bundle that featured a discontinuous sclerenchymatous sheath, with dark deposits surrounding the sheath (Abeysinghe 2024). Vascular bundle shapes varied among species: some displayed oval (*C. cappru-corondo* and *C. dubium*), elongated (*C. citrodorum* and *C. ovalifolium*), or irregular shapes. The vascular bundle in *C. rivulorum* had a "V" shape, while in *C. sinharajaense* and *C. zeylanicum*, the bundles were partially dissected into two segments. *C. litseaefolium* and *C.*

ovalifolium showed vascular bundles partially dissected into three segments. All vascular bundles were open and contained more than ten collateral vessels, arranged as one to five cells in parallel, vertically aligned, and separated by fibers within the bundle (Abeysinghe 2024). The structure, as well as the arrangement of vascular bundles of the *Cinnamomum* species, provided important insights for species differentiation.

5.3.2.3 Root and Stem Anatomy

Stem and root anatomy descriptions are scarce but valuable in systematic studies. Both roots and stems show similar tissue arrangements (Fig. 5.2). The roots of all species were covered by a peridermis that varied in thickness. Moreover, peridermis is a protective tissue found in the stems and roots of plants, particularly those undergoing secondary growth. A TS of *Cinnamomum* stem bark shows a similar arrangement with cortex, pericycle, pericyclic fibers, sclerides, secondary phloem, phloem

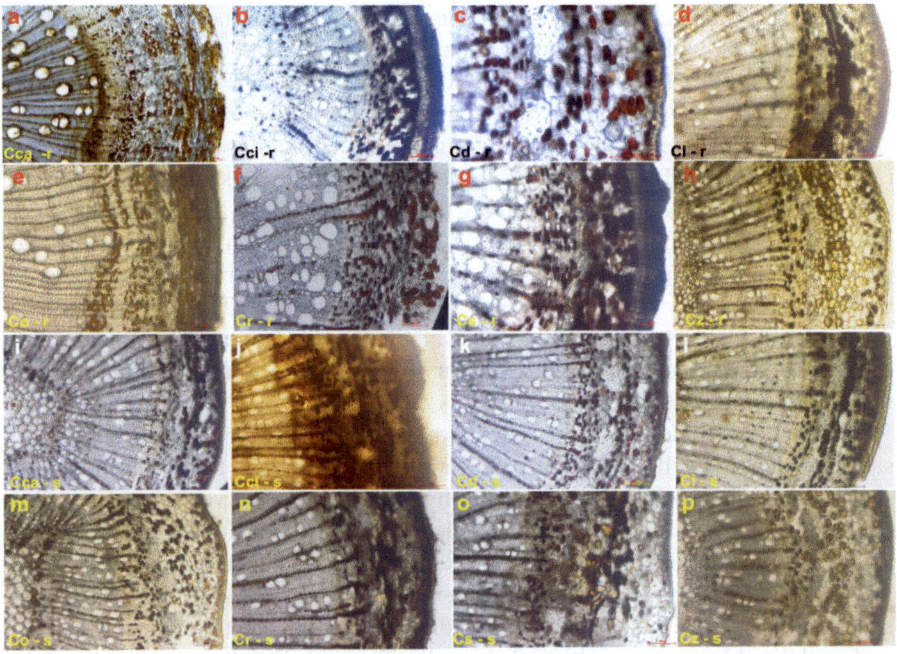

Fig. 5.2 Transverse sections of the root (**a–h**). (**a**) *C. capparu-coronde*; (**b**) *C. citriodorum*; (**c**) *C. dubium*; (**d**) *C. litseafolium*; (**e**) *C. ovalifolium*; (**f**) *C. rivulorum*; (**g**) *C. sinharajaense*; (**h**) *C. zeylanicum*; transverse sections of the stem (**i–p**). (**i**) *C. capparu-coronde*; (**j**) *C. citriodorum*; (**k**) *C. dubium*; (**l**) *C. litseafolium*; (**m**) *C. ovalifolium*; (**n**) *C. rivulorum*; (**o**) *C. sinharajaense*; (**p**) *C. zeylanicum*. Cca—*C. capparu-coronde*; Cci—*C. citriodorum*; Cd—*C. dubium*; Cl—*C. litseafolium*; Co—*C. ovalifolium*; Cr—*C. rivulorum*; Cs—*C. sinharajaense*; Cz—*C. zeylanicum*; r—root; s—stem. Scale bar: 100 μm (unpublished)

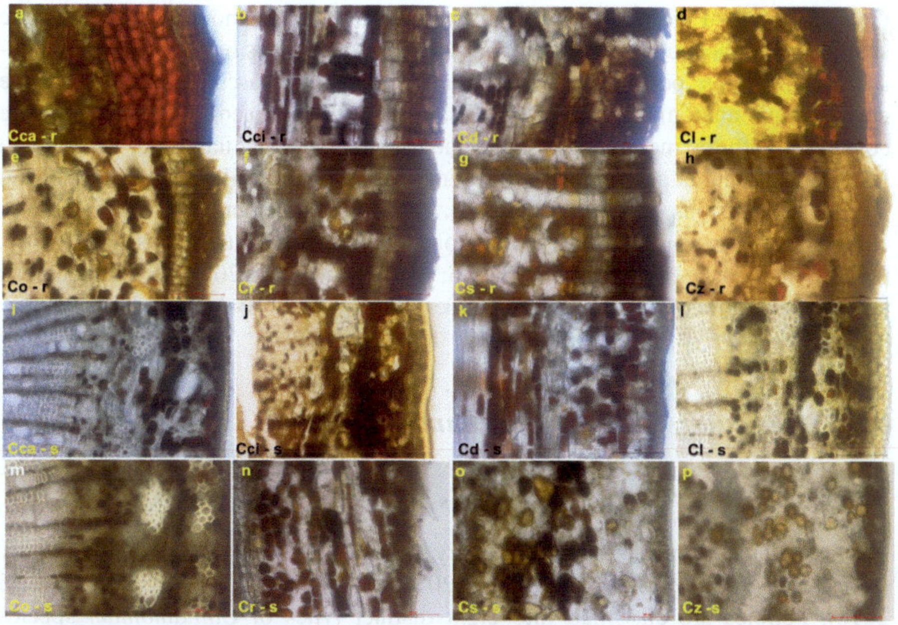

Fig. 5.3 Chemical substances present in the cortex of roots and stems. (**a**–**h**) Transverse sections of the roots. (**a**) *C. capparu-coronde*—dark red substances in five to six layers of rectangular suberized cells; (**b**) *C. citriodorum*—dark brown/black substances in rectangular cells in cortex; (**c**) *C. dubium*—light brown/dark brown substances; (**d**) *C. litseafolium*—dark brown/yellow substances in irregular cells; (**e**) *C. ovalifolium*—low densely spread brown/black substances; (**f**) *C. rivulorum*—dark brown/black/yellowish substances irregular cells; (**g**) *C. sinharajaense*—dark brown substances in irregular cells; (**h**) *C. zeylanicum*—yellow, light brown deposites; (**i**–**p**) transverse sections of the stems. (**i**) *C. capparu-coronde*—dark black substances in irregular cells, and mucilage/oil cells; (**j**) *C. citriodorum*—light brown substances in irregular suberized cells in cortex; (**k**) *C. dubium* coffee brown substances; (**l**) *C. litseafolium*—golden brown deposits; (**m**) *C. ovalifolium*—brown; (**n**) *C. rivulorum*—light brown substances, presence of trichomes; (**o**) *C. sinharajaense* black deposits; (**p**) *C. zeylanicum*—light yellow deposits. Cca—*C. capparu-coronde*, Cci—*C. citirodorum*, Cd—*C. dubium*, Cl—*C. litseafolium*, Co—*C. ovalifolium*, Cr—*C. rivulorum*, Cs—*C. sinharajaense*, Cz—*C. zeylanicum*, r—root, s—stem. Scale bar: 100 μm (unpublished)

parenchyma, phloem fibers, medullary rays, and mucilage cells. In all species, the cortex parenchyma cells of both root and stem bark of the studied taxa contained phenolics and tannins with different colors (dark golden brown/dark red) in different densities and islands of sclerenchyma in the pericycle (Fig. 5.3). Root and stem anatomy of Lauraceae is not reported or is very rare, to compare our results with other results.

Features such as trichomes, cortex structure, pericyclic fibers, scleroids, perforation plates, inter-vessel pitting, ray composition, axial parenchyma distribution, and pit structures are commonly used in taxonomic assessments (Patel 1987; Kumar and Ruba 2018; Swetha et al. 2014; Sun et al. 2015). Among the wide array of root and stem features, the presence (*C. citrodorum*—Fig. 5.2j, *C. dubium*—Fig. 5.2k,

C. ovalifolium—Fig. 5.2m, *C. rivulorum*—Fig. 5.2n) or without trichomes on the stem, cortex layer full of phenolic compounds, ray composition, presence of sclerenchyma groups, perforation plate structure, and inter-vascular pitting were the characteristic anatomical features. The same results had been identified and were consistent with the members of Lauraceae (Shylaja and Manilal 1992; Bakker and Baas 1993; Gang et al. 2021). The presence or absence of oil, mucilage cells, and inorganic inclusions, and secretory cells, and the arrangement of the axial parenchyma have been applied in the evaluation of systematic relationships (Patel 1987; Callado and Costa 1997; Sun et al. 2015; Kumar and Ruba 2018). Suberized cells have been identified in the roots of members of Lauraceae (Bakker and Baas 1993).

5.3.3 Vascular Tissues of *Cinnamomum*

5.3.3.1 Petiolar Vascular Tissue

The vascular system has been a primary focus in the study of petiole anatomy (Kocsis and Borhidi 2003; Noraini et al. 2016; Talip et al. 2017; Long and Oskolski 2018). Petioles are classified into eight distinct types of *Cinnamomum* species depending on the arrangement of vascular tissues (Abeysinghe and Scharaschkin 2019). The various patterns of vascular bundles observed in *Cinnamomum* species—such as open arcs with slightly curved ends directed toward the center, clearly segmented into three portions, partially dissected into three segments, and "C"-shaped open arcs—have facilitated the identification of species differences (Abeysinghe and Scharaschkin 2019). Crescent- or V-shaped vascular bundles have been observed in members of certain Lauraceae (Kamel and Loufty 2001; Coutinho et al. 2006). Different-colored pigmented cells have been observed in the ground tissues of all the studied species (Kamel and Loufty 2001). An open-arch vascular bundle and collateral xylem tissues were documented in *O. duckei* (Coutinho et al. 2006).

5.3.3.2 Root and Stem Vascular Tissue

TS of roots and stems consisted of solitary or multiple vessels with radially arranged vascular systems. Vessel elements were semi-ring porous, transverse to oblique (*C. dubium*—Fig. 5.4c, *C. zeylanicum*—Fig. 5.4h), diffuse porous, composed of vessels predominantly solitary (*C. capparu-coronde*—Fig. 5.4a, *C. citriodorum*—Fig. 5.4b, *C. rivulorum*—Fig. 5.4f) and multiples of two to three (*C. dubium*—Fig. 5.4c, *C. litseafolium*—Fig. 5.4d, *C. ovalifolium*—Fig. 5.4e, *C. sinharajaense*—Fig. 5.4g, and *C. zeylanicum*—Fig. 5.4h). Vessels of most species were unevenly distributed (Fig. 5.4a–h). The results observed in this study are comparable with those of Loutfy (2009) and Sun et al. (2015). The shape of the vessels was mostly round (*C. capparu-coronde*—Fig. 5.4a, *C. citriodorum*—Fig. 5.4b), round/oval (*C. dubium*—Fig. 5.4c,

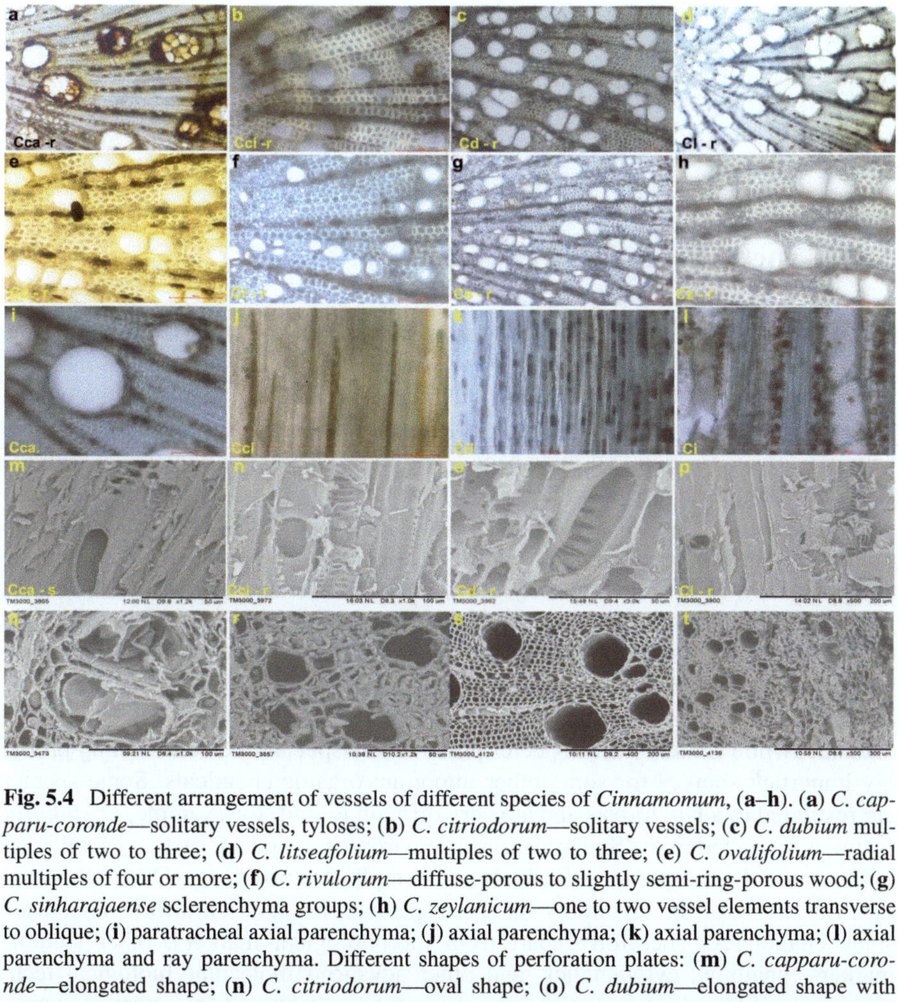

Fig. 5.4 Different arrangement of vessels of different species of *Cinnamomum*, (**a**–**h**). (**a**) *C. capparu-coronde*—solitary vessels, tyloses; (**b**) *C. citriodorum*—solitary vessels; (**c**) *C. dubium* multiples of two to three; (**d**) *C. litseafolium*—multiples of two to three; (**e**) *C. ovalifolium*—radial multiples of four or more; (**f**) *C. rivulorum*—diffuse-porous to slightly semi-ring-porous wood; (**g**) *C. sinharajaense* sclerenchyma groups; (**h**) *C. zeylanicum*—one to two vessel elements transverse to oblique; (**i**) paratracheal axial parenchyma; (**j**) axial parenchyma; (**k**) axial parenchyma; (**l**) axial parenchyma and ray parenchyma. Different shapes of perforation plates: (**m**) *C. capparu-coronde*—elongated shape; (**n**) *C. citriodorum*—oval shape; (**o**) *C. dubium*—elongated shape with scalariform pits; (**p**) *C. litseafolium*—round outline. Crystals were present in different locations of the plant body, (**q**) in ray parenchyma, fibers, and vessel elements; (**r**) in fibers; (**s**) in ray parenchyma; and (**t**) in cortex. Cca—*C. capparu-coronde*, Cci—*C. citriodorum*, Cd—*C. dubium*, Cl—*C. litseafolium*, Co—*C. ovalifolium*, Cr—*C. rivulorum*, Cs—*C. sinharajaense*, Cz—*C. zeylanicum*, r—root, s—stem. Scale bar: 0–30 μm, m, r—50 μm, a–l, n, q—100 μm, p and s—200 μm, and t—300 μm (unpublished)

C. litseafolium—Fig. 5.4d, *C. ovalifolium*—Fig. 5.4e), and round/irregular (*C. rivulorum*—Fig. 5.4f, *C. sinharajaense*—Fig. 5.4g, *C. zeylanicum*—Fig. 5.4h). Vessels are filled with tyloses, which were clear in *C. capparu-coronde* (Fig. 5.4a). Moreover, in other species, tyloses were not observed or were very rare. Large, thick-walled vessels with tylosis are surrounded by thin-walled parenchyma (Fig. 5.4a). Paratracheal parenchyma was observed in *C. capparu-coronde* (Figs. 5.4a, i) and *C. citriodorum* (Fig. 5.4b). Except *C. dubium* (Figs. 5.4c, i) and

C. zeylanicum (Fig. 5.4h), all other species showed uniseriate ray parenchyma (Fig. 5.4j, k). Different shapes of the perforation plates were observed (Fig. 5.4m–p), and various shapes of crystals were observed in different locations of the stems and roots, ray and axial parenchyma (Fig. 5.4q–t), axial parenchyma (Fig. 5.4r), ray parenchyma (Fig. 5.4s), and cortex (Fig. 5.4t).

5.4 Genetic Diversity of *Cinnamomum*

There is a risk of extinction of these valuable germplasms. Considering the *Cinnamomum* species, some of these species (e.g., *C. zeylanicum*) are so morphologically different from others in the field conditions that they can be differentiated into different taxa by looking at morphological characters. Intermediate forms are so intimately connected that interspecific and intraspecific relationships within the *Cinnamomum* have proven more difficult to determine and discriminate species using only morphological characters (Bandusekara et al. 2020). Also, there is quite a variability of Cinnamon in field conditions in terms of shape and size of the leaves, stem diameter, re-juvenile capacity under different soil and environmental conditions, plant height, quality and quantity of oil content of the stem bark, and leaf qualities, and resistance to diseases and major pests. Therefore, these important plant species should be conserved and protected for future breeding programs. Moreover, all *Cinnamomum* species found in Sri Lanka are considered to be endemic, while some of these species possess very special odors, indicating probably important sources for some other important organic chemicals. Some species are restricted to a few forests and are very difficult to find; for example, *C. sinharajaense* is restricted to the Sinharaja forest. Environmental plasticity has a significant influence on the morphology of plants and complicates taxonomy. Due to the fragmentary nature of the taxonomy for these species, conservation efforts have faced some challenges. Therefore, apart from morphological markers, other suitable markers (anatomy, cytogenetics, and other advanced molecular biological techniques) should be used to distinguish, characterize, and breed for future conservation strategies of the *Cinnamomum* species found in Sri Lanka (Bandaranayake and Pushpakumara 2020).

5.4.1 *Molecular Techniques Utilized for Diversity Studies*

It is well known that by using reliable and powerful molecular markers such as amplified fragment length polymorphism (AFLP) sequence-tagged simple sequence repeats (SSRs—Kameyama 2012; Mason 2015), variable number of tandem repeats (VNTR - Zhu et al. 2020), polymerase chain reaction (PCR) target region amplification polymorphism (Hu and Vick 2003), AFLP (Vos et al. 1995), restriction fragment length polymorphism (RFLP), arbitrary primed DNA (Welsh et al. 1992), randomly amplified polymorphic DNA (RAPD; Williams et al. 1990), and sequencing (Adhikari et al. 2017; Hong et al. 2023), there is a possibility to characterize organisms properly at the genetic level. Molecular markers, such as protein gel electrophoresis (isozymes), restriction site mapping, RAPD, AFLP, and PCR-AFLP, have been utilized to clarify the taxonomic status of the genus *Cinnamomum*. With the advancement of technology in molecular biology and biotechnology, there is now a wide array of new techniques to examine the diversity, such as high-throughput DNA sequencing (such as whole-genome sequencing and next-generation sequencing (NGS)) and single-nucleotide polymorphisms (SNPs). Different techniques, along with the above-mentioned molecular markers, sequencing of cpDNA (Chanderbali et al. 2001; Xu et al. 2023), mtDNA, mutations with transposons and mutagenic agents, SNPs, and NGS, have made significant development for the diversity and characterization of *Cinnamomum* species found in Sri Lanka as well as in other countries (Arif et al. 2010; Al-Yasiri and Al-Alwani 2022; Bandaranayake et al. 2023).

5.4.2 Molecular Analysis of Cinnamomum Species

Since the 1980s, with the development of technology, in conjunction with the morphological markers, diversity studies of plant populations have grown at an alarming rate. Allozymes, which are different molecular forms of the same enzyme coding from different alleles of the same gene locus analysis, have been analyzed for over 400 individuals of *C. insularimontanum* and have demonstrated that clear genetic variations exist among different age classes of this species (Chung et al. 2003). Moreover, out of 11 allozyme loci, 9 showed polymorphism in 164 accessions of *C. kanehirae* (Lin et al. 1997). PCR-based markers such as RAPD, SRAP, and inter-simple sequence repeat (ISSR) markers (Joy and Maridass 2008; Ho and Hung 2011; Abeysinghe et al. 2014, 2020; Bandaranayake and Pushpakumara 2020; Hathurusinghe et al. 2023; Liyanage et al. 2020) have been used for diversity studies and examined their relationship. Abeysinghe et al. (2014) have used RAPD and SRAP markers to evaluate the diversity and differentiate *Cinnamomum* species. PCR amplified products were categorized as genus-specific, species-specific, and intraspecific markers. Therefore, using these two techniques, genera and species could be differentiated, and it was possible to examine the diversity and high polymorphism among the species (Abeysinghe et al. 2014). Therefore, RAPD and SRAP techniques can be effectively used in genetic studies for Sri Lankan *Cinnamomum* species (Abeysinghe et al. 2014). Joy and Maridass (2008) also used RAPD markers to examine the diversity of four *Cinnamomum* species: *C. verum*, *C. citronella*, *C. camphora*, and *C. glaucense*. Thirteen RAPD markers were identified, and according to the results, it was possible to clarify the relationship among them. *C. citronella* was more closely related to *C. verum* than the other species (Joy and Maridass 2008).

Morphologically similar 15 accessions of *C. zeylanicum* from various locations in South India were used with 11 RAPD primers. Some primers showed higher levels of diversity among the accessions. Overall findings revealed significant genetic diversity despite their morphological similarities (Sandigawad and Patil 2011). Additionally, ISSR markers have been employed to analyze seedlings derived from the elite variety "Sri Wijaya" of *C. zeylanicum*, which was released by the Department of Export Agriculture in Sri Lanka (Liyanage et al. 2020). Genetic diversity of five *Cinnamomum* species (*C. burmannii, C. cassia, C. loureiroi, C. osmophloeum,* and *C. zeylanicum*) was carried out by employing sequencing the partial ITS2 of ribosomal DNA and showed that *C. burmannii* and *C. osmophloeum* were closely related and shared a more recent common ancestor with *C. cassia.* Unique sequence variation within the ITS2 regions permitted the effective identification of the individual specimens among the five *Cinnamomum* species studied (Lee et al. 2010).

5.4.3 Chloroplast Barcoding

Although a vast number of molecular tools are available, the choice of a proper molecular tool depends on the question being addressed. The sequence information of the chloroplast genome is beneficial in studying evolutionary relationships, diversity studies, and population structure determinations (Le et al. 2025). Compared to the mitochondrial genome, chloroplast genomes are highly conserved, remaining constant in terms of gene content and rearrangements, especially among closely related taxa. These properties, along with the preserved flanking regions, enabled us to design "universal PCR primers" applicable for diversity studies of highly diverse taxa and develop cpDNA probes for phylogenetic studies. A short, variable, and specific DNA sequence qualifies as a DNA barcode when it reflects the distribution of intraspecific and interspecific variations. They are separated by a DNA sequence distance known as the "DNA barcoding gap." Nowadays, DNA barcoding serves as a vital molecular tool for studying plant variability by enabling the regulation of specific genomic regions and the identification that can be routinely sequenced across various samples to distinguish between species, as well as genera and families. The size of the chloroplast genome is about 150 kb in different members of the family Lauraceae (Chen et al. 2017; Zheng et al. 2022; Zhu and Li 2023). The sequence information of the chloroplast genome is beneficial in studying evolutionary relationships, diversity studies, and population structure determinations (Xu et al. 2023; Le et al. 2025). With the advancement of the NGS technology, low cost for sequencing, software development related to sequence assembly, analysis of the chloroplast genome for phylogenetics, and diversity studies have been increased dramatically. Full-length sequencing of cpDNA has led to the identification of candidate regions (*matK*, *rbcL*, *rpoB*, and *rpoC*1) as well as the intergenic spacer region (*trnH-psbA*, *trnL-F*, *atpF-atpH*, and *psbK-psbI*) of the cpDNA for their effective use in distinguishing plant species in phylogenetic analysis in plant species has been discussed (Kuo et al. 2010; Letsiou et al. 2024; Yong et al. 2025; Li et al. 2025). Since they have restriction sites and regional polymorphism, these cpDNA regions are widely used to get data for diversity and phylogenetic studies. cpDNA sequence data are being widely used in population diversity studies (Qin et al. 2025; Kassem 2025).

Non-coding regions evolve much faster than coding regions because of selective constraints. However, both coding and non-coding regions of the chloroplast genome have been employed for a long time, not only in diversity studies, to elucidate the evolutionary and taxonomic relationships in plants in Sri Lanka (Abeysinghe et al. 2014; Chandrasekara et al. 2021; Bandusekara et al. 2025) but also in other countries (Pang et al. 2012; Mishra et al. 2016). Sequences of intrgenic spacer (IGS) and introns in plant chloroplast genomes are very effective in phylogenetic construction at higher taxonomic levels, inferring relationships among species. The *rbcL* and *matK* genes, along with the *trnH-psbA* intergenic region, are commonly employed in diversity and population structure studies and in DNA barcoding (Kress and Erickson 2007; Pang et al. 2012).

Some chloroplast DNA regions (*trnL* intron, *trnT-trnL*, *trnL-trnF*, and intergenic spacer sequences (*trnH-psbA*)), along with the ITS of nuclear rDNA, have been utilized to elucidate the relationship of *Cinnamomum* species found in Sri Lanka (Abeysinghe et al. 2009; Bandaranayake et al. 2023). To solve these problems, not only use the universal barcoding regions of chloroplast DNA (*trnL* intron) but also use the intergenic spacers (*trnT-trnL*, *trnL-trnF*, and *trnH-psbA*) along with the ITS of nuclear rDNA (Abeysinghe et al. 2009; Yang et al. 2022; Bandaranayake et al. 2023). However, there are challenges and limitations to using these candidate regions of cpDNA. Given the limitations and challenges of relying on a single locus, the combination of multiple loci has appeared as a promising approach for achieving accurate species discrimination (Mishra et al. 2016). Doh et al. (2017) utilized the DNA barcode of the 5.8S rDNA from the nuclear DNA of *C. cassia* to identify and evaluate the relationships among *Cinnamomum* species used in traditional medicines in Vietnam, China, Indonesia, Sri Lanka, and Japan. The use of cpDNA sequences for systematic studies of *Cinnamomum* species has been limited to a few markers, such as the *trn*L intron, *trn*L-*trn*F IGS, and *trn*T-*trn*L IGS (Abeysinghe et al. 2009; Bandaranayake et al. 2023). Although cpDNA is extensively used in phylogenetic studies, even non-coding sequences fail to provide phylogenetic information at the species level (Kojoma et al. 2002; Abeysinghe et al. 2009; Bandaranayake et al. 2023). The use of these cpDNA regions in resolving genetic relationships had problematic situations and was unable to solve the genetic relationships they have due to the similarity of these barcodes (Abeysinghe et al. 2009; Chandrasekara et al. 2021; Bandaranayake et al. 2023).

Initially, nrITS regions in fungi, which are not found in cpDNA, were focused on phylogenetics. However, similar functional DNA sequence regions have been searched in plants and found in the chloroplast genome. Slowly evolving rDNA genes and internal and external intergenic spacers are employed at higher taxonomic levels and lower taxonomic levels, respectively, for phylogenetic analyses in plants (Kanazawa et al. 1998). In phylogenetic studies, single-copy nuclear genes and cpDNA provide valuable information. To compensate for this limitation, as well as to obtain additional and independent estimates of the phylogeny of plant species, nuclear rDNA is frequently used in plant systematics (Lantz et al. 2002). *Camphora* Fabr. and *Cinnamomum* Schaeff. belong to the family Lauraceae, and their phylogenetic relationship has been a long-lasting controversy based on plastid data. However, using both plastid (loci *ycf*1, *ycf2*, *trnK* (UUU), *psbJ-psbL*, and *ccsA-ndhD*) and nrDNA data of 22 *Camphora* taxa, 15 *Cinnamomum* taxa, Xu et al. (2025), these barcodes were recognized as potential DNA barcodes for these taxa. Moreover, chromosome-level DNA sequence of some laurales has been completed (*C. camphora*—Chaw et al. 2019; Shen et al. 2021; Wang et al. 2022) along with the complete chloroplast genome of *C. longipetiolatum* (Zheng et al. 2020) and *C. cassia* (Song et al. 2019). The family Lauraceae has been reviewed and emphasized basic relationships within the family, considering the biogeography, taxonomy, and phylogenomic analyses with cpDNA analysis. Results provide robust support for high-level relationships between Lauraceae lineages, leading to the recognition of nine tribes (Li et al. 2025).

5.5 Conclusion

Like any other plant, *Cinnamomum* species also change their morphology with the changing environment. Therefore, the correct plant identification, taxonomic diversity studies, breeding processes, and conservation strategies are rather problematic using only the morphology of plants. Rather than morphological markers, such as anatomy, different molecular tools could be used depending on the problem to be analyzed. This chapter emphasizes the application of anatomical and DNA analysis techniques to identify the *Cinnamomum* species found in Sri Lanka, while also discussing relevant research findings related to Laurels. Traditionally, species identification has relied on morphological traits, but intraspecific diversity, exacerbated by environmental changes, poses additional challenges. Sri Lanka is home to eight *Cinnamomum* species; however, only *C. zeylanicum* is recognized as a multipurpose tree, while the others remain underutilized. Thus, proper identification and characterization of these species are critical for developing sustainable conservation and utilization strategies. Anatomical data provide valuable support for species identification and characterization. Notably, *C. zeylanicum* is distinguished from other *Cinnamomum* species by maintaining minute levels of the toxic compound coumarin, which is found in significantly higher levels in *Cassia* cinnamon. Consequently, further research into the environmental impacts on anatomy and the characterization of chemical and other ergastic substances in all *Cinnamomum* species is vital for their effective utilization in traditional and modern medicine. Beyond its use as a spice, extracts from the leaves and bark of *Cinnamomum* species are employed in the food industry to enhance the shelf life of decaying products through antimicrobial properties and manage pests (insecticidal attacks) during postharvest storage. The potential therapeutic effects that may help prevent various illnesses and disorders. Given the vast possible applications of *C. zeylanicum* in global food, well-being, and cosmetics markets, exploring other species for their utility could prove beneficial. Regarding the molecular characterization of *Cinnamomum* species, it has been noted that DNA barcoding of standard regions has proven inadequate for resolving the closely related phylogenies among these species. Variations in standard barcoding regions such as *rbcL, matK*, and *trnH-psbA*, neither individually nor in combination, are sufficient for distinguishing the *Cinnamomum* species found in Sri Lanka. Therefore, complete genome sequencing is essential to gain a clearer understanding of the diversity patterns within these species. Additionally, transcriptomic studies are crucial for analyzing gene expression, encompassing various genes and the pathways of their gene products in *Cinnamomum* species. These insights could facilitate the enhancement of desirable traits and yields through targeted breeding programs.

References

Abeysinghe PD (2023) LM, micromorphology, SEM of leaf and molecular characterization of the genus *Cinnamomum* (family Lauraceae) found in Sri Lanka: taxonomic significance. In: Saeed OA (ed) Cutting edge research in biology, vol 9. BP International. https://doi.org/10.9734/bpi/cerb/v9/6104A

Abeysinghe PD (2024) Leaf micromorphology and anatomical traits of leaves as potential taxonomic markers for infrageneric classification of *Cinnamomum* (Lauraceae). Cey J Sci 53(3):413–425. https://doi.org/10.4038/cjs.v53i3.8235

Abeysinghe PD, Scharaschkin T (2019) Taxonomic value of the petiole anatomy in the genus *Cinnamomum* (Lauraceae) found in Sri Lanka. Ruh J Sci 10(1):1–17. https://doi.org/10.4038/rjs.v10i1.47

Abeysinghe PD, Wijesinghe KGG, Tachida H, Yoshida T (2009) Molecular characterization of Cinnamon *(Cinnamomum Verum* Presl) accessions and evaluation of genetic relatedness of *Cinnamon* species in Sri Lanka based on *Trn*L intron region, intergenic spacers between *trn*T--*trn*L, *trn*L-*trn*F, *trn*H –*psb*A and nuclear ITS. Res J Agric Biol Sci 5(6):1079–1088

Abeysinghe PD, Samarajeewa NGCD, Li G, Wijesinghe KGG (2014) Preliminary investigation for the identification of Sri Lankan *Cinnamomum* species using randomly amplified polymorphic DNA (RAPD) and sequence-related amplified polymorphic (SRAP) markers. J Natn Sci Founda Sri Lanka 42(3):175–182. https://doi.org/10.4038/jnsfsr.v42i3.7393

Abeysinghe PD, Bandaranayake CGP, Pathirana R (2020) Botany of endemic *Cinnamomum* species of Sri Lanka. In: Senarathna R, Pathirana R (eds) Cinnamon botany, agronomy, chemistry and industrial applications. Springer Nature, Switzerland, pp 85–118. https://link.springer.com/chapter/10.1007/978-3-030-54426-3_4

Abeywikrama (1973) In: Fosberg FR, Clayton WD (eds) A revised handbook to the flora of Ceylon. Amerind Publishing Co. Pvt Ltd, New Delhi, p 105

Adhikari S, Saha S, Biswas A, Rana TS, Bandyopadhyay TK, Ghosh P (2017) Application of molecular markers in plant genome analysis. Nucleus. https://doi.org/10.1007/s13237-017-0214-7

Alves FM, Assis LCS, Quinet A, Batista BJ (2025) A key to the neotropical genera of Lauraceae. Ann Miss Bot Gard 110(1):1–23. https://doi.org/10.3417/2025938

Al-Yasiri Z, Al-Alwani B (2022) Estimation of botanical diversity by molecular marker methods. Biomed Chem Sci 1(3):132–137. https://doi.org/10.48112/bcs.v1i3.200

Andianto IW, Totok KW, Rudi D, Hadiyane A, Fikri MH (2015) Wood anatomy from indonesian genus *Cinnamomum* (Lauraceae) and their identification key. Asian J Plan Sci 14(1):11–19. https://doi.org/10.3923/ajps.2015.11.19

Anu S, Dan M (2020) Taxonomic significance on comparative petiole anatomy of twelve species of *Curcuma* L. (Zingiberaceae) from South India. Plant Archiv 20(1):35–41

Anurag K, Jangid PP, Marimuthu S, Gurav AM, Srikanth N, Mangal AK, Venkateshwarlu B, Shiddamallayya N (2023) Identification and authentication of *Agnimantha* plant species used in Ayurveda on the basis of anatomical and molecular phylogenetic analysis. Plant Sci Today 10(4):26–38. https://doi.org/10.14719/pst.2180

Arif IA, Bakir MA, Khan HA, Al Farhan AH, Al Homaidan AA, Bahkali AH, Sadoon MA, Shobrak M (2010) A brief review of molecular techniques to assess plant diversity. Int J Mol Sci 11(5):2079–2096. https://doi.org/10.3390/ijms11052079

Ariyarathne HBMA, Weerasuriya SN, Senarath WTPSK (2018) Comparison of morphological and chemical characteristics of two selected accessions and six wild species of genus *Cinnamomum* Schaeff. Sri Lankan J Biol 3(1):11–2311. https://doi.org/10.4038/sljb.v3i1.14

Azad R, Senanayake G, Kumara KLW, Ranawaka RAAK, Pushpakumara DKNG, Wijesinghe KGG, Geekiyanage S (2015) Morphological variation within progeny and deviations from mother plant reveal the allele richness in *Cinnamomum verum* germ-plasm: a case study from Deiyandara, Matara collection at the early vegetative stage. Trop Agric Res Exte 18(4):163–167. https://doi.org/10.4038/tare.v18i4.5380

Azad R, Kumara KLW, Senanayake G, Ranawaka RAAK, Pushpakumara DKNG, Geekiyanage S (2018) Flower morphological diversity of cinnamon *(Cinnamomum verum* Presl) in Matara District, Sri Lanka. De Gruyter 3:236–244. https://doi.org/10.1515/opag-2018-0025

Azad R, Jayaprada NVT, Ranaweera SA, Ranawaka RAAK, Jayasekara L, Senanayake G, Hirotoshi T, Geekiyanage S (2024) Diversity and morpho-chemo correlations of Cinnamon (*Cinnamomum verum* J. Presl) bark oil from Sri Lanka. J Agric Food Res 18:101500. https://doi.org/10.1016/j.jafr.2024.101500

Bakker ME, Baas P (1993) Cell walls in oil and mucilage cells. Acta Bot Neerl 42(2):133–139. https://doi.org/10.1111/j.1438-8677.1993.tb00690.x

Bakker ME, Gerritsen AF, van der Schaaf PJ (1992) Leaf anatomy of *Cinnamomum schaeffer* (Lauraceae) with special reference to oil and mucilage cells. Blumea 37:1–30

Bandaranayake PCG, Pushpakumara DKNG (2020) Genetics and molecular characterization of genus *Cinnamomum*. Botany of endemic Cinnamomum species of Sri Lanka. In: Senarathna R, Pathirana R (eds) Cinnamon botany, agronomy, chemistry and industrial applications. Springer Nature, Switzerland, pp 119–140

Bandaranayake PCG, Naranpanawa N, Chandrasekara CHWMRB, Samarakoon H, Lokuge S, Jayasundara S, Bandaranayake AU, Pushpakumara DK, Wijesundara DS (2023) Chloroplast genome, nuclear ITS regions, mitogenome regions, and Skmer analysis resolved the genetic relationship among *Cinnamomum* species in Sri Lanka. PLoS One 18(9):e0291763. https://doi.org/10.1371/journal.pone.0291763

Bandusekara BS, Pushpakumara DKNG, Bandaranayake PCG, Wijesinghe KGG, Jayasinghe GG (2020) Field-level identification of *Cinnamomum* species in Sri Lanka using a morphological index. Tropic Agric Res 31(4):43–53

Bandusekara BS, Pushpakumara DK, Bandaranayake PC, Ranil RH, Wimalasiri KM, Ranawaka RA (2025) Intraspecies diversity of the bioactive compounds of wild and cultivated *Cinnamomum* species in Sri Lanka. BMC Agric 1(1):5. https://doi.org/10.1186/s44399-025-00004-y

Baruah A, Nath SC (2006) Leaf anatomy and essential oil characters of *Cinnamomum pauciflorum* Nees - a potential spice crop from the North-East. Indian J Spices Aromatic Crops 15(1):52–56

Bavappa KVA, Gunaratne LA, Reuttiman RA (1981) Cinnamon cultivation and processing. Department of Minor Export Crops. Ministry of Agricultural Development and Research, Colombo

Biggs MA, Banerjee IA (2025) Targeting breast and gynecologic cancers: the role of natural products with emphasis on cinnamon and its derivatives-advances in nanoscale therapeutics and adjuvant strategies. Macromol 5(1):13. https://doi.org/10.3390/macromol5010013

Binnoubah A, Hamdy R, Raga OG, El-Taher AM, El-Yazied AA, Safhi FA, Elzilal HA, Althobaiti AT, Alshamrani SM, El-Moneim DA, El-Banhawy A (2023) Anatomical and molecular identification of ornamental plant *Ficus* L. Species. Phyton Int J Experim Bot 92(5):1329–1347. https://doi.org/10.32604/phyton.2023.026888

Broadhurst CL, Polansky MM, Anderson RA (2000) Insulin-like biological activity of culinary and medicinal plant aqueous extracts in vitro. J Agric Food Chem 48(3):84952. https://doi.org/10.1021/jf9904517

Callado CH, Costa CG (1997) Wood anatomy of some *Anaueria* and *Beilschmiedia* species (Lauraceae). IAWA J 18(3):247–259

Chanderbali AS, van der Werff H, Renner SS (2001) Phylogeny and historical biogeography of Lauraceae: evidence from the chloroplast and nuclear genomes. Ann Mo Bot Gard 88:104–134. https://doi.org/10.2307/2666133

Chandrasekara RB, Naranpanawa DNU, Bandusekara BS, Pushpakumara G, Wijesundera SA, Bandaranayake PCG (2021) Universal barcoding regions, rbcL, matK and trnH-psbA do not discriminate *Cinnamomum* species in Sri Lanka. PLoS One 16(2):e0245592. https://doi.org/10.1371/journal.pone.0245592

Chaw SM, Liu YC, Wu YW, Wang HY, Lin CY, Wu CS, Ke HM, Chang LY, Hsu CY, Yang HT, Sudianto E (2019) Stout camphor tree genome fills gaps in understanding of flowering plant genome evolution. Nat Plants 5:63–73. https://doi.org/10.1038/s41477-018-0337-0

Chen C, Zheng Y, Liu S, Zhong Y, Wu Y, Li J, Xu LA, Xu M (2017) The complete chloroplast genome of *Cinnamomum camphora* and its comparison with related Lauraceae species. PeerJ 18(5):e3820. https://doi.org/10.7717/peerj.3820

Christophel DC, Rowett AI (1996) Leaf and cuticle atlas of australian leafy Lauraceae. Australian Biological Resources Study, Canberra

Chung M, Nason J, Epperson B, Chung M (2003) Temporal aspects of the fine-scale genetic structure in a population of *Cinnamomum insularimontanum* (Lauraceae). Heredity 90:98. https://doi.org/10.1038/sj.hdy.6800187

Coutinho DF, Agra MF, Basflio IJLD, Barbosa-Filho JM (2006) Morphoanatomical study of the leaves of Ocotea duckei Vattimo (Lauraceae-Lauroideae). Brazi J Pharmaco 16(4):537–544. https://doi.org/10.1590/S0102-695X2006000400017

Dao K, Chen J, Jin P, Dong C, Yang Y, Xu X, Wu J, Xie S, Lin Z, Sun B (2013) A new material of Lindera (Lauraceae) of the Late Pliocene from Tengchong, Yunnan and the genus' biogeography significance. Acta Geol Sin 87(3):690–706. https://doi.org/10.1111/1755-6724.12081

Didehdar M, Chegini Z, Tabaeian SP, Razavi S, Shariati A (2022) *Cinnamomum*: the new therapeutic agents for inhibition of bacterial and fungal biofilm-associated infection. Front Cell Infect Microbiol 12:930624. https://doi.org/10.3389/fcimb.2022.930624

Doh EJ, Kim J-H, Se O, Lee G (2017) Identification and monitoring of Korean medicines derived from *Cinnamomum* spp. by using ITS and DNA marker. Genes Genom 39(1):101–109. https://doi.org/10.1007/s13258-016-0476-51

Elumalai S, Kesavan R, Ramganesh S, Prakasam V, Murugasen R (2010) Comparative study on anti-microbial activities of bark oil extract from *Cinnamomum cassia* and *Cinnamomum zeylanicum*. Biosci Biotechnol Res Asia 7(1):251–258

Gang Z, Liu B, Rohwer JG, Ferguson DK, Yang Y (2021) Leaf epidermal micromorphology defining the clades in *Cinnamomum* (Lauraceae). PhytoKeys 182:125–148. https://doi.org/10.3897/phytokeys.182.67289

Geekiyanage S, Azad R, Ranawaka R, Maddumage RP (2024) Cinnamon: botany, cultivars, and genetic diversity. Cinnamon:23–49. https://doi.org/10.1016/B978-0-443-21820-0.00005-2

Gonawala L, Madhumaali M, Ismail H, Jayasooriya N, Wijekoon N, Rajapakshe S, Erangika H, Amaratunga D, Gunaratna R, Steinbusch WM, Mohan C (2024) Phytochemistry and nutraceutical potential of Ceylon *Cinnamomum* species native to Sri Lanka. Nat Prod Res 3:1–2. https://doi.org/10.1080/14786419.2024.2438269

Gonçalves RDA, Pinheiro AB, Oliveira MA, Do Nascimento RT, Rosalem PF, Garcia VL, Martins AR (2018) Anatomical characters and chemical profile of leaves of three species in Lauraceae family. Rev Bras 28:1–8. https://doi.org/10.1016/j.bjp.2017.11.008

Guo J, Jiang X, Tian Y, Yan S, Liu J, Xie J, Zhang F, Yao C, Hao E (2024) Therapeutic potential of Cinnamon oil: chemical composition, pharmacological actions, and applications. Pharmaceuticals (Basel). https://doi.org/10.3390/ph17121700

Hathurusinghe HABM, Bandusekara BS, Pushpakumar DKNG, Ranawaka RAAK, Bandaranayake PCG (2023) Possibility of utilizing Inter Simple Sequence Repeat regions, bark powder morphology and floral morphometry to characterize the *Cinnamomum* species in Sri Lanka. Trop Agric Res 34(1):65–79. https://doi.org/10.4038/tar.v34i1.8605

Ho K, Hung T (2011) Cladistic relationships within the genus *Cinnamomum* (Lauraceae) in Taiwan based on analysis of leaf morphology and inter-simple sequence repeat (ISSR) and internal transcribed spacer (ITS) molecular markers. Afr J Biotechnol 10(24):4802–4815. https://doi.org/10.5897/AJB10.1852

Hong K, Radian Y, Manda T, Xu H, Luo Y (2023) The development of plant genome sequencing technology and its conservation and application in endangered gymnosperms. Plants 12(23):4006. https://doi.org/10.3390/plants12234006

Hu J, Vick BA (2003) Target region amplification polymorphism: a novel marker technique for plant genotyping. Plant Mol Biol Report 21:289–294. https://doi.org/10.1007/BF02772804

Huang X, Jiang F, Torrico DD (2025) Essential oils from *Cinnamomum longepaniculatum*: a review of their phytochemicals, pharmacological actions, applications, and technology. Nat Prod Commun 20(8):1–13

Hyland BPM (1986) A revision of Lauraceae in Australia (excluding Cassytha). PhD thesis, James Cook University of North Queensland

Ingole SN, Kaikade RS (2015) Study of petiolar anatomy of some medicinal plants mentioned in the Atharvaveda. Int J Res Stud Biosci 3(3):103–106

Jaafarpour M, Hatefi M, Khani A, Khajavikhan J (2015) Comparative effect of cinnamon and Ibuprofen for treatment of primary dysmenorrhea: a randomized double-blind clinical trial. J Clin Diagn Res 9(4):QC04–7. https://doi.org/10.7860/JCDR/2015/12084.5783

John LJ, Shantakumari N (2015) Herbal medicines use during pregnancy: a review from the Middle East. Oman Med J 30:229–236. https://doi.org/10.5001/omj.2015.48

Joy P, Maridass M (2008) Inter-species relationship of *Cinnamomum* species using RAPD marker analysis. Ethnobot Leaf 12:476–480

Kamel EA, Loufty MHA (2001) The significance of cuticular features, petiole anatomy and SDS-PAGE in the taxonomy of the Lauraceae. Pak J Biol Sci 4(9):1094–1100. https://doi.org/10.3923/pjbs.2001.1094.1100

Kameyama Y (2012) Development of microsatellite markers for *Cinnamomum camphora* (Lauraceae). Am J Bot:e1–e3. https://doi.org/10.3732/ajb.1100231

Kanazawa A, Tozuka A, Shimamoto Y (1998) Sequence variation of chloroplast DNA that involves EcoRI and ClaI restriction site polymorphisms in soybean. Genes Genet Syst 73(2):111–119. https://doi.org/10.1266/ggs.73.111

Kassem MA (2025) Comparative analysis of chloroplast genomes across 20 plant species reveals evolutionary patterns in gene content, codon usage, and genome structure. Int J Plant Biol 16:105. https://doi.org/10.3390/ijpb16030105

Khan A, Safdar M, Muzaffar M, Khan A, Khan NK, Anderson RA (2003) Cinnamon improves glucose and lipids of people with type 2 diabetes. Diabetes Care 26:3215–3218. https://doi.org/10.2337/diacare.26.12.3215

Kocsis M, Borhidi A (2003) Petiole anatomy of some Rubiaceae genera. Acta Bot Hungar 45:345–353. https://doi.org/10.1556/ABot.45.2003.3-4.9

Kojoma MK, Kurihara K, Yamada S, Sekita M, Lida SO (2002) Genetic identification of *Cinnamon* spp. based on the trnL-trnF chloroplast DNA. Planta Med 68:94–96. https://doi.org/10.1055/s-2002-20051

Korir NK, Han J, Shangguan L, Fang J (2012) Plant variety and cultivar identification: advances and prospects. Crit Rev Biotechnol 33(2). https://doi.org/10.3109/07388551.2012.675314

Kress WJ, Erickson DL (2007) A two-locus global DNA barcode for land plants: the coding rbcl gene complements the non-coding trnH-psbA spacer region. PLoS One 2(6):e508. https://doi.org/10.1371/journal.pone.0000508

Kumar N, Ruba K (2018) Identification of adulterants by pharmacognostical evaluation in selected medicinal plants. World J Pharmace Med Res 4(2):67–70

Kumarathilake DMHC (2012) Study on the extinction risks, conservation and domestication of endemic wild *Cinnamomum* species in Sri Lanka. Unpublished M.Phil, Thesis, Faculty of Agriculture, University of Ruhuna, Sri Lanka

Kuo DC, Lin CC, Ho KC, Cheng YP, Hwang SY, Lin TP (2010) Two genetic divergence centers revealed by chloroplastic DNA variation in populations of *Cinnamomum kanehirae* Hay. Conserv Genet 11:803–812. https://doi.org/10.1007/s10592-009-9901-5

Lantz H, Andreasen K, Bremer B (2002) Nuclear rDNA ITS sequence data used to construct the first phylogeny of Vanguerieae (Rubiaceae). Plant Syst Evol 230:173–187. https://doi.org/10.1007/s006060200003

Le S, Nguyen HN, Quang MT (2025) The complete chloroplast genome sequence of *Cinnamomum balansae* and its phylogenetic implications. Korean J Plant Taxo 55(2):70–75. https://doi.org/10.11110/kjpt.2025.55.2.70

Lee J, Lee C, Lin M, Ho K (2010) Genetic identification of *Cinnamomum* species based on partial internal transcribed spacer 2 of ribosomal DNA. J Food Drug Anal 18(4):225–231. https://doi.org/10.38212/2224-6614.2232

Letsiou S, Madesis P, Vasdekis E, Montemurro C, Grigoriou ME, Skavdis G, Moussis V, Koutelidakis AE, Tzakos AG (2024) DNA barcoding as a plant identification method. Appl Sci 14(4):1415. https://doi.org/10.3390/app140

Li J, Christophel DC (2000) Systematic relationships within the Litsea complex (Lauraceae): a cladistic analysis based on morphological and leaf cuticle data. Aust Syst Bot 13:1–13. https://doi.org/10.1071/SB98015

Li L, Liu B, Song Y et al (2025) Global advances in phylogeny, taxonomy and biogeography of Lauraceae. Plant Divers 47:341–364. https://doi.org/10.1016/j.pld.2025.04.001

Lin T-P, Cheng Y-P, Huang S-G (1997) Allozyme variation in four geographic areas of *Cinnamomum kanehirae*. J Hered 88:433–438. https://doi.org/10.1093/oxfordjournals.jhered.a023131

Liu X, Zhao G, Tian W, Liu Y, Sun L, Liu Y (2013) Effects of cinnamon additions on volatile flavor compounds of stewed chicken. Food Ferm Indus 39(06):34–40. https://doi.org/10.13995/j.cnki.11-1802/ts.2013.06.042

Liyanage ASU, Senanayake G (2010) The atlas of selected crop wild relatives in Sri Lanka. Department of Agriculture, Sri Lanka, Colombo, pp 11–73

Liyanage NMN, Ranawake AL, Bandaranayake PCG (2020) Cross-pollination effects on morphological, molecular, and biochemical diversity of a selected cinnamon (*Cinnamomum zeylanicum* Blume) seedling population. J Crop Improve 35(1):21–37. https://doi.org/10.1080/15427528.2020.1795769

Long C, Oskolski A (2018) Wood and bark anatomy of Andriana (Heteromorpheae, Apiaceae) with phylogenetic implications. S Afr J Bot 115:138–142. https://doi.org/10.1016/j.sajb.2018.02.003

Loutfy MHA (2001) On the taxonomy of *Apollonias* Nees (Lauraceae). Evidence from general aspects and cuticular features of leaf and fruit. Taeckholmia 21(1):59–80. https://doi.org/10.21608/TAEC.2001.12455

Loutfy MHA (2009) Wood anatomy and its implications on the taxonomy of ApolloniasNEES (Lauraceae). Feddes Repertorium 120(1–2):75–90. https://doi.org/10.1002/fedr.200911195

Lungarini S, Aureli F, Coni E (2008) Coumarin and cinnamaldehyde in cinnamon marketed in Italy: a natural chemical hazard? Food Additi Contamin 11:297–1305. https://doi.org/10.1080/02652030802105274

Mason AS (2015) SSR genotyping. Methods Mol Biol. 1245:77-89. https://doi.org/10.1007/978-1-4939-1966-6_6. PMID:25373750.

Mishra P, Kumar A, Rodrigues V, Shukla AK, Sundaresan V (2016) Feasibility of nuclear ribosomal region ITS1 over ITS2 in barcoding taxonomically challenging genera of subtribe Cassiinae (Fabaceae). PeerJ 4:e2638. https://doi.org/10.7717/peerj.2638

Mohtashamian M, Hajishirkia'ee FA, Fatehi F, Rastegar A (2022) Petiole anatomy: a contribution to the taxonomy of *Acer* L. (Sapindaceae) in Iran. Nordic J Bot:e03417. https://doi.org/10.1111/njb.03417

Nishida S, van der Werff H (2007) Are cuticular characters useful in solving generic relationships of problematic species of Lauraceae? Taxon 56(4):1229–1237. https://doi.org/10.2307/25065914

Nishida S, de Kok R, Yang Y (2016) Cuticular features of *Cryptocarya* (Lauraceae) from Peninsular Malaysia, Thailand and Indo-China and its taxonomic implications. Phytotaxa 244(1):26–44. https://doi.org/10.11646/phytotaxa.244.1.2

Noraini T, Ruzi AR, Ismail BS, Ummu BH, Salwa S, Azi JA (2016) Petiole vascular bundles and its taxonomic value in the tribe Dipterocarpeae (Dipterocarpaceae). Sains Malaysiana 45(2):247–253

Pagliari S, Forcella M, Lonati E, Sacco G, Romaniello F, Rovellini P, Fusi P, Palestini P, Campone L, Labra M, Bulbarelli A, Bruni I (2023) Antioxidant and anti-inflammatory effect of Cinnamon (*Cinnamomum verum* J. Presl) bark extract after in vitro digestion simulation. Foods 12(3):452. https://doi.org/10.3390/foods12030452

Pang X, Liu C, Shi L, Liu R, Liang D, Li H, Cherny SS, Chen S (2012) Utility of the trnH–psbA intergenic spacer region and its combinations as plant DNA barcodes: a meta-analysis. PLoS One 7(11):e48833. https://doi.org/10.1371/journal.pone.0048833

Park BK, Son DC, Ghimire B (2021) Taxonomic relevance of petiole anatomical and micromorphological characteristics of *Clematis* L. (Ranunculaceae) taxa from South Korea. PeerJ 9:e11669. https://doi.org/10.7717/peerj.11669

Patel RN (1987) Wood anatomy of the dicotyledons indigenous to New Zealand 16. Lauraceae. NZ J Bot 25:477–488. https://doi.org/10.1080/0028825X.1987.10410080

Pathirana R, Senaratne R (2020) An introduction to Sri Lanka and its Cinnamon industry. In: Seraratne R, Pathirana R (eds) Cinnamon: botany, agronomy, chemistry and industrial applications. Springer Nature, Switzerland, pp 1–15. https://doi.org/10.1007/978-3-030-54426-3

Purseglove JW (1977) Tropical crops, Dicotyledons. Longmans Group Ltd, London, p 192

Qin Q, Dong Y, Chen J, Wang B, Peng Y, Zhang X, Wang X, Zeng J, Zhong G, Zhang S, Du X (2025) Comparative analysis of chloroplast genomes reveals molecular evolution and phylogenetic relationships within the Papilionoideae of Fabaceae. BMC Plant Biol 25(1):1–20. https://doi.org/10.1186/s12870-025-06138-0

Ranasinghe P, Pigera S, Premakumara GS, Galappaththy P, Constantine GR, Katulanda P (2013) Medicinal properties of "true" cinnamon (*Cinnamomum zeylanicum*): a systematic review. BMC Complement Altern Med 13(1):275–285. https://doi.org/10.1186/1472-6882-13-275

Ro N, Sung P, Haile M, Yoon H, Yu DS, Ko HC, Cho GT, Woo HJ, Chung NJ (2025) A study on the infrageneric classification of *Hordeum* using multiple methods: based on morphological data. Agronomy 15(1):60. https://doi.org/10.3390/agronomy15010060

Sandigawad A, Patil C (2011) Genetic diversity in *Cinnamomum zeylanicum* Blume. (Lauraceae) using random amplified polymorphic DNA (RAPD) markers. Afr J Biotechnol 10:3682–3688

Senaratna LK (2001) A checklist of the flowering plants of Sri Lanka. Published by the National Science Foundation, Sri Lanka, p 197

Senevirathne BS, Jayasinghe MA, Pavalakumar D, Siriwardhana CG (2022) Ceylon cinnamon: a versatile ingredient for futuristic diabetes management. J Future Foods 2(2):125–142. https://doi.org/10.1016/j.jfutfo.2022.03.010

Shan B, Yi-Zhong C, Brooks JD, Corke H (2007) Antibacterial properties and major bioactive components of cinnamon stick (*Cinnamomum burmannii*): activity against foodborne pathogenic bacteria. J Agric Food Chem 55(14):5484–5490. https://doi.org/10.1021/jf070424d

Shen T, Qi H, Luan X, Xu W, Yu F, Zhong Y, Xu M (2021) The chromosome-level genome sequence of the camphor tree provides insights into Lauraceae evolution and terpene biosynthesis. Plant Biotechnol J 20(2):244–246. https://doi.org/10.1111/pbi.13749

Shi G, Xie Z, Li H (2014) High diversity of Lauraceae from the Oligocene of Ningming, South China. Palaeoworld 23:336–356. https://doi.org/10.1016/j.palwor.2014.08.001

Shylaja M, Manilal KS (1992) Bark anatomy of four species of *Cinnamomum* (Lauraceae) from Kerala. J Spices Aromat Crops 1:84–87

Singh G, Maurya S, Delampasona MP, Catalan CAN (2007) A comparison of chemical, antioxidant and antimicrobial studies of cinnamon leaf and bark volatile oils, oleoresins and their constituents. Food Chem Toxicol 45:1650–1661. https://doi.org/10.1016/j.fct.2007.02.031

Song K, He M, Yu J, Guan Y, Bai Y, Xin S, Cao T (2019) Characterization of the chloroplast genome of the family Lauraceae plant species, *Cinnamomum cassia*. Mitochondrial DNA B Resour 4(2):3906–3907. https://doi.org/10.1080/23802359.2019.1687360

Spence C (2024) Cinnamon: the historic spice, medicinal uses, and flavour chemistry. Int J Gastro Food Sci 35:100858. https://doi.org/10.1016/j.ijgfs.2023.100858

Sun J, Wu J, Wang X, Gu J, Gao Z (2015) Comparative wood anatomy of 56 species of Lauraceae from Yunnan. Braz J Bot China 38(3):645–656. https://doi.org/10.1007/s40415-015-0172-8

Swetha VP, Parvathy VA, Sheeja TE, Sasikumar B (2014) DNA barcoding for discriminating the economically important *Cinnamomum verum* from its adulterants. Food Biotechnol 28(3):183–194. https://doi.org/10.1080/08905436.2014.931239

Talip N, Cutler DF, Puad ASA, Ismail BS, Ruzi AR, Juhari AAA (2017) Diagnostic and systematic significance of petiole anatomy in the identification of *Hopea* species (Dipterocarpaceae). S Afr J Bot 111:111–125. https://doi.org/10.1016/j.sajb.2017.03.008

Tang B, Han M, Xu Q, Jin J (2016) Leaf cuticle microstructure of *Machilusmaomingensis* sp. nov. (Lauraceae) from the Eocene of the Maoming Basin, South China. Acta GeologicaSinic 90(5):1561–1571. https://doi.org/10.1111/1755-6724.12802

Tao GF (2001) *Cinnamomum* (Lauraceae) resources in Hubei Province, China. J Wuhan Bot Res 19(6):489–496

Thacker KD, Gavade SK, Lekhak MM, Gondaliya AD, Rajput KS (2021) Comparison of petiole anatomy in *Flemingia* and its potential for delimitation of species. Flora 278:151790. https://doi.org/10.1016/j.flora.2021.151790

The National Red List 2020 (2020) Conservation status of the Flora of Sri Lanka: biodiversity secretariat. Ministry of Environment and the National Herbarium, Department of National Botanical Gardens, pp 79–80

Vos P, Hogers R, Bleeker M, Reijans M, van de Lee T, Hornes M, Frijters A, Pot J, Peleman J, Kuiper M, Zabeau M (1995) AFLP: a new technique for DNA fingerprinting. Nucleic Acids Res 23(21):4407–4414. https://doi.org/10.1093/nar/23.21.4407

Wang L, Chen Q, Zhong Z, Yi Y, Hou W, Wang H (2022) Research on variation of volatile compounds of *Cinnamomum cassia* Presl in different processing stages of stewed beef. Food Sci Technol. https://doi.org/10.1590/fst.0932

Weber JZ (1981) A taxonomic revision of *Cassytha* (Lauraceae) in Australia. J Adelaide Bot Gard 3(3):187–262

Welsh J, Chada K, Dalal SS, Cheng R, Ralph D, McClelland M (1992) Arbitrarily primed PCR fingerprinting of RNA. Nucleic Acids Res 20(19):4965–4970. https://doi.org/10.1093/nar/20.19.4965

Wijesekara ROB, Jayawardena AL, Rajapakse LS (1974) Volatile constituents of leaf, stem and root oils of cinnamon (*Cinnamomum zeylanicum*). J Sci Food Agricul 25:1211–1220

Williams JG, Kubelik AR, Livak KJ, Rafalski JA, Tingey SV (1990) DNA polymorphisms amplified by arbitrary primers are useful as genetic markers. Nucleic Acids Res 18(22):6531–6535. https://doi.org/10.1093/nar/18.22.6531

Wu F, Chen Y, Gao M, Li W, Zhao Y, Wang Y (2025) An updated review on essential oils from Lauraceae plants: chemical composition and genetic characteristics of biosynthesis. Int J Mol Sci 26(12):5690. https://doi.org/10.3390/ijms26125690

Xie C, Guo J, Yan J, Jiang R, Liang A, Che A (2022) The influence of plant morphological structure characteristics on PM2.5 retention of leaves under different wind speeds. Urban For Urban Green 71. https://doi.org/10.1016/j.ufug.2022.127556

Xu Y, Liu Y, Yu Z, Jia X (2023) Complete chloroplast genome sequence of the long blooming cultivar Camellia 'Xiari Qixin': genome features, comparative and phylogenetic analysis. Genes 14(2):460. https://doi.org/10.3390/genes14020460

Xu J, Zhang H, Yang F, Zhu W, Li Q, Cao Z, Song Y, Xin P (2025) Phylogeny of *Camphora* and *Cinnamomum* (Lauraceae) based on plastome and nuclear ribosomal dna data. Int J Mol Sci 26(3):1370. https://doi.org/10.3390/ijms26031370

Yanakiev S (2020) Effects of Cinnamon (*Cinnamomum* spp.) in dentistry: a review. Molecules 25(18):4184. https://doi.org/10.3390/molecules25184184

Yang Y, Zhang L, Liu B, van der Werff H (2012) Leaf cuticular anatomy and taxonomy of *Syndiclis* (Lauraceae) and its allies. Syst Botany 37(4):861–878. https://www.jstor.org/stable/23362704

Yang Y, Zhang L, Liu B, Yang Y, Ferguson D (2022) Phylogeny and taxonomy of *Cinnamomum* (Lauraceae). Ecol Evol 12(10):e9378. https://doi.org/10.1002/ece3.9378

Yang X, Xue T, Gao T, Liang Y, Smets EF, Gadagkar SR, Yu S (2024) An updated infrageneric classification based on phylogenomics and character evolution in Hydrangea (Hydrangeaceae). Taxon 73(2):503–518. https://doi.org/10.1002/tax.13158

Yong Q, Li M, Li Z, Luo C, Zhang J, Bai X (2025) Complete chloroplast genomes of 13 species of the *Impatiens* genus for genomic features and phylogenetic relationships studies. Sci Rep 15(1):4258. https://doi.org/10.1038/s41598-025-87254-7

Zheng Y, Luo Y, Li Y, Wang Y (2020) The complete chloroplast genome sequence of *Cinnamomum longipetiolatum*. Mitochondrial DNA B Resour 5:198–199. https://doi.org/10.1080/23802359.2019.1698994

Zheng Y, Chen Y, Wu Y, Liu X, Wang Y (2022) The chloroplast genome of aromatic plants *Cinnamomum pauciflorum* (Lauraceae). Mitochondrial DNA B 7(4):585–586. https://doi.org/10.1080/23802359.2022.2057245

Zhu H, Li H (2023) Comprehensive analysis of the complete chloroplast genome of *Cinnamomum daphnoides* (Lauraceae), an endangered Island endemic plant. Mol Biotechnol 66(12):3514–3525. https://doi.org/10.1007/s12033-023-00950-5

Zhu L, Wu H, Li H, Tang H, Zhang L, Xu JF, Wang N, Yang L (2020) Short Tandem Repeats in plants: genomic distribution and function prediction. Electron J Biotechnol. https://doi.org/10.1016/j.ejbt.2020.12.003

Chapter 6
Assessing the Richness and Conservation Needs of Sri Lankan Endemic Medicinal Plants

O. W. Ketakela, R. J. Piumal, A. M. K. L. Abeykoon, D. M. R. G. Mayakaduwa, H. S. Kathriarachchi, S. K. M. K. Herapathdeniya, and H. D. Jayasinghe

Abstract Sri Lanka, together with the Western Ghats of India, constitutes one of the world's biodiversity hotspots, renowned for its extraordinary floristic diversity and high levels of endemism. Among its biological wealth, medicinal plants hold exceptional ethnobotanical, cultural, and therapeutic significance. A considerable proportion of these species are endemic, highlighting the ecological uniqueness and global importance. However, this invaluable resource base is increasingly challenged by habitat degradation, overharvesting, invasive species, climate change, and socio-economic pressures. This chapter provides a comprehensive assessment of the taxonomic richness, distribution patterns, and ethnobotanical relevance of Sri Lanka's endemic medicinal flora, while critically examining the threats jeopardizing their persistence. Strategies for conservation and sustainable utilization are discussed, spanning both in-situ and ex-situ approaches, such as protected areas, community-based initiatives, agroforestry cultivation, seed banks, and specialized nurseries. Additionally, the chapter emphasizes the transformative potential of advanced technologies, including tissue culture, cryopreservation, digital herbaria, Geographic Information System-based mapping, and artificial intelligence in strengthening conservation and research capacity. By integrating traditional knowledge systems with modern scientific innovations, the chapter advances a multidimensional framework for safeguarding Sri Lanka's endemic medicinal plant

O. W. Ketakela · R. J. Piumal · A. M. K. L. Abeykoon · D. M. R. G. Mayakaduwa (✉)
H. S. Kathriarachchi
Department of Plant Sciences, Faculty of Science, University of Colombo, Colombo, Sri Lanka
e-mail: ruwani.mayakaduwa@pts.cmb.ac.lk

S. K. M. K. Herapathdeniya
Department of Ayurveda Pharmacology, Pharmaceutics and Community Medicine, Faculty of Indigenous Medicine, University of Colombo, Colombo, Sri Lanka

H. D. Jayasinghe
National Institute of Fundamental Studies, Kandy, Sri Lanka

S. Jha et al. (eds.), *Genetic Diversity in Medicinal Plants*, Sustainable Development and Biodiversity 44, https://doi.org/10.1007/978-981-95-7281-6_6

diversity. Such an approach not only addresses urgent conservation priorities but also ensures the sustainable use of these irreplaceable resources for future generations.

Keywords Endemic · Medicinal plants · Threats and conservation · Taxonomy and diversity

6.1 Introduction

Medicinal plants have been integral to human civilization since antiquity, forming the foundation of both traditional healing practices and modern healthcare systems. Globally, more than 80% of the population in developing countries relies on plant-based remedies as their primary source of healthcare (Sulaiman et al. 2020; Awoke et al. 2024). Endemic medicinal plant species are those confined naturally to specific geographic regions and are not found elsewhere (Hassanpouraghdam et al. 2022). Plants' restricted distribution often results in the evolution of unique biochemical profiles, yielding distinctive bioactive compounds that may not be found elsewhere (Barba-Ostria et al. 2025). These unique compounds make endemic plants particularly valuable in traditional medicine (TM), where they have been used for generations over centuries to treat a variety of ailments, and also in modern pharmaceutical research, where they serve as potential sources for the development of novel drugs (Francioso et al. 2019; Barba-Ostria et al. 2025). Beyond their therapeutic value, medicinal plants play significant roles in nutrition, cultural practices, and socio-economic development, particularly within rural communities (Woldeamanuel et al. 2022).

Sri Lanka, despite its modest geographic extent, is considered one of the world's biodiversity hotspots. Sri Lanka exhibits remarkable biological diversity and harbors the highest concentration of species in the Asian region, owing to varied topography, a range of climatic zones, and a long history of ecological isolation (Gunatilleke et al. 2017). Sri Lanka has a high degree of angiosperm diversity when compared to other South Asian countries. According to *The National Red List 2020* (2020), Sri Lanka possesses 3087 species of angiosperms spanning 186 families, of which 863 (28%) are endemic.

The rich biodiversity, together with the country's long-standing historical, spiritual, and cultural heritage, has fostered a well-established traditional system of medicine, comprising Ayurveda, Siddha, Unani, and Desiya Chikitsa. In Sri Lanka, TM has been practiced for over 3000 years, fulfilling the primary health needs of 60–70% of the rural population (Perera 2012). Ancient Sri Lankan archeological evidence indicates that plants were regularly used by people for curing various ailments even in prehistoric times. According to reports, a total of 1430 plant species from 838 genera and 181 families are thought to have therapeutic potential (Sugathadasa et al. 2008). Figure 6.1 illustrates several endemic medicinal plant species that display distinct morphological characteristics.

Fig. 6.1 Endemic medicinal plants of Sri Lanka. (**a**) *Semecarpus parvifolius*, (**b**) *Monoon acuminatum*, (**c**) *Hydnocarpus venenatus*, (**d**) *Campnosperma zeylanicum*, (**e**) *Holarrhena mitis*, (**f**) *Jeffreycia zeylanica*, (**g**) *Cryptocoryne thwaitesii*, and (**h**) *Elaeodendron glaucum*

Medicinal plants represent profound cultural significance, serving as repositories of indigenous knowledge systems and traditional practices that have been transmitted across generations. They are intimately woven with local traditions, spiritual beliefs, and community identities, thereby representing not only a source of primary healthcare but also a vital component of cultural heritage. However, the sustainability of these resources is increasingly under threat. Unsustainable harvesting, habitat degradation, deforestation, and the accelerating impacts of climate change have placed immense pressure on wild populations, resulting in the decline and endangerment of many species (Mykhailenko et al. 2025).

The urgency of this challenge is particularly acute in the case of endemic medicinal plants, which, by virtue of their restricted geographic distribution, are therefore especially vulnerable to extinction. These species represent irreplaceable reservoirs of unique genetic and phytochemical diversity, with potential to yield novel bioactive compounds of pharmaceutical importance. The loss of these plants would not only diminish biodiversity but also erode cultural heritage and foreclose future opportunities for scientific discovery. Consequently, documenting, studying, and conserving endemic medicinal plants are a matter of both scientific and cultural urgency. Effective conservation must therefore adopt an integrated approach that safeguards ecological integrity while also preserving the traditional knowledge systems.

This chapter aims to provide a comprehensive overview of the richness of Sri Lanka's medicinal flora, highlighting its diversity, historical significance, and enduring role in traditional healthcare systems. It will examine the taxonomic and ecological attributes of endemic medicinal plants, analyze their distribution patterns, and explore the threats faced in the modern era. Furthermore, the chapter highlights the urgent need to conserve these invaluable species in light of escalating threats from habitat loss, overexploitation, and climate change, while emphasizing conservation strategies and the integration of scientific research with technological advancements to promote their sustainable use.

6.2 Historical and Cultural Context: Traditional Medicinal System in Sri Lanka

According to the World Health Organization (WHO 2000), TM is the knowledge, skills, and practices derived from theories, beliefs, and experiences indigenous to different cultures, which are used to maintain health, as well as to prevent, diagnose, improve, or treat physical and mental illnesses. TM systems vary drastically among countries, depending on religious, cultural, historical, and ecological aspects (WHO 2000).

As one of the most historically, spiritually, and culturally diverse countries in the world, Sri Lanka has a rich traditional system of medicine. Sri Lanka, endowed with exceptional biodiversity and a rich cultural heritage, has nurtured an advanced traditional medicinal system for over two thousand five hundred years, influenced by indigenous knowledge, religious traditions, and cross-cultural interactions (Gamage et al. 2021). The Sri Lankan traditional medicinal system is mainly composed of Ayurveda, Sidda, Unani, and Desiya Chikitsa (Weragoda 1980).

Deshiya Chikithsa, also known as *Hela Wedakama*, is the indigenous system of medicine unique to Sri Lanka. Historical records, including Ola-leaf manuscripts and early chronicles such as the *Mahavansa*, document the preparation and use of medicinal formulations for a variety of ailments, highlighting the continuity of these practices from ancient times to the present (Peiris et al. 2017). However, knowledge of these practices has been mainly transmitted across generations through oral traditions, family lineages of practitioners, and ancient manuscripts, ensuring the preservation of invaluable healing wisdom (Peiris et al. 2017).

Ayurveda, which originated in India around 1000 BC, is regarded as the fundamental source of Vedic knowledge on the remedial properties of plants and is considered the most ancient system of medicine. Its principles and practices are preserved in classical texts such as the *Charaka Samhita*, which focuses on internal medicine and therapeutics; the *Sushruta Samhita*, which emphasizes surgical techniques and procedural interventions; and the *Ashtanga Hridaya Samhita*, which integrates both medical and surgical knowledge into a comprehensive system of healthcare (Bhat et al. 2021). In addition to Ayurveda, the Siddha and Unani systems have contributed greatly to strengthening Sri Lanka's rich heritage of plant-based healing practices (Jones and Liyanage 2018). The Unani system, which originated in Greece, was introduced to India and Sri Lanka by Arabs and Persians following the discovery of the sea route to the subcontinent, while the Siddha system developed between the tenth and fifteenth centuries in southern India as a parallel tradition to Ayurveda (Bhat et al. 2021; Weragoda 1980). Both these systems, along with Ayurveda, were later transferred to Sri Lanka through historical invasions, migrations, and cultural exchanges from India; thereafter, they became integrated into local healthcare practices.

According to the WHO, traditional herbal medicines are naturally occurring, plant-derived substances that require minimal or no industrial processing and are used for treating diseases within local or regional healing practices (Tilburt and Kaptchuk 2008). In some cases, a single compound derived from a plant can be highly effective on its own; however, most herbal remedies are complex mixtures that include different parts of one plant or several plants, working together synergistically to enhance the overall therapeutic effectiveness of the preparation. Archeological evidence indicates that in Sri Lanka, herbal medicine has been used to treat a wide range of disorders for over 3000 years (Edwards et al. 2012). Plant and herbal preparations form the cornerstone of both the Ayurveda and Deshiya Chikitsa systems, with approximately 2000 plant species employed in Ayurveda and around 500 species in Deshiya Chikitsa (Weragoda 1980).

Even today, the traditional medicinal system remains relevant, particularly in rural areas, and has been institutionalized through Ayurvedic hospitals, research institutes, and academic programmes. It forms a foundation for ethnopharmacological studies that aim to validate traditional remedies, identify bioactive compounds, and promote sustainable use of plant resources. The historical and cultural context of Sri Lankan TM highlights the deep connection between humans and their environment, emphasizing the importance of preserving both cultural heritage and biodiversity.

6.3 Taxonomy and Diversity of Endemic Medicinal Flora

In Sri Lanka, a substantial body of literature exists on medicinal plants; however, the majority of studies have concentrated on species that are widely used or of high commercial value. To capture the full extent of the country's medicinal plant diversity, several comprehensive compilations have been undertaken, systematically documenting taxa reported as being used or traded in TM. Key contributions include those of Jayaweera (1981, 1982), Sugathadasa et al. (2008), Kankanamalage et al. (2014), and Rajakaruna et al. (2025), as well as curated digital resources such as Barberyn Ayurveda Resorts and University of Ruhuna (2022) and Lakpura LLC (2023). Among these, one of the most significant milestones was the *National Checklist of Medicinal Plants* prepared in 2008 by the Department of Ayurveda in collaboration with the Bandaranayake Memorial Ayurvedic Research Institute. More recently, Rajakaruna et al. (2025) advanced this work through the *Updated Checklist of Medicinal Plants and Conservation Priorities for Sri Lanka*. All referenced sources were thoroughly reviewed, and using these together with *The National Red List 2020*, a comprehensive list of endemic medicinal plants was compiled for this chapter. A total of 166 species across 62 families were confirmed based on their accepted scientific names and verified therapeutic properties and subsequently included in the final compilation (Table 6.1).

Several families, including *Dipterocarpaceae*, *Araceae*, *Annonaceae*, *Anacardiaceae*, *Rubiaceae*, and *Melastomataceae,* had a rich representation in the pool of endemic medicinal plant species. Among these, Dipterocarpaceae emerged as the most dominant family, comprising 18 endemic species, followed by Araceae with 8 species and Annonaceae with 7 species (Fig. 6.2). This taxonomic concentration highlights the ecological and evolutionary significance of these lineages within Sri Lanka's endemic medicinal flora.

Many of Sri Lanka's endemic plant species have been employed by traditional medical practitioners for generations to treat a wide range of disorders. While numerous taxa are incorporated into the traditional systems of medicine, relatively few have been subjected to systematic phytochemical or pharmacological investigation to determine their bioactive constituents and therapeutic potential. This suggests a crucial requirement in undertaking scientific validations of the herbal medicine in order to establish evidence-based medication.

The National Red List 2020 provides the most comprehensive account of the conservation status of the island's flora, encompassing flowering plants (angiosperms), as well as ferns and lycophytes (pteridophytes). The Red List assessment framework follows the globally standardized IUCN Red List Categories and

Table 6.1 A comprehensive list of endemic medicinal plant species of Sri Lanka

Family	Species and conservation status[a]
Acanthaceae	*Rhinacanthus polonnaruwensis* L.H.Cramer (LC)
Achariaceae	*Chlorocarpa pentaschista* Alston (VU), *Hydnocarpus venenatus* Gaertn. (LC), *Trichadenia zeylanica* Thw. (LC)
Anacardiaceae	*Campnosperma zeylanicum* Thwaites (LC), *Mangifera zeylanica* (Blume) Hook. F. (LC), *S. coriaceus* Thwaites (VU), *Semecarpus nigroviridis* Thwaites (LC), *Semecarpus obovatus* Moon ex Thwaites (EN), *Semecarpus parvifolius* Thwaites (LC), *Semecarpus subpeltatus* Thwaites (VU), *Spondias xerophila* Kosterm (VU)
Ancistrocladaceae	*Ancistrocladus hamatus* (Vahl) Gilg (VU)
Annonaceae	*Desmos elegans* (Thw.) Safford (EN), *Goniothalamus gardneri* Hook. f. & Thoms. (VU), *Monoon acuminatum* (Thwaites) B.Xue & R.M.K.Saunders (LC), *Phoenicanthus coriacea* (Thw.) H. Huber (EN), *Polyalthia persicifolia* (Hook.f. & Thomson) Bedd. (EN), *Xylopia championii* Hook. f. & Thoms. (LC), *Xylopia nigricans* Hook. f. & Thoms. (NT)
Apocynaceae	*Holarrhena mitis* (Vahl) R.Br. ex Roem. & Schult. (VU), *Tabernaemontana dichotoma* Roxb. ex Wall. (LC), *Petchia ceylanica* (Wight) Livera (NT), *Wrightia antidysenterica* (L.) R.Br. (VU)
Aponogetonaceae	*Aponogeton jacobsenii* de Wit (CR)
Araceae	*Cryptocoryne alba* de Wit (CR), *Cryptocoryne beckettii* Trimen (VU), *Cryptocoryne bogneri* Rataj (CR), *Cryptocoryne nevillii* sensu auctt. (EN), *Cryptocoryne parva* de Wit (EN), *Cryptocoryne thwaitesii* Schott (EN), *Cryptocoryne undulata* Wendt (CR), *Cryptocoryne walkeri* Schott (CR), *Cryptocoryne wendtii* de Wit (VU), *Cryptocoryne x willisii* Reitz (CR)
Araliaceae	*Schefflera emarginata* (Moon) Harms (VU), *Schefflera heterobotrya* Frodin (VU)
Arecaceae	*Calamus radiatus* Thw(VU), *Calamus zeylanicus* Becc.(EN), *Loxococcus rupicola* (Thw.) H. Wendl. & Drude(EN), *Areca concinna* Thw.(EN), *Oncosperma fasciculatum* Thw. (VU)
Asteraceae	*Gynura hispida* Thw. (CR), *Jeffreycia zeylanica* (L.) H.Rob., S.C.Keeley & Skvarla (LC)
Balsaminaceae	*Impatiens repens* Moon ex Wight (EN)
Burseraceae	*Canarium zeylanicum* (Retz.) Blume (VU)
Centroplacaceae	*Bhesa ceylanica* (Arn. ex Thwaites) Ding Hou (LC)
Celastraceae	*Elaeodendron glaucum* (Rottb.) Pers. (LC), *Kokoona zeylanica* Thw. (EN)
Calophyllaceae	*Calophyllum moonii* Wight (VU), *Calophyllum tomentosum* Wight (VU), *Calophyllum walkeri* Wight (VU), *Mesua stylosa* (Thw.) Kosterm. (CR), *Mesua thwaitesii* Planch. & Triana (LC)
Clusiaceae	*Garcinia echinocarpa* Thwaites (NT), *Garcinia quaesita* Pierre (LC), *Garcinia zeylanica* Roxb. (EN)
Connaraceae	*Connarus championii* Thwaites (NT)
Crypteroniaceae	*Axinandra* zeylanica Thw. (NT)
Dilleniaceae	*Acrotrema walkeri* Wight ex Thwaites (VU), *Schumacheria castaneifolia* Vahl (VU)
Dioscoreaceae	*Dioscorea koyamae* Jayasuriya (EN), *Dioscorea trimenii* Prain & Burkill (EN)

(continued)

Table 6.1 (continued)

Family	Species and conservation status[a]
Dipterocarpaceae	*Cotylelobium scabriusculum* (Thwaites) Brandis (EN), *Dipterocarpus glandulosus* Thw (EN), *Dipterocarpus zeylanicus* Thw. (NT), *Hopea cordifolia* (Thw.) Trimen (EN), *Hopea discolor* Thw. (EN), *Hopea jucunda* Thw. (VU), *Hopea modesta* (A.DC.) Kosterm. (VU), *Shorea congestiflora* (Thw.) Ashton (VU), *Shorea cordifolia* (Thw.) Ashton (VU), *Shorea disticha* (Thw.) Ashton (VU), *Shorea megistophylla* Ashton (VU), *Shorea oblongifolia* Thw. (VU), *Shorea ovalifolia* (Thw.) Ashton (CR), *Shorea stipularis* Thw. (VU), *Shorea zeylanica* (Thw.) Ashton (EN), *Vateria copallifera* (Retz.) Alston (NT), *Vatica obscura* Trimen (VU), *Vatica paludosa* Kosterm. (EN)
Ebenaceae	*Diospyros atrata* Alston (EN), *Diospyros attenuata* Thw. (EN), *Diospyros pegophila* I.M.Turner (VU), *Diospyros oppositifolia* Thw. (EN), *Diospyros quaesita* Thw. (EN)
Elaeocarpaceae	*Elaeocarpus montanus* Thw. (VU)
Ericaceae	*Rhododendron arboreum* subsp. zeylanicum (Booth) Tagg (VU)
Eriocaulaceae	*Eriocaulon ceylanicum* Körnicke (EN)
Euphorbiaceae	*Agrostistachys hookeri* (Thwaites) Trimen (LC), *Macaranga digyna* (Wight) Muell. Arg. (VU)
Fabaceae	*Adenanthera aglaosperma* Alston (VU), *Cynometra zeylanica* Kosterm. (NT), *Derris parviflora* Benth. (LC), *Dialium ovoideum* Thw. (VU), *Humboldtia laurifolia* Vahl (LC)
Gentianaceae	*Exacum trinervium* (L.) Druce (NT), *Swertia zeylanica* (Griseb.) Walker ex Clarke (CR)
Lamiaceae	*Premna procumbens* Moon (LC)
Lauraceae	*Actinodaphne stenophylla* Thw. (VU), *Cinnamomum capparu-coronde* Blume (VU), Cinnamomum verum J. Presl (VU), *Baratranthus nodiflorus* (Thwaites) Tiegh. (EN)
Malvaceae	*Cullenia ceylanica* (Gardner) K.Schum. (LC), *Pterygota thwaitesii* (Mast.) Alston (VU), *Pityranthe verrucosa* Thwaites (LC)
Melastomataceae	*Memecylon capitellatum* L. (LC), *Memecylon rivulare* K.Bremer (VU), *Memecylon rostratum* Thwaites (NT), *Memecylon sylvaticum* Thw. (NT), *Osbeckia octandra* DC. (LC)
Meliaceae	*Dysoxylum championii* Hook.f. & Thomson ex Thwaites (VU)
Monimiaceae	*Hortonia floribunda* Wight ex Arn (EN)
Moraceae	*Artocarpus gomezianus* subsp. zeylanicus Jarrett (NT), *Artocarpus nobilis* Thw. (LC), *Allaeanthus zeylanicus* Thwaites (VU), *Ficus fergusonii* (King) T.B.Worth. ex Corner (LC)
Myristicaceae	*Horsfieldia iryaghedhi* (Gaertn.) Warb. (VU)
Myristicaceae	*Myristica dactyloides* Gaertn. (LC)
Myrtaceae	*Eugenia rotundata* (Trimen) Trimen (NT), *Syzygium cordifolium* (Wight) Walp. (EN), *Syzygium cylindricum* (Wight) Alston (VU), *Syzygium firmum* Thwaites (EN), *Syzygium umbrosum* Thwaites (EN)
Nepenthaceae	*Nepenthes distillatoria* L. (VU)
Orchidaceae	*Dendrobium maccarthiae* Thwaites (EN), *Ipsea speciosa* Lindl. (EN), *Podochilus falcatus* Lindl. (VU), *Podochilus saxatilis* Lindl. (NT), *Zeuxine regia* (EN)

(continued)

Table 6.1 (continued)

Family	Species and conservation status[a]
Pandanaceae	*Pandanus ceylanicus* Solms (VU)
Pentaphylacaceae	*Adinandra lasiopetala* (Wight) Choisy (EN)
Peraceae	*Chaetocarpus coriaceus* Thwaites (LC)
Phyllanthaceae	*Aporosa lanceolata* (Tul.) Thw. (LC), *Bridelia moonii* Thw. (LC), *Cleistanthus pallidus* (Thw.) Muell. Arg. (LC), *Margaritaria cyanosperma* (Gaertn.) Airy Shaw (VU)
Pittosporaceae	*Pittosporum ceylanicum* Wight (NT)
Plantaginaceae	*Adenosma camphorata* Hook.f (EN)
Poceae	*Ochlandra stridula* Thw. (LC)
Polygalaceae	*Xanthophyllum zeylanicum* Meijden (LC)
Primulaceae	*Ardisia willisii* Mez (VU)
Putranjivaceae	*Putranjiva zeylanica* (Thw.) Muell. Arg. (NT)
Rhamnaceae	*Ziziphus linnaei* M.A.Lawson (NT), *Ziziphus lucida* Moon ex Thwaites (CR)
Rhizophoraceae	*Carallia calycina* Benth. (VU)
Rosaceae	*Prunus walkeri* (Wight) Kalkman (LC)
Rubiaceae	*Diplospora erythrospora* (Thwaites) Bedd. (VU), *Diyaminauclea zeylanica* (Hook. f.) Ridsd. (EN), *Gardenia crameri* Tirveng.(EN), *Gardenia fosbergii* Tirveng. (VU), *Knoxia zeylanica* L. (NT), *Scyphostachys coffeoides* Thwaites (EN)
Rutaceae	*Micromelum minutum* var. ceylanicum B.C.Stone (LC)
Salicaceae	*Dovyalis hebecarpa* (Gardn.) Warb. (EN), *Scolopia pusilla* Willd. (LC)
Sapindaceae	*Allophylus zeylanicus* L. (LC)
Sapotaceae	*Isonandra compta* Dubard (EN), *Madhuca clavata* Jayas. (CR), *Madhuca fulva* (Thwaites) J.F.Macbr. (VU), *Palaquium grande* (Thw.) Engler (VU), *Palaquium thwaitesii* Trimen (VU)
Stemonuraceae	*Stemonurus apicalis* (Thw.) Miers (NT)
Symplocaceae	*Symplocos coronata* Thwaites (VU)
Theaceae	*Polyspora ceylanica* (Wight) Orel, Peter G.Wilson, Curry & Luu (EN)

[a]The conservation status of the species according to The National Red List 2020 is indicated within brackets after each species (*CR* critically endangered, *EN* endangered, *VU* vulnerable, *NT* near threatened, *LC* least concern)

Criteria, which classify species into nine categories: not evaluated (NE), data deficient (DD), least concern (LC), near threatened (NT), vulnerable (VU), endangered (EN), critically endangered (CR), extinct in the wild (EW), and extinct (EX) (Walter and Gillett 1998). Under this system, species designated as CR, EN, and VU are considered nationally threatened, and their identification is regarded as the first and most critical step in advancing conservation measures.

Analysis of the 166 endemic medicinal plant species recorded revealed that 111 species (67%) are highly threatened, with 12 species (7.2%) classified as CR, 42 species (25.3%) as EN, and 57 species (34.3%) as VU. This high proportion of threatened taxa emphasizes the urgent need for targeted conservation interventions to prevent extinction and minimize global biodiversity loss. In addition, 11.4% of

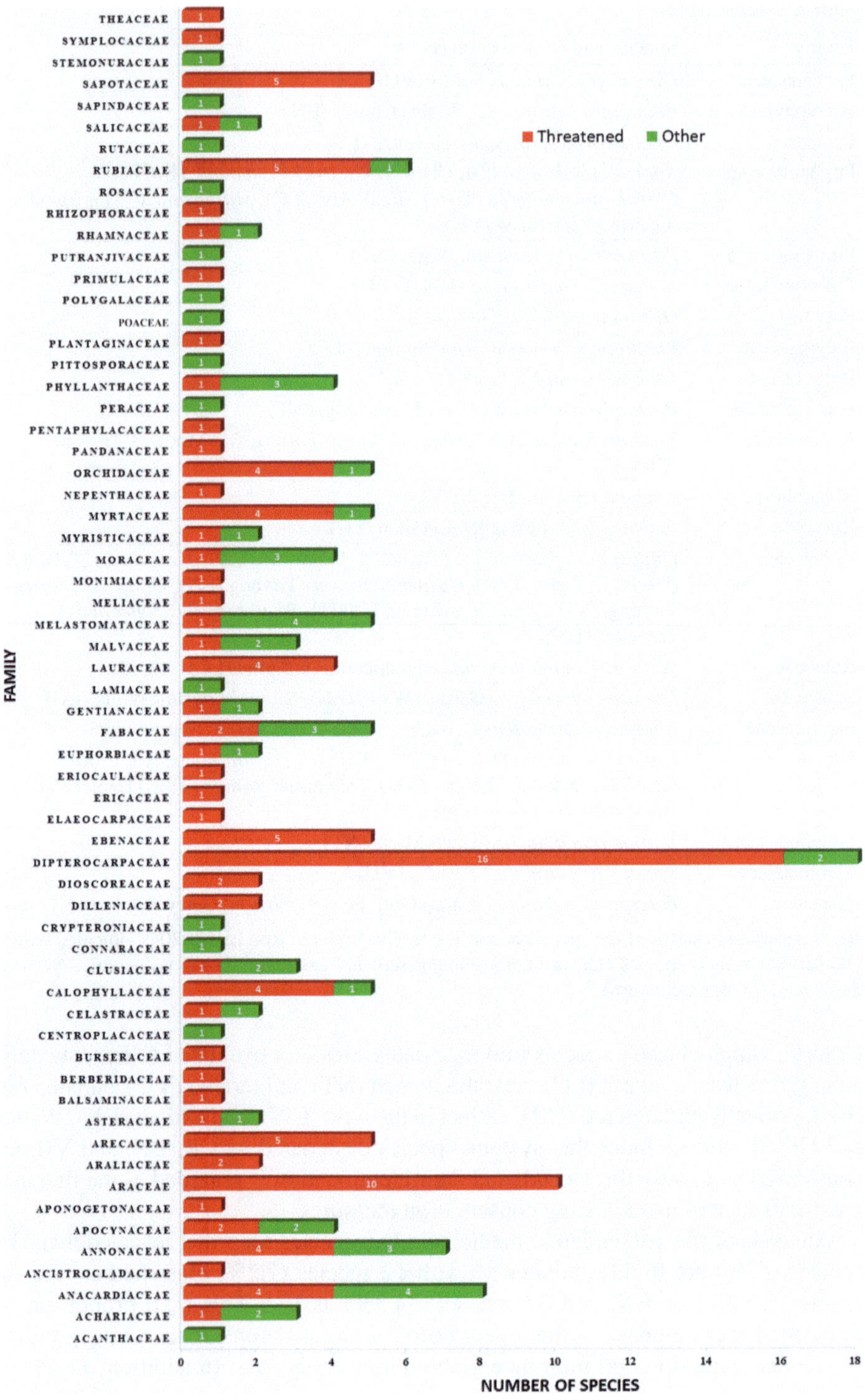

Fig. 6.2 Illustration of the number of endemic medicinal plant species recorded within each plant family

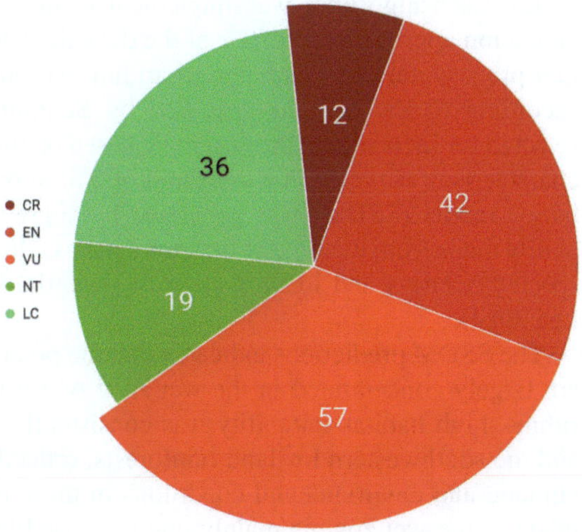

Fig. 6.3 Extent of threatened endemic medicinal plant species in Sri Lanka

endemic medicinal plants were classified as NT, while 21.7% were placed in the LC category (Fig. 6.3). Although these taxa are not immediately threatened, proactive conservation measures are recommended to reduce the risk of future decline. Notably, endemic species categorized as EX, EW, DD, or NE had no recorded therapeutic potential, reflecting knowledge gaps or extinction-driven losses in Sri Lanka's ethnomedicinal heritage.

6.4 Ecological and Geographical Distribution

The distribution of plant species is mainly shaped by a variety of biotic and abiotic factors. The plant occurrence and habitat suitability are mainly influenced by abiotic factors, including rainfall, humidity, temperature, and light intensity. An ensemble species distribution model (ESDM) was performed to predict the potential distribution of endemic medicinal plant species across Sri Lanka. By integrating predictions from multiple modeling algorithms, the ESDM enhances predictive accuracy and reduces uncertainties associated with individual models (Chung and Lee 2024).

Bioclimatic variables for each sampling point were derived from the *WorldClim* database (www.worldclim.org), which provides 19 standardized climate layers at 30-inch spatial resolution averaged over the period 1970–2000 (Hijmans et al. 2017). Species occurrence data were obtained from the *Global Biodiversity Information Facility* (https://www.gbif.org/). Spatial data were processed and mapped at a resolution of 100 × 100 m (excluding bioclimatic layers) using *ArcGIS Desktop* v.10.8.1 (ESRI Inc., Redlands, CA, USA).

The ESDM for each endemic plant species was modeled using 19 environmental predictors, employing multiple algorithms, including random forests, MaxEnt, and support vector machines (Meyer et al. 2019; Hijmans et al. 2017; Liaw and Wiener

2002). Each algorithm was implemented and evaluated using a hold-out cross-validation approach, with 70% of the data allocated to a training set and the validation process iterated twice per algorithm. All other algorithm settings were tuned according to the defaults specified by Schmitt et al. (2017). Algorithms were selected for inclusion in the ensemble based on the area under the receiver operating characteristic curve, with a threshold of 0.8. Subsequently, a stacked species distribution model (SSDM) was generated by aggregating the continuous habitat suitability maps from the ESDM for each species, allowing for the assessment of overall species richness and identification of potential biodiversity hotspots (Chung and Lee 2024).

The SSDM predictions indicated that the potential habitats of the studied species are largely concentrated in the wet zone of Sri Lanka (Fig. 6.4). Across all algorithms, high habitat suitability was consistently observed in the central highlands and the southwestern lowland rainforests, reflecting the concentration of favorable climatic and environmental conditions in these regions. The high degree of endemism in the wet zone is mainly due to favorable conditions like adequate rainfall and humidity. Moreover, Sri Lanka's unique geo-evolutionary history and

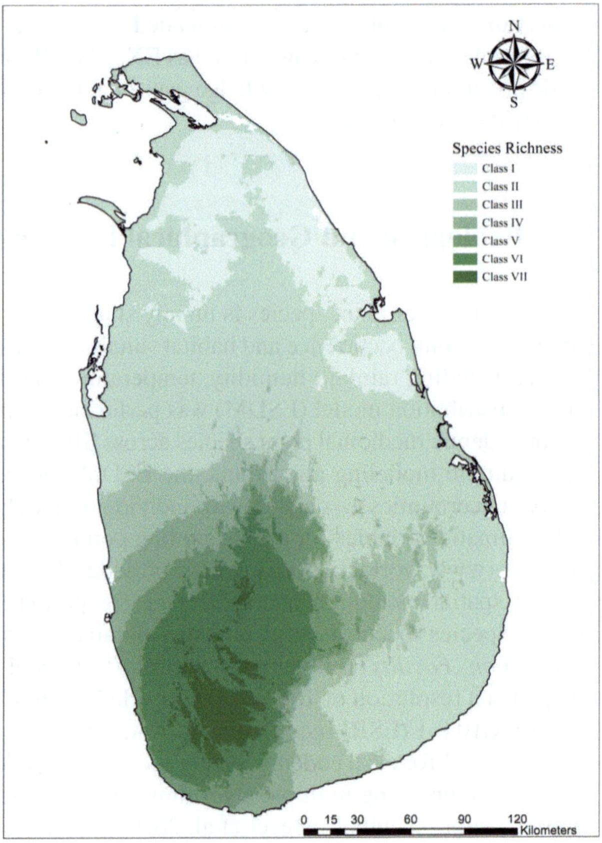

Fig. 6.4 SSDM predicting the potential distribution of endemic medicinal plant species across Sri Lanka showing species class richness categories: Class I (<14.0), Class II (14.0–19.0), Class III (19.0–24.0), Class IV (24.0–29.0), Class V (29.0–34.0), Class VI (34.0–39.0), and Class VII (39.0–44.0)

pronounced climatic and topographic variation have facilitated the persistence of numerous geographically relict species, particularly within the Sinharaja rainforest, a United Nations Educational, Scientific and Cultural Organization (UNESCO) World Heritage Site recognized as a global biodiversity hotspot (Pushpakumara et al. 2012).

6.5 Threats to Medicinal Flora

Overharvesting and unsustainable use are among the most serious threats to medicinal plant species (Mykhailenko et al. 2025; Ssenku et al. 2022). The growing demand for both traditional remedies and commercially valuable herbal products has led to the excessive collection of wild plants, particularly targeting roots, rhizomes, bark, and entire individuals, resulting in rapid depletion of wild populations (Khakurel et al. 2024; Groner et al. 2022). The lack of large-scale cultivation or domestication further intensifies the reliance on wild resources, placing additional strain on naturally occurring populations (Liu et al. 2019). As a result, overexploitation not only diminishes the availability of medicinal plants but also threatens the long-term survival of species and the ecological balance of the habitats occupied (Lorite 2024).

Habitat destruction and fragmentation pose significant threats to medicinal plant diversity, with natural populations increasingly impacted by deforestation, agricultural expansion, urbanization, and changes in land use (Mykhailenko et al. 2025). These activities not only diminish the extent of natural habitats but also fragment them, isolating plant populations and disrupting ecological connectivity (Collins et al. 2017).

Climate change intensifies these challenges by disrupting the ecological processes critical to the survival of medicinal plants, as rising temperatures, irregular precipitation, drought, and extreme climatic events threaten the distribution, survival, and chemical composition of medicinal plants (Mykhailenko et al. 2025; Kougioumoutzis et al. 2024). Some plants are being forced to migrate to higher altitudes or latitudes in search of suitable conditions, while others, with limited adaptive capacity, face the risk of extinction (Cahyaningsih et al. 2021). Climate-induced stress also influences the phytochemical composition of plants, potentially diminishing the potency or altering the efficacy of their medicinal properties (Mansinhos et al. 2024). Thus, climate change not only threatens the persistence of species but also compromises their therapeutic value.

Pollution, particularly the contamination of soil and water with heavy metals such as cadmium, lead, mercury, and arsenic from industrial discharges, mining, and intensive agricultural practices, poses a significant threat to the survival and safe use of medicinal plants (Çolak et al. 2023). These pollutants are readily taken up by medicinal plants via their roots and accumulate in various tissues, including those routinely utilized in traditional and pharmaceutical applications (Mandal and Dhal 2023).

Adding to these pressures is the spread of invasive alien species, which aggressively compete with native medicinal flora for resources. These invasive plants frequently alter soil conditions and ecological processes, making habitats less suitable for indigenous species (Dar et al. 2023). Current climate change has facilitated the spread of invasive species, giving them a competitive advantage over native flora within the natural environments (Giejsztowt et al. 2020).

Collectively, these threats create a precarious situation for medicinal plants and not only compromise the richness and diversity of these species but also undermine ongoing efforts to safeguard their ecological, cultural, and pharmacological significance. Endemic medicinal plants, confined to narrow ecological niches and often limited in distribution, are particularly vulnerable and face disproportionately high risks from these pressures. Addressing these challenges requires urgent, integrated strategies that combine scientific research, policy enforcement, sustainable utilization, and effective conservation measures to ensure the long-term survival of medicinal plants.

6.6 Conservation Strategies for Protecting Endemic Medicinal Plant Species

Implementing effective conservation strategies is essential, both to secure the survival of these plants in their natural habitats and to ensure their continued availability for future generations. Such strategies should incorporate both in-situ approaches, which safeguard plants within their natural ecosystems, and ex-situ measures, which preserve species outside their native ranges, together forming a comprehensive framework for protecting plant biodiversity (Fig. 6.5).

6.6.1 In-Situ Conservation

According to the Convention on Biological Diversity (1992), in-situ conservation refers to the conservation and protection of ecosystems and natural habitats, as well as the maintenance and recovery of viable populations of species in their natural habitats. Natural reserves and wild nurseries are the most common in-situ conservation sites.

Most medicinal plants are endemic, and their medicinal properties are usually a result of secondary metabolites produced in response to environmental stimuli, which are not always expressed outside of their natural habitats (Coley et al. 2003; Figueiredo and Grelle 2009). Therefore, in-situ conservation is the most appropriate method of conserving such medicinal plants (Gepts 2006). Moreover, in-situ conservation enhances the amount of diversity that can be preserved and promotes the link between resource conservation and sustainable use (Long et al. 2003; Forest et al. 2007).

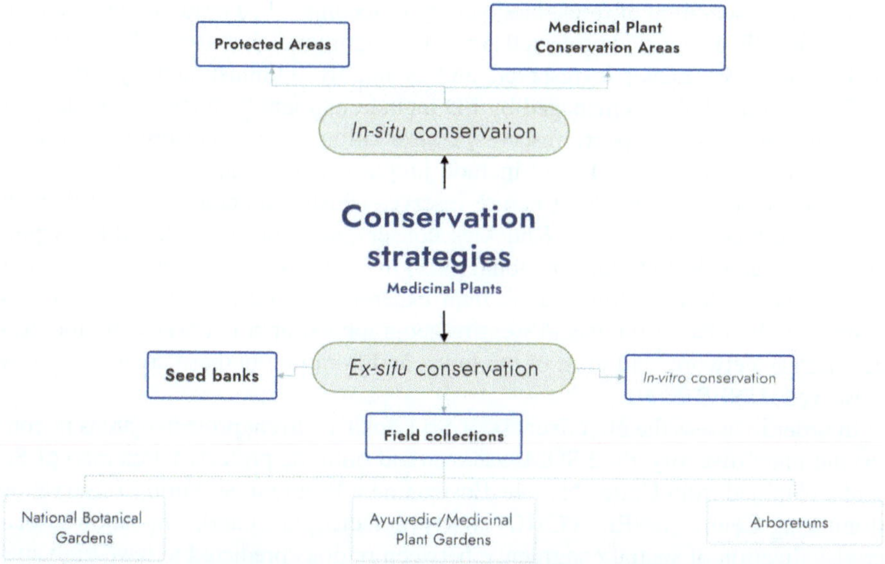

Fig. 6.5 Schematic diagram representing different conservation strategies implemented or could be implemented

Worldwide, in-situ conservation efforts have focused on establishing protected areas and promoting ecosystem-oriented conservation rather than species-oriented conservation (Jianzhang et al. 2012). As a result, this method of in-situ conservation would allow the conservation of species along with their natural communities and the associated network of relationships.

In Sri Lanka, the Ministry of Environment serves as the national focal point for biodiversity conservation, providing leadership in policy formulation, implementation, monitoring, and resource mobilization. Under its purview, the Wildlife and Forest Resources Conservation Division coordinates the activities of the Department of Forest Conservation (DFC) and the Department of Wildlife Conservation (DWLC), which share responsibility for the management and protection of the country's natural forest resources. Together with the Forest Department (FD), these institutions oversee the conservation and sustainable management of multiple categories of natural forests, which constitute critical habitats for endemic biodiversity, including medicinal plant species.

In addition to these primary agencies, several other institutions contribute to in-situ conservation. The Central Environmental Authority, the Mahaweli Authority, and local administrative offices such as Divisional Secretariats play supplementary but significant roles in safeguarding forest ecosystems and ensuring compliance with environmental regulations (Pushpakumara et al. 2012). Collectively, these agencies form the institutional backbone for the in-situ conservation of Sri Lanka's endemic medicinal flora, highlighting the importance of coordinated, multi-stakeholder governance in biodiversity conservation.

In-situ conservation of forest biodiversity is accomplished through the establishment of legally designated protected areas. It is significant that over 26% of the total land area of Sri Lanka is reserved and is mainly administrated by DFC and DWLC. Natural habitats managed by FD include national heritage and wilderness areas, international biosphere reserves, conservation forests, and reserved forests, while those managed by DWLC include jungle corridors, national parks, nature reserves, sanctuaries, and strict nature reserves (Pushpakumara et al. 2012). Four forests (Singharaja rainforest, Knuckles conservation forest, Peak wildness protected area, and Horton-plains national park) in Sri Lanka have been designated as Natural World Heritage Sites due to their exceptional biodiversity and high endemism. Nearly 14% of the island's entire geographical area is covered by the protected areas network, and most of the fauna and flora within these areas are highly conserved (UNEP 2009).

In order to assess the effectiveness of Sri Lanka's current protected areas in conserving plant diversity, the SSDM was overlaid onto the protected area map of Sri Lanka obtained from United Nations Environment Programme World Conservation Monitoring Centre (UNEP-WCMC) (www.protectedplanet.net). This allowed better visualization of spatial congruence between regions predicted to have high species richness and the boundaries of existing protected areas (Fig. 6.6). By integrating the SSDM with the protected area network, it was possible to identify gaps where regions of high conservation value fall outside formally protected areas, highlighting potential targets for future conservation efforts. The overlay also provided insights into how well endemic and threatened species are confined within the current system of reserves, offering a quantitative basis for prioritizing areas for protection, restoration, or sustainable management.

This shows that endemic medicinal plant species are highly concentrated within the protected areas of the wet zone of Sri Lanka. Moreover, the distribution analysis showed that the majority of data points were concentrated within Sri Lanka's wet zone, and the majority of endemic plant species lie within Sri Lanka's tropical rainforests. Only about 18% of the total protected area falls within the biologically rich wet zone (Padmalal et al. 2006). It is clear that, although Sri Lanka has an extensive protected areas network, there are critical gaps in biodiversity conservation (e.g., the protected areas network of Sri Lanka fails to adequately include the floristically rich wet zones). This poses a major threat to biological diversity, particularly to endemic medicinal plant species. Hence, these spatial analyses could be used for guiding evidence-based conservation planning.

In addition, the Sri Lanka Conservation and Sustainable Use of Medicinal Plants Project—a project funded by the Global Environment Facility in 1998—has established Medicinal Plant Conservation Areas (MPCAs) (IUCN 2004). MPCAs have been established in Bibila, Ritigala, Rajawaka, Naula, and Kanneliya, representing different climatic zones. This project successfully implemented in-situ conservation by establishing a collection of medicinal plant species from the wild. *Vateria copallifera*, *Garcinia quaesita*, and *Cinnamomum zeylanicum* are a few of the endemic medicinal plant species present within these MPCAs (Russell-Smith et al. 2006).

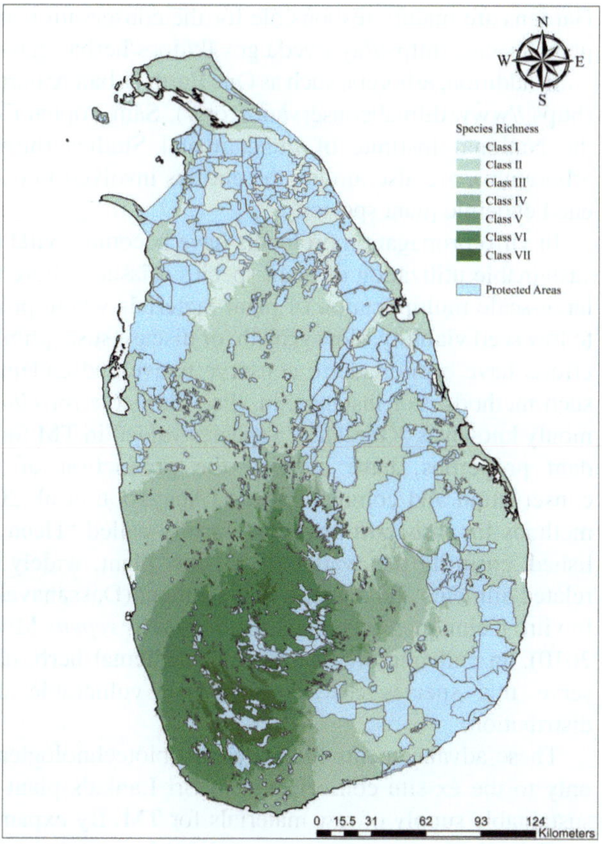

Fig. 6.6 SSDM overlaid onto the protected area network map of Sri Lanka, showing species class richness categories: Class I (<14.0), Class II (14.0–19.0), Class III (19.0–24.0), Class IV (24.0–29.0), Class V (29.0–34.0), Class VI (34.0–39.0), and Class VII (39.0–44.0)

6.6.2 *Ex-Situ Conservation*

The Department of National Botanical Gardens (DNBG) is mainly responsible for conserving the flora of Sri Lanka and currently holds over 8000 species (Pushpakumara et al. 2012). The DNBG has five botanical gardens, mainly to represent all major climatic zones of Sri Lanka, located at Peradeniya (representing the mid-country wet zone), Hakgala (representing the montane zone), Henerathgoda—Gampaha (representing the low country wet zone), Sitawaka—Awissawella (representing the low country wet zone), and Mirijjawila—Hambantota District (representing the dry zone). Furthermore, Ganewatta Medicinal Plant Garden, under the purview of DNBG, contains an extensive and unique collection of medicinal plants and is mainly dedicated to ex-situ conservation of medicinal plants. A rich collection of preserved plant specimens is being maintained in the National Herbarium in Peradeniya (https://en.botanicgardens.gov.lk/).

In addition, the Department of Ayurveda primarily oversees seven Ayurvedic Herbal Gardens, including those in Bathgoda—Haldummulla, Girandurukotte, Pattipola, Nawinna, Pallekele, Pinnaduwa, and Kanneliya. These Ayurvedic Herbal

Gardens are mainly responsible for the conservation and propagation of medicinal plant species (https://ayurveda.gov.lk/docs/herbal-gardens/).

In addition, arboreta, such as One Earth Urban Arboretum—Dilmah Conservation (https://www.dilmahconservation.org/), Sam Popham's Arboretum, maintained by the National Institute of Fundamental Studies (https://www.nifs.ac.lk/facilities/arboretum), are also among the entities involved in ex-situ conservation of threatened endemic plant species.

In vitro propagation protocols have become a vital tool in the conservation and sustainable utilization of plant species. Tissue culture techniques enable rapid and large-scale multiplication of plant materials where propagation is challenging due to low seed viability, slow growth, or disease susceptibility. In Sri Lanka, significant efforts have been made to conserve native and endemic medicinal plants through such methods. For instance, in vitro protocols for *Garcinia quaesita* Pierre (commonly known as "Goraka"), a species valued in TM for its anti-obesity and antioxidant properties, have enabled the production of healthy plantlets for both conservation and commercial use (Abeykoon et al. 2021). Similarly, propagation methods for *Osbeckia octandra*, locally called "Heen Bovitiya," have been established, ensuring the availability of this plant, widely used in Ayurveda for liver-related ailments and diabetes management (Dassanayake et al. 2018). Furthermore, in vitro techniques developed for *Impatiens repens* Moon (Herath and Wijesundara 2010), an endemic ornamental and medicinal herb, have provided means to conserve this species that is otherwise vulnerable due to its limited natural distribution.

These advancements illustrate how biotechnological approaches contribute not only to the ex-situ conservation of Sri Lanka's plant biodiversity but also to the sustainable supply of raw materials for TM. By expanding such protocols to other medicinally important species, Sri Lanka can strengthen its TM systems while ensuring the long-term protection of its unique plant heritage.

6.7 Legislation Framework Toward Conservation of Endemic Medicinal Plant Species in Sri Lanka

There are two national policies and 15 regulatory frameworks in Sri Lanka that are directly or indirectly involved in biodiversity conservation (MoERE 2014; Gunatilleke et al. 2017). In addition, Sri Lanka has ratified eight international conventions associated with the conservation of biodiversity (Gunatilleke et al. 2017).

Mainly, the Forest Ordinance No. 16 of 1907, the Fauna and Flora Protection Ordinance (FFPO) No. 2 of 1937 and its amendments, and the Convention on International Trade in 18 Endangered Species of Wild Fauna and Flora (CITES) are directly involved in the conservation of endemic plant species in Sri Lanka (UNEP 2009). CITES: Being a CITES member, Sri Lanka gains support from global collaboration in regulating the increasing threat of international trade to endangered wild fauna and flora. As a result, international trade in endangered endemic plant species is forbidden. The Fauna and Flora Protection (Amendment) Act, No. 22 of 2009: Sri Lanka's legal framework for biodiversity conservation is mainly based on FFPO and its amendments (Miththapala 2016; Gunatilleke et al. 2017). Mainly, the FFPO uses two strategies to promote species conservation: the establishment and regulation of "Protected Areas" in Sri Lanka and the listing of "protected species" (Gunatilleke et al. 2017). The main purpose of declaring an area as a protected area is to limit the extent and character of human activities, as well as to protect all biota. Thereby, protected areas enable conservation of faunal and floral species within their natural habitats (UNEP 2009). No one is permitted to uproot, destroy, cause any harm to, or destroy a tree that has been designated as a "protected tree" under Schedule VIII of the Fauna and Flora Protection (Amendment Act), No. 22 of 2009, for the duration of the schedule. Thereby, endemic plant species listed as "protected species" by the FFPO are conserved. According to Schedule VIII of FFPO, at least 58 endemic medicinal plant species have been declared as protected plant species. Furthermore, these species are banned from export.

6.8 Sustainable Utilization, Future Prospects, and Innovative Practices in Medicinal Plant Conservation

Future strategies must respect traditional heritage while incorporating evidence-based scientific validation to ensure the safety, efficacy, and sustainability of medicinal plant resources and their products. Therefore, conservation has to be reinforced both biologically by protecting vulnerable species from overexploitation and culturally, by preserving the traditional knowledge that is useful for responsible resource management.

The conservation of endemic medicinal plants must remain a priority due to their irreplaceable ecological and therapeutic value. Effective conservation requires a combination of strategies, including strengthening legal protection, incorporating regular assessment of endemic plants, and establishing gene banks. It is also important to maintain and manage protected areas, which safeguard critical habitats and support in-situ conservation of species in their natural ecosystems. Identifying and designating protected species, along with ongoing monitoring and habitat management, ensures that endemic medicinal plants are highly conserved.

Sustainable use is especially critical for endemic medicinal plants, which are geographically restricted and therefore more vulnerable to extinction. Many of these species are highly valued in TM systems but face intense pressure from habitat loss and overharvesting. To ensure their survival, site-specific management plans and community-based conservation initiatives must prioritize endemic species. Adopting *Good Agricultural and Collection Practices*, promoting cultivation under agroforestry models, and establishing dedicated endemic plant nurseries and seed banks can significantly reduce extraction pressure on wild populations (Chen et al.

2016). A key approach to sustainable utilization is the domestication and cultivation of medicinal plants (Bariotakis et al. 2023). Shifting from wild collection to controlled cultivation reduces extraction pressure on natural ecosystems and preserves wild populations, while standardized cultivation practices ensure consistent plant material quality essential for therapeutic efficacy, commercial production, and research applications (Marcelino et al. 2023).

Advances in biotechnology and digital sciences offer unprecedented opportunities to conserve endemic medicinal plants. Biotechnological interventions have opened new horizons in conservation and utilization. Techniques such as micropropagation and cryopreservation enable the rapid and large-scale propagation of endangered species, ensuring the preservation of genetic diversity (Chatterjee and Ghosh 2020). Tissue culture techniques are valuable for conserving threatened species by enabling rapid propagation of rare or endangered plants and producing genetically uniform, disease-free material for medicinal and research use (Kaur et al. 2022; Unal et al. 2023; Tagimanova et al. 2024). Cryopreservation complements this by providing long-term storage of plant germplasm under ultra-low temperatures (El Merzougui et al. 2023; Zeng et al. 2024). Molecular marker-assisted selection and genomic studies further enhance the ability to identify and conserve unique genotypes, while also facilitating the discovery of novel bioactive compounds (Hussain and Nisar 2020). In addition, advances in metabolomics and synthetic biology hold potential to reduce dependence on wild populations by enabling the biosynthesis of medicinally important phytochemicals in vitro (Marchev et al. 2020; Wawrosch and Zotchev 2021).

In the future, advanced digital platforms will play a crucial role in strengthening the sustainable utilization and conservation of endemic medicinal plants. Geographic Information System-based habitat mapping, coupled with species distribution modeling, provides powerful tools to identify priority regions for conservation, restoration, and sustainable management (Kaky and Gilbert 2016; Wu et al. 2019; Tafesse et al. 2023). These approaches not only depict areas of high species richness and endemism but also allow predictive assessments under future climate change scenarios, thereby guiding proactive interventions (Porfirio et al. 2014). Similarly, remote sensing technologies ranging from satellite imagery to drone-based surveys enable the detection of habitat changes, forest degradation, and land-use dynamics in real time, offering early warning systems that can inform policy and conservation strategies (Sivaraj et al. 2020; Ding et al. 2023; Meng et al. 2023).

Today, artificial intelligence (AI) is rapidly transforming biodiversity research by enabling the automated classification and identification of species from digital images, herbarium specimens, and even community-sourced photographs (Wäldchen and Mäder 2018; Little et al. 2020). The integration of AI into public platforms, such as mobile applications and online biodiversity databases, empowers local communities, students, and naturalists to actively contribute to scientific monitoring. Promoting AI-assisted mobile tools can instantly identify medicinal plant species from photographs taken in the field, enabling local users to document populations and detect threats. AI-driven tools can significantly enhance taxonomic precision, reduce misidentification, and support large-scale biodiversity monitoring at lower costs (Huixian 2020).

In parallel, blockchain-enabled traceability systems are emerging as innovative solutions for ensuring ethical and sustainable trade of flora (Bandara and Katukurunda 2024). By creating tamper-proof digital records of supply chains, blockchain technology ensures that medicinal plants entering global markets are harvested legally, sustainably, and in compliance with international laws (Li et al. 2022).

6.9 Conclusion

Endemic medicinal plants represent a unique and irreplaceable facet of global biodiversity, embodying profound ecological value while also serving as reservoirs of traditional heritage and traditional knowledge. Their restricted distribution, however, renders them exceptionally vulnerable to pressures such as overexploitation, habitat degradation, and the accelerating impacts of climate change. These threats underscore the urgent need for targeted, evidence-based conservation strategies. Safeguarding these species requires a multidimensional approach that integrates traditional knowledge systems with scientific innovation, while prioritizing endemic taxa within both in-situ and ex-situ conservation frameworks. Through coordinated action spanning ecological research, community engagement, policy support, and technological advancement, societies can ensure the long-term protection and sustainable use of Sri Lanka's endemic medicinal plant diversity for future generations.

References

Abeykoon RMGBS, Attanayake CIM et al (2021) *In vitro* propagation of *Garcinia quaesita* Pierre. Int Res J Adv Eng Sci 6(3):5–9

Awoke A, Gudesho G, Akmel F, Shanmugasundaram P (2024) Traditionally used medicinal plants for human ailments and their threats in Guraferda District, Benchi-Sheko zone, Southwest Ethiopia. J Ethnobiol Ethnomed 2024(1):82

Bandara KM, Katukurunda KG (2024) Exploring the Blockchain's Green Revolution in medicinal plant supply chains (Blossoming Trust)–a review. J Agric Value Addit 7(2):35

Barba-Ostria C, Guamán-Bautista J, Tosi-Vélez AA, Puente-Pineda JA, Cedeño-Zambrano MA, Teran E, Guamán LP (2025) Recent advances in the therapeutic potential of bioactive molecules from plants of Andean origin. Nutrients 17(11):1749

Barberyn Ayurveda Resorts and University of Ruhuna. [Online] Ayurvedic medicinal plants of Sri Lanka (2022). Available from: http://www.ayurvedicmedicinalplantssrilanka.org/. Accessed 19 Aug 2025

Bariotakis M, Georgescu L, Laina D, Koufaki M, Souma M, Douklias S, Giannakakis KA, Chouli KN, Paoli L, Loppi S, Karousou R (2023) Climate change dependence in ex situ conservation of wild medicinal plants in Crete, Greece. Biology 12(10):1327

Bhat MN, Singh B, Surmal O, Singh B, Shivgotra V, Musarella CM (2021) Ethnobotany of the Himalayas: safeguarding medical practices and traditional uses of Kashmir regions. Biol 10(9):851

Cahyaningsih R, Phillips J, Brehm J, Gaisberger H, Maxted N (2021) Climate change impact on medicinal plants in Indonesia. Glob Ecol Conserv 30:e01752

Chatterjee T, Ghosh B (2020) Micropropagation of medicinal plants: a review. Int J Econ Plants 7(2):066–072

Chen SL, Yu H, Luo HM, Wu Q, Li CF, Steinmetz A (2016) Conservation and sustainable use of medicinal plants: problems, progress, and prospects. Chin Med 11(1):37

Chung OS, Lee JK (2024) Assessing the distribution and richness of mammalian species using a stacking species distribution model in a temperate forest. Animals 14(5):759

Çolak S, Yilmaz ŞB, Öztekin E (2023) Bioaccumulation factors of heavy metal (loid) s in some medicinal and aromatic plants species: example of Zonguldak/Türkiye. Water Air Soil Pollut 234(8):522

Coley PD, Heller MV, Aizprua R, Araúz B, Flores N, Correa M, Gupta M, Solis PN, Ortega-Barría E, Romero LI, Gómez B (2003) Using ecological criteria to design plant collection strategies for drug discovery. Front Ecol Environ 1(8):421–428

Collins CD, Banks-Leite C, Brudvig LA, Foster BL, Cook WM, Damschen EI, Andrade A, Austin M, Camargo JL, Driscoll DA, Holt RD (2017) Fragmentation affects plant community composition over time. Ecography 40(1):119–130

Convention on Biological Diversity (CBD) (1992) Convention on biological diversity. United Nations, Rio de Janeiro (Brazil)

Dar MA, Ahmad M, Singh R, Kohli RK, Singh HP, Batish DR (2023) Invasive plants alter soil properties and nutrient dynamics: a case study of *Anthemis cotula* invasion in Kashmir Himalaya. Catena 226:107069

Dassanayake DM, Krishnarajah SA, Eeswara JP (2018) Development of an effective propagation protocol for *Osbeckia octandra* (L.) DC.(Heen Bovitiya). Trop Agric Res 29(2):123

Department of Ayurveda. Herbal Gardens [Online]. Available from: https://ayurveda.gov.lk/docs/herbal-gardens/. Accessed 2 Oct 2025

Department of National Botanic Gardens. [Online]. Available from: https://en.botanicgardens.gov.lk/. Accessed 2 Oct 2025

Dilmah Conservation. Home – Dilmah Conservation [Online]. Available from: https://www.dilmahconservation.org/. Accessed 2 Oct 2025

Ding R, Luo J, Wang C, Yu L, Yang J, Wang M, Zhong S, Gu R (2023) Identifying and mapping individual medicinal plant *Lamiophlomis rotata* at high elevations by using unmanned aerial vehicles and deep learning. Plant Methods 19(1):38

Edwards S, Da-Costa-Rocha I, Lawrence MJ, Cable C, Heinrich M (2012) Use and efficacy of herbal medicines: part 1-historical and traditional use. Pharmac J 289(7717):161–162

El Merzougui S, Benelli C, El Boullani R, Serghini MA (2023) The cryopreservation of medicinal and ornamental geophytes: application and challenges. Plants 12(11):2143

Fick SE, Hijmans RJ [Online] WorldClim 2: Global climate data. Available from: https://www.worldclim.org/. Accessed 2025 Aug 15

Figueiredo MS, Grelle CE (2009) Predicting global abundance of a threatened species from its occurrence: implications for conservation planning. Divers Distrib 15(1):117–121

Forest F, Grenyer R, Rouget M, Davies TJ, Cowling RM, Faith DP, Balmford A, Manning JC, Procheş Ş, Van Der Bank M, Reeves G (2007) Preserving the evolutionary potential of floras in biodiversity hotspots. Nature 445(7129):757–760

Francioso A, Franke K, Villani C, Mosca L, D'Erme M, Frischbutter S, Brandt W, Sanchez-Lamar A, Wessjohann L (2019) Insights into the phytochemistry of the cuban endemic medicinal plant *Phyllanthus orbicularis*: Fideloside, a novel bioactive 8-C-glycosyl 2, 3-dihydroflavonol. Molecules 24(15):2855

Gamage DG, Dharmadasa RM, Abeysinghe DC, Wijesekara RG, Prathapasinghe GA, Someya T (2021) Emerging herbal cosmetic production in Sri Lanka: identifying possible interventions for the development of the herbal cosmetic industry. Scientifica 2021(1):6662404

Gepts P (2006) Plant genetic resources conservation and utilization: the accomplishments and future of a societal insurance policy. Crop Sci 46(5):2278–2292

Giejsztowt J, Classen AT, Deslippe JR (2020) Climate change and invasion may synergistically affect native plant reproduction. Ecology 101(1):e02913

Global Biodiversity Information Facility (GBIF) [Online] GBIF: The Global Biodiversity Information Facility. Available from: https://www.gbif.org/. Accessed 15 Aug 2025

Groner VP, Nicholas O, Mabhaudhi T, Slotow R, Akçakaya HR, Mace GM, Pearson RG (2022) Climate change, land cover change, and overharvesting threaten a widely used medicinal plant in South Africa. Ecol Appl 32(4):e2545

Gunatilleke N, Pethiyagoda R, Gunatilleke S (2017) Biodiversity of Sri Lanka. J Nat Sci Found Sri Lanka 2017:36

Hassanpouraghdam MB, Ghorbani H, Esmaeilpour M, Alford MH, Strzemski M, Dresler S (2022) Diversity and distribution patterns of endemic medicinal and aromatic plants of Iran: implications for conservation and habitat management. Int J Environ Res Public Health 19(3):1552

Herath HM, Wijesundara DS (2010) Large-scale in vitro propagation of *Impatiens repens* Moon., a critically endangered medicinal plant in Sri Lanka. J Hortic Sci Biotechnol 85(3):211–214

Hijmans RJ, Phillips S, Leathwick J, Elith J (2017) dismo: Species distribution modeling. R package version 1(4):1

Huixian J (2020) The analysis of plants image recognition based on deep learning and artificial neural network. IEEE Access 8:68828–68841

Hussain H, Nisar M (2020) Assessment of plant genetic variations using molecular markers: a review. J Appl Biol Biotechnol 8(5):99–109

International Union for Conservation of Nature (IUCN) (2004) Final report of Sri Lanka: conservation and sustainable use of medicinal plants. The World Conservation Union, Colombo (Sri Lanka)

Jayaweera DMA (1981) Medicinal plants (indigenous and exotic) used in Ceylon, vol I–III. The National Science Council of Sri Lanka, Colombo (Sri Lanka)

Jayaweera DMA (1982) Medicinal plants (indigenous and exotic) used in Ceylon, vol IV–V. The National Science Council of Sri Lanka, Colombo (Sri Lanka)

Jianzhang M, Ke R, Kun C (2012) Research and practice on biodiversity *in situ* conservation in China: progress and prospect. Biodivers Sci 20(5):551

Jones M, Liyanage C (2018) Traditional medicine and primary health care in Sri Lanka: policy, perceptions, and practice. Asian Rev World Histor 6(1):157–184

Kaky E, Gilbert F (2016) Using species distribution models to assess the importance of Egypt's protected areas for the conservation of medicinal plants. J Arid Environ 135:140–146

Kankanamalage TN, Dharmadasa RM, Abeysinghe DC, Wijesekara RG (2014) A survey on medicinal materials used in traditional systems of medicine in Sri Lanka. J Ethnopharmacol 155(1):679–691

Kaur N, Kaur N, Saggoo MI (2022) Conservation strategies for medicinal plants in the face of environmental challenges. In: Environmental challenges and medicinal plants: sustainable production solutions under adverse conditions. Springer International Publishing, Cham, pp 461–485

Khakurel D, Uprety Y, Karki S, Khadka B, Poudel BD, Ahn G, Cha JY, Kim WY, Lee SH, Rajbhandary S (2024) Assessing the risks to valuable medicinal plants in Nepal from human activities and environmental factors. Glob Ecol Conserv 51:e02860

Kougioumoutzis K, Tsakiri M, Kokkoris IP, Trigas P, Iatrou G, Lamari FN, Tzanoudakis D, Koumoutsou E, Dimopoulos P, Strid A, Panitsa M (2024) Assessing the vulnerability of medicinal and aromatic plants to climate and land-use changes in a Mediterranean biodiversity hotspot. Land 13(2):133

Lakpura LLC (2023) [Online] Ayurvedic medicinal plants of Sri Lanka. Available from: https://lk.lakpura.com/pages/ayurvedic-medicinal-plants. Accessed 16 Aug 2025

Li D, Gong Y, Zhang X, Huang M (2022) An exploratory study on the design and management model of traditional Chinese medicine quality safety traceability system based on blockchain technology. Secur Commun Netw 2022(1):7011145

Liaw A, Wiener M (2002) Classification and regression by random Forest. R News 2(3):18–22
Little DP, Tulig M, Tan KC, Liu Y, Belongie S, Kaeser-Chen C, Michelangeli FA, Panesar K, Guha RV, Ambrose BA (2020) An algorithm competition for automatic species identification from herbarium specimens. Appl Plant Sci 8(6):e11365
Liu H, Gale SW, Cheuk ML, Fischer GA (2019) Conservation impacts of commercial cultivation of endangered and overharvested plants. Conserv Biol 33(2):288–299
Long CL, Li H, Ouyang Z, Yang X, Li Q, Trangmar B (2003) Strategies for agrobiodiversity conservation and promotion: a case from Yunnan, China. Biodivers Conserv 12(6):1145–1156
Lorite J (2024) Overharvesting is the leading conservation issue of the endangered flagship species *Artemisia granatensis* Boiss. Diversity 16(12):744
Mandal K, Dhal NK (2023) Bioaccumulation of industrial heavy metals and interactive biochemical effects on two tropical medicinal plant species. Environ Sci Pollution Res 30(15):43860–43871
Mansinhos I, Gonçalves S, Romano A (2024) How climate change-related abiotic factors affect the production of industrial valuable compounds in Lamiaceae plant species: a review. Front Plant Sci 15:1370810
Marcelino S, Hamdane S, Gaspar PD, Paço A (2023) Sustainable agricultural practices for the production of medicinal and aromatic plants: evidence and recommendations. Sustainability 15(19):14095
Marchev AS, Yordanova ZP, Georgiev MI (2020) Green (cell) factories for advanced production of plant secondary metabolites. Crit Rev Biotechnol 40(4):443–458
Meng J, You X, Zhang X, Shi T, Zhang L, Chen X, Zhao H, Xu M (2023) Remote sensing application in chinese medicinal plant identification and acreage estimation—a review. Remote Sens 15(23):5580
Meyer D, Dimitriadou E, Hornik K, Weingessel A, Leisch F, Chang CC, Lin CC (2019) e1071: misc functions of the department of statistics, probability theory group (formerly: E1071), TU Wien. R package version 1(2)
Ministry of Environment and renewable energy (MoERE) (2014) Sri Lanka's fifth national report to the convention on biological diversity 2014. Ministry of Environment and Renewable Energy, Colombo (Sri Lanka)
Miththapala S (2016) Conservation revisited. Ceylon J Sci (Biological Sciences) 44(2):1
Mykhailenko O, Jalil B, McGaw LJ, Echeverría J, Takubessi M, Heinrich M (2025) Climate change and the sustainable use of medicinal plants: a call for "new" research strategies. Front Pharmacol 15:1496792
National Institute of Fundamental Studies. NIFS Sam Popham's Arboretum [Online]. Available from: https://www.nifs.ac.lk/facilities/arboretum. Accessed 2 Oct 2025
Padmalal UK, Dela J, Gunawardena HG, Ratnayake HD, Bambaradeniya C, Wijesinghe L (2006) Addendum to the biodiversity conservation in Sri Lanka-a framework for action chapter report 2: *In-Situ* conservation
Peiris R, Karunarathna H, Samaratunga U, Kaldera H, Wewalwala S (2017) Significant characteristics of traditional medicine in Sri Lanka: a review based on sources of available evidence
Perera PK (2012) Current scenario of herbal medicine in Sri Lanka. In: Proceedings of the 4th annual herbal international summit cum exhibition on medicinal & aromatic products, spices and finished products (hi-MAPS). NSIC, Okhla India
Plants of the World Online. [Online] Available from: https://powo.science.kew.org/. Accessed 1 Oct 2025
Porfirio LL, Harris RM, Lefroy EC, Hugh S, Gould SF, Lee G, Bindoff NL, Mackey B (2014) Improving the use of species distribution models in conservation planning and management under climate change. PLoS One 9(11):e113749
Pushpakumara DK, Bandara KM, Weerawardena ND (2012) The country report of forest genetic resources of Sri Lanka for the State of World's Forest Genetic Resources (SOW-FGR)'. FAO and Forest Department, Colombo, Sri Lanka and Rome
Rajakaruna RW, Yakandawala DM, Jayasuriya KM (2025) Preserving Sri Lanka's indigenous healing heritage: an updated checklist of medicinal plants and conservation priorities. Ceylon J Sci 54(1):65

Russell-Smith J, Karunaratne NS, Mahindapala R (2006) Rapid inventory of wild medicinal plant populations in Sri Lanka. Biol Conserv 132(1):22–32

Schmitt S, Pouteau R, Justeau D, De Boissieu F, Birnbaum P (2017) ssdm: an r package to predict distribution of species richness and composition based on stacked species distribution models. Methods Ecol Evol 8(12):1795–1803

Sivaraj N, Venkateswaran K, Pandravada SR, Dikshit N, ThirupathiReddy M, Rajasekharan PE, Ahlawat SP, Rao VR (2020) Geospatial technologies for threatened medicinal plant conservation. Conservation and utilization of threatened medicinal plants. Springer International Publishing, Cham, pp 229–274

Ssenku JE, Okurut SA, Namuli A, Kudamba A, Tugume P, Matovu P, Wasige G, Kafeero HM, Walusansa A (2022) Medicinal plant use, conservation, and the associated traditional knowledge in rural communities in Eastern Uganda. Trop Med Health 50(1):39

Sugathadasa KSS, Jeevandara PM et al (2008) A checklist of medicinal plants in Sri Lanka. Bandaranaike Memorial Ayurvedic Research Institute, Department of Ayurveda and ICRAF Sri Lanka Program

Sulaiman SSM, Sadaf AM, Gul B, Begum M (2020) Ethnoecological, elemental, and phytochemical evaluation of five plant species of Lamiaceae in Peshawar, Pakistan. Scientifica 2020(1):2982934

Tafesse B, Bekele T, Demissew S, Dullo BW, Nemomissa S, Chala D (2023) Conservation implications of mapping the potential distribution of an Ethiopian endemic versatile medicinal plant, Echinops kebericho Mesfin. Ecol Evol 13(5):e10061

Tagimanova D, Raiser O, Danilova A, Turzhanova A, Khapilina O (2024) Micropropagation of rare endemic species *Allium microdictyon* Prokh. threatened in Kazakhstani Altai. Horticulture 10(9):943

The National Red List 2020 (2020) Conservation status of the flora of Sri Lanka. Colombo (Sri Lanka): Biodiversity Secretariat, Ministry of Environment and the National Herbarium, Department of National Botanical Gardens

Tilburt JC, Kaptchuk TJ (2008) Herbal medicine research and global health: an ethical analysis. Bull World Health Organ 86:594–599

Unal BT, Turker H, Ozturk M (2023) Ex-situ conservation of medicinal and aromatic plants using in vitro techniques. Plants as medicine and aromatics. CRC Press, pp 13–22

UNEP-WCMC and IUCN (2025) Protected area profile for Sri Lanka; World Database on Protected Areas (WDPA). UNEP-WCMC, Cambridge (UK). [Online] Available from: https://www.protectedplanet.net/country/LKA. Accessed 2 Sept 2025

United Nations Environment Programme (UNEP) (2009) Judges and environmental law: a handbook to the Sri Lankan judiciary. United Nations Environment Programme, Colombo 5 (Sri Lanka)

Wäldchen J, Mäder P (2018) Machine learning for image based species identification. Methods Ecol Evol 9(11):2216–2225

Walter KS, Gillett HJ (eds) (1998) 1997 IUCN red list of threatened plants. IUCN

Wawrosch C, Zotchev SB (2021) Production of bioactive plant secondary metabolites through in vitro technologies—status and outlook. Appl Microbiol Biotechnol 105(18):6649–6668

Weragoda PB (1980) The traditional system on medicine in Sri Lanka. J Ethnopharmacol 2(1):71–73

Woldeamanuel MM, Geda MK, Mohapatra S, Bastia TK, Rath P, Panda AK (2022) Ethnobotanical study of endemic and non-endemic medicinal plants used by indigenous people in environs of Gullele botanical garden Addis Ababa, central Ethiopia: a major focus on Asteraceae family. Front Pharmacol 13:1020097

World Health Organization (2000) General guidelines for methodologies on research and evaluation of traditional medicine. World Health Organization, Geneva. (WHO/EDM/TRM/2000.1)

Wu J, Li X, Huang L, Meng X, Hu H, Luo L, Chen S (2019) A new GIS model for ecologically suitable distributions of medicinal plants. Chin Med 14(1):4

Zeng L, Sun Z, Fu L, Gu Y, Li R, He M, Wei J (2024) Cryopreservation of medicinal plant seeds: strategies for genetic diversity conservation and sustainability. Plants 13(18):2577

[illegible]

Chapter 7
Classical and Contemporary Approaches to Assessing Genetic Diversity, Including Case Studies of Selected Endangered Medicinal Plants in India

Anrini Majumder

Abstract India is endowed with a rich diversity of medicinal plants. These plants form the backbone of the traditional systems of medicine, and many medications have been semi-synthesized from lead compounds isolated from these plants. This rich natural resource, however, is disappearing rapidly due to various factors. Therefore, its conservation and management are of utmost importance. For planning effective conservation and breeding strategies, it is essential to have a thorough understanding of a species' genetic makeup and diversity across different populations. It is a well-established fact that gene flow among populations and the percentage of polymorphism are key factors ensuring the survival of a species. There are very few studies on cytogenetic diversity and its correlation with phytochemicals at inter- or intraspecific levels in medicinal plants. This review provides a brief account of various cytogenetic approaches, including DNA markers, to assess the cytogenetic character and genetic diversity of some endangered medicinal plants. This approach may enable the identification of highly valuable genotypes in medicinal plant populations that should be prioritized for conservation.

Keywords Cytogenetics · DNA markers · Endangered species · Genetic diversity · Medicinal plants

7.1 Introduction

The global biodiversity hotspots are biogeographic regions with significant levels of biodiversity that are threatened with destruction (Jaisankar et al. 2018). The vast expanse of the Indian landscape has resulted in the formation of 15 agro-climatic zones, harboring a rich phyto-diversity of 10.45% of the entire global floral diversity

A. Majumder (✉)
Department of Botany, Harimohan Ghose College, Kolkata, West Bengal, India

S. Jha et al. (eds.), *Genetic Diversity in Medicinal Plants*, Sustainable Development and Biodiversity 44, https://doi.org/10.1007/978-981-95-7281-6_7

(Gowthami et al. 2021). Since time immemorial, medicinal plants have been used by mankind for therapeutic purposes, and ancient Indian texts such as the Vedas, Upanishads, and Samhitas have documented medicinal plants for treating various ailments (Che et al. 2024). Along with Allopathy, Homoeopathy, and Naturopathy, the Ayurvedic, Unani, and the Siddha systems of medicine are still well practiced in India as the traditional systems in addition to the different community-specific ethnomedicinal practices (Gowthami et al. 2021; Rizvi et al. 2022). These traditional and indigenous medicines using medicinal plants have the potential of being used either directly or indirectly for the treatment of many life-threatening diseases. Many pharmaceutically active secondary metabolites isolated from these plants have also been used as lead compounds for the semi-synthesis of drugs by the pharmaceutical industry (Kaundal and Kumar 2025). Unfortunately, a large percentage of these plants are facing a threat of extinction because of various factors, viz., degradation of population, reduction of reproductive or seed germination capacity, climate change, depletion of pollinators, destruction of habitat because of natural calamities as well as anthropogenic activities such as deforestation, urbanization, pollution, invasion of land for agricultural and industrial purposes, grazing by livestock, overexploitation for collection of bioactive ingredients and other domestic purposes, unskilled harvesting, and loss of genetic diversity due to increased genetic drift and inbreeding (Gao et al. 2020; Gowthami et al. 2021). Thus, various measures are being taken to protect these natural treasures from becoming extinct.

To design effective in-situ and ex-situ conservation and breeding programs, it is crucial to have a deep understanding of a species' genetic makeup and the range of genetic variation within and among populations. The interplay of several factors is responsible for creating the genetic profile of a plant population and for genetically distinguishing different populations of a single species (Vaishnav et al. 2023). Plant form, reproductive strategies, seed-dispersal processes, pollination mode, geographic distribution range, and anthropogenic activities are other factors that govern a species' genetic diversity (Gamba and Muchhala 2020; De Kort et al. 2021). The survival of a species depends on its genetic makeup, a diverse gene pool, gene flow, and the level of polymorphism between populations (Tomar et al. 2021; Negi et al. 2022).

This review summarizes cytogenetic diversity in medicinal plants in general and discusses some endangered medicinal plants of India that may assist in their conservation for future use.

7.2 Cytogenetic and Phytochemical Diversity of Medicinal Plants

Plant chromosomes exhibit wide diversity in number and structure at the intra- and interspecific levels. Data on genetic diversity at the chromosome level, such as ploidy, structure, size, mitotic and meiotic behavior, aberrations, structural variations, and banding patterns (https://bsi.gov.in/page/en/database-on-genome-related-information-of-indian-plants), are crucial not only for planning breeding and conservation strategies but also in understanding evolutionary trends among plants, interspecies relationships, and reproductive behavior, as well as in understanding variation of metabolite profiles. Karyotype analysis using classical and molecular cytogenetic techniques is an effective tool for assessing genetic diversity at the chromosome level. Some species exhibit polyploidy, a trait that is significant for evolution, adaptation, and the development of desired characteristics. The effects of polyploidy are variable, which may be desirable or undesirable. To a large extent, diverse cytotypes give rise to new chemotypes through variations in phytochemical profile and content. Production of plant secondary metabolites, one of the defining characteristics of medicinal plants, might be affected by ploidy level, leading to increases or decreases in bioactive phytochemicals. There are very few studies on cytogenetic diversity and its correlation with phytochemicals at inter- or intraspecific levels, as revealed in Table 7.1.

7.3 Molecular Markers in Assessing Genetic Diversity of Medicinal Plants

The last few decades have witnessed an extensive use of DNA-based molecular markers for species identification, tracing their origin, understanding evolutionary relationships, gene mapping, studying population genetics, and detecting polymorphism in plants (Nadeem et al. 2018; Negi et al. 2022; Bidyananda et al. 2024). These markers have also proven to be valuable tools for plant breeding and conservation schemes (Nadeem et al. 2018; Kumar et al. 2024). These can be classified as non-PCR, ((polymerase chain reaction) hybridization-based restriction fragment length polymorphism (RFLP) markers and the PCR-based marker systems, such as randomly amplified polymorphic DNA (RAPD), amplified fragment length polymorphism (AFLP), simple sequence repeats (SSRs), inter-simple sequence repeats (ISSRs), and so on, which are extensively used in plant genomic research (Kumar et al. 2009). RAPD analysis is rapid, easy to analyze, with 10-mer primers of arbitrary nucleotide sequences to amplify random segments of genomic DNA. However, its limitations include low reproducibility and dominant expression of alleles. For species identification, codominant, reproducible, reliable, sequence-based markers or sequence-characterized amplified region (SCAR) created by converting the RAPD amplicons might be helpful (Kiran et al. 2010; Sairkar et al. 2016). Microsatellites or SSRs are very effective multipurpose molecular markers in population genetics, genome mapping, and analysis of genetic diversity, taxonomic studies, and so on because of their abundance, reproducibility, codominant and polymorphic nature, and genome-wide distribution (Hosseinzadeh-Colagar et al. 2016; Kalariya et al. 2021). ISSR marker techniques are fast, simpler than the SSRs, economic, and do not require any prior knowledge of DNA sequences, and the

Table 7.1 Reports on the correlation of phytochemicals with natural or induced variation in ploidy level in selected species of medicinal plants

Plant species	Family	Chromosome number	Correlation with phytochemicals	References
Inula racemosa Hook. f.	Asteraceae	$n = 10$, $n = 20$	An upsurge in the quantity of isoalantolactone and alantolactone was noticed with increasing ploidy levels	Jeelani et al. (2022)
Rauvolfia spp.	Apocynaceae	$2n = 22$ in *R. serpentina* (3 populations), $2n = 22$ in *R. vomitoria*, $2n = 66$ in *R. canescens*	Variation in reserpine content (0.23–0.41%) in populations of *R. serpentina* reported. *R. vomitoria* showed 0.40% while hexaploid *R. canescens* showed 0.27% reserpine	Banerjee and Sharma (1989)
Artemisia annua L.	Asteraceae	$2n = 18$	Artemisinin content increased from 39% to 56% in induced tetraploids	Lin et al. (2011)
			Increase in the average level of artemisinin in induced autotetraploids from 42.0% to 63.6%.	Xia et al. (2018)
Centella asiatica (L.) Urb.	Apiaceae	$2n = 18$	Total triterpenoids increased significantly in induced tetraploids	Thong-on et al. (2014)
Datura stramonium L.	Solanaceae	$2n = 24$	Higher alkaloid content and scopolamine/hyoscyamine ratio in induced tetraploids	Berkov and Philipov (2002)
Eclipta alba (L.) Hassk.	Asteraceae	$2n = 22$	Higher accumulation of wedelolactone in induced tetraploid plants	Salma et al. (2018)
Linum album Kotschy ex Boiss.	Linaceae	$2n = 30$	1.39- and 1.23-fold enhancement of podophyllotoxin production in leaves and stems of induced tetraploids, respectively	Javadian et al. (2017)
Papaver somniferum L.	Papaveraceae	$2n = 22$	Significant enhancement in morphine content in induced tetraploids	Mishra et al. (2010)
Thymus vulgaris L.	Lamiaceae	$2n = 30$	Essential oil, thymol, carvacrol, γ-terpinene, and linalool content increased in induced tetraploid plants	Mohammadi et al. (2023)
			41.11% higher essential oil yield with enhanced thymol and γ-terpinene content in induced tetraploids	Gupta et al. (2024)
Zingiber officinale Rosc.	Zingiberaceae	$2n = 22$	Essential oil content is significantly higher in induced tetraploids	Prasath et al. (2022)

results are reproducible (Kurane et al. 2009). These have been the markers of choice for fingerprinting of germplasms of wild as well as crop plants, phylogenetic studies, gene tagging, genome mapping, evolutionary relationships, and studying intra- and interspecific genetic diversity in a wide range of plant species (Kulhari et al. 2015; Rathore et al. 2016; Thakur et al. 2016; Sarma and Tanti 2017; Negi et al. 2022). Directed Amplification of Minisatellite DNA (DAMD) is another reproducible PCR-based marker system revealing polymorphism due to minisatellites and has been used to detect genetic diversity in a number of plants of medicinal relevance (Bhattacharyya et al. 2015; Panapitiya and Welikala 2018; Saleh 2019; Vaishnav et al. 2023). Genome sequencing is one of the major milestones of biological research. It is an immensely important technique for conservation of species, as it provides information regarding genetic diversity of domestic and wild species, integrates genetic diversity through breeding programs to preserve species existence, and assists in genetic rescue schemes to protect and improve the viability of threatened populations of species (Ralls et al. 2020; Schneider 2022). Areas such as plant responses to stress, evolutionary dynamics of species, improvement of agronomic traits, and so on are also within the purview of this technique (Thottathil et al. 2016; Hong et al. 2023).

7.4 Case Studies

7.4.1 Aconitum heterophyllum

Variously known as *Atis* or *Ativisha* or the Greenish Himalayan Monkshood, *Aconitum heterophyllum* Wall. ex Royle ($2n = 16$) belonging to the family Ranunculaceae, is an important medicinal herb (Jeelani et al. 2015; Wani et al. 2022). Biennial in nature, it is commonly found in the alpine and subalpine Himalayan regions at altitudes of 2000–5000 m above mean sea level (amsl), with distributions in the Indian states of Jammu and Kashmir, Himachal Pradesh, and Uttarakhand (Jeelani et al. 2015; Wani et al. 2022). The tubers of the plant are rich in secondary metabolites such as aconitine, mesaconitine, hypaconitine, atisine, isoatisine, hetisine, heteratisine, and heterophylline, in addition to flavonoids and saponins, which account for its therapeutic potential against a wide range of diseases of the immune, digestive, and nervous systems of human beings (Murti and Khorana 1968; Mori et al. 1989; Malhotra et al. 2014). The only non-toxic member of the genus, *A. heterophyllum,* has notable hepatoprotective, analgesic, diuretic, antipyretic, astringent, aphrodisiac, antioxidant, carminative, antibacterial, antifungal, insecticidal, antiviral, anticancer, and anti-inflammatory activities (Tomar et al. 2025). The whole plant has to be uprooted for harvesting the tubers (Tomar et al. 2025). Therapeutic demand resulting in indiscriminate uprooting of whole plants for tuber collection from the wild, coupled with other factors such as patchy, limited distribution, low population density, reduced genetic diversity, long seed dormancy, poor germination, high rate of mortality at the seedling stage, and cultivation in small scale has led to the depletion of natural populations of this species and subsequently its categorization as an "endangered species" according to the IUCN (International Union for Conservation of Nature) (Jeelani et al. 2015; Gowthami et al. 2021; Negi et al. 2022; Tomar et al. 2025). The plant has been prioritized by the National Medicinal Plant Board (NMPB) as one of the 32 threatened medicinal plants of India to be conserved (Gowthami et al. 2021).

Various authors have undertaken studies to understand the genetic composition and investigate genetic diversity among populations of *A. heterophyllum* collected from multiple locations in India. Jeelani et al. (2015) studied 21 natural populations of *A. heterophyllum* collected from Kashmir and the Ladakh Himalayas. They assessed the collected populations of *A. heterophyllum* cytologically, morphologically, and ecologically to examine cytogenetic diversity in relation to the distribution patterns governed by various ecological factors. Although wide variation was observed in morphological features at the intrapopulation level, chromosome analysis revealed that all populations were diploid ($2n = 16$). However, intraspecific genetic variation was revealed by the presence of meiotic anomalies in some of the populations. Although 10 populations showed normal meiosis and microsporogenesis with high pollen fertility, 11 populations exhibited abnormalities at various meiosis stages, leading to abnormal microsporogenesis (Jeelani et al. 2015). Studies also revealed an uneven, low-density distribution of *A. heterophyllum* populations and that genetic diversity was associated with ecological conditions (Jeelani et al. 2015).

Inter- and intrapopulation genetic composition of different populations of *A. heterophyllum* from the Himalayan state of Uttarakhand was studied by Negi et al. (2022) using ISSR markers. Leaves, rhizomes, and roots from 25 individuals from seven different populations were analyzed using five primer combinations. The average polymorphism percentage per primer was 78. Results indicated that genetic variation at the intrapopulation level was 35%, whereas it was 65% among populations, indicating genetic variation among populations. Different populations were related as revealed by the UPGMA (unweighted pair group method with arithmetic mean) dendrogram, with the populations being divided into two major clusters. The authors hypothesized that the smaller population size in the region and limited gene flow among the populations of this species, because of overexploitation, endemism, and disintegrated habitats, were responsible for the lower genetic diversity and polymorphism of this species (Negi et al. 2022).

ISSR markers were similarly used to assess the genetic diversity of eight germplasm accessions of *A. heterophyllum* maintained at the Nature Interpretation Site, Baniyakund, in Rudraprayag district, Uttarakhand (Chauhan et al. 2023). Although distinct variations were noted among the plants collected from the Garhwal and Kumaun regions, accessions of each population were highly similar, indicating low intrapopulation genetic diversity. Kumar et al. (2023) also studied the genetic diversity and population structure of *A. heterophyllum* from the North Western Himalayan states of Uttarakhand and Himachal Pradesh using microsatellite markers on 36 accessions representing eight populations of the species. Narrow genetic diversity at the species level was observed, along with higher genetic differentiation and low gene flow, due to population fragmentation, suggesting possible reasons for the decline of populations of this immensely important medicinal plant. The authors recommended developing in-situ conservation strategies to improve gene flow between populations (Kumar et al. 2023).

7.4.2 Aristolochia indica

Aristolochia indica L. (Aristolochiaceae, $2n = 12$) is a woody, climbing, perennial herb or shrub. Also known as the Indian Birthwort in English and *Ishwarmul* in Hindi, the plant is well-documented in the Siddha, Ayurveda, Unani, and other indigenous systems of medicine (Sarma and Tanti 2015; Gautam and Bansal 2022; Sreelakshmi et al. 2024). Considered as an antidote against snakebites and insects, the plant has been reported to relieve stiffness and soreness in the joints and in the treatment of dry cough, bowel disorders, biliousness, intermittent fever, epilepsy, schizophrenia, leprosy, ulcers, wounds, postpartum infections, and so on. The plant also exhibits abortifacient, antimicrobial, antibacterial, anti-inflammatory, antidiabetic, and anticancer activities (Gautam and Bansal 2022; Sreelakshmi et al. 2024; Neethu et al. 2024). Phytochemical analysis of extracts from different plant parts has revealed the presence of steroids, glycosides, fatty acids, alkaloids, flavonoids, saponins, tannins, triterpenoids, and volatile oils in addition to anthocyanidins, aristolochic acid, ceryl alcohol, β-sitosterol, and rutin (Sati et al. 2011; Pattar and Jayaraj 2012). Factors such as overexploitation due to its medicinal importance, habitat fragmentation, and other anthropogenic activities have resulted in the depletion of *A. indica* populations in the wild, resulting in its enlistment as a rare and endangered species in Assam, Madhya Pradesh, and the Eastern Ghats of India (Sarma et al. 2018; Gautam and Bansal 2022; Gajarajan et al. 2023; Neethu et al. 2024).

Sarma and Tanti (2015) conducted karyomorphological analyses of *A. indica* along with two other species (*A. saccata* Wall. and *A. cathcartii* Hook. f.) collected from Kamrup, Barpeta, and Karbi Anglong districts of Assam. Different parameters, such as somatic chromosome number, total chromosome length, volume of chromosomes, arm ratio (length of long arm/length of short arm), and centromeric position, were studied from fresh root tip preparations from the collected plants. Studies revealed that the somatic cells of *A. indica* had 12 chromosomes with lengths ranging from 1.36 to 3.06 μm and volumes of 0.004–0.176 μm^3; arm ratios ranged from 1 to 1.5 with a prevalence of median and metacentric chromosomes (six median and six metacentric chromosomes), suggesting a primitive nature. The karyotypic formula was depicted as $M_6 + m_6 = 2n = 12$. Although Sharma and Varma (1959) reported a pair of chromosomes with secondary constriction, no such chromosome was detected by Sarma and Tanti (2015), prompting the authors to suggest that habitat and climatic conditions are responsible for such karyological variations (Sarma and Tanti 2015).

In a subsequent study, the authors (Sarma and Tanti 2017) analyzed the genetic diversity and phylogenetic relationship of *A. indica* with three other species (*A. saccata* Wall., *A. tagala* Cham., and *A. cathcartii* Hook. f.) from Assam using a panel of eight ISSR-based primers. Binary data from banding patterns generated by the ISSR markers was used to construct a consensus tree, which depicted that *A. indica* and *A. cathcartii* were 62% similar at the genetic level and were included in the same cluster, while the other two species were included in separate clusters, indicating polymorphism at the genetic level. The authors pointed out that the diverse genetic stock of *Aristolochia* might be identified using such studies, paving the way for conservation of such stocks on a preference basis.

7.4.3 Commiphora wightii

Indian bdellium or *Commiphora wightii* (Arn.) Bhandari (syn. *C. mukul* Hook. ex Stocks, $2n = 26$) of the family Burseraceae is a small, perennial tree or shrub, with distributions in the arid and semi-arid zones of North Western India, particularly in the states of Gujarat and Rajasthan (Sobti and Singh 1961; Sairkar et al. 2016; Bishoyi et al. 2018; Brindavanam et al. 2022). Commonly known as the *Guggul* or *Guggal* in Hindi, the plant displays an array of biological activities, viz., anti-inflammatory, antirheumatic, hypocholesterolemic, hypolipidemic, antifertility, anticancer, hepatoprotective, neuroprotective, antioxidant, antimicrobial, antimalarial, antiarthritic, and cardioprotective activities, being used in Allopathy, the Ayurvedic, and Unani systems of medicine (Satyavati 1990; Sairkar et al. 2016; Kunnumakkara et al. 2018; Bishoyi et al. 2018). The plant contains a plethora of metabolites, including amino acids, fatty acids, organic acids, phenolic acids, pregnane derivatives, steroids, sterols, sugars, sugar alcohols, terpenoids, and tocopherol in the leaves, stem, roots, latex, and fruits (Bhatia et al. 2015). Upon injury, the stems and leaves of the plant ooze out a fragrant, bitter oleo-gum-resin known as *Guggul* (Myrrh), having guggulsterone as the major bioactive ingredient responsible for the plant's various therapeutic properties, which solidifies to a brownish resin in contact with the hot environment (Sairkar et al. 2016; Bishoyi et al. 2018; Tomar et al. 2021). *Guggul* is also used for perfumery, calico-printing, and in fabric-dying industries (Brindavanam et al. 2022). Local tribals extract the resin from the main trunk by "tapping," resulting in infection by bacterium *Xanthomonas axonopodis,* causing gummosis and subsequent death of the plant (Samanta et al. 2012; Bishoyi et al. 2018). Reckless, illegal harvesting to extract the bioactive resin for medicinal, religious, and domestic purposes, associated with factors like poor seed setting and germinating ability of the plant, slow growth rate, and lack of proper cultivation methods have led to the categorization of *C. wightii* as "critically endangered" in the IUCN Red Data list (Reddy et al. 2012; Gowthami et al. 2021). Thus, initiatives were taken to characterize the different germplasms of this plant collected from different habitats to adopt conservation and breeding strategies.

Kulhari et al. (2015) used ISSR markers for the first time, as well as RAPD markers, to study genetic diversity among *C. wightii* germplasms from the Indian states of Rajasthan and Haryana. Of the 100 markers used (50 each of RAPD and ISSR), RAPD markers were superior to the ISSR ones, detecting 74.16% polymorphism, compared to 62.52% polymorphism detected by the ISSR primers in 52 accessions of the species. The collected accessions were separated into two major clusters as revealed by the UPGMA dendrogram constructed, with one accession from Bikaner of Rajasthan being not included in any of the clusters. Wide genetic variation could be detected within the species, data of which the authors claimed was applicable for planning conservation of germplasm, plant breeding, and population genetics studies.

In addition to being "critically endangered," because of the similarities of gum *Guggul* with gum resins of other species at the intra- and intergeneric level, there are chances of adulteration in commercial samples. Therefore, to identify and discriminate between *C. wightii* and closely related *C. myrrha*, Sairkar et al. (2016) developed species-specific SCAR markers. These markers are characterized by their locus specificity, targeting specific regions of the genome as these are polymorphic in nature, codominance, abundance, cheapness, ease of usage, rapidity, and high reproducibility (Sairkar et al. 2016). For the purpose of their study, 28 accessions of these two species of *Commiphora* were collected by the authors from different regions of the states of Madhya Pradesh, Maharashtra, Gujarat, and Rajasthan. After screening of 20 RAPD primers, two species-specific RAPD amplicons (1 kb amplicon of OPD-02 and 0.6 kb amplicon of OPD-08) were sequenced following selection, gel purification, and cloning. Four SCAR markers, namely, Sc1P, Sc1Pm, Sc2P, and Sc2Pm, were developed using sequences of 979 and 590 nucleotides (Genebank accession numbers K90051 and K90052), of which Sc1Pm was highly specific for *C. wightii* amplifying 910 bp amplicon, whereas primer Sc2P distinguished between the two *Commiphora* species, producing 491 bp amplicon for *C. wightii* and 1200 bp amplicon for *C. myrrha.* Notably, 491 and 570 bp amplicons for *C. wightii* and *C. myrrha*, respectively, were produced by Sc2Pm. The study proved that *C. wightii* species could be identified and authenticated using SCAR markers.

In an attempt to analyze the genetic diversity of *C. wightii* populations in relation to reproductive behavior, molecular markers (RAPD and ISSR) were used by Bishoyi et al. (2018) on fresh and young leaves of 73 accessions collected from different geographical regions of Gujarat (the only state in India with a distribution of sexual populations) and Rajasthan. These accessions were males, females (either apomicts or sexuals), or hermaphrodites. Out of 140 RAPD primers screened, 24 produced reproducible bands with a total of 185 DNA amplicons in all the accessions, with an average of 80% polymorphism, whereas 16 out of 35 ISSR primers screened produced 128 clear amplified bands with 87 polymorphic, 37 monomorphic, and 4 unique bands; the average polymorphism was 68.5%. Results indicated that instead of being governed by geographical locations, the pattern of genetic diversity could be correlated to the reproductive behavior of the populations, with low diversity being noted in the apomictic populations of Rajasthan, whereas the highest diversity, along with sexual forms, was seen in the Gujarat populations. Based on their studies, the authors speculated that populations of *C. wightii* distributed in parts of Gujarat and Rajasthan might have originated from ancestral stocks occurring in specific regions of Gujarat (Jamnagar, Dwarka, and Kutchh). The authors also hypothesized that the primitive sexual forms might have transformed into the apomictic forms because of adverse environmental conditions prevailing in the newly adapted areas (Bishoyi et al. 2018).

In a similar study, Tomar et al. (2021) conducted vast surveys in forest areas of Rajasthan to map the population distribution, density, and genetic diversity of *C. wightii* populations and analyze the relationship among geographical and genetic distances and agroclimatic zones. To assess the genetic diversity, DNA was isolated from fresh, juvenile leaves of 96 genotypes, and 6 out of a total of 15 RAPD primers were selected on the basis of the generation of a high number of scorable, polymorphic bands. Surveys revealed that an average of 21.9 *Guggul* plants per ha were distributed over the entire state of Rajasthan, with regions such as Sawai Madhopur, Jhunjhunu, Barmer, and Jalore being recognized as the hotspots of Guggul populations. Altitude differences contributed to the variations in population densities because of higher anthropogenic activities at lower altitudes. The percentage of polymorphism as generated by the RAPD primers ranged from 75 to 100; the data also revealed that genetic diversity was higher within populations than among populations of *C. wightii* in four agroclimatic zones of Rajasthan and indicated a moderate gene flow between the zones. The authors suggested dispersal of seeds and other vegetative propagules by birds or anthropogenic activities to long distances as probable reasons for moderate gene flow, which again led to the fragmentation of populations (Tomar et al. 2021).

ISSR markers were also used to assess genetic diversity in *C. wightii* populations in conservation sites (Medicinal Plants Conservation Areas or MPCAs developed for long-term, non-interventional conservation of *Guggul*) of the Kachchh region of Gujarat (Brindavanam et al. 2022). A total of 40 ISSR primers were tested on 38 accessions of *Guggul* collected from five sites (four MPCA sites and one wildlife sanctuary—Narayan Sarovar Wildlife Sanctuary or NSWS), of which 14 primers produced 49 scorable bands, including 20 polymorphic ones. Results indicated that accessions of the MPCAs had narrow genetic diversity. However, the accessions of NSWS were genetically distant to the populations of the MPCAs, making it a potential site for the development of the fifth MPCA aimed at conserving all the genotypes of the Kachchh region.

7.4.4 Gymnema sylvestre

Commonly known as the *Gurmar* or *Madhunashini* (the sugar killer) due to its antidiabetic properties, *Gymnema sylvestre* (Retz.) R. Br. ex Schult. ($2n = 22$, Asclepiadaceae) is a perennial, slow-growing liana, found primarily in the tropical peninsular India and the Andaman and Nicobar Islands, with extensions till Uttar

Pradesh and Bihar (Thakur et al. 2012; Verma et al. 2016; Polaiah et al. 2023; Vaishnav et al. 2025). Mentioned in the Indian Pharmacopoeia as an official antidiabetic drug because of its stimulatory effect on pancreatic cells and potentiality to treat Type I and Type II diabetes, according to the Ayurveda, it is also used traditionally in the treatment of dyspepsia, constipation, jaundice, hemorrhoids, renal and vesicle calculi, cardiopathy, asthma, bronchitis, amenorrhea, and leukoderma (Verma et al. 2016; Verma and Dhawan 2017). It has also been demonstrated to exhibit hypolipidemic, hepatoprotective, neuroprotective, immunomodulatory, cardioprotective, anti-inflammatory, antioxidant, anticancerous, antibacterial, and antiviral properties and used for maintaining normal blood pressure, prevention of dental caries, cataract, and so on (Kalariya et al. 2021; Vaishnav et al. 2023). Herbal formulations based on *Gymnema,* either alone or in combination with other plants, are preferred choices over other plant-derived antidiabetic formulations, and among other constituents such as gymnemasaponins, gymnemosides, gymnemanol, gymnemagenin, and gymnemasin, along with several glycosides, essential oils, alkaloids, flavonoids, valepotriates, and so on, gymnemic acid is the primary bioactive principle exhibiting the antidiabetic activity by stimulating the release of insulin from the pancreas (Verma et al. 2016; Jinu et al. 2019; Kalariya et al. 2021; Vaishnav et al. 2023). Listed among the 242 most traded medicinal plants in India, it has seen surging commercial demand, leading to overexploitation and unscientific harvesting from the wild (Vaishnav et al. 2023). Other factors such as loss of habitat, stress of climatic and environmental change, invasion by exotic species, slow natural regeneration of the species, poor seed germination, and lack of large-scale, sustainable cultivation protocols have led to the categorization of this species as near threatened (NT)/vulnerable(VU)/endangered (EN) in several states of India, as declared by the Conservation Assessment and Management Prioritization evaluation (Verma et al. 2016; Gowthami et al. 2021, Vaishnav et al. 2023, 2025). The NMPB has selected *G. sylvestris* as one of 32 threatened medicinal plants in India for conservation (Gowthami et al. 2021).

Identification of elite genotypes with desirable characteristics is a prerequisite for designing crop improvement and conservation programs. A number of reports indicate that efforts have been directed toward elucidating the genetic makeup of different *G. sylvestre* populations. Genetic diversity and population genetic structure of 117 accessions of *G. sylvestre* collected from Karnataka, Maharashtra, and Goa in the Western Ghats region and from West Bengal in Eastern India using ISSR and RAPD markers were studied by Rathore et al. (2016). A high degree of genetic diversity at the species level was noted with both types of markers. Both the seven selected ISSR and six selected RAPD primers produced 100% polymorphic bands. The ISSR markers generated 45.86% rare, 52.63% shared, and 1.51% similar bands, whereas the figures were 38.23%, 58.59%, and 3.13%, respectively, for the RAPD marker-generated bands. Thus, the markers helped in the identification of particular accessions through the rare bands and the resembling accessions through the shared and similar bands. At the species rank, 97.08% and 98.46% bands were polymorphic based on ISSR and RAPD fingerprinting, respectively. Variations were also noted in polymorphism according to the geographical regions, with the ISSR and

RAPD markers depicting 58.49% and 83.05% polymorphism for the Eastern Indian region accessions, respectively, whereas both sets of primers revealed 100% polymorphism for the Western Ghats accessions. While genetic differentiation was high in the species, gene flow was low, indicating that genetically distinct populations should be conserved. The populations were separated by geographic barriers and anthropogenic activities, affecting gene flow through seed and pollen dispersal. The authors speculated that habitat fragmentation, low seed dispersal, and geographical isolation might be some of the reasons for low gene flow in populations. The dendrogram created by UPGMA suggested that most of the populations branched according to their geographical location. The authors also presumed that in spite of the wide distribution of *G. sylvestre,* the fragmented habitats and small-sized isolated populations were the reasons for a high degree of genetic differentiation among the populations (Rathore et al. 2016).

Genetic and chemical fingerprinting using ISSR markers and high-performance liquid chromatography (HPLC) analysis, respectively, of 14 accessions of *G. sylvestre*, collected from different locations of Central India and maintained at Central Institute of Medicinal and Aromatic Plants (Lucknow, Uttar Pradesh, India), were carried out by Verma et al. (2016) with an aim of selecting high metabolite yielding, elite accessions or improved genotypes. 15 polymorphic of 40 ISSR markers produced scorable, unambiguous bands of which 12 were anchored-type dinucleotide repeat, one anchored-trinucleotide repeat, and one non-anchored tetra nucleotide repeat. Notably, 75.89% alleles were polymorphic, indicating a high degree of polymorphism or genetic variation among the 14 genotypes analyzed. Two of the primers tested revealed 100% polymorphism, indicating their potentiality of being used for identifying *G. sylvestre* accessions. Also, nine unique accession-specific bands were generated by the ISSR primers, which the authors claimed could be used as identifiers for protecting specific accessions and to detect any mixing up of cultivars. Variability was also detected among the accessions in the content of gymnenic acid, as revealed by the HPLC analysis of the alcoholic extracts of powdered leaf samples. The phylogenetic tree constructed indicated that the high-yielding accessions were closely related to each other. The authors suggested that for the identification of high-yielding accessions or genotypes, the polymorphic ISSR markers could be used as molecular probes (Verma et al. 2016).

In a similar study, seven accessions of *G. sylvestre* were collected from Tamil Nadu, India, and subjected to RAPD and HPLC analysis (Jinu et al. 2019). Notably, 13 of 100 oligonucleotide primers screened revealed DNA polymorphism among the populations, with an average of 54.2% polymorphism. A dendrogram constructed on the basis of the data obtained from RAPD analysis could divide the seven accessions into four major groups. The authors noticed that accessions from different geographical origins were clustered together. Notable variation in gymnemic acid content was also recorded among the accessions, with accession number L7 possessing the maximum amount of gymnemic acid. Also, the highest level of genetic diversity was with L7 as revealed by the UPGMA dendrogram, suggesting a positive correlation between the genetic make-up and gymnemic acid content of the accessions, for selection of potential genotypes for large-scale cultivation.

Kalariya et al. (2021) investigated the genetic diversity at the molecular level of 25 genotypes of *G. sylvestre* collected from different locations of India, using the efficient Expressed Sequence Tags-SSR (EST-SSR) markers. The authors designed 40 pairs of primers, with 27 producing polymorphic loci, and identified a total of 4448 EST-SSR loci. Different types of EST-SSR repeat sequence motifs (di-, tri-, tetra-, penta-, and hexanucleotide) were also identified, with trinucleotide motif repeats being the most predominant. Based on the results, a dendrogram was built, which revealed that the 25 genotypes could be clustered into two main groups, with genotypes from the same state being distributed in separate clusters. The authors presumed that the disorganized distribution of the genotypes might be due to the migration of *G. sylvestre* populations to different places.

In a subsequent study, a comprehensive and cumulative analysis of genetic variation and population genetic structure of 118 accessions of 11 populations collected from the Indian states of Uttar Pradesh, Madhya Pradesh, Chhattisgarh, Maharashtra, Telangana, Karnataka, and Kerala, that is, spanning a vast region of the peninsular India, was done using eight DAMD and 17 ISSR markers (Vaishnav et al. 2023). High degree, that is, 82.58% and 79.93% of polymorphism was detected with the DAMD and ISSR markers, respectively, while 80.89% was the cumulative value of polymorphism, indicating that *G. sylvestre* has a broad genetic base. Studies revealed significant genetic diversity at the species level, while low diversity was detected at the population level, which could be attributed to the differences in genetic diversity of different populations, with an average of 45.12% polymorphism per population. Most of the diversity was because of genetic diversity within populations. Region-wise, maximum genetic diversity was noted in the populations of South India, followed by Central India, with the North Indian populations exhibiting the least genetic diversity. The genetic identities and genetic distances between 11 populations of *G. sylvestre* were consistent with geographical diversity. The authors presumed that various factors, such as the long-lived, perennial, liana-like nature of *G. sylvestre*, its outbreeding capability because of remarkable entomophily, the presence of comose seed with a wind dispersal mechanism, and wide distribution over the entire peninsular India, are responsible for its genetic diversity. They concluded that populations with maximum genetic diversity should be given priority for conservation purposes, whereas attempts must be made to increase the diversity in the low variation populations. The authors also proposed that the genetically distinct populations identified through their study might serve as potential genetic resources valuable for future conservation programs of *G. sylvestre* (Vaishnav et al. 2023).

Next-generation sequencing (NGS) is a technology used for parallel sequencing of DNA and RNA and for detecting variants and/or mutants. Using this technology, hundreds or thousands of genes or the whole genome can be sequenced in a short period of time. The development of NGS has revolutionized genomics by providing cost-effective, high-throughput techniques for the identification of markers in plants (Qin 2019; Panahi et al. 2024). Polaiah et al. (2023) reported, for the first time, the development of novel genomic SSR markers using NGS to assess the genetic diversity of 26 genotypes of *G. sylvestre* collected from diverse regions of India and maintained at the germplasm field gene bank of the ICAR-Directorate of Medicinal and Aromatic Plants Research, Anand, Gujarat. Notably, 58 of the 100 SSR primers synthesized were amplified, yielding alleles ranging from one to six. The UPGMA dendrogram revealed that the genotypes could be grouped into three clusters, which were not according to their geographic origin. According to the authors, selected divergent genotypes could be used for breeding programs (Polaiah et al. 2023).

7.4.5 Lilium polyphyllum

Lilium polyphyllum D. Don ex Royle or the White Himalayan Lily (syn. *L. punctatum* Jacquem. ex Duch., Liliaceae, $2n = 24$) is a perennial, bulbous herb, native in the Himalayan region, with few populations in restricted pockets of the Indian states of Uttarakhand, Himachal Pradesh, and Jammu and Kashmir (Rana and Samant 2011; Sourabh et al. 2015; Kumar et al. 2021). Known as *Ksirakakoli*, the bulb of this herb is known to have soothing, astringent, and anti-inflammatory properties, and preparations from the bulb are used as expectorant, galactogogue, aphrodisiac, diuretic, antipyretic, in the treatment of cough, bronchitis, burning sensation, fevers, hematemesis, and rheumatalgia (Rana and Samant 2011; Sourabh et al. 2015; Bhardwaj and Sankhyan 2021; Kumar et al. 2021). The plant is reported to contain various phenolic phytochemicals such as catechin, caffeic acid, chlorogenic acid, ferulic acid, gallic acid, and kaempferol, together with terpenes like linalool, caryophyllene, and humulene, which are responsible for its various therapeutic effects (Giri et al. 2017; Bhardwaj and Sankhyan 2021). The entire plant must be uprooted to collect the bulbs. Reckless exploitation of the species for trade, degradation of habitats, changing climatic conditions, and invasion by other species have resulted in the depletion of its natural population, and this species is listed as "critically endangered" as per the IUCN Red List (Rana and Samant 2011; Gowthami et al. 2021; Dhyani et al. 2022).

In a multidimensional approach, Kumar et al. (2021) studied the morphological characters, chromosome count, karyomorphology, pollen grains, and seed surface sculptures through the light microscope and scanning electron microscope from new populations of the species collected from wild habitats in Sechu Tuan Nalla Wildlife Sanctuary in Chamba district of Himachal Pradesh, an Indian state. Mitotic studies of somatic cells from the ovary revealed $2n = 24$, and the karyotype consisted of 2 metacentric, 2 submetacentric, 4 acrocentric, and 16 telocentric chromosomes (Kumar et al. 2021). Other morphometric parameters such as the total chromosome length and the arm ratio ranged from 4.93 to 12.44 μm and from 1.43 to 13.25, respectively, while the centrometric index varied from 5.8 to 41.1. The authors claimed that the overall karyotype of this *Lilium* species was consistent with that of other *Lilium* species (Kumar et al. 2021).

7.4.6 Nepenthes khasiana

The Indian pitcher plant or *Nepenthes khasiana* Hook. f. (Nepentheceae, $2n = 80$) is endemic to North East India, nestled in the misty Garo, Khasi, and Jaintia Hills of Meghalaya and Dima Hasao of Assam, states of India (Devi et al. 2019; Betlu 2016; Bhattacharya et al. 2024). Also known as *Tiew-rakot* (the demon flower or the devouring plant) by the local Khasi people and *Ghatparni* in Hindi, the unopened pitcher fluids of the sole *Nepenthes* species found in India is a reservoir of medicinal compounds that have been used by tribals for treating ophthalmic disorders (cataract and night blindness), gastrointestinal and gynecological problems, as well as skin infections (Ghosh and Ghosh 2012). Leaf juices of this dioecious vine have been traditionally used in the treatment of diabetes and urinary disorders (Ghosh and Ghosh 2012). A recent GC-MS (gas chromatography-mass spectrometry) study for phytochemical profiling of the methanolic extract of the pitcher fluid revealed 22 compounds, including palmitic acid, stearic acid, hexadecanoic acid, 2,4-di-tert-butylphenol, triethyl citrate, and so on, which were coronavirus inhibitors, in addition to displaying hypocholesterolemic, anticancer, antioxidant, anti-inflammatory, diuretic, and diaphoretic activities (Bhattacharya et al. 2024). However, this immensely important species is categorized as endangered in the IUCN Red List of Threatened Species and is included in the Negative List of Exports of the Government of India because of destruction of habitat as a result of shifting cultivation practised by the local population, overexploitation for medicinal and ornamental purposes, anthropogenic activities like coal mining, and unsustainable harvesting; it is also included in Appendix I of the Convention on International Trade in Endangered Species of Wild Fauna and Flora, which prohibits trade of this species (Devi et al. 2015; Betlu 2016; Gowthami et al. 2021; Bhattacharjee et al. 2024; Konhar et al. 2025).

Devi et al. (2012) first reported a chromosome analysis of the Indian pitcher plant, *N. khasiana*, collected from Jarain, Meghalaya, and documented the chromosome number as $2n = 80$ without any variations. Based on chromosome number reports on the different species of the genus *Nepenthes* ($x = 5, 10$), Devi et al. (2012) suggested the taxa studied as polyploid.

Rapid clonal propagation of plants with desirable traits has been a major area of interest in plant biotechnology for decades, a method that is also applied for safeguarding the existence of endangered plants. The process is, however, often limited by the formation of tissue-culture-generated somaclonal variants, resulting in the development of plants with undesirable traits. Thus, maintenance of genetically uniform micropropagated plants is important for conservation and for commercial propagation purposes. Devi et al. (2015) assessed the genetic fidelity of three *N. khasiana* in vitro generations using cytogenetics and heterochromatin patterns. Mitotic studies on root tips showed that 76.66% of first-generation cells had normal chromosome numbers ($2n = 80$), while others varied ($2n = 76, 78$). Fluorescent staining with chromomycin A3 (CMA) and 4′,6-diamidino-2-phenylindole (DAPI) revealed that similar to the mother plant, regenerants from the three subsequent generations had less GC-rich regions (CMA^+ sites) compared to the AT-rich regions ($DAPI^+$ sites), with little variation noted in the first generation plants, indicating an abundance of AT bases in the heterochromatin regions of *N. khasiana*. Thus, cytological differences were noted in micropropagated plantlets of *N. khasiana*, which were morphologically identical (Devi et al. 2015).

In another effort, Bhattacharjee et al. (2024) reported the formation of in vitro pitchers and compared the effects of a natural cytokinin (2-isopentenyl adenine or 2iP) with those of other cytokinins on in vitro regeneration. 2iP could regenerate shoots at high percentages (83.33 and 93.33, respectively) from both nodal explants and in vitro raised axillary buds. Shoots developed indirectly via callus formation also. The regenerants were polymorphic as revealed by ISSR and start codon targeted (SCoT) markers.

Complete chloroplast (cp) and mitochondrial (mt) genomes of *N. khasiana* were first reported by Konhar et al. (2025), who claimed that these were crucial for comprehending its genetic diversity and the distinctive adaptations. The length of the cp genome was 156,914 bp with a GC content of 37%. There were two inverted repeat regions of 25,193 bp each: one large single-copy region of 87,237 bp and one small single-copy region of 19,291 bp, with 131 genes, including 86 protein-coding genes, 79 of which were unique, 37 tRNA-coding genes (29 were unique), and 8 rRNA-coding genes (4 were unique). The mt genome (900,001 bp) had 84 genes, including 50 protein (36 unique), seven rRNA (3 unique), and 27 tRNA (16 unique) coding genes. The datasets were useful for cp and mt genome comparative analysis.

7.5 Concluding Remarks

India has a rich diversity of medicinal plants. Parts of these plants or compounds isolated from them are either used in the traditional systems of medicine or by the pharmaceutical industry for the preparation of drugs. However, many of these plants are disappearing at an alarming rate because of various factors, natural as well as anthropogenic, rendering their enlistment as endangered/threatened/vulnerable species in the IUCN Red Data Book. Thus, conservation and management of this rich medicinal phytodiversity are of utmost importance. To frame conservation programs, either ex situ or in situ, it is crucial to have a thorough knowledge of the genetic constitution of any species and the range of genetic diversity existing within and among populations of the same species. Wild populations of any species are storehouses of improved genetic material (Zhang et al. 2016). All existing populations of a species are important as loss of any population has a direct and detrimental effect on its genetic diversity, worsening its chances of extinction through inbreeding depression (Negi et al. 2022).

Instead of aiming to prevent the extinction of an entire species, efforts should focus on conserving genetically unique populations, as a species' survival and adaptability largely depend on the genetic diversity within its populations. The assessment of plant diversity has evolved from traditional methods to advanced techniques in molecular genetics that allow for the identification of highly valuable genotypes in populations of medicinal plants that should be prioritized for conservation.

This review assessed the genetic makeup and diversity at intra- and interpopulation levels of some endangered medicinal plants in India, using traditional cytological techniques and advanced DNA-based marker methods. To date, cytological studies have proven to be powerful tools, providing insights into the population structures of both endangered and common plant species. Chromosomal analyses, including determining chromosome numbers and structures, as well as examining mitotic and meiotic behavior and anomalies, are crucial for understanding the cytogenetic composition of species, their relationships within different taxa, and laying a foundation for future genetic improvement programs. Although various types of DNA-based molecular markers have been developed to identify elite genotypes, several gaps remain in their application for conservation, breeding programs, and the identification of high-metabolite-producing genotypes, as these markers have their own limitations. As demonstrated by the case studies in the review, only some molecular markers are widely employed in plant genotyping due to their theoretical and technical advantages, ease of application, and straightforward analysis of outcomes. Further innovation in molecular marker technology and -omics approaches is essential for ensuring the survival of endangered plant species.

References

Banerjee N, Sharma AK (1989) Chromosome constitution and alkaloid content in *Rauwolfia* L. (Apocynaceae). Cytologia 54:723–728

Berkov S, Philipov S (2002) Alkaloid production in diploid and autotetraploid plants of *Datura stramonium*. Pharm Biol 40(8):617–621

Betlu ALS (2016) *Nepenthes khasiana* Hook. f. (Nepenthaceae) from Assam. Curr Sci 3(8):1311

Bhardwaj P, Sankhyan S (2021) Medicinal importance of an endangered ayurvedic medicinal herb - *Lilium polyphyllum*. Ijppr Human 21(4):112–120

Bhatia A, Sidhu SP, Roy R, Nautiyal CS (2015) Metabolic profiling of *Commiphora wightii* (guggul) reveals a potential source for pharmaceuticals and nutraceuticals. Phytochemistry 110:29–36

Bhattacharjee S, Washmin N, Borah T, Sarkar A, Mudoi KD, Saikia SP, Verma JS, Banik D (2024) Conspectus on endangered carnivorous pitcher plant *Nepenthes khasiana* Hook. f. emphasizing *in-vitro* regeneration, pitcher development, and stability in genetic makeup. S Afr J Bot 167:270–284

Bhattacharya A, Verma AK, Rao S, Momin SG, Radhakrishnan P, Sarkar P (2024) Unlocking the phytochemicals in unopened pitcher fluids of *Nepenthes khasiana* -a GC–MS study. Intell Pharm 2:284–290

Bhattacharyya P, Kumaria S, Tandon P (2015) Applicability of ISSR and DAMD markers for phyto-molecular characterization and association with some important biochemical traits of *Dendrobium nobile*, an endangered medicinal orchid. Phytochemistry 117:306–316

Bidyananda N, Jamir I, Nowakowska K, Varte V, Vendrame WA, Devi RS, Nongdam P (2024) Plant genetic diversity studies: insights from DNA marker analyses. Int J Plant Biol 15(3):607–640

Bishoyi AK, Kavane A, Sharma A, Geetha KA, Samantaray S, Maiti S (2018) Molecular marker analysis of genetic diversity in relation to reproductive behaviour of *Commiphora wightii* populations distributed in Gujarat and Rajasthan states of India. S Afr J Bot 117:141–148

Brindavanam NB, Goraya GS, Singh SP, Kumar A, Tiwari A, Sarvepalli BN, Raturi PP (2022) Genetic diversity in *Commiphora wightii* (Arn.) Bhandari (Guggul): an assessment of populations in conservation sites of Kachchh region (Gujarat) of India. Pharm J 14(4):379–387

Chauhan J, Purohit VK, Kumar A, Ranjan RK, Dwivedi H, Kumar VV, Prasad P, Nautiyal MC (2023) Assessment of genetic diversity of *Aconitum heterophyllum* Wall. ex. Royle in Uttarakhand, India using ISSR markers. Med Plants Int J Phytomed Relat Ind 15(2):419–424

Che CT, George V, Ijinu TP, Pushpangadan P, Andrae-Marobela K (2024) Traditional medicine. In: Mccreath SB, Clement YN (eds) Pharmacognosy, 2nd edn. Academic Press, pp 11–28

De Kort H, Prunier JG, Ducatez S, Honnay O, Baguette M, Stevens VM, Blanchet S (2021) Life history, climate and biogeography interactively affect worldwide genetic diversity of plant and animal populations. Nat Commun 12:516

Devi SP, Kumaria S, Rao SR, Tandon P (2015) Genetic fidelity assessment in micropropagated plants using cytogenetical analysis and heterochromatin distribution: a case study with *Nepenthes khasiana* Hook. f. Protoplasma 252(5):1305–1312

Devi SP, Kumaria S, Sharma PR, Khoyumthem P, Tandon P (2019) *Nepenthes khasiana* Hook f., an endangered tropical pitcher plant of India. Indian J Tradit Knowl 18(1):68–75

Devi SP, Rao SR, Kumaria S, Tandon P (2012) Mitotic chromosome studies in *Nepenthes khasiana*, an endemic insectivorous plant of Northeast India. Cytologia 77(3):381–384

Dhyani A, Nautiyal BP, Nautiyal MC (2022) Seedling performance, phenotypic traits and growing media for domestication of *Lilium polyphyllum.* Plants People Planet 4:685–695

Gajarajan M, Shareef Khan M, Sridharan MC, Dhamodharan R, Jayaraman P, Saraswathy A, Pazhanivel S (2023) *Aristolochia indica* L. is endangered species in Red Data Book, but growing in South India. Int J Curr Sci 13(1):332–338

Gamba D, Muchhala N (2020) Global patterns of population genetic differentiation in seed plants. Mol Ecol 29:3413–3428

Gao JG, Liu H, Wang N, Yang J, Zhang XL (2020) Plant extinction excels plant speciation in the Anthropocene. BMC Plant Biol 20:430

Gautam S, Bansal YK (2022) Indian Birthwort (*Aristolochia indica* L.): an important endemic and endangered plant with immense medicinal potential. J Em Technol Inn Res 9(13):b1–15

Ghosh D, Ghosh S (2012) India's only pitcher plant in peril. Sci Rep 49(6):53–56

Giri L, Belwal T, Bahukhandi A, Suyal R, Bhatt ID, Rawal RS, Nandi SK (2017) Oxidative DNA damage protective activity and antioxidant potential of Ashtvarga species growing in the Indian Himalayan region. Ind Crop Prod 102:173–179

Gowthami R, Sharma N, Pandey R, Agrawal A (2021) Status and consolidated list of threatened medicinal plants of India. Genet Resour Crop Evol 68:2235–2263

Gupta N, Bhattacharya S, Dutta A, Tauchen J, Landa P, Urbanová K, Houdková M, Fernández-Cusimamani E, Leuner O (2024) Synthetic polyploidization induces enhanced phytochemical profile and biological activities in *Thymus vulgaris* L. essential oil. Sci Rep 14:5608

Hong K, Radian Y, Manda T, Xu H, Luo Y (2023) The development of plant genome sequencing technology and its conservation and application in endangered gymnosperms. Plants (Basel) 12(23):4006

Hosseinzadeh-Colagar A, Haghighatnia MJ, Amiri Z, Mohadjerani M, Tafrihi M (2016) Microsatellite (SSR) amplification by PCR usually led to polymorphic bands: evidence which shows replication slippage occurs in extend or nascent DNA strands. Mol Biol Res Commun 5(3):167–174

Jaisankar I, Velmurugan A, Swarnam TP, Singh AK (2018) Hotspots: an introduction and role in conservation. In: Sivaperuman C, Venkataraman K (eds) Indian hotspots. Springer, Singapore, pp 1–21

Javadian N, Karimzadeh G, Sharifi M, Moieni A, Behmanesh M (2017) *In vitro* polyploidy induction: changes in morphology, podophyllotoxin biosynthesis, and expression of the related genes in *Linum album* (Linaceae). Planta 245(6):1165–1178

Jeelani SM, Shahnawaz M, Gupta AP, Lattoo SK (2022) Phytochemical diversity in relation to cytogenetic variability in *Inula racemosa* Hook. f., an endangered medicinal plant of Himalayas. Chem Biodivers 19(12):e202200486

Jeelani SM, Siddique MAA, Rani S (2015) Variations of morphology, ecology and chromosomes of *Aconitum heterophyllum* Wall., an endangered Alpine medicinal plant in Himalayas. Caryologia 68(4):294–305

Jinu U, Thiyagarajan M, Javed S, Alatar AA, Venkatachalam P (2019) Detection of molecular and phytochemical variations in Tamilnadu landraces of *Gymnema sylvestre* genotype: an important anti-diabetic plant. J Agric Sci Tech 21(6):1521–1534

Kalariya KA, Poojara L, Minipara D, Saran PL, Meena RP, Macwan S, Manivel P (2021) Development and validation of EST-SSR markers in *Gymnema sylvestre* R. Br J Plant Physiol Pathol 9:5

Kaundal R, Kumar D (2025) Current demands for standardization of Indian medicinal plants: a critical review. Med Drug Discov 27:100211

Kiran U, Khan S, Mirza KJ, Ram M, Abdin MZ (2010) SCAR markers: a potential tool for authentication of herbal drugs. Fitoterapia 81(8):969–976

Konhar R, Dash D, Biswal DK (2025) First report of the chloroplast and mitochondrial genomes of the Indian pitcher plant, *Nepenthes khasiana* Hook. f. BMC Res Notes 18:45

Kulhari A, Singh R, Chaudhary A, Dhawan AK, Kalia RK (2015) Assesment of genetic variability through ISSR and RAPD markers in *Commiphora wightii* (Arn.) Bhandari. Acta Physiol Plant 37:113

Kumar A, Tiwari A, Narendran P, Raturi PP (2023) Genetic diversity and population structure of *Aconitum heterophyllum*- an endangered plant from North-western Himalayas using microsatellite markers. Int J Phytomed Relat Ind 15(2):390–401

Kumar P, Gupta VK, Misra AK, Modi DR, Pandey BK (2009) Potential of molecular markers in plant biotechnology. Plant Omics 2(4):141–162

Kumar P, Singh H, Singh SK (2021) Karyomorphological and SEM studies in newly discovered populations of a critically endangered medicinal plant — *Lilium polyphyllum* D. Don ex Royle from a high altitude protected area of north-western Himalaya. Microsc Res Tech 2021:1–11

Kumar R, Das SP, Choudhury BU, Kumar A, Prakash NR, Verma R, Chakraborti M, Devi AG, Bhattacharjee B, Das R, Das B, Devi HL, Das B, Rawat S, Mishra VK (2024) Advances in genomic tools for plant breeding: harnessing DNA molecular markers, genomic selection, and genome editing. Biol Res 57:80

Kunnumakkara AB, Banik K, Bordoloi D, Harsha C, Sailo BL, Padmavathi G, Roy NK, Gupta SC, Aggarwal BB (2018) Googling the guggul (*Commiphora* and *Boswellia*) for prevention of chronic diseases. Front Pharmacol 9:686

Kurane J, Shinde V, Harsulkar A (2009) Application of ISSR marker in pharmacognosy: current update. Pharmacogn Rev 3(6):216–228

Lin X, Zhou Y, Zhang J, Lu X, Zhang F, Shen Q, Wu S, Chen Y, Wang T, Tang K (2011) Enhancement of artemisinin content in tetraploid *Artemisia annua* plants by modulating the expression of genes in artemisinin biosynthetic pathway. Biotechnol Appl Biochem 58(1):50–57

Malhotra N, Kumar V, Sood H, Singh TR, Chauhan RS (2014) Multiple genes of mevalonate and non-mevalonate pathways contribute to high aconites content in an endangered medicinal herb, *Aconitum heterophyllum* Wall. Phytochemistry 108:26–34

Mishra BK, Pathak S, Sharma A, Trivedi PK, Shukla S (2010) Modulated gene expression in newly synthesized auto-tetraploid of *Papaver somniferum* L. S Afr J Bot 76:446–452

Mohammadi V, Talebi S, Ahmadnasab M, Mollahassanzadeh H (2023) The effect of induced polyploidy on phytochemistry, cellular organelles and the expression of genes involved in thymol and carvacrol biosynthetic pathway in thyme (*Thymus vulgaris*). Front Plant Sci 14:1228844

Mori T, Ohsawa T, Murayama M (1989) Studies on *Aconitum* species. VIII. Components of “Kako-Bushi-Matsu”. Heterocycles 29(5):873–884

Murti BSR, Khorana ML (1968) Identification of aconites and estimation of alkaloids. Indian J Pharm 30(9):206–208

Nadeem MA, Nawaz MA, Shahid MQ, Doğan Y, Comertpay G, Yıldız M, Hatipoğlu R, Ahmad F, Alsaleh A, Labhane N, Özkan H, Chung GFS (2018) DNA molecular markers in plant breeding: current status and recent advancements in genomic selection and genome editing. Biotechnol Equip 32(2):261–285

Neethu R, Tandel T, Thakur N, Ahmad A, Borkar R, Pingle R (2024) In-vitro assays to investigate anti-inflammatory, anti-diabetic and anti- microbial activity of herbal extract of *Aristolochia indica*- a review. World J Pharm Res 13(11):1860–1874

Negi RK, Nautiyal P, Rawat JM, Bhatia R (2022) Assessment of genetic diversity of *Aconitum heterophyllum* Wall. ex Royle using ISSR markers: an endangered medicinal herb of Himalaya. Vegetos 35:942–952

Panahi B, Jalay HM, Hamid R (2024) Using next-generation sequencing approach for discovery and characterization of plant molecular markers. Curr Plant Biol 40:100412

Panapitiya PDDM, Welikala N (2018) Development of RAPD, DAMD and ISSR markers for authentication of medicinal plant *Cassia auriculata* and its adulterant *Cassia surattensis*. Int J Environ Agric Biotech 3(5):1906–1912

Pattar PV, Jayaraj M (2012) Pharmacognostic and phytochemical investigation of whole plant of *Aristolochia indica* L.- a threatened medicinal plant. Int J Pharm Sci Rev Res 15(2):95–98

Polaiah AC, Damor PR, Reddy RN, Manivel P, Shivakumara KT, Suthar MK, Thondaiman V, Manjesh GN, Bindu KH, Kumar J (2023) Development of genomic SSR markers in *Gymnema sylvestre* (Retz.) R.Br. ex Sm. using next generation DNA sequencing and their application in genetic diversity analysis. J Appl Res Med Aromat Plants 34:100455

Prasath D, Nair RR, Babu PA (2022) Effect of colchicine induced tetraploids of ginger (*Zingiber officinale* Roscoe) on cytology, rhizome morphology, and essential oil content. J Appl Res Med Aromat Plants 31:100422

Qin D (2019) Next-generation sequencing and its clinical application. Cancer Biol Med 16(1):4–10

Ralls K, Sunnucks P, Lacy RC, Frankham R (2020) Genetic rescue: a critique of the evidence supports maximizing genetic diversity rather than minimizing the introduction of putatively harmful genetic variation. Biol Conserv 251:108784

Rana MS, Samant SS (2011) Population biology of *Lilium polyphyllum* D. Don ex Royle—a critically endangered medicinal plant in a protected area of Northwestern Himalaya. J Nat Conserv 19(3):137–142

Rathore PK, Madihalli S, Hegde S, Hegde HV, Bhagwat RM, Gupta VS, Kholkute SD, Jha TB, Roy S (2016) Assessment of genetic diversity of *Gymnema sylvestre* (Retz.) R.Br. from Western Ghats and Eastern India, India. J Biodivers Environ Sci 9(1):82–92

Reddy CS, Meena SL, Krishna PH, Charan PD, Sharma KC (2012) Conservation threat assessment of *Commiphora wightii* (Arn.) Bhandari—an economically important species. Taiwania 57(3):288–293

Rizvi SAA, Einstein GP, Tulp OL, Sainvil F, Branly R (2022) Introduction to traditional medicine and their role in prevention and treatment of emerging and re-emerging diseases. Biomolecules 12:1442

Sairkar PK, Sharma A, Shukla NP (2016) SCAR marker for identification and discrimination of *Commiphora wightii* and *C. myrrha*. Mol Biol Int 2016:1482796

Saleh B (2019) Molecular characterization using Directed Amplification of Minisatellite-region DNA (DAMD) marker in *Ficus sycomorus* L. (Moraceae). Open Agric J 13:74–81

Salma U, Kundu S, Hazra AK, Ali MN, Mandal N (2018) Augmentation of wedelolactone through in vitro tetraploid induction in *Eclipta alba* (L.) Hassk. Plant Cell Tissue Organ Cult 133:289–298

Samanta JN, Mandal K, Maiti S (2012) A novel pathovar of *Xanthomonas axonopodis* causes gumming of Guggal (*Commiphora wightii*). Eur J Plant Pathol 135:115–125

Sarma B, Baruah PS, Tanti B (2018) Habitat distribution modeling for reintroduction and conservation of *Aristolochia indica* L. - a threatened medicinal plant in Assam, India. J Threat Taxa 10(11):12531–12537

Sarma B, Tanti B (2015) Karyomorphology of three species of *Aristolochia*—rare and endemic medicinal plants of Assam, India. Caryologia 68(2):154–158

Sarma B, Tanti B (2017) Analysis of genetic diversity of certain species of *Aristolochia* using ISSR-based molecular markers. Curr Life Sci 3(4):47–53

Sati H, Sati B, Saklani S, Bhatt PC, Mishra AP (2011) Phytochemical and pharmacological potential of *Aristolochia indica*: a review. Res J Pharm Biol Chem Sci 2(4):647–654

Satyavati GV (1990) Use of plant drugs in Indian traditional systems of medicine and their relevance to primary health care. In: Wagner H, Farnsworth NR (eds) Economic and medicinal plant research. 4. Academic Press, London, UK, pp 39–56

Schneider H (2022) Integrating genomics and conservation to safeguard plant diversity. Integr Conserv 2:10–18

Sharma AK, Varma B (1959) The somatic chromosomes of five species of *Aristolochia*. Phyton Rev Internac Bot Expt Argentina 12:101–108

Sobti SN, Singh SDA (1961) Chromosomal survey of medicinal plants part 1. Proc Indian Acad Sci 54b:138–144

Sourabh P, Thakur J, Uniyal PL, Pandey AK (2015) Biology of *Lilium polyphyllum* - a threatened medicinal plant. Med Plants - Int J Phytomed Relat Ind 7(2):158–166

Sreelakshmi KP, Arya A, Delna MS, Balakrishnan MK, Sandra PD (2024) Comprehensive review on *Aristolochia indica* and *Boerhavia diffusa*: ethnomedical uses, phytochemistry and pharmacological effects. Intl Res J 11(5):a424–a431

Thakur GS, Sharma R, Sanodiya BS, Pandey M, Prasad GBKS, Bisen PS (2012) *Gymnema sylvestre*: an alternative therapeutic agent for management of diabetes. J Appl Pharm Sci 2(12):1–6

Thakur J, Dwivedi MD, Sourabh P, Uniyal PL, Pandey AK (2016) Genetic homogeneity revealed using SCoT, ISSR and RAPD markers in micro propagated *Pittosporum eriocarpum* Royle- an endemic and endangered medicinal plant. PLoS One 11(7):0159050

Thong-on W, Arimatsu P, Pitiporn S, Soonthornchareonnon N, Prathanturarug S (2014) Field evaluation of in vitro-induced tetraploid and diploid *Centella asiatica* (L.) Urban. J Nat Med 68:267–273

Thottathil GP, Jayasekaran K, Othman AS (2016) Sequencing crop genomes: a gateway to improve tropical agriculture. Trop Life Sci Res 27(1):93–114

Tomar UK, Singhal H, Gaur A, Saini LS (2021) Population density, genetic diversity and hot spots of *Commiphora wightii (*Arnott) Bhandari in Rajasthan State. J Appl Res Med Aromat Plants 25:100323

Tomar S, Thakur S, Kanwal KS, Bhatt ID, Puri S (2025) Population assessment and habitat suitability modelling of endangered medicinal plant, *Aconitum heterophyllum* Wall. ex Royle in the western Himalaya. Sci Rep 15:33794

Vaishnav K, Tiwari V, Durgapal A, Meena B, Rana TS (2023) Estimation of genetic diversity and population genetic structure in *Gymnema sylvestre* (Retz.) R.Br. ex Schult. populations using DAMD and ISSR markers. J Genet Eng Biotechnol 21:42

Vaishnav K, Maurya A, Durgapal A, Rana TS (2025) An integrative analysis of population genetics and species distribution modeling of *Gymnema sylvestre* in India. J Appl Res Med Aromat Plants 46:100631

Verma AK, Dhawan SS (2017) Chromosomal fragmentation: a possible marker for the selection of high gymnemic acid yielding accessions of *Gymnema sylvestre* R. Br Pharmacogn Mag 13:S481–S483

Verma AK, Dhawan SS, Singh S, Bharati KA, Jyotsana (2016) Genetic and chemical profiling of *Gymnema sylvestre* accessions from central India: its implication for quality control and therapeutic potential of plant. Pharmacogn Mag 12:S407–S413

Wani TA, Kaloo ZA, Dangroo NA (2022) *Aconitum heterophyllum* Wall. ex Royle: a critically endangered medicinal herb with rich potential for use in medicine. J Integr Med 20:104–113

Xia J, Ma YJ, Wang Y, Wang JW (2018) Deciphering transcriptome profiles of tetraploid *Artemisia annua* plants with high artemisinin content. Plant Physiol Biochem 130:112–126

Zhang H, Mittal N, Leamy LJ, Barazani O, Song BH (2016) Back into the wild-Apply untapped genetic diversity of wild relatives for crop improvement. Evol Appl 10(1):5–24

Part II
Threats and Conservation of Medicinal Plant Genetic Diversity

Chapter 8
Medicinal Plants from Southern Brazilian Grasslands (*Campos Sulinos*): Threatened Species and Current Knowledge About Their Genetic Diversity

Liliana Essi, Yago Luís Nóbrega Mendes, Marina Scheuer, Henrique da Rosa Ferreira, Hulia Juana Scherer, Milena de Melo Hettwer, and Táfinis Sturbelle Schiavon

Abstract This review covers the medicinal plants of Brazil's southern grasslands (*Campos Sulinos*), highlighting their historical, ecological, and scientific importance. It defines the concept of medicinal plants and their cultural and therapeutic relevance from the time of ancient civilizations onward, demonstrating that the use of plant species in traditional medicine was essential for the development of modern pharmaceuticals and still represents the primary therapeutic resource for a large portion of the world's population. The study addresses the botanical richness of Brazil—home to over 40,000 native plant species—and emphasizes the southern grasslands (grassland formations of the Pampa and Atlantic Forest biomes) as ecosystems that are highly biodiverse, yet threatened by land-use changes and limited scientific attention. The search revealed 271 threatened plant species in these grasslands, of which 24 have reported medicinal uses. Among these, families such as Asteraceae and Arecaceae predominate, but most lack genetic studies—a key factor in guiding conservation strategies. Few species, among them *Araucaria angustifolia, Butia eriospatha, Hesperozygis ringens,* and *Colletia paradoxa*, have been studied for genetic diversity. These studies reveal low variability and population risks resulting from fragmentation and overexploitation. The chapter concludes by highlighting the urgent need to expand ethnobotanical and genetic research, recognize traditional knowledge lost to colonization, and integrate environmental and cultural

L. Essi (✉)
Department of Biology, Universidade Federal de Santa Maria, Santa Maria, Brazil
e-mail: liliana.essi@ufsm.br

Y. L. N. Mendes · M. Scheuer · H. da Rosa Ferreira · H. J. Scherer · M. de Melo Hettwer
Post Graduate Program in Agrobiology – Universidade Federal de Santa Maria, Santa Maria, Brazil

T. S. Schiavon
Graduate in Biological Sciences – Universidade Federal de Santa Maria, Santa Maria, Brazil

S. Jha et al. (eds.), *Genetic Diversity in Medicinal Plants*, Sustainable Development and Biodiversity 44, https://doi.org/10.1007/978-981-95-7281-6_8

conservation efforts to ensure the sustainable use of the medicinal flora of the *Campos Sulinos*.

Keywords Atlantic Forest · Pampa · Endangered plants · Red List · Nature conservation · Ethnobotany · Genetic variability

8.1 Introduction

A medicinal plant can be defined as any plant that, when administered in any form and by any route to humans or other animal species, exerts some type of pharmacological action on them (Silva and Carvalho 2004). It is also possible to label a species as a "plant with medicinal properties" rather than a "medicinal plant." However, we herein use the term "medicinal plant" as defined by the Brazilian National Policy on Medicinal Plants and Phytotherapeutics (*Política Nacional de Plantas Medicinais e Fitoterápicos*—Brasil 2006) and the World Health Organization (WHO); that is, as a plant that contains substances with medicinal properties in its parts (such as the leaves, flowers, roots, or bark). These substances can have therapeutic effects on the human body and are used to treat or alleviate diseases, symptoms, or health conditions (Brasil 2006).

The use of plants to treat maladies is one of the oldest therapeutic practices in the world. It is a widespread practice in folk medicine and has served as the basis and inspiration for the isolation and development of countless pharmaceuticals. Some examples include pilocarpine, extracted from the leaves of *jaborandi* (*Pilocarpus jaborandi* Holmes) and used in the treatment of glaucoma; the alkaloids vincristine and vinblastine, extracted from the leaves of the *vinca* (*Catharanthus roseus* (L.) G.Don) and used in the treatment of cancer (Guerra and Nodari 2001); and the widely used analgesic and anti-inflammatory acetylsalicylic acid, extracted from *Salix* (for a historical overview about the development of salicylates, see Rainsford 2004). Although there are multiple isolated substances available—either of natural or synthetic origin—that could replace the traditional uses of plants nowadays, medicinal plants are often the only accessible treatment option for certain communities, either due to the difficulty of accessing conventional medicine or the cost of medication (Battisti et al. 2013).

In addition to therapeutic practice, the use of medicinal plants is itself part of people's cultural history, passed down from generation to generation. For instance, Brazil is enormously rich in medicinal plants and their uses due to both the multicultural origins of Brazilian people (Native Indian tribes, enslaved African peoples, European colonizers and immigrants, etc.) and the mega diversity of plants. According to Flora e Funga do Brasil (2025), Brazil hosts more than 40,000 native plant species, distributed across six main phytogeographic domains or biomes: the Amazon, Caatinga, Cerrado, Atlantic Forest, Pampa, and Pantanal.

These biomes include forest and grassland natural formations. The forest biomes, such as the Amazon and Atlantic Forest, are the best known abroad, whereas the

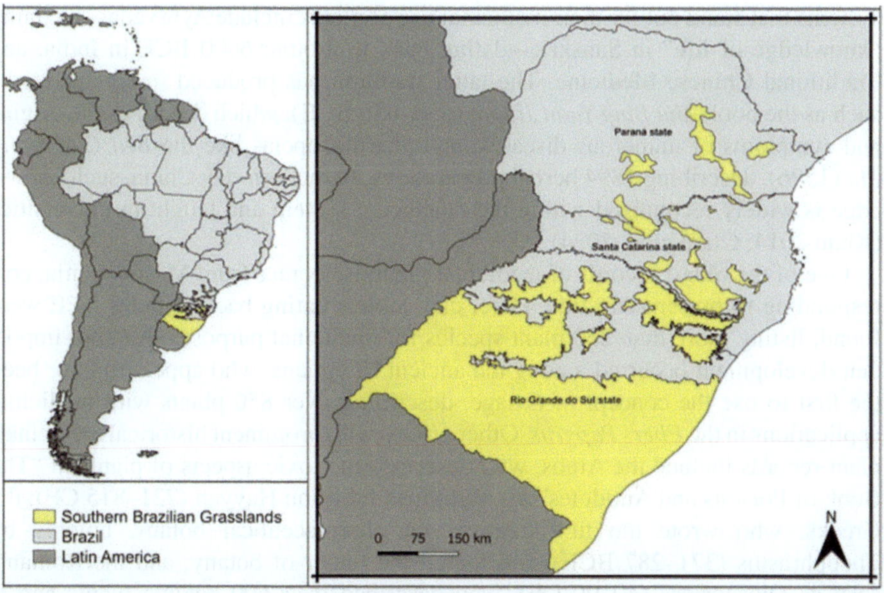

Fig. 8.1 Geographic location of the southern Brazilian grasslands (highlighted in yellow) within South America. The left panel shows their position in relation to other regions from Brazil (light gray) and to other Latin American countries (dark gray), while the right panel details their distribution within southern Brazil

grassland biomes have only started to receive international attention more recently (Ellwanger et al. 2022; Andrade et al. 2023), despite also being very rich in native and endemic species. Southern Brazil features campestral formations belonging to the Atlantic Forest biome and the Pampa biome. Together, they compose the so-called southern Brazilian grasslands (*Campos Sulinos*, see Fig. 8.1). In the next sections, we present a historical background on medicinal plants worldwide and then focus on the *Campos Sulinos*. We present an ecological characterization, conservation status data, and finally a synthesis of the genetic studies conducted on the threatened medicinal plants that inhabit this region.

8.2 Medicinal Plants Worldwide: Historical Background

From the time of the earliest human civilizations, *Homo sapiens* has used various plant parts to prevent, alleviate, treat, or cure pathological conditions. In fact, archeological evidence has revealed that the use of medicinal plants is the oldest method of medication employed by humans and that these plants' medicinal properties were likely already part of the knowledge of prehistoric peoples (Khan 2014; Hardy 2021).

Different human populations, throughout various historical periods and across different geographic regions, have used medicinal plants in their cultures. In Asia, systems that stand out for their medicinal use of plants include Ayurveda—meaning "knowledge of life" in Sanskrit—dating back to around 6000 BCE in India, and Traditional Chinese Medicine. The latter tradition has produced important works such as the book *Zhu Bing Yuan Ji Lun* (550–630 BCE), which describes the origins and symptoms of numerous diseases, and pharmacopeias like the *Ben Cao Gang Mu* (1596), describing 1894 herbal preparations. In modern-day China, such knowledge is widely recognized within the healthcare system and taught in universities (Khan 2014; Claro et al. 2024).

One of the oldest records of medicinal plant use comes from Mesopotamia, corresponding to present-day Iraq. Here, clay tablets dating back to 5000 BCE were found, listing more than 250 plant species for medicinal purposes. Another important development occurred among the ancient Egyptians, who appear to have been the first to use the concept of dosage, describing over 850 plants with medicinal applications in the *Ebers Papyrus*. Other groups with prominent historical medicinal-plant records include the Arabs, who described the toxic aspects of plants in "The Book of Poisons and Antidotes" by Abu Musa Jabir ibn Hayyan (721–815 CE); the Greeks, who wrote the first treatises on pharmaceutical botany, notably by Theophrastus (371–287 BCE), considered the father of botany; and the Romans, such as Dioscorides (40–90 CE), who described over 600 healing plants and is regarded as an important writer in ancient phytomedicine as well as the father of pharmacognosy (Claro et al. 2024).

The use of plants with medicinal properties in Western cultures derives mainly from Graeco-Roman traditions, preserved by the Arabs during the Dark Ages and Middle Ages (fifth to fifteenth centuries), and combined with herbs from Chinese, Indian, and various Western cultures' traditions, initially applied in an empirical way. Nevertheless, it was only from the nineteenth century onward that the discovery of plant-derived medicines began to occur rationally, leading to the emergence of many well-known compounds such as morphine, nicotine, caffeine, cocaine, and capsaicin, derived from poppy, tobacco, coffee, coca leaf, and peppers, respectively. Other essential compounds for modern medicine also emerged, such as paclitaxel, obtained from species of the genus *Taxus* L. (Atanasov et al. 2015).

Throughout Latin America, which is composed of 20 independent countries, plants serve as the primary means of healthcare through the incorporation of traditional knowledge (Salazar-Gómez and Alonso-Castro 2022). Within this geographic region, Brazil holds the greatest plant biodiversity on the planet, besides a wide variety of ethnic groups that make use of this vegetation, including for therapeutic purposes. Due to factors such as their easy accessibility, fewer side effects, and low financial cost, herbal medicines are widely used in traditional indigenous and rural communities, which account for 25% of Brazil's territorial occupation (Tribess et al. 2015; Santos et al. 2018; Melro et al. 2020).

Over time, medicinal plant knowledge has lost its mystical character and begun to be scientifically studied, giving rise to the development of new medicines as a result of advances in science and the growing demand for less-invasive treatments. However, this generational knowledge has been increasingly lost due to the interference of external social groups—including in the impact of European colonization on native indigenous peoples—and because of its fragile nature, as this information is mainly transmitted orally and then entrusted to memory (Tribess et al. 2015; Melro et al. 2020).

By the end of the twentieth century, there was a growing global popularization of herbal medicine, and the large amount of documented information on ethnobotany and traditional uses facilitated the development of new drugs from medicinal plants (Atanasov et al. 2015). The term "ethnobotany" was originally coined in 1895 by American botanist John William Harshberger and is nowadays understood as the relationship between humans and plants (Ijinu et al. 2022).

Currently, traditional knowledge accounts for approximately 40–80% of medicines developed, and its importance is recognized and promoted by the WHO (WHO 2023; Claro et al. 2024). The use of plants for medicinal purposes is an important part of healthcare in both developing and developed nations, employed among 80% of the global population (Dar et al. 2017; Ijinu et al. 2022).

It is estimated that around 70,000 plant species have medicinal potential for the treatment of human diseases. Due to its historical background and, above all, its mega diversity, Brazil is a country with a long-standing tradition in the use of medicinal plants; this has sparked interest among Brazilian researchers in developing herbal medicines from native plants. However, due to worsening climate change and other anthropogenic impacts, many plant species have become threatened, making it urgent to implement conservation efforts and conduct research in various fields (taxonomy, genetics, pharmacology, etc.) on such species (Atanasov et al. 2015; Dutra et al. 2016).

8.3 The Southern Brazilian Grasslands: Ecological Context

The grasslands of southern Brazil are natural areas—witnesses of ancient ecosystems—dominated by grasses and rich in biodiversity (Silveira et al. 2020). Occurring in the Pampa and Atlantic Forest biomes, their plant communities are shaped by environmental gradients such as altitude (from sea level to over 900 m a.s.l.) and by differences in management, including the use of fire and grazing, which are very common in the region (IBGE 2004; Menezes et al. 2022).

The region's phytoecological diversity is characterized by distinct classifications (Fig. 8.2). The *Instituto Brasileiro de Geografia e Estatistca* (IBGE 2004) refers to the southern grasslands as the Steppe region, an international term for grassland formations in temperate zones with constant rainfall. These formations include the *Campanha Gaúcha*, dominated by cespitose grasses of the genera *Stipa* and *Agrostis*, rhizomatous grasses of the genera *Paspalum* and *Axonopus*, and legumes and composites. The phanerophytes are primarily spiny and deciduous species. Additionally, there is the Espinilho Park, a Steppe Savannah in the floodplain of southwestern Rio Grande do Sul featuring genera such as *Proposis* and *Acacia*. On the Santa Catarina and Paraná plateaus, the *Campos Gerais* forms mosaics with forests of *Araucaria angustifolia* Kuntze. The climate is predominantly humid

Fig. 8.2 Some of the phytoecological diversity in the southern Brazilian grasslands: (**a**) Highland grassland (HIG): >700 m a.s.l., gentle relief, deep low-fertility soils, dominated by clump-forming grasses and Fabaceae; (**b**) shallow-soil grassland (SSG): 30–400 m a.s.l., gentle relief, very shallow stony soils, dominated by small tufted grasses, Asteraceae, Fabaceae and stress-tolerant species; (**c**) coastal subtropical forest (CSF): sub-montane/montane, undulate relief, deep to shallow soils, diverse forest species; (**d**) sandy grassland (SAG): 30–400 m a.s.l., gentle relief, deep sandy soils, dominated by Poaceae, Fabaceae, Asteraceae, Cactaceae and drought-adapted endemic species; (**e**) inland sub-montane grassland (ISG): 30–400 m a.s.l., gentle relief, deep soils, prostrate grasses in lower stratum, Andropogoneae in upper stratum, Fabaceae, and Asteraceae in disturbed areas; and (**f**) Bush grassland (BUG): 400–500 m a.s.l., undulate relief, shallow to deep low-fertility soils, dominated by tussock-forming grasses, Fabaceae, shrubs, and endemic Cactaceae. (Adapted from Hasenack et al. (2023). Image credit: Marina Scheuer)

subtropical, characterized by hot summers (Cfa) at lower altitudes and temperate summers (Cfb) at higher altitudes, without pronounced dry seasons (Alvares et al. 2013). More recent research complements this view, identifying three (Andrade et al. 2019) to five (Menezes et al. 2022) distinct groups based on their floristic, soil, and climatic characteristics.

The biodiversity of the grasslands is remarkable, with the Pampa harboring 3642 species of vascular plants and 9% of Brazil's total biodiversity despite representing just over 2% of the national territory. The families with the greatest species richness are Asteraceae (476 species), Poaceae (423), Fabaceae (292), Orchidaceae (162), and Cyperaceae (155) (Andrade et al. 2023). However, this diversity is at risk due to the current rate of land-use conversion, which has caused the homogenization of the remaining landscape (Staude et al. 2018). The Pampa lost 30.3% of its natural vegetation between 1985 and 2024, while the area dedicated to forestry increased 15.7-fold in the same period (Mapbiomas 2025).

Grassland plant communities are shaped by vegetation management, but their intensity and frequency can lead to ecosystem degradation and loss of resilience. The exclusion of grazing, for example, promotes the accumulation of dead biomass and the dominance of only a few plant species, reducing diversity. On the other hand, overgrazing also causes degradation, replacing quality forage plants with ruderal species. Additionally, the introduction of exotic foraging species, which may have a high competitive advantage, drives degradation by altering the floristic composition of grasslands (Andrade et al. 2015).

Studies on the restoration of open ecosystems such as grasslands are still incipient, in contrast to the attention given to forests—only 1% of ecological restoration studies in Brazil focus on the Pampa, indicating a significant knowledge gap (Guerra et al. 2020). Nevertheless, only 2.58% of the grassland area in Rio Grande do Sul State is protected in Conservation Units (Brandão et al. 2008). The use of plants for medicinal purposes is a historical and cultural practice in the region. However, the neglect of these biomes and the focus on forest restoration models hinder the appreciation of native flora and the development of techniques to preserve the ethnobotanical richness of the southern grasslands (Silveira et al. 2020).

8.4 Threatened Medicinal Plant Species: A Regional Diagnosis

The evaluation of extinction risk is an important tool for monitoring biodiversity loss and establishing conservation strategies. Brazil has a National Plant Red List that is published periodically by the Ministry of the Environment. In addition, some Brazilian states produce their own Red Lists, making it possible to monitor species conservation locally. However, evaluating more than 40,000 plant species poses a real challenge.

To date, only 9445 plant species in Brazil have been assessed for extinction risk, and around 3209 species are currently on the Brazilian Red List (Brasil 2022). In the context of conservation, these are significant data, as they point to the biodiversity crisis driven by habitat loss, fragmentation, land-use changes, and various other anthropogenic impacts (Fagundes et al. 2025; Menezes et al. 2025; Le Roux et al. 2019). Regionally, the Red Lists of the three southern Brazilian states where *Campos Sulinos* are situated are unfortunately not properly updated (for Santa Catarina and Rio Grande do Sul States, the last official lists were published in 2014, and for Paraná State in 1992). Thus, the conservation status of species in *Campos Sulinos* needs closer analysis.

How many formally evaluated medicinal plants are threatened by extinction in the southern Brazilian grasslands? How many of these have already been studied from a genetic point of view? These were the questions our research group sought to answer by conducting a literature review using the current official list of endangered plant species for Brazil (BRASIL 2022). We selected those species occurring in southern grasslands, categorized them by degree of threat, and then searched for articles on genetic diversity for endangered plant species, as well as records of any potential or confirmed medicinal uses for them, within the national database Periódicos Capes. Our results are presented and discussed below.

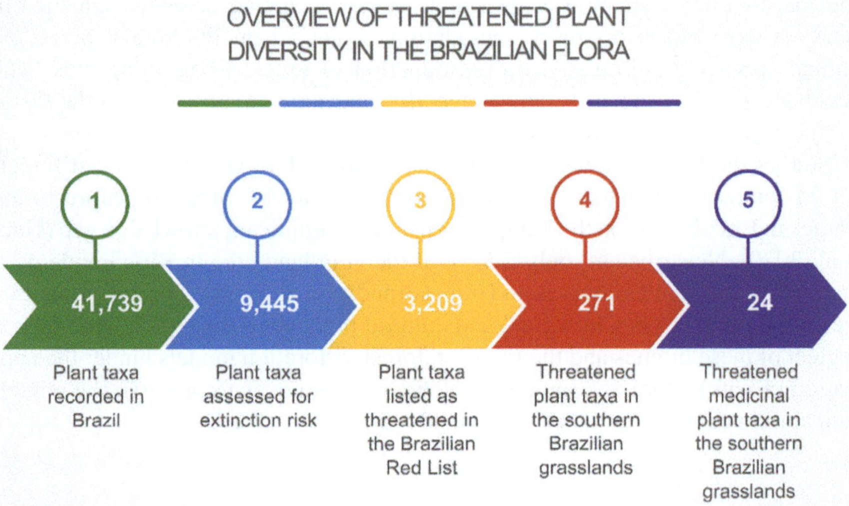

Fig. 8.3 Overview of threatened plant diversity in Brazil and in the southern Brazilian grasslands, summarizing the number of native plant taxa recorded in Brazil, those assessed for extinction risk, and those officially listed as threatened in the Brazilian Red List. Emphasis is given to the number of threatened plant taxa occurring in the southern Brazilian grasslands, including medicinal species. (Infographic adapted by the authors based on data from the CNCFlora (Centro Nacional de Conservação da Flora, https://cncflora.jbrj.gov.br/) and the Brazilian Red List of Threatened Species)

Focusing on the southern Brazilian grasslands, which include the Pampa grasslands and the high-altitude grasslands associated with the Atlantic Forest domain, we obtained 271 recorded threatened plant species (Fig. 8.3). Of these, only 24 have documented and scientifically confirmed medicinal uses (Fig. 8.3), posing the question of whether this accurately represents the total number of native medicinal plant species in the southern grasslands.

Before we continue our discussion of recorded threatened species, it is important to briefly contextualize the local knowledge about medicinal plants. An important point to highlight is that current written knowledge about the use of medicinal plants in Brazil is heavily influenced by immigrants—especially European immigrants—who brought many exotic medicinal plants for their own use, such as chamomile, mint, artichoke, and others (Paulet et al. 2019; Corrêa and Scheffer 2014). These immigrants included only a few native plants in their medicinal practices, without properly documenting the rich uses of plants among the original peoples. The indigenous peoples of the region corresponding to the southern grasslands have a tradition of transmitting knowledge orally. Colonization by Europeans—which resulted in acculturation, the extinction of tribes and their native languages, and the loss of knowledge among the native peoples, among other devastating effects on local cultures—gradually led to the loss of indigenous knowledge about the use of native plants. A major study published in 2021 (Cámara-Leret and Bascompte 2021) warned that numerous indigenous languages in three regions of the world with high biocultural diversity were at risk of disappearing, and with them a vast and unique knowledge of medicinal plants. No similar research has been conducted for the southern grassland region of Brazil, and this information is difficult to retrieve, as there are no longer isolated indigenous communities in the region, and the existing communities have already suffered a significant impact from acculturation, loss of territories, and possibly a significant loss of traditional knowledge.

Therefore, our approach takes into account only what has already been recorded in the scientific literature, but there is still much to be illuminated regarding the traditional knowledge about medicinal plants native to the southern grasslands.

Along with the degradation of sociocultural knowledge is environmental degradation—as pointed out in Sect. 8.3—which is the main cause of the threat of extinction for medicinal plant species. Our findings showed that most medicinal species threatened with extinction are categorized as endangered (EN), as can be seen in Fig. 8.4.

Taxonomically, the 24 threatened medicinal plants collated in this study are distributed across 14 different botanical families, among which the Asteraceae—one of the most species-rich families in the southern grasslands (Andrade et al. 2023)—and Arecaceae, with genus *Butia*, comprise the highest numbers of species. The other families with more than one representative were Amaryllidaceae, Lamiaceae, Lauraceae, and Rhamnaceae, while the remaining families were represented by only one species, as shown in Fig. 8.5.

When we searched for genetic studies on threatened medicinal species, we found many species on which no studies had been conducted, as shown in Fig. 8.6. This result is consistent with what was observed for threatened Brazilian plant species in general by Essi et al. (2020), who identified a clear lack of population genetic studies for species at risk of extinction.

Categorizing the species by family highlights Asteraceae as having the fewest genetic diversity studies, with all four of its threatened medicinal species lacking published research (Fig. 8.7). This is also in accordance with Essi et al. (2020), who found Asteraceae to be the most neglected botanical family in terms of genetic diversity studies in Brazil. The complete list of species, ordered by family, is available in Table 8.1.

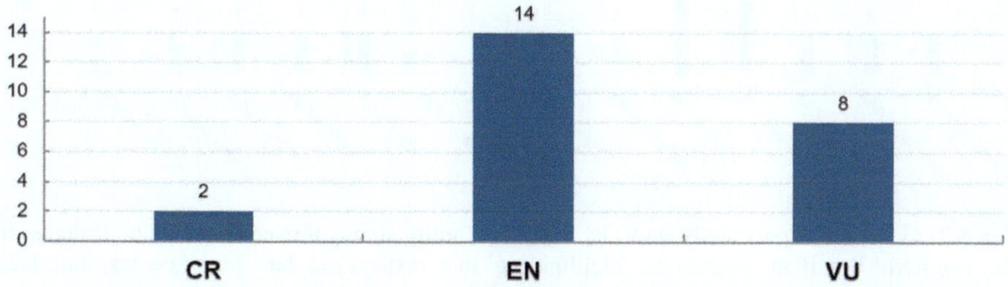

Fig. 8.4 Distribution of threat categories among 24 medicinal plant species, with 2 species classified as Critically Endangered (CR), 14 as Endangered (EN), and 8 as Vulnerable (VU), according to the Brazilian National Plant Red List

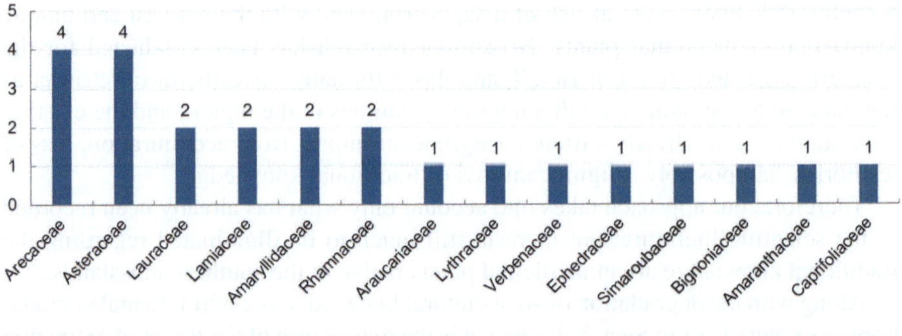

Fig. 8.5 Distribution of threatened medicinal plant species per botanical family in the southern Brazilian grasslands, with Asteraceae and Arecaceae represented by four species each; Amaryllidaceae, Lauraceae, Lamiaceae, and Rhamnaceae by two species each; and the other families by one species each

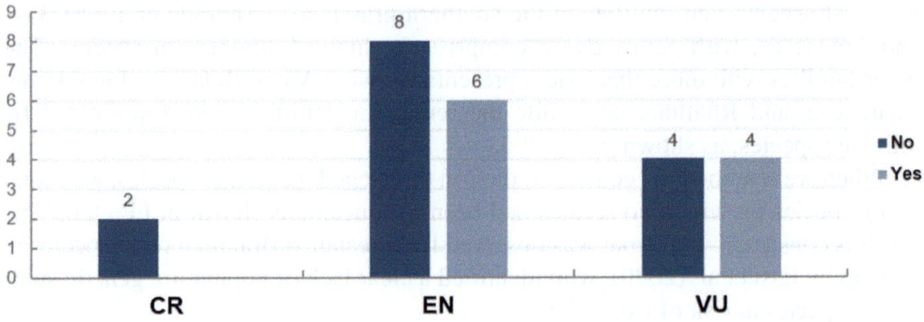

Fig. 8.6 Genetic diversity study status by threat category among threatened medicinal species in the southern Brazilian grasslands, with six studies for endangered (EN), and four for vulnerable (VU) species, while two CR, eight EN, and three VU species have not been studied. Categorization according to the Brazilian National Plant Red List

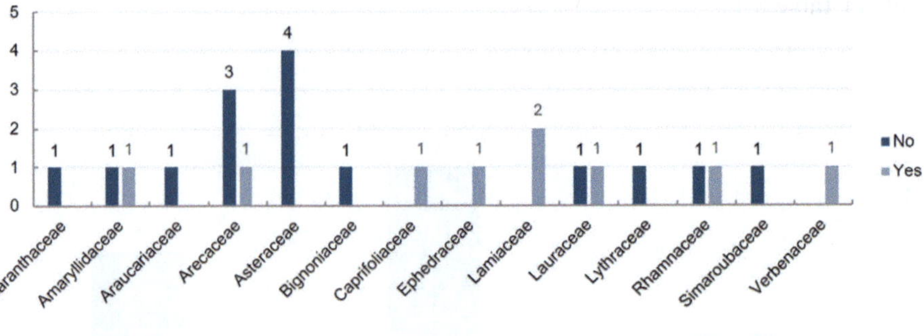

Fig. 8.7 Genetic diversity study status by botanical family among threatened medicinal species in the southern Brazilian grasslands, highlighting that Asteraceae has four species that lack genetic studies

Table 8.1 Threatened medicinal plant species from the southern Brazilian grasslands, with information on their threat category and the presence or absence of published genetic diversity studies

	Family	Species	Threat category	Genetic diversity study
1	Amaranthaceae	*Alternantherahirtula*(Mart.) R. E. Fr. (Fig. 8.8b)	VU	No
2	Amaryllidaceae	*Hippeastrum breviflorum* Herb.	EN	No
3	Amaryllidaceae	*Zephyranthes candida* (Lindl.) Herb.	EN	No
4	Araucariaceae	*Araucaria angustifolia* (Bertol.) Kuntze	EN	Yes
5	Arecaceae	*Butiaeriospatha*(Mart. exDrude) Becc.	VU	Yes
6	Arecaceae	*Butia exilata* Deble & Marchiori	CR	No
7	Arecaceae	*Butia microspadix* Burret	VU	No
8	Arecaceae	*Butia yatay* (Mart.) Becc.	VU	No
9	Asteraceae	*Mikaniaduseniii*B.L.Rob.	CR	No
10	Asteraceae	*Mikania variifolia* Hieron.	EN	No
11	Asteraceae	*Perezia multiflora*Less.(Fig. 8.8c)	EN	No
12	Asteraceae	*Stevia leptophylla* Sch.Bip. ex Baker	EN	No
13	Bignoniaceae	*Anemopaegma arvense* (Vell.) Stellfeldex de Souza	EN	Yes
14	Caprifoliaceae	*Valeriana glechomifolia*F.G.Mey.	EN	No
15	Ephedraceae	*Ephedra tweediana* C.A.Mey.	VU	No
16	Lamiaceae	*Cunila incana* Benth.	EN	Yes
17	Lamiaceae	*Hesperozygis ringens* (Benth.) Epling	VU	Yes
18	Lauraceae	*Ocotea catharinensis* Mez	VU	Yes
19	Lauraceae	*Ocotea porosa* (Nees & Mart.) Barroso	EN	Yes
20	Lythraceae	*Cuphealindmaniana*KoehneexBacig.	EN	No
21	Rhamnaceae	*Colletia paradoxa* (Spreng.) Escal.	EN	Yes
22	Rhamnaceae	*Discaria americana* Gilliesex Hook.	VU	No
23	Simaroubaceae	*Castela tweedii* Planch.	EN	No
24	Verbenaceae	*Aloysia hatschbachii* Moldenke	EN	Yes

8.5 Current Knowledge on Genetic Diversity of Threatened Medicinal Species and Implications for Conservation

When a plant species is threatened, its population size declines, and its genetic diversity usually reduces, resulting in consequences for its genetic health, causing isolation and making it vulnerable to environmental, demographic, and genetic events (Viegas et al. 2011). In this context, the maintenance of good levels of genetic diversity provides conditions for adaptation, evolution, and survival and is important for the sustainability of species (Rajora and Mosseller 2001).

The conservation of a species depends on several factors, among the most crucial of which is the maintenance of genetic diversity. Therefore, studies evaluating genetic diversity and genetic variability are essential, as they provide support for defining priority conservation areas and implementing effective management strategies (Kamada et al. 2009; Ramalho et al. 2016).

Fig. 8.8 Some medicinal plants from *Campos Sulinos* threatened with extinction: (**a**) *Araucaria angustifolia* (habit); (**b**) *Alternanthera hirtula* (inflorescence); (**c**) *Perezia multiflora* (flower heads); (**d**, **e**) *Hesperozygis ringens*; (**e**) habit and (**d**) detail of a flower. (Photo credits: (**a**, **b**, **d**, **e**) by Liliana Essi, and (**c**) by Yago Nóbrega Mendes)

Our review found that, unfortunately, few threatened medicinal species occurring in the grasslands of southern Brazil have been studied in regard to their genetic diversity or variability (Table 8.1), demonstrating the need for research in this area. From the 24 threatened medicinal species sourced from the literature, only nine had already been studied from a genetic perspective—some only preliminarily. In addition, no genetic studies were recovered for any of the four threatened Asteraceae species.

The Asteraceae family has approximately 23,000 species worldwide (Chiqueti and Tiuman 2021) and is important for ornamental, medicinal, aromatic, and edible purposes. Several genera from this family have medicinal uses. For instance, the genus *Mikania* Willd. is used for fever, rheumatism, and colds; however, only 12% of its 430 species have undergone pharmacological study (Rufatto et al. 2012).

Some families with several threatened species in the National Red List (Brasil 2022) are less commonly employed in Brazilian traditional medicine and are better known for other popular uses; this includes Orchidaceae, with eight threatened species in *Campos Sulinos*, which are mainly known for their ornamental use. However, for some other families whose members are commonly used in folk medicine, we did not recover specific records in the literature regarding threatened species' medicinal uses. This was the case for Poaceae (29 threatened species in *Campos Sulinos*), Apiaceae (13 species), Bromeliaceae and Fabaceae (11 species each), and Myrtaceae and Solanaceae (8 species each).

Among the medicinal species for which the search recovered at least one genetic diversity study, different plant families were represented: Araucariaceae, Arecaceae, Bignoniaceae, Lamiaceae, Lauraceae, Rhamnaceae, and Verbenaceae. However, only four species for which genetic studies were retrieved are endemic to Brazil, and for those species with broader distribution ranges, several recovered studies did not include populations from *Campos Sulinos*. In the next paragraphs, we briefly present some of the species for which genetic studies were conducted.

Butia (Becc.) Becc. is a palm genus with 21 species native to Brazil, 12 of which are endemic (Heiden and Sant'Anna-Santos 2025); these are popularly known as butiás. *Butia* species are generally used as food, *in natura* or in the preparation of sweets, jellies, and drinks. According to Aguiar et al. (2014), the fruits of *Butia* are rich in carotenoids, phenolic compounds, fatty acids, ascorbic acid, and minerals. They are also a source of unsaturated fatty acids (Lopes et al. 2012). A review by Schneider et al. (2017) compiled a list of therapeutic uses for *Butia* species, such as antimicrobial, anti-inflammatory, and antioxidant activity. From the four threatened *Butia* species in the *Campos Sulinos*, we only recovered genetic studies for *B. eriospatha* (Mart. ex Drude) Becc. (Nazareno et al. 2011; Nazareno and dos Reis 2014). ***Butia eriospatha*** (Mart. ex Drude) Becc. is a species endemic to the Mata Atlânticabiome grasslands, popularly known as *butiá-da-serra* (Heiden and Sant'Anna-Santos 2025). It has several economic uses, including as timber, fibers for craftsmanship, oil extracted from seeds, and ornamental use (Amorin and Bicalho 2021). According to Franco and Fontana (1997), the tea of the flowers is used in folk medicine to treat *amarelão* (jaundice/anemia) and as a sedative, or to improve sleep. The pulp of the fruit, consumed in its natural state, helps eliminate uric acid. The species suffers from several pressures, such as habitat destruction, international illegal trade, and overexploitation (Amorin and Bicalho 2021). Nazareno et al. (2011) developed microsatellite markers for the species from 50 *B. eriospatha* samples from two populations from Santa Catarina state, which enabled investigations into the genetic diversity, spatial genetic structure, mating system, and population dynamics of the species. Two years later, Nazareno and dos Reis (2014) investigated the patterns of demography, natural regeneration, and herbivory, as well as the levels of genetic diversity, of *B. eriospatha* using microsatellite loci sampled from four populations in South Brazil. Their results revealed that cattle grazing in *B. eriospatha* population areas severely affected the species' demographic structure. They found that the species presented no recruitment and a mortality rate of 2.0%, pointing to an extremely high risk of local extinction for the species—with a greater than 50% reduction in the effective population size—in the next 40 years (Nazareno and dos Reis 2014).

Anemopaegma arvense (Vell.) Stellfeld ex de Souza, popularly known as *catuaba*, is a species cited in some works as endemic to the Cerrado, with bioactive agents effective in the treatment of cancer, malaria, and HIV (Batistini 2006; Batistini et al. 2009). According to CNC Flora (dos Santos Filho and Messina 2012), the species is found in Brazil (in the north, northeast, central-west, southeast, and south, in the State of Paraná), Bolivia, and Paraguay. The species is being commercialized to produce cosmetics and perfumery (dos Santos Filho and Messina 2012).

Batistini et al. (2009) evaluated the genetic diversity of seven populations of *A. arvense* found in the State of São Paulo, using random amplified polymorphic DNA markers. They found significant genetic variability within and between subpopulations. Genetic variability among subpopulations was not spatially structured. The authors suggest that an in situ conservation strategy is best for this species, keeping it in its own area of natural occurrence, with local adaptations favoring the establishment of the species in this natural habitat. They also emphasized that only one of the studied populations was in a protected area (Mogi-Guaçu Biological Reserve), so germplasm bank conservation could be a strategy for conserving genetic variants in unprotected areas (Batistini et al. 2009).

Cunila incana Benth. is a herbaceous to shrubby aromatic species with occurrence in southern Brazilian grasslands (Flora e Funga do Brasil 2025) and northeast Argentina (Xifreda and Mallo 2006). The species is used for essential oil extraction (Agostini et al. 2010), and species of this genus more broadly are used in folk medicine as stimulants, aromatics, antispasmodics, emmenagogues, antithermics, and in the treatment of chronic coughs and respiratory infections (Agostini et al. 2008). Eleven South American species of the genus, including one population of *C. incana*, were evaluated by ISSR (Inter-Simple Sequence Repeats) marker by Agostini et al. (2008). The study contributed to the knowledge of the taxonomy of botanical sections in *Cunila*, but a genetic population approach focused on the conservation and sustainable use of the species is still lacking.

Hesperozygis ringens (Benth.) Epling (Figs. 8.8d, e), popularly known as *espanta-pulga*, is an aromatic shrubby plant endemic to the Rio Grande do Sul State. It is found in sandy and rocky soils of the Pampa biome and has been increasingly studied in recent years due to its multiple potential applications. Leite et al. (2024) published a review on the current knowledge about the species, presenting several of its described uses or potential uses, such as antiparasitic and allelopathic properties against some species of pests, parasites, and weeds, and notable anesthetic and bactericidal activity for controlling fish parasitic microorganisms—properties of great interest to the fish-farming industry. Two important studies were conducted to evaluate the genetic diversity of *H. ringens*, both using ISSR markers and similar populations (Fracaro and Echeverrigaray 2006; Leite et al. 2023). The recovered studies found low genetic diversity values, although the genetic diversity remained stable when comparing the two studies. The species is not protected in any Conservation Unit, and its great risk of extinction comes from the intense degradation of its natural areas of occurrence. A small ex-situ collection, representing a single natural population, is maintained in the Santa Maria Botanical Garden (in the Universidade Federal de Santa Maria, Rio Grande do Sul State). This collection should be expanded to include representatives of other natural populations. The collection is currently mainly used for environmental education purposes, with the aim of raising awareness among the population about the importance of preserving Pampa species.

Colletia paradoxa (Spreng.) Escal., popularly known as *curro*, *quina-do-campo* or *quina-do-rio-grande*, is a thorny shrub growing in open areas in southern Brazil, Uruguay, and Argentina. In Argentina, it occurs in large clusters called *currales*. The species is used in folk medicine, with the roots and bark of the branches used to reduce fever. Bastos-Záchia et al. (2021) investigated the antioxidant and antimycobacterial activity of extracts of *C. paradoxa.* They found significant amounts of phenolic compounds, including quercetin, quercitrin, and rutin, besides considerable antioxidant and antimycobacterial activity. A recent study conducted by Taques et al. (2025) evaluated the effects of different LED lighting systems on the photosynthetic and antioxidant metabolism of *C. paradoxa* cultivated in vitro. They recommended combining red and blue LEDs for the efficient micropropagation of *C. paradoxa*, as they found this reduced light-induced stress in its shoots. The genetic diversity of six natural populations of *C. paradoxa* was assessed through ISSR markers by Bastos-Záchia et al. (2021). Their results showed high genetic variance within populations and a good number of migrants ($N_m = 1.0465$). The species is considered threatened mainly due to habitat loss, either due to the expansion of agriculture in its areas of occurrence or due to the reduction in habitat quality due to invasive species.

Aloysia hatschbachii Moldenke is an aromatic shrub endemic to the state of Paraná, found in rocky fields. The genus *Aloysia* is commonly used in folk medicine, and according to Segecin (2008), this species is also used in conventional medicine, although its specific uses were not mentioned. Araujo (2020) highlighted the potential medicinal use of its essential oil, which they characterized, finding sesquiterpenes such as β-elemene and germacrene B (22.6%), monoterpenes, and phenylpropanoids in low concentrations. Araujo (2020) also pointed out that the high concentration of sesquiterpenes found (above 84%) indicates potential antimicrobial action. In terms of genetic diversity, a Ph.D. study conducted by Chagas

(2023) evaluated a total of 150 individuals from different subpopulations with isoenzyme markers. The subpopulations showed low genetic diversity and an excess of homozygotes, especially in the most isolated subpopulation; in addition, the author found greater genetic variability among individuals than between subpopulations. The low genetic diversity was attributed to anthropogenic effects, with short-term conservation measures being needed (Chagas 2023).

Although herbs and shrubs are more typically associated with the southern grasslands, there are also trees present, mainly close to rivers and other water bodies. The genus *Ocotea* Aubl., family Lauraceae, with 178 trees and shrub species recorded for Brazil (*Ocotea* in Flora e Funga do Brasil 2025), has several species with different uses, especially as timber or medicine. The genus includes the endangered ***Ocotea porosa*** (Nees & Mart.) Barroso and the vulnerable ***Ocotea catharinensis*** Mez. For *O. porosa*, known as *imbuia*, research was conducted on its properties, revealing cytotoxicity against human cancer cells, mouse cancer cells, and non-tumoral mouse cells, as well as insecticidal activity of the essential oil against susceptible bedbugs through topical application (Brustolin et al. 2020). For *O. catharinensis*, Meira and da Silva (2015) performed analytical tests on plant extracts and demonstrated the presence of alkaloids, proteins and amino acids, condensed tannins, and foamy saponins, as well as volatile coumarins. Tests also demonstrated the toxicity of the species in an assay with *Artemia salina* (Meira and da Silva 2015). Both species were included in a microsatellite-based study (Martins et al. 2015) that investigated the patterns and levels of genetic diversity and inbreeding of three *Ocotea* species to define priority populations for conservation management actions focusing on population enhancement and ex situ germplasm collections. Martins et al. (2015) found high genetic diversity for each of the species, with moderate genetic differentiation among populations. According to the authors, most populations displayed significant inbreeding and isolation by distance.

Another important tree in the *Campos Sulinos* is ***Araucaria angustifolia*** (Bertol.) Kuntze (Fig. 8.8a), a gymnosperm species that was exploited for decades as a source of timber, but also for its edible seeds and medicinal applications (Freitas et al. 2009; Michelon et al. 2012; Souza et al. 2014; Peralta et al. 2016). The species is currently classified as Endangered (EN) (Brasil 2022) due to intense logging, which has caused the degradation and fragmentation of its populations (Inza et al. 2018; Souza et al. 2020; Rimoli and Lima 2024). It is one of the best-known species in terms of genetic diversity, thanks to pioneering research which helped to diagnose its genetic decline. Several groups have published important research characterizing and monitoring its natural populations (Auler et al. 2002; Inza et al. 2018; de Sousa et al. 2020; da Costa et al. 2021), which supported the conservation planning and sustainable use of the species. For instance, the species is protected against timber exploitation, the use of the seeds is allowed only in a restricted period (during the closed season—*defeso*—collecting, transporting, and trading its seeds is forbidden), planting is incentivized, the importance of conservation is frequently addressed in schools, and new cultivars are even being developed for commercial purposes, reducing the pressure on natural populations. In this context, *A. angustifolia* represents a good example of how the development of studies on its genetic diversity has been fundamental to the development of effective guidelines for its conservation and sustainable use.

8.6 Conclusion

Formal knowledge about medicinal plant species native to Brazilian *Campos Sulinos*, as well as their conservation status, is still insufficient. Many studies focus on introduced exotic species, and little research has been conducted on the genetic diversity of threatened medicinal plants from the region. The scenario is even more worrying for the endemic medicinal plants of these grasslands, with very few studies having been conducted on these species. Even species frequently mentioned in the literature, such as *Ephedra tweediana*, lack studies focused on their population genetic diversity and conservation. Genetic diversity studies with long-term monitoring, such as those conducted for *A. angustifolia*, are essential for guiding appropriate strategies for the conservation and sustainable use of the genetic heritage of the area.

Brazil has several challenges: to restore its indigenous people's knowledge about native medicinal plant species, to invest more heavily in basic studies about such species (general biology, propagation, sustainable use, and genetic diversity or variability), and to plan more effective conservation policies to prevent species extinction. Another challenge is to invest in the conservation of all the diverse biomes found in this continental country, including our less well-known but species-rich and beautiful *Campos Sulinos*.

Acknowledgments We would like to thank the Coordination for the Improvement of Higher Education Personnel (CAPES) for the master's scholarships awarded to Yago Luís Nóbrega Mendes, Hulia Juana Scherer, and Milena de Melo Hettwer, as well as the National Council for Scientific and Technological Development (CNPq) for the master's scholarships awarded to Marina Scheuer and Henrique da Rosa Ferreira. We would also like to thank the Universidade Federal de Santa Maria for the Scientific Initiation scholarship (FIPE/UFSM) awarded to Táfinis Sturbelle Schiavon.

References

Agostini G, Echeverrigaray S, Souza-Chies TT (2008) Genetic relationships among South American species of *Cunila* D. Royen ex L. based on ISSR. Plant Syst Evol 274:135–141. https://doi.org/10.1007/s00606-008-0037-8

Agostini G, Souza-Chies TT, Agostini F, Atti-Serafni L, Echeverrigaray S (2010) Essential oil composition of *Cunilaincana* Benth. (Lamiaceae). J Essent Oil Res 22:433–434. https://doi.org/10.1080/10412905.2010.9700365

Aguiar MCS, Silvério FO, de Pinho GP, Lopes PSN, Fidêncio PH, Ventura SJ (2014) Volatile compounds from fruits of *Butia capitata* at different stages of maturity and storage. Food Res Int 62:1095–1099. https://doi.org/10.1016/j.foodres.2014.05.039. Get rights and content

Alvares CA, Stape JL, Sentelhas PC, Gonçalves JLM, Sparovek G (2013) Köppen 's climate classification map for Brazil. Meteorol Z 22(6):711–728. https://doi.org/10.1127/0941-2948/2013/0507

Amorim E, Bicalho M (2021) *Butiaeriospatha* (Arecaceae). Lista Vermelha da Flora Brasileira: Centro Nacional de Conservação da Flora/ Instituto de Pesquisas Jardim Botânico do Rio de Janeiro. Available at https://proflora.jbrj.gov.br/html/Butia%20eriospatha_2021.html

Andrade BO, Koch C, Boldrini II, Martin EV, Hasenack H, Hermann JM, Kollmann J, Pillar VD, Overbeck GE (2015) Grassland degradation and restoration: a conceptual framework of stages and thresholds illustrated by southern Brazilian grasslands. Nat Conserv 13(2):95–104. https://doi.org/10.1016/j.ncon.2015.08.002

Andrade BO, Bonilha CL, Overbeck GE, Vélez-Martin E, Rolim RG, Bordignon SAL et al. (2019) Classification of South Brazilian grasslands: Implications for conservation. Appl Veg Sci 22(1): 168–184. https://doi.org/10.1111/avsc.12413

Andrade BO, Dröse W, Aguiar CA de, Aires ET, Alvares DJ, Barbieri RL, Carvalho CJB de, Bartz M, Becker FG, Bencke GA, Beneduzi A, Silva JB, Blochtein B, Boldrini II, Boll PK, Bordin J, Silveira RMB da, Martins MB et al (2023) 12,500+ and counting: biodiversity of the Brazilian Pampa. Front Biogeogr 15(2). https://doi.org/10.21425/F5FBG59288

Araujo GM (2020) Caracterização da morfoanatômica, fenologia e composição química do óleo essencial de *Aloysiahatschbachii* (Verbenaceae). Dissertação. Mestrado em Agronomia: Agricultura e Ambiente. Universidade Federal de Santa Maria. Available at https://repositorio.ufsm.br/handle/1/21092. Accessed 15 Oct 2025

Atanasov AG, Waltenberger B, Pferschy-Wenzig EM, Linder T, Wawrosch C, Uhrin P, Temml V, Wang L, Schwaiger S, Heiss EH, Rollinger JM, Schuster D, Breuss JM, Bochkov V, Mihovilovic MD, Kopp B, Bauer R, Dirsch VM, Stuppner H (2015) Discovery and resupply of pharmacologically active plant-derived natural products: a review. Biotechnol Adv 33(8):1582–1614. https://doi.org/10.1016/j.biotechadv.2015.08.001

Auler NMF, Reis MSD, Guerra MP, Nodari RO (2002) The genetics and conservation of *Araucaria angustifolia*: I. Genetic structure and diversity of natural populations by means of non-adaptive variation in the state of Santa Catarina, Brazil. Genet Mol Biol 25:329–338. https://doi.org/10.1590/S1415-47572002000300014

Bastos-Záchia NR, Piana M, dos Santos SF, da SJR, Athayde ML, Essi L, de Campos MMA (2021) Phytochemical analysis and evaluation of antioxidant and antimycobacterial activity of *Colletia paradoxa* from Brazil. Nat Prod Res 35(5):802–806. https://doi.org/10.1080/14786419.2019.1598987

Batistini AP (2006) Diversidade morfológica, genética e química de populações naturais de Anemopaegma arvense (Vell.) Stellf. xii, 83 f. Tese (doutorado) - Universidade Estadual Paulista, Faculdade de Ciências Agrárias e Veterinárias. Available at https://repositorio.unesp.br/entities/publication/b6e318f0-3c95-4af8-a831-1531c876c43c. Acessed 15 Oct 2025

Batistini AP, Telles MPC, Bertoni BW, Coppede J da S, Môro FV, França S de C, Pereira AMS (2009) Genetic diversity of natural populations of *Anemopaegma arvense* (Bignoniaceae) in the Cerrado of São Paulo State, Brazil. Genet Mol Res 8(1):52–63. Available at: https://repositorio.unesp.br/entities/publication/446e04d0-f757-4a6b-9bc2-59b8fdf3e9fa. Accessed 13 April 2025

Battisti C, Garlet TMB, Essi L, Horbach RK, de Andrade A, Badke MR (2013) Plantas medicinais utilizadas no município de Palmeira das Missões, RS, Brasil. Rev Bras Bioc 11(3):338–348. Available at https://seer.ufrgs.br/index.php/rbrasbioci/article/view/115518. Accessed 13 April 2025

Brandão T, Trevisan R, Both R (2008) Unidades de Conservação e os Campos do Rio Grande do Sul. Rev Bras Bioc 5(S1):843–845. Available at https://seer.ufrgs.br/index.php/rbrasbioci/article/view/115909. Accessed 15 Oct 2025

Brasil (2006) Ministério da Saúde. Política Nacional de Plantas Medicinais e Fitoterápicos. Brasília. https://bvsms.saude.gov.br/bvs/publicacoes/politica_nacional_fitoterapicos.pdf. Accessed 29 Jul 2025

Brasil (2022) Ministério do Meio Ambiente. Portaria MMA n° 148, de 7 de junho de 2022. Diário Oficial da União: seção 1, Brasília, DF, p. 74, 8 Jun 2022. https://www.gov.br/icmbio/pt-br/assuntos/centros-de-pesquisa/aves-silvestres/arquivos/portaria-148-2022.pdf. Accessed 29 Jul 2025

Brustolin LJR, Monteiro LM, Almeida VP, Raman V, Maia BHLNS, Casapula I, Paludo KS, Bussade JE, Renman JU, Kahn IA, Farago PV, Budel JM (2020) *Ocotea porosa*: anatomy and histochemistry of leaves and stems, chemical composition, cytotoxicity and insecticidal activities of essential oil. Braz Arch Bioltechnol 63. https://doi.org/10.1590/1678-4324-2020190082

Cámara-Leret R, Bascompte J (2021) Language extinction triggers the loss of unique medicinal knowledge. Proc Natl Acad Sci 118(24):e2103683118. https://doi.org/10.1073/pnas.2103683118

Centro Nacional de Conservação da Flora (CNCFlora). Brazilian flora conservation portal. Jardim Botânico do Rio de Janeiro. Available at https://cncflora.jbrj.gov.br/. Accessed 29 Jul 2025

Chiqueti J, Tiuman TS (2021) Atividade anti-Candida de extratos e frações de espécies da família Asteraceae. In: XXVI Seminário de Iniciação Científica e Tecnológica da UTFPR. Available at https://eventos.utfpr.edu.br//sicite/sicite2021/paper/viewFile/7927./4354. Acessed 07 Ago 2025

Claro AE, Silva RS, Pereira LM, Souza MF (2024) Historical use of medicinal plants and future potential from phytotherapy to phytochemicals. Ann Bot (Roma) 14(1). https://doi.org/10.13133/2239-3129/18564

Corrêa C, Scheffer MC (2014) As plantas medicinais, aromáticas e condimentares e a agricultura familiar. Hortic Bras 32(3):376–376. https://doi.org/10.1590/S0102-05362014000300023

da Costa NCF, Stedille LIB, Lauterjung MB, Montagna T, Candido-Ribeiro R, Bernardi AP, Nodari RO (2021) Spatiotemporal variation in mating system and genetic diversity of Araucaria angustifolia: implications for conservation and seed collection. For Ecol Manag 481:118716. https://doi.org/10.1016/j.foreco.2020.118716

Dar RA, Shahnawaz M, Qazi PH (2017) General overview of medicinal plants: a review. J Phytopharmacol 6(6):349–351

das Chagas KPT (2023) Autoecologia e diversidade genética de *Aloysiahatschbachii*Moldenke (Verbenaceae), uma espécie em perigo de extinção. PhD Thesis. Universidade Federal do Paraná, Setor de Ciências Agrárias. Programa de Pós-Graduação em Engenharia Florestal. Curitiba. Available at https://acervodigital.ufpr.br/xmlui/handle/1884/87654. Accessed 15 Oct 2025

de Meira TSO, da Silva AFG (2015) Estudo Fitoquímico e Avaliação de Atividades Biológicas das Folhas de *Ocoteaodorifera* e *Ocoteacatharinensis*. Anais do ENIC1(4). Available at: https://anaisonline.uems.br/index.php/enic/article/view/1765

de Sousa VA, Reeves PA, Reilley A, de Aguiar AV, Stefenon VM, Richards CM (2020) Genetic diversity and biogeographic determinants of population structure in *Araucaria angustifolia* (Bert.) O.Ktze. Conserv Genet 21(2):217–229. https://doi.org/10.1007/s10592-019-01242-9

dos Santos Filho LAF, Messina T (2012) *Anemopaegma arvense* (Bignoniaceae). Lista Vermelha da Flora Brasileira: Centro Nacional de Conservação da Flora/ Instituto de Pesquisas Jardim Botânico do Rio de Janeiro. Available at https://proflora.jbrj.gov.br/html/Anemopaegma%20arvense_2012.html. Accessed 10 Oct 2025

Dutra RC, Campos MM, Santos ARS, Calixto JB (2016) Medicinal plants in Brazil: pharmacological studies, drug discovery, challenges and perspectives. Pharmacol Res 112:4–29. https://doi.org/10.1016/j.phrs.2016.01.021

Ellwanger JH, Ziliotto M, Chies JAB (2022) Protect Brazil's overlooked Pampa biome. Science 377(6607):720. https://www.science.org/doi/10.1126/science.ade1838. Acessed 11 April 2025

Essi L, Lima MFRC, Leite LG, Wolf MM (2020) Threatened and understudied: the lack of genetic data of endangered Brazilian plant species. Ciênc Nat 42:e28–e28. https://doi.org/10.5902/2179460X40965

Fagundes ACA, Menezes IS, Rocha DSB, Funch LS, de Siqueira JA (2025) Habitat loss and climate change threaten the survival of endangered *Dyckia* (Bromeliaceae) species in the Caatinga domain. Plant Ecol 226:497–511. https://doi.org/10.1007/s11258-025-01510-4

Fracaro F, Echeverrigaray S (2006) Genetic variability in *Hesperozygisringens* Benth. (Lamiaceae), an endangered aromatic and medicinal plant of Southern Brazil. Biochem Genet 44(11–12):471–482. https://doi.org/10.1007/s10528-006-9044-z

Franco IJ, Fontana VL (1997) Ervas & plantas: a medicina dos simples. Imprimax, Erechim, p 177

Freitas AM, Almeida MTR, Andrighetti-Fröhner CR, Cardozo FTGS, Barardi CRM, Farias MR, Simões CMO (2009) Antiviral activity-guided fractionation from *Araucaria angustifolia* leaves extract. J Ethnopharmacol 126(3):512–517. https://doi.org/10.1016/j.jep.2009.09.005

Guerra MP, Nodari RO (2001) Biodiversidade: aspectos biológicos, geográficos, legais. In: Farmacognosia: da planta ao medicamento. v 6. Editora UFRGS, Porto Alegre, pp 13–28

Guerra A, Reis LK, Borges FLG, Ojeda PTA (2020) Ecological restoration in Brazilian biomes: identifying advances and gaps. For Ecol Manag 458:117802. https://doi.org/10.1016/j.foreco.2019.117802

Hardy K (2021) Paleomedicine and the evolutionary context of medicinal plant use. Rev Bras 31(1):1–15. https://doi.org/10.1007/s43450-020-00107-4

Hasenack H, Weber EJ, Boldrini II, Trevisan R, Flores CA, Dewes H (2023) Biophysical delineation of grassland ecological systems in the State of Rio Grande do Sul, Southern Brazil. Iheringia, Série Botânica 78. https://doi.org/10.21826/2446-82312023v78e2023001

Heiden G, Sant'Anna-Santos BF *Butiain*Flora e Funga do Brasil.Jardim Botânico do Rio de Janeiro. Available at https://floradobrasil.jbrj.gov.br/FB15703. Accessed 13 Oct 2025

IBGE - Instituto Brasileiro de Geografia e Estatística (2004) Mapa de vegetação do Brasil. https://geoftp.ibge.gov.br/informacoes_ambientais/vegetacao/mapas/brasil/vegetacao.pdf. Accessed 20 Aug 2025

Ijinu TP, George V, Pushpangadan P (2022) History of research on medicinal plants in India. In: Dev S, Srivastava A, Sharma V (eds) Medicinal and aromatic plants of India, vol 1. Springer International Publishing, Cham, pp 35–61

Inza MV, Aguirre NC, Torales SL, Pahr NM, Fassola HE, Fornes LF, Zelener N (2018) Genetic variability of *Araucaria angustifolia* in the Argentinean Parana Forest and implications for management and conservation. Trees 32:1135–1146

Kamada T, Picoli EAT, Alfenas AC, Cruz CD, Vieira RF, Otoni WC (2009) Diversidade genética de populações naturais de *Pfaffiaglomerata* (Spreng.) Pedersen estimadapormarcadores RAPD. Acta Sci Agron 31(3):403–409

Khan H (2014) Medicinal plants in light of history: recognized therapeutic modality. JBECAM 19(3):216–219

Le Roux JJ, Hui C, Castillo ML, Iriondo JM, Keet JH, Khapugin AA et al (2019) Recent anthropogenic plant extinctions differ in biodiversity hotspots and coldspots. Curr Biol 29(17):2912–2918

Leite LG, Leonardo JLSH, Mazzocato AC, Essi L (2023) Genetic diversity of *Hesperozygisringens* (Benth.) Epling (Lamiaceae): current situation on natural populations of Rio Grande do Sul, Brazil. Rev Bras Bot 46(1):61–68. https://doi.org/10.1007/s40415-023-00869-w

Leite LG, Wolf MM, Mazzocato AC, Essi L (2024) Hesperozygis ringens (Benth.) Epling (Lamiaceae): uma revisão. Rev Fitos 18: e1199–e1199. https://doi.org/10.32712/2446-4775.2024.1199

Lopes RM, da Silva JP, Vieira RF, da Silva DB, Gomes IS, Agostoni-Costa TS (2012) Composition of fat acids in pulp of native fruits from the Brazilian savanna. Rev Bras Frutic 34(2):635–640

MAPBIOMAS (2025) Coleção 10 de Mapas de Cobertura e Uso da Terra do Brasil. https://brasil.mapbiomas.org/infograficos/. Acessed 12 Ago 2025

Martins EM, Lamont RW, Martinelli G, Lira-Medeiros CF, Quinet A, Shapcott A (2015) Genetic diversity and population genetic structure in three threatened *Ocotea* species (Lauraceae) from Brazil's Atlantic Rainforest and implications for their conservation. Conserv Genet 16(1):1–14

Melro JCL, Souza IM, Barbosa Júnior J, Rocha EA, Alves RR (2020) Ethnodirigid study of medicinal plants used by the population assisted by the "Programa de Saúde da Família" (Family Health Program) in Marechal Deodoro-AL, Brazil. Braz J Biol 80(2):410–423

Menezes LS, Ely CV, Lucas DB, Minervini-Silva GH, Vélez-Matin E, Hasenack H, Trevisan R, Boldrini II, Pillar VD, Overbeck GE (2022) Reference values and drivers of diversity for South Brazilian grassland plant communities. An Acad Bras Ciênc 94(1). https://doi.org/10.1590/0001-3765202220201079

Menezes AAD, Braga EJB, Iganci J (2025) Where Endemism meets urgency: native Cactaceae and the conservation crisis in the subtropical South America Pampa. Diversity 17(6):397. https://doi.org/10.3390/d17060397

Michelon F, Branco CS, Calloni C, Giazzon I, Agostini F, Spada PKWDS, Salvador M (2012) *Araucaria angustifolia*: a potential nutraceutical with antioxidant and antimutagenic activities. Curr Nutr Food Sci 8(3):155–159. https://doi.org/10.2174/157340112802651103

Nazareno AG, Zucchi MI, Reis MS (2011) Microsatellite markers for Butia eriospatha (Arecaceae), a vulnerable palm species from the Atlantic Rainforest of Brazil. Am J Bot 98: e198–e200. https://doi.org/10.3732/ajb.1100064

Nazareno AG, dos Reis MS (2014) At Risk of Population Decline? An Ecological and Genetic Approach to the Threatened Palm Species Butia eriospatha (Arecaceae) of Southern Brazil. J Hered 105: 120–129. https://doi.org/10.1093/jhered/est065

Ocoteain Flora e Funga do Brasil. Jardim Botânico do Rio de Janeiro. Disponível em: https://floradobrasil.jbrj.gov.br/FB8440. Accessed 13 Oct 2025

Paulert R, Zonetti P, Cordeiro J, Kozera C, Stefanello S (2019) Cultivo de plantas medicinais: integração do conhecimento tradicional e científico. In: Stadnik MJ, Velho AC, Zorrilla SE (eds) Desenvolvimento Sustentável na Produção Agroalimentar. Florianópolis, UFSC, pp 73–88. Available at https://labfitop.paginas.ufsc.br/livros/. Accessed 13 Oct 2025

Peralta RM, Koehnlein EA, Oliveira RF, Correa VG, Corrêa RC, Bertonha L, Ferreira IC (2016) Biological activities and chemical constituents of *Araucaria angustifolia*: an effort to recover a species threatened by extinction. Trends Food Sci Technol 54:85–93. https://doi.org/10.1016/j.tifs.2016.05.013

Rainsford KD (2004) History and development of the salicylates. In: Rainsford KD (ed) Aspirin and related drugs. CRC Press, London, pp 1–23. https://doi.org/10.1201/9780203646960

Rajora OP, Mosseler A (2001) Challenges and opportunities for conservation of forest genetic resources. Euphytica 118:197–212. https://doi.org/10.1023/A:1004150525384

Ramalho AB, Rossi AAB, Dardengo JFE, Zortéa K, Évelin M, Tiago AV, Martins KC (2016) Diversidade genética entre genótipos de*Bertholletia excelsa* por meio de marcadores moleculares ISSR. Floresta 46(2):207–214. https://doi.org/10.5380/rf.v46i2.41970

Rimoli PAR, Lima de S (2024) Possibilidades de mudança: movimentos que podem inverter o atual estágio de declínio das araucárias no Brasil. Revista Mirante, Anápolis (GO) 17(2):333–363

Rufatto LC, Gower A, Schwambach J, Moura S (2012) Genus *Mikania*: chemical composition and phytotherapeutical activity. Rev Rev Bras Farmacogn 22(6). https://doi.org/10.1590/S0102-695X2012005000099

Salazar-Gómez A, Alonso-Castro AJ (2022) Medicinal plants from Latin America with wound healing activity: ethnomedicine, phytochemistry, preclinical and clinical studies—a review. Pharmaceuticals 15(9):1095. https://doi.org/10.3390/ph15091095

Santos MO, Rodrigues AP, Guarçoni EA, Tavares TB, Pires M (2018) Medicinal plants: versatility and concordance of use in the Caatinga area, Northeastern Brazil. An Acad BrasCiênc 90:2767–2779. https://doi.org/10.1590/0001-3765201820170594

Schneider LR, dos Santos DC, Campos AD, Lund RG (2017) The phytochemistry and pharmacology of *Butia* sp.: a systematic review and an overview of the technological monitoring process. Phytother Res 31(10):1495–1503. https://doi.org/10.1002/ptr.5883

Segecin S (2008) Morfoanatomia das Folhas Cotiledonares, Protofilo e Metafilo de *Aloysiahatschbachii*Moldenke (Verbenaceae). Akrópolis- Revista de Ciências Humanas da UNIPAR 3(11):21–29. Available at: https://revistas.unipar.br/index.php/akropolis/article/view/1642. Accessed 12 Oct 2025

Silva MC, Carvalho JCT (2004) Plantas medicinais. In: Carvalho JCT (ed) Fitoterápicos anti-inflamatórios: aspectos químicos, farmacológicos e aplicações terapêuticas. Tecmedd, Ribeirão Preto, pp 39–41

Silveira FAO, Arruda AJ, Bond W, Durigan G, Fidelis A, Kirman K, Oliveira RS, Overbeck GE, Sansevero JBB, Siebert F, Siebert SJ, Young TP, Buisson E (2020) Myth-busting tropical grassybiomerestoration. Restor Ecol 28(5):1067–1073. https://doi.org/10.1111/rec.13202

Souza CM Jr, Shinbo JZ, Rosa M, Parente L (2020) Reconstructing three decades of land use and land cover changes in brazilian biomes with landsat archive and earth engine. Remote Sens 12(17):2735. https://doi.org/10.3390/rs12172735

Souza MO, Branco CS, Sene J, DallAgnol R, Agostini F, Moura S, Salvador M (2014) Antioxidant and antigenotoxic activities of the Brazilian pine *Araucariaangustifolia* (Bert.) O.Kuntze. Antioxidants 3(1):24–37. https://doi.org/10.3390/antiox3010024

Staude IR, Vélez-Matin E, Andrade BO, Podgaiski LR, Boldrini II, Mendonça M Jr, Pillar VD, Overbeck GE (2018) Local biodiversity erosion in south Brazilian grasslands under moderate levels of landscape habitat loss. J Appl Ecol 55(3):1241–1251. https://doi.org/10.1111/1365-2664.13067

Taques RCV, Souza RCT, Giloni-Lima PC, de Lima VA, Ribas LLF (2025) Light quality affects photosynthesis and antioxidant metabolism of *Colletia paradoxa* cultivated in-vitro. Plant Cell Tissue Organ Cult 161(1):20

Tribess B, Pintarelli GM, Bini LA, Camargo A, Silva CV, Gasper AL, Rego F, Hanazaki N (2015) Ethnobotanical study of plants used for therapeutic purposes in the Atlantic Forest region, Southern Brazil. J Ethnopharmacol 164:136–146. https://doi.org/10.1016/j.jep.2015.02.005

Viegas MP, Silva CLSP, Lacerda C, Moreira JP, Cardin LT, Azevedo VCR, Ciampi AY, Freitas MLM, Moraes MLT, Sebbenn AM (2011) Diversidade genética e tamanho efetivo de duas populações de *Myracrodruon urundeuva* Fr. All., sob conservação ex situ. Revista Árvore 35(4):769–779. https://doi.org/10.1590/S0100-67622011000500002

World Health Organization (WHO) (2023) Traditional medicine has a long history of contributing to conventional medicine and continues to hold promise. Newsroom – Feature stories, Geneva, 10 Aug. https://www.who.int/news-room/feature-stories/detail/traditional-medicine-has-a-long-history-of-contributing-to-conventional-medicine-and-continues-to-hold-promise. Accessed 29 Jul 2025

Xifreda CC, Mallo AC (2006) El Género *Cunila* (Lamiaceae, Mentheae) en Argentina. *Darwiniana*. Nueva Serie 44(1):298–308. https://doi.org/10.14522/darwiniana.2014.441.299

Chapter 9
Strengthening Medicinal Plant Resources: Genetic Diversity for Sustainable Production, Accurate Identification, and Conservation

Marina Silalahi, Nabela Fikkirya, Endang C. Purba, and Anisa Anggraeni

Abstract The global demand for medicinal plants has increased significantly due to their widespread use in therapeutic applications. However, a major challenge in utilizing these plants is the variability in the concentration of bioactive compounds, which complicates the standardization of medicinal plants for consistent therapeutic use. To address this issue and meet the growing demand, strategies to improve the quality and quantity of medicinal plants are essential. This includes advancements in breeding and cultivation, the development of superior genotypes, polyploidy induction, and the use of gene mapping techniques. Furthermore, to combat the risks of misidentification, genetic-based plant identification methods, such as multilocus barcoding, genome-based DNA barcodes, mini-barcoding, and meta-barcoding, can be employed to ensure accurate species authentication. However, genetic-based identification itself faces challenges, including genetic mutations, recombination, habitat fragmentation, and climate change, which threaten both the genetic integrity and adaptability of medicinal plants. To mitigate these risks and preserve genetic diversity, conservation strategies such as tissue culture technology, cryopreservation, gene banking, and ecological niche modeling are crucial.

M. Silalahi (✉)
Biology Education Department, Universitas Kristen Indonesia, Jakarta, Indonesia
e-mail: marina.silalahi@uki.ac.id

N. Fikkirya
Aquatic Resource Management, Faculty of Fisheries and Marine Science, Universitas Jenderal Soedirman, Purwokerto, Indonesia

E. C. Purba
Institute of Ecology and Evolutionary Biology, National Taiwan University, Taipei, Taiwan

A. Anggraeni
Department of Economic Plants and Biotechnology, Yunnan Key Laboratory for Wild Plant Resources, Kunming Institute of Botany, Chinese Academy of Sciences, Kunming, China

University of Chinese Academy of Science, Huairou, Beijing, China

S. Jha et al. (eds.), *Genetic Diversity in Medicinal Plants*, Sustainable Development and Biodiversity 44, https://doi.org/10.1007/978-981-95-7281-6_9

Collectively, these integrated methodologies offer a comprehensive framework for the sustainable use, conservation, and reliable production of medicinal plant resources, thereby facilitating the long-term availability of plant-based therapeutics.

Keywords Bioactive compounds · Genetic diversity · Misidentification · Sustainable use · Conservation technologies

9.1 Introduction

Studies suggest that between 12.5% and 18% of the world's total plant species—approximately 422,000 species—are known to contain bioactive compounds with medicinal properties (Rao et al. 2004; Rajeswara et al. 2012). These compounds have attracted considerable attention due to their diverse pharmacological effects, ranging from pain relief and anti-inflammatory properties to the treatments of complex conditions such as motion sickness and glaucoma (Rao et al. 2004). As a result, the global demand for medicinal plants (MPs) has increased significantly, reflecting their growing recognition as vital bioresources in the healthcare industry. Since 1971, over 20,000 plant species have been utilized and traded as raw materials for medicinal purposes, with the global market value of these plants surpassing $60 billion (Mohammadi and Saghaian 2022). According to projections by the World Health Organization (WHO), the market value of MPs could reach an estimated $5 trillion by 2050, underscoring the expanding economic significance of this sector (Salim and Munadi 2017). This rapid growth highlights not only the increasing reliance on plant-based resources for therapeutic applications but also the essential role MPs play in both healthcare systems and bioresource industries.

Despite the therapeutic potential of MPs, considerable variation exists in the concentration and efficacy of their bioactive compounds, even within the same species (Bagdonaite et al. 2012). For example, *Hypericum* species have been shown to exhibit varying concentrations of bioactive compounds, influenced by both internal genetic factors and external environmental conditions (Cirak and Radusiene 2019). This variability complicates the standardization of MPs for therapeutic use and can jeopardize the safety and efficacy of plant-based medicines (Vijayakumar et al. 2019; Jamdade et al. 2022). Furthermore, such variations in bioactive compounds can also contribute to the misidentification of plant species, as morphological similarities between species may mask differences in their chemical composition. Misidentification, in turn, increases the risk of using toxic substitutes, which can severely undermine the therapeutic value of MPs. A well-documented example of this risk is the substitution of *Stephania tetrandra* S. Moore with *Aristolochia fangchi* Y.C. Wu ex L.D. Chow, and S.M. Hwang, resulting in nephrotoxicity and cancer due to the ingestion of toxic aristolochic acids (Ekor 2014). These challenges underscore the critical need for accurate and reliable plant identification methods to safeguard the safety and efficacy of MP products (Bhelkar and Bolavarapu 2024; Lankasena et al. 2024).

To address these challenges, Zhu et al. (2022) propose a robust, technology-driven framework that integrates advanced genetic-based identification, machine learning algorithms, and the establishment of a standardized reference database for precise species verification. However, genetic-based identification itself is not without challenges (Rajeswara et al. 2012), as internal factors such as genetic mutations and recombination, combined with environmental pressures like habitat fragmentation and climate change, threaten the survival of individual species and erode their genetic diversity (Rajeswara Rao 2016; Niazian 2019; Sharma et al. 2020). This genetic erosion significantly impedes the adaptive capacity of medicinal plant species, further exacerbating concerns about their long-term availability. For instance, the unsustainable harvesting of *Scutellaria baicalensis* Georgi roots in China has caused a sharp decline in its population (Yuan et al. 2010), while *Clivia miniata* (Lindl.) Verschaff., commonly traded in South Africa's informal medicinal markets, has seen its population decrease of over 40% in the past 90 years (Groner et al. 2022). Given these challenges, this chapter will comprehensively examine strategies for enhancing MP production to meet global demand, explore accurate and reliable identification methods, assess key factors contributing to genetic vulnerability, and identify effective conservation strategies to preserve genetic diversity within MP populations. By addressing these key issues, this research will contribute to the sustainable use and conservation of MPs, ensuring their availability for future generations and supporting the continued growth of the global MP market.

9.2 Improvement of Medicinal Plant Production

To meet the growing global demand for MPs, it is crucial to improve both their quality and quantity. Several strategies can be employed to achieve this objective, including advanced breeding techniques, optimized cultivation methods, the development of superior genotypes, polyploidy induction, and gene mapping. These approaches are designed not only to enhance the productivity of MPs but also to boost their therapeutic potential by increasing the concentration of bioactive compounds. The following sections provide a detailed explanation of each of these methods.

9.2.1 Breeding

Plant breeding, encompassing both conventional and modern techniques, aims to enhance the quality and quantity of crop yields while simultaneously improving plant resilience to biotic stresses (e.g., pests and diseases) and abiotic stresses (e.g., environmental conditions). Conventional breeding approaches, however, are often limited by challenges such as prolonged breeding cycles, seasonal dependencies, and unpredictable outcomes. To overcome these limitations, biotechnological breeding methods are increasingly being explored. These advanced techniques include tissue culture, protoplast fusion, and other genetic engineering strategies. In the context of MPs, breeding efforts primarily focus on enhancing the production of secondary metabolite bioactive compounds that contribute to the plant's medicinal properties. Biotechnology-based breeding methods offer the potential to achieve these improvements more rapidly than traditional approaches (Bajaj and Ishimaru 1999; Niazian 2019). Furthermore, these methods allow for the precise modification of the genetic composition of MPs to achieve specific medicinal or economic goals (Niazian 2019; Hasan et al. 2024). For instance, techniques such as polyploidy induction and chromosome doubling can promote the development of larger plant tissues, while mutation breeding—through the application of physical and chemical mutagens—can facilitate the generation of novel plant varieties (Hasan et al. 2024; Parthasarathy et al. 2025).

9.2.2 Cultivation

The cultivation of MPs plays a vital role in ensuring a sustainable supply for the production of plant-based medicines and herbal remedies, thus reducing the harvest pressure on wild populations (Yuan et al. 2010; Rajeswara Rao 2016). It also offers a promising alternative to wild collection, helps preserve genetic diversity, and provides high-quality raw materials for various industries, including pharmaceuticals, cosmetics, and food. Moreover, cultivated MPs are preferred over wild-harvested plants in the herbal industry due to their predictable quality and consistent composition of secondary metabolites (Dajic-Stevanovic and Pljevljakusic 2015). Cultivation can be carried out using conventional methods, such as planting on agricultural land, or through biotechnological techniques, like tissue culture. However, conventional continuous cropping (CC) of MPs can lead to significant economic losses due to negative effects on soil health, including changes in microbial communities and nutrient availability. For instance, continuous cultivation of *Panax notoginseng* (Burkill) F. H. Chen in Southwest China has resulted in reduced plant quality and increased disease incidence, including microbial infections from *Fusarium oxysporum* and *Phaeosphaeria rousseliana* (Dong et al. 2016; Zhang et al. 2019). These challenges are primarily attributed to the depletion of soil nutrients and allelopathic effects, which hinder the growth and sustainability of MPs. To address these challenges, tissue culture offers a more controlled environment, ensuring more consistent cultivation outcomes and improving varieties with stable success (Parthasarathy et al. 2025).

9.2.3 Superior Genotypes and Polyploidy

The synthesis and accumulation of secondary metabolites in MPs is a highly complex process, influenced by both internal genetic factors and external environmental conditions. Internally, genetic circuits involving regulated genes and enzymes play a significant role, while external factors such as light, temperature, water availability, and salinity can impact the production of these metabolites (Li et al. 2020). One approach to addressing these complexities is the genetic manipulation of MPs using in vitro plant regeneration systems. This method enables the engineering of metabolic pathways that lead to the production of secondary metabolites (Gómez-Galera et al. 2007). A key technique in this regard is polyploidy induction, which involves the acquisition of three or more complete sets of chromosomes in an organism (Iannicelli et al. 2020). Polyploidy results in larger plant sizes and can significantly enhance the production of secondary metabolites (Salma et al. 2017; Iannicelli et al. 2020). The generation of synthetic polyploids, as part of plant breeding strategies, has led to the development of new and improved cultivars with increased metabolic output (Iannicelli et al. 2020). Commonly used antimitotic chemicals in this process include colchicine, trifluralin, and oryzalin, which facilitate polyploidy induction by disrupting normal cell division. Inducing polyploidy in medicinal plants can therefore be a promising strategy for improving plant vigor and enhancing the yield of desired secondary metabolites (Salma et al. 2017).

9.2.4 Gene Mapping

Gene mapping is a crucial tool in plant genetics that enables the identification and localization of specific genes responsible for important traits in medicinal plants (Ge et al. 2012; Shang et al. 2020). By pinpointing genes associated with secondary metabolite biosynthesis, gene mapping can help improve the yield and quality of bioactive compounds used in pharmaceuticals and other industries (Wu et al. 2022; Guo et al. 2024; Martínez-Chávez et al. 2024). Modern gene mapping techniques, such as genome-wide association studies (GWAS) (He and Gai 2023) and quantitative trait locus (QTL) mapping (Powder 2020), allow researchers to associate specific genomic regions with desirable traits like increased production of alkaloids, flavonoids, or terpenoids. These methods provide a deeper understanding of the genetic mechanisms underlying secondary metabolite production, which can inform breeding programs aimed at enhancing these traits. Furthermore, gene mapping plays a critical role in developing genetically modified plants that are tailored to produce higher levels of active compounds or possess improved disease resistance (Aziz et al. 2022). As the availability of genomic resources and sequencing technologies continues to grow, gene mapping will remain an essential tool for advancing plant breeding strategies and ensuring the sustainable production of medicinal plants.

9.3 Genetic Diversity-Based Plant Identification

DNA barcoding, a molecular technique first introduced by Paul Hebert in 2003, has fundamentally transformed the field of plant species identification by integrating molecular genetics, high-throughput sequencing technologies, and bioinformatics (Kress 2017). Unlike traditional morphological methods, DNA barcoding provides a robust, reliable approach to species identification that is independent of the developmental stage or environmental conditions. This makes it particularly advantageous for identifying plants at any stage of growth, including juveniles, which are often difficult to classify using morphological traits alone. Its consistency allows for the development of highly specific, repeatable markers that can be used across a broad range of plant taxa. As of April 2025, the Barcode of Life Data Systems (BOLD) database has catalogued approximately 362,416 plant species, covering a vast array of taxonomic families and genera, including both medicinal and non-medicinal species. Thus, DNA barcoding represents a powerful tool in plant taxonomy, with far-reaching implications for biodiversity conservation, ecological research, and the safe application of plant-derived products in the health sector (Table 9.1).

In the early stages, barcoding efforts predominantly utilized single-locus DNA markers, a strategy known as the monolocus approach. Among the most commonly used markers in plant DNA barcoding are *matK*, *rbcL*, *trnH-psbA*, and ITS (Reddy et al. 2022). This approach gained popularity due to its relatively low cost and rapid processing times, making it a practical and efficient tool for species identification and authentication, particularly in large-scale studies or routine plant screenings. For example, Krawczyk et al. (2014) demonstrated that the combination of *trnH-psbA* and *matK* was highly effective in authenticating species in the genus *Lamium*, providing reliable species resolution. In contrast, a study by Xiong et al. (2022) revealed that the ITS locus showed the highest success rate for distinguishing species within the genus *Ardisia*. The utility of the ITS region as a robust barcode marker is further supported by several studies. Chen et al. (2010) reported an identification success rate of 92.7% using ITS2, while Liu et al. (2019) achieved 92.15% accuracy for identifying poisonous medicinal plants from the Chinese Pharmacopoeia. Similarly, Zhang et al. (2015) found that ITS2 provided the highest identification efficiency in the genus *Uncaria*, with a remarkable success rate of 94.4% based on Best Close Match analysis. However, the monolocus approach is not without limitations, the most prominent being its inability to reliably distinguish species within closely related genera. In such cases, the use of a single marker can lead to taxonomic ambiguities, resulting in inconsistencies in classification data (Larraín et al. 2019). In light of these limitations, alternative methodologies, such as multilocus barcoding, genome-based DNA barcode markers, mini-barcoding, and meta-barcoding, are increasingly employed to enhance the resolution and accuracy of species identification, as discussed below.

Table 9.1 The selected medicinal plants and the specific application of DNA barcodes in plant identification

Study group	Sample strategy	Success rate	Reference
Orchid genera in Asia	422 species represent all five subfamilies, including Apostasioideae, Vanilloideae, Cypripedioideae, Orchidoideae and Epidendroideae 6986 sequences from five single barcodes (ITS, ITS2, *matK*, *rbcL*, *trnH-psbA*, and their seven combinations)	Single locus: ITS performed the best barcode Multilocus: ITS + *matK* exhibited the most efficient barcode	Raskoti and Ale (2021)
Order Caryophyllales in United Arab Emirates	20 species, 13 species (36 individuals) collected from the field and 7 species from NCBI gene bank	ITS2 was more successful in distinguishing between examined species	Jamdade et al. (2022)
Various medicinal plant species	6600 individuals, 4800 species, and 753 genera Used seven candidate DNA barcodes, including psbA-trnH, matK, *rbcL*, *rpoC1*, *ycf5*, ITS2, and ITS	The species level, ITS2 can provide an identification success rate of 92.7%.	Chen et al. (2010)
Poisonous medicinal plants based on Chinese Pharmacopoeia	106 species, 27 families, 65 genera Four commonly used regions as candidate DNA barcodes, including ITS2, psbA-trnH, matK, and rbcL	ITS2 exhibited highest correct identification efficiency using BLAST or distance method at species level (92.15%)	Liu et al. (2019)
Various medicinal plants based on Ayurvedic Pharmacopoeia of India (API)	347 medicinal plants, 308 genera, 112 families, and 45 orders Used *rbcL* DNA barcode	The *rbcL* DNA barcode differentiates 319 species (85%) with pairwise divergences ranging between 0.2% and 29.9%	Vassou et al. (2016)
Selected of local medicinal plants of Kingdom of Bahrain	29 medicinal plants Used *rbcLa*, *matK*, and ITS2 DNA barcode	rbcLa stands out for highest sequencing success rate (97%) and highest accuracy for species-level identification (89.28%)	Pathak et al. (2018)
Indonesian medicinal plant	61 species and 30 families Used four DNA barcoding regions, including ITS2, *matK*, *rbcL*, and trnL	*mat*K stands out for highest accuracy for species-level identification (31.15%)	Cahyaningsih et al. (2022)

(continued)

Table 9.1 (continued)

Study group	Sample strategy	Success rate	Reference
Selected medicinal plants recommended by WHO	257 samples and 8 species Used three DNA barcoding regions, including *matK*, *rbcL*, and ITS2	*rbcL* showed the highest overall efficiency in PCR amplification and sequencing, working for 228 of 257 samples (88.72%) *mat*K was successful in 209 samples (81.32%) ITS2 had the lowest success rate, working for 185 samples (71.98%)	Palhares et al. (2015)
Selected medicinal plants from Pakistan	78 species Used three DNA barcoding regions, including *rbc*L, *mat*K, and intergenic spacer region *psb*A-*trn*H	*psbA-trnH* was less effective, grouping fewer species (only 25 successful sequences out of 78) *mat*K and *rbcL* are validated as the most useful barcodes for studying large plant populations and evolutionary relationships	Khan Shinwari et al. (2018)
Uncaria genera in China	54 specimens and 10 Uncaria species Used five DNA barcoding regions, including nrITS, nrITS2, *matk*, *rbcL*, and *trnH-psbA*	ITS2 showed 100.0% amplification and sequencing efficiency and highest ability of species identification on the basis of the Best Close Match (94.4%)	Zhang et al. (2015)
Tulipa edulis (Miq.) Baker	1 species Used three DNA barcoding regions, including *matK*, *psbA trnH*, and *rbcL*	*mat*K sequence information enabled the correct species identification, performed well in PCR amplification efficiency, sequencing efficiency, and differential intra- and interspecific divergences	Ma et al. (2014)
Astragalus genera in China	Used four DNA barcoding regions, including ITS, *mat*K, *rbc*L, and *trn*H–*psb*A	The ITS region showed a high success rate of PCR amplification and bidirectional sequencing, as well as a high discriminatory ability (72%). The *mat*K region also possessed relatively perfect university and had a high discriminatory ability (61%) The *rbc*L region showed good university but low discriminatory ability (44%) *trn*H–*psb*A region was not suitable for identifying *Astragalus* species	Zhang and Jiang (2020)

(continued)

Table 9.1 (continued)

Study group	Sample strategy	Success rate	Reference
Lamiaceae Family	72 samples collected from 314 species Four candidate barcodes (*matK, rbcL*, ITS2, and *psbA-trnH*)	ITS2 successfully identified 78.3% at the species level and 100% at the genus level	Han et al. (2012)
Selected medicinal plants in Pakistan	12 species *rbcL, matK*, and *psbA-trnH*	*rbcL* is a good barcode for *Acacia nilotica*, with a 100% match to multiple vouchers (representing different subspecies), but it is not useful for distinguishing *Anethum graveolens* and *Foeniculum vulgare* *psbA-trnH* spacer does not appear to be a good barcode for *Acacia nilotica* or *Linum usitatissimum*	Schori and Showalter (2011)
Common herbal plants collected from Peninsular Malaysia	602 samples and 112 species *rbcL* and *psbA-trnH*	*rbcL*: Used as the primary barcode, showing: High amplification and sequencing success. ~87.5% species resolution across 112 herbal species. *trnH-psbA*: Used as a second-tier barcode to improve species discrimination: Exhibited high sequence divergence and insertion/deletion rates. Increased species resolution to 96.4% when combined with *rbcL*	Tnah et al. (2019)

9.3.1 Multilocus DNA Barcoding

A multilocus DNA barcoding can enhance the accuracy of plant species identification by integrating several molecular markers, such as *rbcL* and *trnH-psbA*, *trnH-psbA* and ITS, as well as *matK* and *trnH-psbA*. This multilocus approach is particularly beneficial when a single locus fails to provide sufficient resolution for accurate species differentiation. Jiamahate et al. (2025) identified *trnH-psbA* and ITS as the most widely used molecular markers for the authentication of flowering plants (angiosperms), as both loci exhibit substantial discriminatory power for species identification. The combination of these markers has been proven highly effective in various studies, such as the work by Bolson et al. (2015), which demonstrated that *trnH-psbA* and ITS provided 91% species-specific clusters in Maximum Likelihood analysis when applied to threatened commercial woody angiosperms from the southern Brazilian Atlantic Rainforest. This high level of resolution underscores the efficacy of this marker combination in enhancing species-level accuracy, particularly for complex and endangered plant species. In addition to *trnH-psbA* and ITS, other marker combinations have also proven successful in improving identification accuracy. Raskoti and Ale (2021), for example, showed that the combination of ITS and *matK* was the most efficient multilocus barcode for identifying 422 orchid species across Asia. On the other hand, studies by Khan Shinwari et al. (2018) and Palhares et al. (2015) indicated that the combination of *matK* and *rbcL* provided superior accuracy for species-level identification, particularly for plants with closely related species. These findings suggest that the combination of multiple loci can substantially enhance taxonomic resolution, making it an essential tool for plant species authentication, especially when dealing with difficult-to-distinguish species. Despite the success of the multilocus approach in improving identification accuracy, it does have certain limitations. One of the primary challenges is the potential for minimal amplification success, especially when plant DNA is degraded or poorly preserved (Li et al. 2015). In some cases, the amplification of one or more markers may fail, leading to incomplete or unreliable results. Moreover, the cost and complexity associated with sequencing multiple markers can be prohibitive in large-scale applications. Nevertheless, the multilocus DNA barcoding method remains one of the most promising tools for plant species authentication, offering a higher degree of confidence in distinguishing species and improving the reliability of medicinal plant products on the market.

9.3.2 Genome-based DNA Barcode Markers

Since the advent of next-generation sequencing (NGS) technology in 2010, there has been a significant increase in the use of genome-based DNA barcode markers, often referred to as super-barcodes, in plant species identification (Batovska et al. 2017). These genome-based markers involve the utilization of entire chloroplast genomes or other genomic sequences to identify species, expanding upon traditional single- and multilocus DNA barcoding methods (Yu et al. 2021). Chloroplast genomes, in particular, have shown promise in surpassing the limitations of single-locus barcodes, offering higher resolution in species identification due to their relatively slow mutation rate and maternal inheritance, which enhances the detection of genetic variation across species (Wu et al. 2021). This approach provides a more comprehensive means of species identification, particularly for complex plant groups that may be difficult to distinguish using traditional markers. Despite the advantages, the application of super-barcodes is not without challenges. One of the primary obstacles is the high cost of sequencing entire genomes, which makes it a resource-intensive method, particularly for large-scale studies or routine identification tasks. Additionally, the complexity of genome annotation and the requirement for advanced bioinformatics tools and technical expertise pose significant barriers to its widespread adoption, as not all laboratories possess the necessary infrastructure or capacity to handle such data. Furthermore, the utility of chloroplast genome-based super-barcoding is limited in taxa with a history of extensive hybridization, where species boundaries may be unclear due to gene flow between species. For example, research by Yang et al. (2017) demonstrated that chloroplast genes alone are insufficient to resolve species identity in the genus *Quercus*, a complex group with a history of hybridization. Similarly, Yan et al. (2018) found that while universal chloroplast DNA (cpDNA) could effectively distinguish between the subgenera *Cerris* (Old World clade) and *Quercus* (New World clade) within *Quercus*, it was unable to resolve species-level differences. This limitation arises from factors such as divergence, hybridization, and incomplete lineage sorting, leading to low interspecific variation and high intraspecific variation within the genus. The super-barcoding approach, while facing certain challenges, is increasingly being acknowledged as a transformative tool in molecular taxonomy, biodiversity conservation, and species authentication, particularly for medicinal plants (Yu et al. 2021).

9.3.3 Mini-Barcoding

Mini-barcoding is an innovative modification of the traditional DNA barcoding approach, specifically designed to overcome the challenges associated with degraded or damaged DNA samples, which are often encountered in herbarium specimens, processed herbal products, or other biological materials that have undergone deterioration over time (Gao et al. 2019). This technique utilizes shorter DNA fragments, typically ranging from 100 to 250 base pairs (bp), which still contain sufficient genetic information for effective species differentiation, despite their reduced length compared to full-length barcode markers (Yu et al. 2020). Mini-barcoding has proven particularly valuable for applications in which obtaining high-quality DNA is difficult, such as with processed medicinal products or ancient herbarium specimens (Dormontt et al. 2018; Jiao et al. 2018; Gao et al. 2019). One of the key advantages of mini-barcoding is its high efficiency in PCR amplification, which makes it an ideal solution for laboratories with limited technical resources or when dealing with degraded DNA samples (Yu et al. 2021). The study by Parveen et al. (2016) highlighted the application of mini-barcoding in the authentication of herbal ingredients in commercially available products containing species, such as *Ginkgo biloba* L., *Serenoa repens* (W. Bartram) Small, and *Harpagophytum procumbens* (Burch.) DC. ex Meisn. Although mini-barcoding offers lower resolution compared to full-length DNA barcodes, it remains a highly effective and practical tool for various applications, including medicinal plant authentication, quality control of natural products, molecular conservation, and forensic investigations (Dubey et al. 2011; Raclariu-Manolică et al. 2021; Zhu et al. 2022). For instance, targeting variable regions of markers, such as *trnL*, has been shown to be particularly successful in distinguishing species even within complex DNA mixtures, where traditional barcoding methods might struggle (Ming et al. 2017). This makes mini-barcoding a promising alternative for situations where traditional DNA barcoding methods might not be feasible, offering a cost-effective and efficient solution for species identification in the context of degraded samples or difficult-to-handle materials.

9.3.4 Meta-Barcoding

Meta-barcoding represents an innovative molecular approach that combines traditional DNA barcoding with NGS technologies, enabling the simultaneous identification of multiple species within complex, mixed samples, such as soil, water, or herbal products (Seethapathy et al. 2019). This technique uses universal primers to target conserved DNA regions, such as ITS, *rbcL*, or *trnL*, which are then amplified and analyzed in parallel during a single high-throughput sequencing run (Travadi et al. 2023). The primary advantage of meta-barcoding lies in its ability to detect a broad range of species from fragmented or mixed DNA, without the need for isolating individual organisms. This makes it especially valuable for environments or samples where direct species isolation is not feasible or where the DNA is highly degraded (Arulandhu et al. 2017). In the plant context, meta-barcoding has become an indispensable tool for a variety of applications, including biodiversity assessments, ecological monitoring, and the authentication of raw materials in herbal products and foods (Bruno et al. 2019; Liang et al. 2025). By enabling the identification of species within a sample, even if they are present in low abundance, meta-barcoding helps detect contaminants or unauthorized plant species, including invasive alien species that could pose ecological or economic threats (Ruppert et al. 2019). For instance, research by Urumarudappa et al. (2022) demonstrated the efficacy of meta-barcoding in identifying plant species listed in the National List of Essential Medicines (NLEM) of Thailand, achieving accurate identification down to the genus and family levels. This highlights its potential for quality control in medicinal plant authentication, offering a powerful tool for ensuring the authenticity of herbal medicines and preventing adulteration. However, despite its significant advantages, meta-barcoding does have certain limitations. One of the major challenges is the potential for PCR amplification bias, which can affect the representation of species within a sample (Skelton et al. 2023). This bias arises due to variations in the efficiency of amplification among different DNA regions, which can skew results. Additionally, the success of meta-barcoding is highly dependent on the quality and completeness of reference databases. Short DNA barcodes, which are commonly used in meta-barcoding, may lack sufficient resolution for accurately identifying certain genera. In contrast, conventional DNA barcodes, such as *matK*, *rbcL*, *trnH-psbA*, and *ITS2*, typically exceed 500 base pairs and may be less suitable for degraded DNA samples, such as those found in processed herbal products (Yu et al. 2020). This limitation further underscores the importance of developing more robust and comprehensive reference databases to improve the accuracy and reliability of meta-barcoding in plant species identification.

9.4 Determinants of Genetic Vulnerability

The genetic diversity of medicinal plants is influenced by a complex interplay of intrinsic and extrinsic factors. Intrinsic factors, which are endogenous to the organism, include its mutational events, recombination, and inheritance patterns, which together govern the genetic variation within populations and modulate phenotypic expression. These factors are pivotal in shaping the adaptive potential and evolutionary trajectory of plant species. In contrast, extrinsic factors are primarily shaped by the environmental conditions in which a plant exists (Halder and Jha 2023). Environmental perturbations, such as climate fluctuations and habitat degradation (Lee et al. 2018), can enhance the adaptability of medicinal plants to changing conditions; however, they also pose a threat to genetic diversity, which is inherently sensitive to disturbances in environmental stability (Nag et al. 2015). Among extrinsic factors, habitat fragmentation stands out as a critical determinant, as it disrupts ecological interactions, alters species distributions, impairs gene flow, and ultimately erodes genetic diversity (Lee et al. 2018). A detailed exploration of the various factors influencing the genetic diversity of medicinal plants is provided below.

9.4.1 Genetic Mutation

Genetic mutations can occur both naturally and through deliberate human intervention. Natural mutations arise spontaneously due to various factors, including environmental influences such as radiation, chemicals, and other mutagenic agents (Udage 2021). In contrast, intentional mutations are induced to develop new plant varieties with desirable traits, such as enhanced active compound content, increased disease resistance, improved environmental tolerance, and optimized agronomic characteristics like plant size, leaf morphology, or flowering time (Dong and Ronald 2019). These induced mutations are typically achieved through methods such as radiation, chemical treatments, or genetic engineering. Despite their considerable advantages in advancing plant breeding, these methods also introduce potential risks. Induced genetic changes can lead to unintended consequences, including the emergence of undesirable traits, the development of new pests or diseases, and potential plant mortality (Zhao and Wolt 2017). Furthermore, genetic mutations can create difficulties in plant propagation, particularly when they result in seedless fruit or increased susceptibility to environmental stressors and toxins. Such mutations may also induce genetic instability, which can cause the loss of desired traits in subsequent generations, thereby complicating the long-term maintenance of these traits.

9.4.2 Recombination

Whereas genetic mutations generate new alleles through direct alterations in the DNA sequence, genetic recombination involves the exchange or rearrangement of genetic material during meiosis, producing novel genetic combinations that can spread throughout a population. Although genetic recombination—particularly when applied through genetic engineering—offers advantages such as increased yields, improved nutritional quality, and greater resilience to environmental stress, it also carries potential ecological risks. For example, the introduction of transgenic plants with herbicide resistance may lead to the emergence of new, persistent weed species that are difficult to manage, potentially increasing pesticide use. Moreover, genetic modifications in cultivated plants can disrupt ecological balance, especially if transgenic varieties outcompete native species and reduce biodiversity (Ijaz and Haq 2019; Zou et al. 2024). These considerations highlight the need for careful assessment and monitoring of the ecological impacts associated with the use of genetic recombination and genetic engineering in plant populations.

9.4.3 Cross-Pollination

Cross-pollination in medicinal plants poses a significant threat to the integrity of local plant varieties, particularly those with unique genetic characteristics. When genetically distinct lines or species interbreed, there is a risk of altering or diluting desirable traits that have been carefully selected within local varieties. Such genetic mixing can result in the loss of key bioactive compounds or modifications to pharmacological properties that are critical for the therapeutic efficacy of these plants. Additionally, the introduction of foreign genes may compromise the genetic purity of local varieties, weakening their distinctiveness and potentially contributing to the erosion of valuable genetic resources that have adapted to specific ecological conditions over time. The impact of cross-pollination can be especially detrimental for rare or endangered local varieties, as the infusion of nonlocal genes can blur original gene pools, leading to the extinction of unique genetic lines. Consequently, conserving local medicinal plant varieties becomes increasingly challenging, as cross-pollination may reduce their genetic diversity and ecological resilience, ultimately threatening the sustainability of traditional medicine practices that depend on these plants.

9.4.4 Phenological Changes

Changes in phenology, or the timing of life-cycle events such as flowering, fruiting, and seed germination, can have profound effects on the genetic material of medicinal plants. These phenological shifts are often driven by environmental factors such as temperature, precipitation, and seasonal variation (Woldearegay 2020; Hassan et al. 2024). Such changes can alter the expression of genes related to key traits, including secondary metabolite production, which are crucial for the plant's medicinal value. Moreover, shifts in phenology can influence genetic diversity by affecting gene flow within and between populations. For example, if phenological events like flowering and pollination occur out of sync with the optimal timing for cross-pollination, this can lead to reduced genetic exchange and, consequently, a decline in genetic variability (Wolf et al. 2017). This reduced genetic diversity makes the plant population more susceptible to environmental stresses and diseases, thus compromising the long-term sustainability of medicinal plant resources. Additionally, phenological changes may result in the loss of specific genetic traits that have evolved to confer advantages in particular ecological niches, ultimately affecting the plant's ability to adapt to future climate conditions.

9.4.5 Geographic Distribution

The bioactive composition of medicinal plants is profoundly impacted by several extrinsic factors, including geographical location (Kadu et al. 2012). This factor not only influences the plant's bioactive compound profile but also has significant implications for its genetic diversity. For instance, a study by Al-Hmadi et al. (2021) illustrates that the bioactive composition of *Pimpinella lutea* essential oils varies significantly depending on the growing location in Tunisia. Different regions, such as North Bizerta, South Bizerta, and Tabarka, exhibit distinct profiles of major compounds, with these variations attributed to local soil conditions, climate, and altitude. These environmental factors alter the plant's metabolic pathways, which in turn lead to variability in the types and quantities of bioactive compounds produced. This metabolic plasticity, while beneficial for adapting to different environments, can also result in genetic shifts, especially when plants undergo selective pressures from varying climates and cultivation practices.

9.4.6 Habitat Fragmentation

Habitat fragmentation leads to the spatial isolation of plant species, disrupting ecological connectivity and gene flow between populations (Lee et al. 2018). This isolation can significantly reduce genetic diversity by altering population size and increasing genetic drift, which hinders the ability of populations to adapt to environmental changes. Several factors contribute to habitat fragmentation, including human activities such as deforestation for agriculture and plantations, as well as natural disturbances like fires and storms. Among these, forest fires—whether intentional or accidental—are considered a primary driver of fragmentation. Fires significantly contribute to habitat fragmentation through a range of scale-dependent mechanisms. First, they generate edges by altering the structure of habitats, which in turn intensifies edge effects, such as increased exposure to wind, sunlight, and invasive species. Second, fires degrade habitat patches by reducing vegetation cover, altering microclimatic conditions, and compromising the availability of resources, making these habitats less suitable for species survival and reproduction. Finally, fires disrupt landscape-scale connectivity by breaking up continuous habitats, creating barriers to species movement, and isolating populations, which hampers gene flow and exacerbates the genetic isolation of MPs (Driscoll and Smith 2021).

9.4.7 *Climate Change*

Climate change indicators, including fluctuations in temperature, alterations in precipitation patterns, an increased frequency of extreme weather events (Ahmadi-Lahijani and Moori 2023; Sudhakaran 2024), and rising atmospheric CO_2 concentrations (Applequist et al. 2020; Ahmadi-Lahijani and Moori 2023), represent significant drivers of metabolic changes in MPs. These environmental stressors modulate key metabolic pathways, thereby influencing the physiological responses of plants. Consequently, these alterations affect the biosynthesis of secondary metabolites, which are crucial for the therapeutic efficacy of MPs (Mishra et al. 2016). Disruptions in these metabolic pathways can lead to diminished quality and safety of MPs, while also adversely impacting plant growth and biomass production—factors that are integral to their pharmacological properties. These ongoing changes highlight the critical need for the development of adaptive strategies to preserve the integrity and sustainability of medicinal plant resources in the face of a rapidly changing climate.

9.5 Conservation of Genetic Diversity Within Medicinal Plant Populations

In response to the challenge described in the previous chapter, biotechnology-based conservation has emerged as a crucial step in the genetic conservation of plants, especially for those that cannot be preserved through traditional methods. Biotechnology offers valuable tools for plant conservation, including tissue culture, genetic banks, and cryopreservation (Păunescu 2009; Cruz-Cruz et al. 2013), which together support the long-term conservation and restoration of plant biodiversity. These techniques are particularly effective for conserving endangered, rare, and economically significant plants, including tuber crops (Cruz-Cruz et al. 2013; Dewi et al. 2016).

9.5.1 *Tissue Culture Technology*

Advancements in biotechnology, particularly tissue culture, have significantly enhanced the conservation, propagation, and stability of medicinal plants. Tissue culture facilitates the rapid clonal propagation of plants under controlled sterile conditions, ensuring the consistent genetic integrity of plant material, which is crucial in overcoming challenges such as genetic mutations, recombination, and the unintended effects of cross-pollination. Furthermore, tissue culture offers the advantage of slow-growth in vitro storage, allowing for the long-term conservation of rare, endangered, and economically important species, effectively mitigating the risks associated with habitat fragmentation and climate change (Cruz-Cruz et al. 2013; Kahat 2023). By manipulating growth conditions, tissue culture can also optimize the production of secondary metabolites, such as alkaloids, flavonoids, and terpenoids, which are vital for the therapeutic efficacy and pharmacological properties of medicinal plants (Rani and Kumar 2017). This technique not only guarantees a reliable and continuous supply of bioactive compounds but also enhances their quality, potency, and safety by minimizing the risk of disease, pathogen contamination, and genetic instability. Moreover, the controlled environment of tissue culture eliminates the unpredictability of phenological changes, ensuring uniform growth and the maintenance of desired traits over time. The sustained availability of these bioactive metabolites supports pharmaceutical development, providing a stable source of raw materials for drug innovation and therapeutic formulation. While tissue culture offers substantial potential in terms of increasing productivity and ensuring the sustainable supply of medicinal plants, its successful application in the industry requires rigorous quality control, standardized protocols, and safety assessments to ensure the long-term genetic integrity and efficacy of plant-derived products (Anushi et al. 2023).

9.5.2 *Cryopreservation for Long-Term Conservation of Medical Plants*

Cryopreservation, as an advanced method for the preservation of medicinal plants, offers significant benefits in maintaining the genetic diversity and long-term viability of plant species, especially those facing environmental and anthropogenic pressures (Păunescu 2009; Cruz-Cruz et al. 2013). While genetic mutations, recombination, and cross-pollination can complicate plant breeding and genetic conservation, cryopreservation mitigates these issues by preserving plant genetic material at ultralow temperatures, thus maintaining genetic integrity over time and reducing the risk of undesirable genetic changes. This is particularly critical in the face of climate change and habitat fragmentation, where the natural habitats of many medicinal plants are increasingly altered or destroyed. Cryopreservation offers solutions to this challenge by enabling the rapid multiplication and conservation of endangered species, including those that are impacted by phenological changes or shifting reproductive patterns due to climate change (Sun et al. 2024). For example, cryopreservation has been successfully applied to a variety of plant species, including *Peganum harmala* L. and *Tanacetum ulutavicum* Tzvelev, maintaining the viability of seeds and tissues from diverse plant types, including wild species and geophytes, which are particularly vulnerable to fragmentation and environmental stresses. By preserving these species under controlled conditions, both tissue culture and cryopreservation can counteract genetic erosion and ensure the future availability of valuable medicinal plants for both research and pharmaceutical applications (Gavrilkova et al. 2019).

9.5.3 Gene Banking

Gene banking has become a cornerstone in modern plant genetic resource management, playing a critical role in preserving genetic diversity across species used in food, agriculture, and medicine (Aubry 2023). Gene banks store, maintain, and document genetic material from over 70,000 plant species worldwide, including medicinal and aromatic plants, ensuring that their genetic resources are safeguarded for future use (Lohwasser and Weise 2020). This preservation is especially crucial for protecting the genetic integrity of endangered or rare medicinal plants, which are susceptible to the effects of genetic mutation, recombination, and cross-pollination that may compromise their unique traits. Additionally, gene banks mitigate the impacts of phenological changes and habitat fragmentation by ensuring that genetically diverse populations are maintained despite environmental disruptions. By offering a repository of genetic material, gene banks provide critical resources for breeding efforts aimed at enhancing traits such as disease resistance, drought tolerance, and resilience to climate change. Through genetic conservation, gene banks contribute to the sustainable use of medicinal plants, ensuring that bioactive compounds, vital for pharmaceutical development, remain accessible even as external pressures like climate change and habitat loss pose challenges to plant populations. As the global demand for medicinal plants increases, gene banks are indispensable in securing both the genetic integrity and availability of plant-based medicinal resources for future generations.

9.5.4 Ecological Niche Modeling

Climate change and overexploitation represent two of the most pressing challenges for the future conservation of medicinal plants. To address these challenges, Ecological Niche Modeling (ENM), particularly when combined with big data and artificial intelligence (AI), offers innovative solutions for the identification and conservation of medicinal plants. ENM has emerged as a powerful tool for predicting plant distributions based on environmental data, even when available data is limited (Kaky and Gilbert 2017). By linking species observations to environmental variables, ENMs help identify ecological niches and forecast the geographical range of species under various environmental scenarios (Chauvier et al. 2022). Recent advancements in ENM, particularly in studies involving medicinal plants in Egypt, have allowed for the identification of biodiversity hotspots, both present and future, and have provided insights into how climate change may influence species richness patterns (Kaky and Gilbert 2017). These findings highlight the crucial role of ENMs in enhancing the efficiency and effectiveness of medicinal plant conservation strategies, especially in the context of habitat fragmentation and climate change-induced challenges.

9.6 Conclusion

In summary, ensuring the sustainable future of medicinal plants requires an integrated strategy that simultaneously improves production, strengthens accurate species identification, addresses genetic vulnerabilities, and applies effective conservation measures. By combining improved breeding, cultivation, superior genotypes, polyploidy, and gene mapping with advanced DNA barcoding and genome-based techniques, it is possible to maintain the quality and authenticity of medicinal resources. Moreover, recognizing and mitigating genetic risks through targeted conservation actions—such as tissue culture, cryopreservation, gene banking, and ecological modeling will help safeguard genetic diversity and secure the long-term availability of these valuable plants. Ultimately, this comprehensive approach supports not only the ecological sustainability of medicinal plant populations but also their critical role in advancing global public health.

References

Ahmadi-Lahijani MJ, Moori S (2023) Current status of medicinal plants in perspective of environmental challenges and global climate changes. In: Aftab T (ed) Environmental challenges and medicinal plants: sustainable production solutions under adverse conditions. Springer International Publishing, pp 1–28

Al-Hmadi H, Mokni ER, Joshi RK, Ashour ML, Hammami S (2021) The impact of geographical location on the chemical compositions of *Pimpinella lutea* Desf. growing in Tunisia. Appl Sci 11(16):7739

Anushi JS, Rathod M, Mishra G, Kumari VLP, Baksh H, Saxena S, Prasad L (2023) Plant tissue culture for medical therapy: unlocking the potential of medicinal plants. Curr J Appl Sci Technol 42(46):7–22

Applequist WL, Brinckmann JA, Cunningham AB, Hart RE, Heinrich M, Katerere D, Van Andel T (2020) Scientists' warning on climate change and medicinal plants. Planta Med 86(1):10–18

Arulandhu AJ, Staats M, Hagelaar R, Voorhuijzen MM, Prins TW, Scholtens I, Costessi A, Duijsings D, Rechenmann F, Gaspar FB, Barreto Crespo MT, Holst-Jensen A, Birck M, Burns

M, Haynes E, Hochegger R, Klingl A, Lundberg L, Natale C, Kok E (2017) Development and validation of a multi-locus DNA metabarcoding method to identify endangered species in complex samples. GigaScience 6(10):gix080

Aubry S (2023) Genebanking plant genetic resources in the postgenomic era. Agric Hum Values 40(3):961–971

Aziz M, Brini F, Rouached H, Masmoudi K (2022) Genetically engineered crops for sustainably enhanced food production systems. Front Plant Sci 13:1027828. https://doi.org/10.3389/fpls.2022.1027828

Bagdonaite E, Mártonfi P, Repčák M, Labokas J (2012) Variation in concentrations of major bioactive compounds in *Hypericum perforatum* L. from Lithuania. Ind Crops Prod 35(1):302–308

Bajaj YPS, Ishimaru K (1999) Genetic transformation of medicinal plants. In: Bajaj YPS (ed) Transgenic medicinal plant. Springer, Berlin, Heidelberg, pp 1–29

Batovska J, Cogan NOI, Lynch SE, Blacket MJ (2017) Using next-generation sequencing for DNA barcoding: capturing allelic variation in ITS2.G3. G3 (Bethesda) 7(1):19–29. https://doi.org/10.1534/g3.116.036145

Bhelkar NS, Bolavarapu S (2024) Machine learning techniques for medicinal plants identification using leaf features: review. Front Health Inform 13(7):520–526. www.healthinformatics-journal.com

Bolson M, De Camargo Smidt E, Brotto ML, Silva-Pereira V (2015) ITS and trnH-psbA as efficient DNA barcodes to identify threatened commercial woody angiosperms from Southern Brazilian Atlantic Rainforests. PLoS One 10(12):e0143049

Bruno A, Sandionigi A, Agostinetto G, Bernabovi L, Frigerio J, Casiraghi M, Labra M (2019) Food tracking perspective: DNA metabarcoding to identify plant composition in complex and processed food products. Genes 10(3):248. https://doi.org/10.3390/genes10030248

Cahyaningsih R, Compton LJ, Rahayu S, Brehm JM, Maxted N (2022) DNA barcoding medicinal plant species from Indonesia. Plants 11(10):1375. https://doi.org/10.3390/plants11101375

Chauvier Y, Descombes P, Guéguen M, Boulangeat L, Thuiller W, Zimmermann NE (2022) Resolution in species distribution models shapes spatial patterns of plant multifaceted diversity. Ecography 2022(10):e05973. https://doi.org/10.1111/ecog.05973

Chen S, Yao H, Han J, Liu C, Song J, Shi L, Zhu Y, Ma X, Gao T, Pang X, Luo K, Li Y, Li X, Jia X, Lin Y, Leon C (2010) Validation of the ITS2 region as a novel DNA barcode for identifying medicinal plant species. PLoS One 5(1):e8613

Cirak C, Radusiene J (2019) Factors affecting the variation of bioactive compounds in *Hypericum* species. Biol Futur 70(3):198–209

Cruz-Cruz C, González-Arnao M, Engelmann F (2013) Biotechnology and conservation of plant biodiversity. Resources 2(2):73–95

Dajic-Stevanovic Z, Pljevljakusic D (2015) Challenges and decision making in cultivation of medicinal and aromatic plants. In: Máthé Á (ed) Medicinal and aromatic plants of the world: scientific, production, commercial and utilization aspects. Springer Netherlands, Dordrecht, pp 145–164. https://doi.org/10.1007/978-94-017-9810-5_8

Dewi N, Dewi IS, Roostika I (2016) Pemanfaatan Teknik Kultur In Vitro untuk Konservasi Plasma Nutfah Ubi-ubian. J AgroBiogen 10(1):34–44

Dong OX, Ronald PC (2019) Genetic engineering for disease resistance in plants: recent progress and future perspectives. Plant Physiol 180(1):26–38

Dong L, Xu J, Feng G, Li X, Chen S (2016) Soil bacterial and fungal community dynamics in relation to *Panax notoginseng* death rate in a continuous cropping system. Sci Rep 6:31802

Dormontt EE, van Dijk KJ, Bell KL, Biffin E, Breed MF, Byrne M, Caddy-Retalic S, Encinas-Viso F, Nevill PG, Shapcott A, Young JM, Waycott M, Lowe AJ (2018) Advancing DNA barcoding and metabarcoding applications for plants requires systematic analysis of herbarium collections-an Australian perspective. Front Ecol Evol 6:134

Driscoll KP, Smith DM (2021) Development of riparian and groundwater-dependent ecosystem assessments for national forests in the western US. Sustainability 13(8):4488

Dubey B, Meganathan PR, Haque I (2011) DNA mini-barcoding: an approach for forensic identification of some endangered Indian snake species. Forensic Sci Int Genet 5(3):181–184

Ekor M (2014) The growing use of herbal medicines: issues relating to adverse reactions and challenges in monitoring safety. Front Neurol 4:177

Gao Z, Liu Y, Wang X, Wei X, Han J (2019) DNA mini-barcoding: a derived barcoding method for herbal molecular identification. Front Plant Sci 10:987. Frontiers Media S.A

Gavrilkova Y, Orlova S, Dodonova A, Yerastenkova M, Verzhuk V, Sitnikov M, Radchenko OY, Filipenko G, Murashev S, Pavlov A (2019) Cryopreservation as a Method to Preserve Some Fruit and Berry Crops and Wild Medicinal Plants. Problems of Cryobiology and Cryomedicine. 29(1):044–057. https://doi.org/10.15407/CRYO29.01.044

Ge Y, Wang T, Wang N, Wang Z, Liang C, Ramchiary N, Choi SR, Lim YP, Piao ZY (2012) Genetic mapping and localization of quantitative trait loci for chlorophyll content in Chinese cabbage (*Brassica rapa* ssp. pekinensis). Sci Hortic 147:42–48

Gómez-Galera S, Pelacho AM, Gené A, Capell T, Christou P (2007) The genetic manipulation of medicinal and aromatic plants. Plant Cell Rep 26(10):1689–1715

Groner VP, Nicholas O, Mabhaudhi T, Slotow R, Akçakaya HR, Mace GM, Pearson RG (2022) Climate change, land cover change, and overharvesting threaten a widely used medicinal plant in South Africa. Ecol Appl 32(4):e2545

Guo M, Lv H, Chen H, Dong S, Zhang J, Liu W, He L, Ma Y, Yu H, Chen S, Luo H (2024) Strategies on biosynthesis and production of bioactive compounds in medicinal plants. Chin Herb Med 16(1):13–26

Halder M, Jha S (2023) Medicinal plants and bioactive phytochemical diversity: a fountainhead of potential drugs against human diseases. In: Jha S, Halder M (eds) Medicinal plants: biodiversity, biotechnology and conservation. Springer Nature, Singapore, pp 39–93

Han JP, Shi LC, Chen XC, Lin YL (2012) Comparison of four DNA barcodes in identifying certain medicinal plants of Lamiaceae. J Syst Evol 50(3):227–234

Hasan N, Laskar RA, Farooqui SA, Naaz N, Sharma N, Budakoti M, Joshi DC, Choudhary S, Bhinda MS (2024) Genetic improvement of medicinal and aromatic plant species: breeding techniques, conservative practices and future prospects. Crop Des 3(4):100080

Hassan T, Gulzar R, Hamid M, Ahmad R, Waza SA, Khuroo AA (2024) Plant phenology shifts under climate warming: a systematic review of recent scientific literature. Environ Monit Assess 196(1):36

He J, Gai J (2023) Genome-wide association studies (GWAS). In: Shavrukov Y (ed) Plant genotyping: methods and protocols. Springer US, pp 123–146

Iannicelli J, Guariniello J, Tossi VE, Regalado JJ, Di Ciaccio L, van Baren CM, Pitta Álvarez SI, Escandón AS (2020) The "polyploid effect" in the breeding of aromatic and medicinal species. Sci Hortic 260:108854

Ijaz S, Haq I (2019) Recombinant DNA technology. Cambridge Scholars Publishing

Jamdade R, Mosa KA, El-Keblawy A, Al Shaer K, Al Harthi E, Al Sallani M, Al Jasmi M, Gairola S, Shabana H, Mahmoud T (2022) DNA barcodes for accurate identification of selected medicinal plants (Caryophyllales): toward barcoding flowering plants of the United Arab Emirates. Diversity 14(4):262

Jiamahate A, Bozorov TA, Wang J, Zhang J, Zhang H, Wang X, Yang H, Zhang D (2025) Insights from DNA Barcodes-based phylogenetic analysis of medicinal plants and estimation of their conservation status: a case study in the Tianshan Wild Forest, China. Plants 14(1):99

Jiao L, Yu M, Wiedenhoeft AC, He T, Li J, Liu B, Jiang X, Yin Y (2018) DNA barcode authentication and library development for the wood of six commercial *Pterocarpus* species: the critical role of xylarium specimens. Sci Rep 8(1):1945

Kadu CAC, Parich A, Schueler S, Konrad H, Muluvi GM, Eyog-Matig O, Muchugi A, Williams VL, Ramamonjisoa L, Kapinga C, Foahom B, Katsvanga C, Hafashimana D, Obama C, Vinceti B, Schumacher R, Geburek T (2012) Bioactive constituents in *Prunus africana*: geographical variation throughout Africa and associations with environmental and genetic parameters. Phytochemistry 83:70–78

Kahat PM (2023) Conservation of threatened medicinal plants – a futuristic approach. Int J Multidiscip Res 5(2):1–5

Kaky E, Gilbert F (2017) Predicting the distributions of Egypt's medicinal plants and their potential shifts under future climate change. PLoS One 12(11):e0187714

Khan Shinwari Z, Jan SA, Talha Khalil A, Khan A, Ali M, Qaiser M, Zahra NB (2018) Identification and phylogenetic analysis of selected medicinal plant species from Pakistan: DNA barcoding approach. Pak J Bot 50(2):553–560

Krawczyk K, Szczecińska M, Sawicki J (2014) Evaluation of 11 single-locus and seven multilocus DNA barcodes in *Lamium* L. (Lamiaceae). Mol Ecol Resour 14(2):272–285

Kress WJ (2017) Plant DNA barcodes: applications today and in the future. J Syst Evol 5(4):291–307. https://doi.org/10.1111/jse.12254

Lankasena N, Nugara RN, Wisumperuma D, Seneviratne B, Chandranimal D, Perera K (2024) Misidentifications in ayurvedic medicinal plants: convolutional neural network (CNN) to overcome identification confusions. Comput Biol Med 183:109349. https://doi.org/10.1016/j.compbiomed.2024.109349

Larraín MA, González P, Pérez C, Araneda C (2019) Comparison between single and multi-locus approaches for specimen identification in *Mytilus mussels*. Sci Rep 9(1):19714. https://doi.org/10.1038/s41598-019-55855-8

Lee S, Choi J, Lee B, Yu J, Lim C (2018) Genetic diversity and structure of an endangered medicinal herb: implications for conservation. AoB Plants 10(2):ply021

Li X, Yang Y, Henry RJ, Rossetto M, Wang Y, Chen S (2015) Plant DNA barcoding: from gene to genome. Biol Rev 90(1):157–166. https://doi.org/10.1111/brv.12104

Li Y, Kong D, Fu Y, Sussman MR, Wu H (2020) The effect of developmental and environmental factors on secondary metabolites in medicinal plants. Plant Physiol Biochem 148:80–89. https://doi.org/10.1016/j.plaphy.2020.01.006

Liang X, Yang X, Sha N, Wang J, Qiu G, Chang M (2025) Application of eDNA metabarcoding technology to monitor the health of aquatic ecosystems. Water 17(8):1109. https://doi.org/10.3390/w17081109

Liu M, Li X-W, Liao B-S, Luo L, Ren Y-Y (2019) Species identification of poisonous medicinal plant using DNA barcoding. Chin J Nat Med 17(8):585–590. https://doi.org/10.1016/S1875-5364(19)30060-3

Lohwasser U, Weise S (2020) Genetic resources of medicinal and aromatic plants. In: Medicinal, aromatic and stimulant plants. Springer, Cham, pp 1–205. https://doi.org/10.1007/978-3-030-38792-1_1

Ma HL, Zhu ZB, Zhang XM, Miao YY, Guo QS (2014) Species identification of the medicinal plant *Tulipa edulis* (Liliaceae) by DNA barcode marker. Biochem Syst Ecol 55:362–368. https://doi.org/10.1016/j.bse.2014.03.038

Martínez-Chávez LA, Hernández-Ramírez MY, Feregrino-Pérez AA, Esquivel Escalante K (2024) Cutting-edge strategies to enhance bioactive compound production in plants: potential value of integration of elicitation, metabolic engineering, and green nanotechnology. Agronomy 14(12):2822. https://doi.org/10.3390/agronomy14122822

Ming S, Gang-Qiang D, Ya-Qin Z, Xia L, Wei S (2017) Identification of processed Chinese medicinal materials using DNA mini-barcoding. Chin J Nat Med 15(7):481–486

Mishra P, Kumar A, Nagireddy A, Mani DN, Shukla AK, Tiwari R, Sundaresan V (2016) Barcoding: an efficient tool to overcome authentication challenges in the herbal market. Plant Biotechnol J 14(1):8–21

Mohammadi H, Saghaian S (2022) Factors affecting consumption of different forms of medicinal plants: the case of licorice. Agriculture 12(9):1453. https://doi.org/10.3390/agriculture12091453

Nag A, Ahuja P, Sharma R (2015) Genetic diversity of high-elevation populations of an endangered medicinal plant. AoB Plants 7:plu076

Niazian M (2019) Application of genetics and biotechnology for improving medicinal plants. Planta 249:953–973

Palhares RM, Drummond MG, Dos Santos Alves Figueiredo Brasil B, Cosenza GP, Das Graças Lins Brandão M, Oliveira G (2015) Medicinal plants recommended by the world health organization: DNA barcode identification associated with chemical analyses guarantees their quality. PLoS One 10(5):0127866

Parthasarathy M, Anand AK, Rakesh Rao KR, Ramakrishnan R, Sudheer WN, Nagella P (2025) Plant breeding advances in turmeric (*Curcuma longa* L.). In: biodiversity and genetic improvement of herbs and spices. Springer Nature, Cham, pp 435–469

Parveen I, Gafner S, Techen N, Murch SJ, Khan IA (2016) DNA barcoding for the identification of botanicals in herbal medicine and dietary supplements: strengths and limitations. Planta Med 82(14):1225–1235. Georg Thieme Verlag. https://doi.org/10.1055/s-0042-111208

Pathak MR, Mohamed AAM, Farooq M (2018) DNA barcoding and identification of medicinal plants in the Kingdom of Bahrain. Am J Plant Sci 09(13):2757–2774. https://doi.org/10.4236/ajps.2018.913200

Păunescu A (2009) Biotechnology for endangered plant conservation: a critical overview. Rom Biotechnol Lett 14:4095–4103

Powder KE (2020) Quantitative trait loci (QTL) mapping. In: Shi XM (ed) eQTL analysis: methods and protocols. Springer US, pp 211–229. https://doi.org/10.1007/978-1-0716-0026-9_15

Raclariu-Manolică AC, Anmarkrud JA, Kierczak M, Rafati N, Thorbek BLG, Schrøder-Nielsen A, de Boer HJ (2021) DNA metabarcoding for quality control of basil, oregano, and paprika. Front Plant Sci 12:665618. https://doi.org/10.3389/fpls.2021.665618

Rajeswara Rao BR (2016) Genetic diversity, genetic erosion, conservation of genetic resources, and cultivation of medicinal plants. In: Ahuja MR, Jain SM (eds) Genetic diversity and erosion in plants: case histories. Springer International Publishing, pp 357–407. https://doi.org/10.1007/978-3-319-25954-3_11

Rajeswara R, Syamasundar K, Rajput D, Nagaraju G, Adinarayana G (2012) Biodiversity, conservation, and cultivation of medicinal plants. J Pharmacogn 3(2):59–62. https://doi.org/10.13140/2.1.2927.4247

Rani A, Kumar S (2017) Tissue culture as a plant production technique for medicinal plants: a review. Int Conf Innov Res Sci Technol Manag 6:784–795

Rao MR, Palada MC, Becker BN (2004) Medicinal and aromatic plants in agroforestry systems. Agrofor Syst 61–62(1–3):107–122. https://doi.org/10.1023/B:AGFO.0000028993.83007.4b

Raskoti BB, Ale R (2021) DNA barcoding of medicinal orchids in Asia. Sci Rep 11(1):23651. https://doi.org/10.1038/s41598-021-03025-0

Reddy VD, Mehandi S, Janeja HS, Saxena K, Prakash S (2022) Concept on plant DNA barcodes and their application in identification of plants. Biol Forum 14(2a):360

Ruppert KM, Kline RJ, Rahman MS (2019) Past, present, and future perspectives of environmental DNA (eDNA) metabarcoding: a systematic review in methods, monitoring, and applications of global eDNA. Glob Ecol Conserv 17:e00547. Elsevier B.V. https://doi.org/10.1016/j.gecco.2019.e00547

Salim Z, Munadi E (2017) Info Komoditi Tanaman Obat, vol 1. Badan Pengkajian dan Pengembangan Perdagangan

Salma U, Kundu S, Mandal N (2017) Artificial polyploidy in medicinal plants: advancement in the last two decades and impending prospects. J Crop Sci Biotechnol 20(1):9–19. https://doi.org/10.1007/s12892-016-0080-1

Schori M, Showalter A (2011) DNA barcoding as a means for identifying medicinal plants of Pakistan. Pak J Bot 43:1–4. http://www.ncbi.nlm.nih.gov/genbank/

Seethapathy GS, Raclariu-Manolica AC, Anmarkrud JA, Wangensteen H, de Boer HJ (2019) DNA metabarcoding authentication of ayurvedic herbal products on the European market raises concerns of quality and fidelity. Front Plant Sci 10:68. https://doi.org/10.3389/fpls.2019.00068

Shang J, Kong S, Li N, Wang J, Zhou D, Li N, Ma S (2020) Genetic mapping and localization of major QTL for bitterness in melon (*Cucumis melo* L.). Sci Hortic 266:109286. https://doi.org/10.1016/j.scienta.2020.109286

Sharma M, Thakur R, Sharma M, Kumar Sharma A, Kumar Sharma A (2020) Changing scenario of medicinal plants diversity in relation to climate change: a review. Plant Arch 20(2):4389–4400

Skelton J, Cauvin A, Hunter ME (2023) Environmental DNA metabarcoding read numbers and their variability predict species abundance, but weakly in non-dominant species. Environ DNA 5(5):1092–1104. https://doi.org/10.1002/edn3.3

Sudhakaran G (2024) Impact of climate change on the yield of medicinal plants in recent years. Nat Prod Res 39(12):3614–3615

Sun Z, Zeng L, Wei J, He M, Li R, Gu Y, Fu L (2024) Cryopreservation of medicinal plant seeds: strategies for genetic diversity conservation and sustainability. Plants 13(18):2577. https://doi.org/10.3390/plants13182577

Tnah LH, Lee SL, Tan AL, Lee CT, Ng KKS, Ng CH, Nurul Farhanah Z (2019) DNA barcode database of common herbal plants in the tropics: a resource for herbal product authentication. Food Control 95:318–326. https://doi.org/10.1016/j.foodcont.2018.08.022

Travadi T, Shah AP, Pandit R, Sharma S, Joshi C, Joshi M (2023) A combined approach of DNA metabarcoding collectively enhances the detection efficiency of medicinal plants in single and polyherbal formulations. Front Plant Sci 14:1169984. https://doi.org/10.3389/fpls.2023.1169984

Udage AC (2021) Introduction to plant mutation breeding: different approaches and mutagenic agents. J Agric Sci 16(3):466–483. https://doi.org/10.4038/jas.v16i03.9472

Urumarudappa SKJ, Tungphatthong C, Jaipaew J, Pornputtapong N, Pakdeesattayapong D, Vimolmangkang S, Sukrong S (2022) Development of a DNA barcode library of plants in the Thai herbal pharmacopoeia and monographs for authentication of herbal products. Sci Rep 12(1):9624. https://doi.org/10.1038/s41598-022-13287-x

Vassou SL, Nithaniyal S, Raju B, Parani M (2016) Creation of reference DNA barcode library and authentication of medicinal plant raw drugs used in Ayurvedic medicine. BMC Complement Altern Med 16(Suppl 1):186

Vijayakumar AS, Jeyaraj M, Kolanchinathan P, Selvakumar M, Praveenkumar D (2019) Moleculer identification of the medicinal plant *Justicia gendarussa* (Burm) F. using its gene. Res J Life Sci Bioinform Pharm Chem Sci 5(1):663–674

Woldearegay M (2020) Climate change impacts on the distribution and phenology of plants: a review. Trop Plant Res 7(1):196–204

Wolf AA, Zavaleta ES, Selmants PC (2017) Flowering phenology shifts in response to biodiversity loss. Proc Natl Acad Sci USA 114(13):3463–3468

Wu L, Wu M, Cui N, Xiang L, Li Y, Li X, Chen S (2021) Plant super-barcode: a case study on genome-based identification for closely related species of *Fritillaria*. Chin Med 16(1):52

Wu Z, Singh SK, Lyu R, Pattanaik S, Wang Y, Li Y, Yuan L, Liu Y (2022) Metabolic engineering to enhance the accumulation of bioactive flavonoids licochalcone A and echinatin in *Glycyrrhiza inflata* (Licorice) hairy roots. Front Plant Sci 13:932594

Xiong C, Sun W, Wu L, Xu R, Zhang Y, Zhu W, HEJ, Panjwani, Liu Z, Zhao B (2022) Evaluation of four commonly used DNA barcoding loci for *Ardisia* species identification. Front Plant Sci 13:860778

Yan M, Xiong Y, Liu R, Deng M, Song J (2018) The application and limitation of universal chloroplast markers in discriminating east Asian evergreen oaks. Front Plant Sci 9:569

Yang J, Vázquez L, Chen X, Li H, Zhang H, Liu Z, Zhao G (2017) Development of chloroplast and nuclear DNA markers for Chinese oaks (*Quercus* subgenus *Quercus*) and assessment of their utility as DNA barcodes. Front Plant Sci 8:816

Yu X, Tan W, Gao H, Miao L, Tian X (2020) Development of a specific mini-barcode from plastome and its application for qualitative and quantitative identification of processed herbal products using DNA metabarcoding technique: a case study on Senna. Front Pharmacol 11:585687. https://doi.org/10.3389/fphar.2020.585687

Yu J, Wu X, Liu C, Newmaster S, Ragupathy S, Kress WJ (2021) Progress in the use of DNA barcodes in the identification and classification of medicinal plants. Ecotoxicol Environ Saf 208:111691

Yuan QJ, Zhang ZY, Hu J, Guo LP, Shao AJ, Huang LQ (2010) Impacts of recent cultivation on genetic diversity pattern of a medicinal plant, *Scutellaria baicalensis* (Lamiaceae). BMC Genet 11(1):29

Zhang D, Jiang B (2020) Species identification in complex groups of medicinal plants based on DNA barcoding: a case study on *Astragalus* spp. (Fabaceae) from southwest China. Conserv Genet Resour 12(3):469–478

Zhang ZL, Song MF, Guan YH, Li HT, Niu YF, Zhang LX, Ma XJ (2015) DNA barcoding in medicinal plants: testing the potential of a proposed barcoding marker for identification of *Uncaria* species from China. Biochem Syst Ecol 60:8–14

Zhang Y, Zheng Y, Xia P, Xun L, Liang Z (2019) Impact of continuous *Panax notoginseng* plantation on soil microbial and biochemical properties. Sci Rep 9(1):13205

Zhao H, Wolt JD (2017) Risk associated with off-target plant genome editing and methods for its limitation. Emerg Top Life Sci 1(2):231–240

Zhu S, Liu Q, Qiu S, Dai J, Gao X (2022) DNA barcoding: an efficient technology to authenticate plant species of traditional Chinese medicine and recent advances. Chin Med 17(1):112

Zou M, Shabala S, Zhao C, Zhou M (2024) Molecular mechanisms and regulation of recombination frequency and distribution in plants. Theor Appl Genet 137(4):86

Chapter 10
Impact of Cultivation and Domestication of Medicinal Plants on Genetic and Phytochemical Diversity

Gülnur Ekşi Bona and Mehmet Bona

Abstract Medicinal plants play a vital role in healthcare and the pharmaceutical industry, yet they face growing threats from habitat degradation, overexploitation, and unsustainable practices. While cultivation and domestication can aid conservation and ensure a steady supply, these processes may also diminish genetic diversity and alter phytochemical compositions, potentially reducing adaptive traits found in wild populations. Conservation strategies, both in-situ and ex-situ—such as field gene banks, cryopreservation, and sustainable harvesting—combined with advances in molecular genetics and biotechnology, provide valuable tools for maintaining genetic resilience and metabolomic diversity. Integrating traditional conservation approaches with modern breeding and genomic technologies is essential to achieve a balance between sustainable use and biodiversity protection. This review summarizes the current understanding of how cultivation and domestication affect the genetic diversity and phytochemistry of medicinal plants, while also highlighting future directions for conservation, research, and policy development. Moreover, this research emphasizes the rich diversity of Turkey's medicinal plants, details key approaches and conditions necessary for their sustainable utilization and conservation, and demonstrates their therapeutic value through selected examples.

Keywords Cultivation · Domestication · Genetic diversity · Phytochemical diversity · Medicinal plants

G. E. Bona (✉)
Department of Pharmaceutical Botany, Faculty of Pharmacy, İstanbul University-Cerrahpaşa, Istanbul, Turkey
e-mail: gulnur.bona@iuc.edu.tr

M. Bona
Department of Botany, Faculty of Science, İstanbul University, Istanbul, Turkey

S. Jha et al. (eds.), *Genetic Diversity in Medicinal Plants*, Sustainable Development and Biodiversity 44, https://doi.org/10.1007/978-981-95-7281-6_10

10.1 Introduction

Genetic diversity refers to the range of genetic differences among individuals within a species or population, and it plays a central role in driving natural selection and evolutionary processes. It is vital for sustaining a healthy population since it preserves a variety of genes that may provide resistance against pests, diseases, and other stress factors. It also supports the ability of individuals to cope with both biological and environmental challenges (Salgotra and Chauhan 2023). Genetic variation equips plant species with the ability to evolve, remain resilient, and adapt to changing environmental conditions (Ashry 2013). Genetic diversity shifts over time due to factors such as genetic drift, gene flow, mutation, and selection. Both natural and artificial selection strongly influence the choice of superior genotypes, which in turn alters the genetic and genotypic composition of populations (Ray et al. 2015).

Over the last hundred years, a steady global decline in genetic diversity within wild populations has been observed (Hoban et al. 2021; Puffenberger 2021). Many species are diminishing mainly due to habitat destruction and insufficient conservation measures (Ceballosa et al. 2017; Khoury et al. 2019; Leigh et al. 2019). Major causes of genetic loss include climate change, habitat fragmentation, overexploitation, overgrazing, and urbanization (Hoban et al. 2021; Salgotra and Chauhan 2023).

Since the advent of agriculture, people have practiced the selection, cultivation, and preservation of seeds from locally adapted plants, commonly referred to as landraces (Panis et al. 2020). This practice persisted until the time of Gregor Mendel in the twentieth century, whose work laid the foundation for breeding programs aimed at producing high-yielding and stress-resistant crop varieties. By the mid-twentieth century, this progress laid the groundwork for the "Green Revolution," which greatly boosted agricultural output. However, it also resulted in the displacement of landraces and the spread of monoculture farming. Consequently, more than 75% of plant genetic diversity and around 90% of traditional crop varieties were lost (Salgotra and Chauhan 2023).

For centuries, medicinal plants have been vital to human health and healing. Across different cultures, they have been used to treat a wide range of illnesses. Their therapeutic value comes from their diverse chemical components, which include bioactive compounds with pharmacological properties. These plants not only serve as natural remedies but have also been the foundation for the development of many modern medicines. The rising global demand for medicinal plants—driven by population growth, urbanization, and a growing interest in alternative medicine—has placed heavy pressure on their populations. This situation is further aggravated by climate change, habitat loss, overharvesting, and unsustainable collection practices (Shukla 2023).

The Neolithic Revolution, post-domestication expansion of agriculture to new regions, secondary domestications of new crops, the Silk Road, and the Green Revolution have had a significant impact on humanity and plant biodiversity. These can be considered major milestones in the history of agriculture. It is estimated that

Earth hosts approximately 500,000 species of higher plants, with over 350,000 already identified and described (Willis 2017b).

Preagricultural human societies are believed to have used around 7000 plant species for food, yet only a small portion of the vast plant kingdom has ever been domesticated. Roughly 2500 plant species have been partially domesticated, while about 250 have been fully domesticated (Meyer et al. 2012; Gaut et al. 2018; Smýkal et al. 2018; Von Wettberg et al. 2020; Diamond 2005; Meyer et al. 2012; Meyer and Purugganan 2013; Salman-Minkov et al. 2016; Gaut et al. 2018; Fernie and Yan 2019; Alseekh et al. 2021; Gutierrez and Grillo 2022). Crop wild relatives serve as the most important reservoir of genetic diversity for crop improvement and have been widely used to introduce traits such as disease and pest resistance, as well as tolerance to environmental stresses (Vavílov et al. 1992; Hajjar and Hodgkin 2007; Warschefsky et al. 2014; Castañeda-Álvarez et al. 2016; Smýkal et al. 2018; Coyne et al. 2020; Von Wettberg et al. 2020).

Domestication is a coevolutionary process based on mutualism, where the domesticator creates and manages an environment that ensures the survival and reproduction of the domesticated species, in return for resources or services (Purugganan 2022). Human-directed domestication started about 12,000 years ago in the Middle East and later spread to different parts of the world (Purugganan and Fuller 2009; Xu 2010; Meyer and Purugganan 2013; Fernie and Yan 2019; Wang et al. 2020). Domesticated plants have undergone extended periods of selective breeding and isolation, resulting in new cultivars and varieties tailored to human requirements (Doebley et al. 2006). The domestication of wild plants may have led to a decline in their genetic diversity (Allaby et al. 2019). It is crucial to study the genetic diversity of wild plant populations, as they offer vast potential for developing improved breeding varieties that can enhance the genetic base of cultivated plants. Anatolia is recognized as a cradle of agriculture, with plant breeding and domestication practices dating back millennia. During the Neolithic period (c. 8000–6000 BCE), inhabitants began to collect and domesticate key crops such as wheat (*Triticum* L. spp.), barley (*Hordeum* L. spp.), chickpea (*Cicer arietinum L.*), and lentil (*Lens* Mill. spp.), applying early forms of selective cultivation. These initial practices not only enhanced agricultural productivity but also laid the foundation for the preservation and development of genetic diversity. Consequently, plant breeding and domestication activities in Anatolia extend back approximately 10,000 years, underscoring its historical significance in both agricultural and biological contexts (Ekşi et al. 2020). Plant breeding is essential for improving crop yield, quality, and resistance to biotic and abiotic stresses, thereby supporting food security and sustainable agriculture. However, intensive breeding can reduce genetic diversity, increasing vulnerability to pests, diseases, and environmental changes, while the focus on high-yield varieties may threaten local landraces and traditional cultivars. Recent advances in molecular technologies, especially genome sequencing, have revealed how human selection has influenced numerous loci during and after crop domestication. Additionally, population-level molecular studies, combined with archaeological evidence, help trace the demographic history and origins of cultivated crops (Von Wettberg et al. 2018; Smýkal et al. 2018).

The cultivation of medicinal plants has been practiced for over 2000 years (Xiao 1991; Olsen and Gross 2008). Around 12% of all documented plant species worldwide are medicinal, forming a vital foundation for traditional medicine and herbal products (Schippmann et al. 2002). The modern global food system is rooted in thousands of years of selective breeding, where early farmers chose plant species and genotypes with desirable traits for cultivation and nutrition. This long process led to crop domestication and the creation of landraces adapted to local environments. The rediscovery of Mendel's inheritance principles in 1900, combined with Darwin's theory of natural selection, laid the foundation for modern plant breeding and the development of today's crop varieties (Hallauer 2011). Since the early 1900s, the creation of high-yielding and uniform commercial crop varieties—initially in temperate regions and later, during the Green Revolution, in tropical and subtropical areas—has significantly boosted global annual yields of major crops such as maize, millet, rice, sorghum, and wheat (Evenson and Gollin 2003).

Crop domestication is shaped by human selection, farming practices, and agricultural environments. This process often reduces genetic diversity in regions tied to desirable traits—such as nonshattering seeds or improved taste. In addition, cultivation practices shrink effective population sizes, making crops more vulnerable to genetic drift and the random loss of alleles (Cowling et al. 2009; Gross and Olsen 2010; Olsen and Wendel 2013).

10.2 Cultivation of Medicinal Plants

Medicinal plants play a vital role in sustainable development by creating employment, enhancing public health, ensuring food security, protecting the environment, safeguarding genetic diversity, and supporting economic progress. Owing to these diverse contributions, their value has risen both nationally and internationally. As their role in pharmaceutical supply chains expands, medicinal plants have become an important indicator of global development. Because their cultivation greatly benefits rural communities and offers strong export potential, it is essential that production aligns with scientific guidelines and legal regulations (Dalir et al. 2025). While medicinal plants are widely acknowledged as valuable economic and ecological assets, most research has centered on their medicinal properties or commercial export potential, leaving their role in supporting rural communities relatively overlooked. Actively engaging local populations in cultivation and incorporating sustainable farming practices are essential steps toward maximizing their benefits (Dalir et al. 2025).

To overcome the difficulties associated with cultivating medicinal and aromatic plants, the Turkish Ministry of Agriculture and Forestry initiated the National Research Project of Medicinal and Aromatic Plants in 1990. The project's primary objective is to develop new plant varieties capable of supplying high-quality raw materials for diverse industries. Research institutions throughout Turkey identify species suited to local ecological conditions and carry out breeding, variety enhancement, and quality assessments to determine their active compounds. So far, 41 cultivars belonging to 15 species have been registered, and cultivation areas are steadily increasing (Karık et al. 2016). In 2018, nearly 300,000 tons of medicinal plants were produced across 100,000 ha, of which 50,000 tons were exported. Worldwide, Turkey is the third-largest producer, following China and India. Around 50 medicinal plant species are cultivated on about 90,000 ha, with anise, cumin, poppy, and thyme accounting for 90% of the total area. Other species grown include cardamom, carob, clove, coriander, fenugreek, laurel, licorice, linden, oregano, poppy seed, rosemary, saffron, spearmint, sumac, tea, and turmeric. Since the development of medicinal and aromatic plant varieties in Turkey began later than for other crops, the number of species examined and varieties registered remains limited. However, continued breeding and cultivation efforts are expected to increase the number of registered varieties and support standardized high-quality production (Karık et al. 2016).

The World Health Organization (1998) reports that medicinal plants serve as the primary source of healthcare for 80% of people in developing nations, and that at least 25% of medicines listed in modern pharmacopoeias are still derived from plants (Kingston 2011). The majority of medicinal plants in trade are harvested from natural habitats, with only a small portion being cultivated. As the global demand for medicinal and aromatic plants continues to rise, many species are becoming threatened or endangered (Willis 2017a; Akalın et al. 2020). Turkey is recognized as a biodiversity hotspot in the Northern Hemisphere. Its flora includes 1220 genera and 9753 species from 154 families (11,466 taxa when subspecies are included). Approximately 1500 species are identified as medicinal plants, with about 350 collected from the wild for domestic use and around 100 for export (Güner et al. 2012; Tuzlacı 2016). In Turkey, commonly cultivated medicinal and aromatic plants include anise, black cumin, black tea, caper, coriander, cumin, dill, fennel, flax, hemp, lavender, lemon balm, mint, poppy, red pepper, rose, sage, stinging nettle, and thyme (Akalın et al. 2020; Akalın 2023; TUIK 2024) (Table 10.1).

With the exception of products used traditionally, medicinal plants must be licensed before they can be sold as herbal medicinal products, and this licensing depends on having an approved therapeutic indication. Products without an indication are authorized by the Ministry of Agriculture and Forestry, whereas those with an indication require approval from the Ministry of Health and must comply with the applicable directives and regulations. Although various cultivation projects have been carried out by research institutes and medicinal plant botanical gardens, broader and more comprehensive efforts are still required. The Turkey Seed Gene Bank, operated by the Biodiversity and Genetic Resources Department of the Field Crops Central Research Institute, was opened in 2010 and had preserved 122,000 accessions by 2015 (Akalın et al. 2020). In addition, ex-situ conservation activities are carried out in several botanical gardens, and 144 locations have been designated as Important Plant Areas, along with other protected regions and biodiversity hotspots (Özhatay 2006; Özhatay et al. 2010; Akalın et al. 2013; Koçyiğit and Demirci 2018). The Ministry of Agriculture and Forestry has assessed the conservation status of medicinal plants in Turkey in line with WHO, IUCN, and WWF criteria (Akalın et al. 2020).

Table 10.1 Medicinal and aromatic plants cultivated and exported in Turkey, along with their phytochemical compositions and therapeutic applications

Species	Family	Phytochemicals	Medicinal uses	Production 2024 (TUIK 2024)	References
Camellia sinensis L. (black tea)	Theaceae	Flavonoids and phenolic acids	Anticancer, antimicrobial, and anti-inflammatory	145 tons	Boudou et al. (2025)
Capsicum annuum L. (red pepper)	Solanaceae	Capsaicinoids, phenolics and fatty acids, flavonoids, vitamins, and essential oils	Anti-inflammatory, antifungal, antibacterial, antioxidant, and anticancer	346 tons	Jang et al. (2024); Tunç et al. (2024)
Orchis Tourn. Ex (L.) ssp. (salep plant)	Orchidaceae	Polysaccharides, fibers, proteins, steroids, and flavonoids	Antidiabetic, antitumor, and anti-inflammatory	147 tons	Pokhrel et al. (2026)
Papaver somniferum L. (poppy)	Papaveraceae	Narcotic alkaloids, flavonoids, and phenolics	Antioxidant, antimicrobial, anti-inflammatory, anti-depressant, analgesic, and sedative	8.5 tons	Aalinezhad et al. (2025); Nazari et al. (2025)
Cuminum cyminum L. (cumin)	Apiaceae	Terpenoids, aldehyde, phenolic acids, and essential oil	Antidiabetic, anti-inflammatory, immunomodulatory, antimicrobial, antioxidant, and insecticidal	17 tons	Msegued Ayam et al. (2025)
Capparis spinosa L. (caper bush)	Capparaceae	Phenolic acids, flavonoids, phytosterols, and alkaloids	Antibacterial, hepatoprotective, and anti-diabetic	23 tons	Babaei Rad et al. (2025)

Urtica dioica L. (stinging nettle)	Urticaceae	Polyphenols, alkaloids, essential oils, flavonoids, and sterols	Antioxidant, antimicrobial, and antiulcer, anti-proliferative, anti-inflammatory, analgesic, immunomodulatory, hypotensive, anticancer immunomodulatory, and cardioprotective	0.4 tons	Dakhli et al. (2025)
Cannabis sativa L. (hemp)	Cannabaceae	Essential oils, fatty acids, waxes, flavonoids, polyphenols, and terpenophenol	Anti-inflammatory, antimicrobial, and antioxidant	556 tons	El Oihabi et al. (2025)
Mentha ssp. L. (mint)	Lamiaceae	Essential oils and phenolics	Antimicrobial, antiviral, antioxidant, immunomodulatory, anticancer, insecticidal, anti-allergic and antihypertensive	234 tons	Kar et al. (2025)
Thymus L. ssp. (thyme)	Lamiaceae	Polyphenols, flavonoids, and essential oils	Antioxidant antimicrobial, anti-inflammatory, sedative, immunomodulatory, antiseptic, spasmolytic, and antidiabetic	24 tons	Djermane et al. (2025)
Rosa x*damascena* Herrm. (rose)	Rosaceae	Essential oils, flavonoids, polyphenols, and organic acids	Antimicrobial, antioxidant, anti-inflammatory, analgesic, anxiolytic, and neuroprotective effects	19 tons	Lalovski et al. (2025)

(continued)

Table 10.1 (continued)

Species	Family	Phytochemicals	Medicinal uses	Production 2024 (TUIK 2024)	References
Pimpinella aromatica M. Bieb. (anise)	Apiaceae	Flavonoids, essential oils, polyphenols, tannins, steroids, terpenoids, and saponins	Antimicrobial, antioxidant, anti-inflammatory, anticonvulsant, antispasmodic, estrogenic, cytotoxic, insecticidal, repellent, hepatoprotective, anesthetic, nephroprotective, anxiolytic, antihypertensive, hypolipidemic, antiparasitic, antidiabetic, local hypnotic, antidepressant, anti-cholinesterase, immunomodulatory, anticancer, laxative effects, and increased milk production, prevented osteoporosis, and minimized premenstrual syndrome	5 tons	İlktaç et al. (2024); Al-Snafi (2024)
Nigella sativa L. (black cumin)	Ranunculaceae	Essential oils, alkaloids, polyphenols, fatty acids, and phytosterols	Antioxidant, antimicrobial, immunomodulatory, anti-inflammatory, neuroprotective, cardioprotective, anticancer, bronchodilator, and gastroprotective	4.5 tons	Abbas et al. (2024)

Calluna vulgaris (L.) Hull (heather)	Ericaceae	Flavonoids, phenols, and steroids	Antioxidant, antimicrobial, cytotoxic, genoprotective, cardioprotective, antiproliferative, expectorant, antiseptic, choleretic, anti-inflammatory, and antimicrobial	2 tons	Altuntaş et al. (2025)
Foeniculum vulgare Mill. (fennel)	Apiaceae	Polyphenols, flavonoids, alkaloids and essential oils	Anti-inflammatory, antidiabetic, antioxidant, anti-inflammatory, antispasmodic, antiseptic, antithrombotic, antitumor, chemopreventive, cytoprotective, hepatoprotective, hypoglycemic, estrogenic, carminative, diuretic, and analgesic	1 tons	Ncube and Gupta (2025); Moumen et al. (2025)
Lupinus albus L. (white lupin)	Fabaceae	Flavonoids, alkaloids, and phenolic compounds, and fatty acids	Hypocholesterolemic, hypoglycemic, antibacterial, immunomodulatory, anti-inflammatory, antihypertensive, and angiotensin-converting enzyme inhibitory	104 tons	Tian et al. (2024); Abd-Elsamei et al. (2025); Hunie Tesfa et al. (2025)

(continued)

Table 10.1 (continued)

Species	Family	Phytochemicals	Medicinal uses	Production 2024 (TUIK 2024)	References
Lavandula angustifolia Mill. (lavender)	Lamiaceae	Essential oils, phenolics, phytosterols, phenolic acids, flavonoids, and polyphenols.	Antioxidant, anticancer, anti-inflammatory, hepatoprotective, DNA protective, antiaging, antiallergic, and antimicrobial, insecticidal, antispasmodic, analgesic, hypotensive, antiseptic, and tonic	9 tons	Manzoor et al. (2025); Sevindik et al. (2025)
Melissa officinalis L. (lemon balm)	Lamiaceae	Flavonoids, phenolic acids, polyphenols, and essential oil	Antioxidant, anti-alzheimer, anti-inflammatory, cardioprotective, hypolipidemic, vasorelaxant, antidepressant, anxiolytic, hypnotic, antispasmodic, antitumor, antimicrobial, antithyroid, and analgesic	234 tons	Zafar and Zafar (2025)
Salvia officinalis L. (sage)	Lamiaceae	Essential oils, phytosterols, flavonoids, and phenolic acids	Antimicrobial, antioxidant, and anti-inflammatory	2 tons	Anagnostou et al. (2025)
Coriandrum sativum L. (coriander)	Apiaceae	Flavonoids and essential oils	Antioxidant, anti-inflammatory, antimicrobial, hypoglycemic, hypolipidemic, and neuroprotective	1 tons	Agarwal et al. (2025)
Humulus lupulus L. (hops)	Cannabaceae	Phenolic compounds, phenolic acids, and flavonoids	Antimicrobial, antioxidant, anti-inflammatory, antiaging, moisturizing, and UV-protective	759 tons	Silva et al. (2025)

Allium tuncelianum (Kollmann) Özhatay, B. Mathew & Siraneci is a medicinal plant endemic to Turkey. This species is also used by local people as an alternative to *Allium sativum* (common garlic) (Figs. 10.1 and 10.2). In recent years, *A. tuncelianum* has become increasingly recognized, and its cultivation has begun. As a result, the pressure on natural populations caused by uncontrolled collection from the wild has been reduced. In a country like Turkey, where nearly one-third of the plant species are endemic, conserving these plants is of great importance to ensure the sustainability of their medicinal uses. Continuous efforts to protect this and other endemic species, as well as to ensure their controlled cultivation, will not only guarantee the sustainability of these rare species but also contribute to the national economy. More extensive efforts are needed to enhance agronomic practices for cultivated medicinal plants and to domesticate wild species with medicinal value. If not, many endemic species native to Turkey may become depleted because of uncontrolled harvesting and sale in local markets. Both the loss and degradation of natural habitats caused by human activity, as well as unregulated collection, present

Fig. 10.1 *Allium tuncelianum* (Kollmann) Özhatay, B. Mathew & Siraneci. (**a–c**) Plants from the habitat and (**d–f**) plants from a cultivation area in Tunceli province in Turkey. The plant photographs in the figure (Prof. Mehmet Koyuncu appears in photograph **f**) were taken by Prof. Mehmet Koyuncu

Fig. 10.2 *Allium tuncelianum*. (**a1–a2**) General appearance, (**b**) early stage of plant, (**c**) spathe, (**d**) flower, (**e**) transverse section of the flower, and (**f**) pistil. (Drawn by Gülnur Ekşi Bona)

serious threats to biodiversity. Promoting the cultivation of medicinal species and shifting toward standardized production are key strategies to safeguard and preserve these endemic plants.

This review evaluates how domestication and cultivation influence the diversity of medicinal plants, focusing on biotechnological applications, conservation approaches, genetic impacts, and phytochemical variations. By combining ethnobotanical, molecular, and phytochemical insights, it emphasizes strategies that balance pharmaceutical standardization with the preservation of ecological and evolutionary resilience.

10.3 Impact of Plant Domestication and Cultivation

Plant domestication is an ongoing evolutionary process in which wild species are altered through genetic changes to become useful cultivated forms (Van Tassel et al. 2020). Many changes occur during the domestication of plants, including the development of larger fruits or grains, stronger plants, reduced seed dispersal and dormancy, decreased bitterness, increased central stem growth, altered photoperiod sensitivity, and synchronized flowering (Doebley et al. 2006). Through next-generation sequencing, plant breeders now have access to comprehensive genomic and transcriptomic data for these plants. Genomic and transcriptomic data, when integrated with proteomic and metabolomic information, can uncover previously unknown natural products in medicinal plants (Niazian 2019; Hasan et al. 2024).

10.3.1 Positive Outcomes of Domestication and Cultivation

While stricter regulations and sustainable wild-harvesting practices can help protect certain species, expanding the domestic cultivation of medicinal plants offers a more sustainable long-term solution. Cultivation also enables the application of biotechnology to address challenges in herbal medicine production, such as species extract inconsistency, genetic and phenotypic variation, misidentification, as well as the presence of toxic compounds and contaminants. Cultivation provides a means to maximize yield while ensuring consistency and high-quality standards in the final product (Canter et al. 2005).

Phytochemicals are naturally occurring plant compounds with biological activity that offer health benefits beyond the effects of basic nutrients like macronutrients and micronutrients (Hasler and Blumberg 1999). Phytochemicals are plant-derived compounds that help protect plant cells against drought, environmental stresses, pathogen attacks, pollution, and UV radiation (Gibson et al. 1998; Mathai 2000). It is now well established that phytochemicals contribute to the protection of human health. Phytochemicals can be categorized based on their chemical composition, physical properties, and protective roles (Meagher and Thomson 1999).

Phytochemicals are stored in various parts of plants, including the flowers, fruits, roots, seeds, and stems (Costa et al. 1999). The levels of phytochemicals vary among plants, depending on the environmental conditions, species, and variety (King and Young 1999; Saxena et al. 2013).

Pathway engineering in medicinal plants aims to boost the production of active phytochemicals and is a well-recognized target for genetic modification, though it poses significant challenges. A major limitation is that the metabolic pathways responsible for synthesizing active compounds are still not well understood, and only a small number of genes involved in crucial enzymatic or regulatory steps have been identified. Nonetheless, some cases of pathway engineering have successfully produced valuable improvements for medicinal plant breeding (Ferreira and Duke 1997; Charlwood and Pletsch 2002; Canter et al. 2005).

The direct modification of DNA to influence gene expression in medicinal plants holds significant potential for growth. As long as a trait is linked to one or a few genes, it can, in principle, be modified. While the main focus of trait modification in medicinal plants is enhancing the levels of active compounds, it is also essential to improve fundamental agronomic traits such as development, growth, resistance to both biotic and abiotic stresses, stability, and uniformity for successful crop development (Bajaj 1998; Bernath 2002; Dubey and Guerra 2002). Biotechnological research on medicinal plants is already well advanced, with significant efforts focused on altering biosynthetic pathways to generate drug precursors, food ingredients, and pesticides. For instance, in Mentha species, these pathways have been engineered to adjust essential-oil production in trichomes and to improve resistance against fungal infections and abiotic stresses (Veronese et al. 2001; Canter et al. 2005). Similar to transgenic food crops, where traits such as disease, herbicide, and pest resistance were prioritized, these same traits have also been among the earliest targets in medicinal plant biotechnology. Herbicide-resistant transgenic *Atropa* plants have been reported for bialaphos and glufosinate (Saito et al. 1992), while *Panax ginseng* resistant to Basta was developed through the introduction of the phosphinothricin acetyltransferase (Choi et al. 2003; Canter et al. 2005).

Growing medicinal plants offers potential to enhance the bioactivity, consistency, overall biomass yield, purity, and quality of the raw materials (Xego et al. 2016). The sustainable production of many innovative plant-based products is hindered by irregular supplies of quality raw materials; however, cultivation can enhance biomass yield and ensure consistency by adjusting growing conditions, without compromising the plants' bioactivity. For instance, cultivating *Pelargonium sidoides* under well-watered conditions greatly enhanced total biomass, while maintaining stable levels of its active compounds (Mofokeng et al. 2020). Extracts of *P. sidoides* from greenhouse-acclimatized plants exhibited antimicrobial and antioxidant properties comparable to those of wild-collected plants, supporting the feasibility of large-scale cultivation (Moyo et al. 2013; Mofokeng et al. 2022). The bulbs of cultivated *Bowiea volubilis* showed biological activities comparable to those of wild-harvested specimens sold in markets, with no major variations observed (Masondo et al. 2013).

Controlled cultivation systems enable the regulation of phenotypic variation in the levels of medicinally valuable compounds at harvest, with the goal of enhancing potency, lowering toxin content, and ensuring greater consistency and predictability of extracts. The compounds of interest are typically secondary metabolites, which plants often produce as adaptations to changing light and temperature conditions, environmental stress, herbivory, or infections. For instance, *Mentha piperita* grown in shaded conditions produces less essential oil and has a reduced menthol concentration compared to plants grown in full light. *Papaver somniferum* cultivated in cooler conditions tends to accumulate higher levels of morphine, although its overall alkaloid content is lower than that of plants grown in warmer conditions. The accumulation of secondary metabolites is also influenced by factors such as changes in soil pH and nutrient levels, interactions with soil microorganisms, and water availability (McChesney 1999; Canter et al. 2005).

Cultivating herbs allows the use of both traditional and biotechnological breeding approaches at the genetic level to enhance yield, consistency, and adjust potency or toxicity. Target traits such as seed production and viability can often be improved by selecting vigorous, fertile genotypes, which also helps develop plant populations well adapted to the provided growing conditions. Artificial selection is often time-consuming, but the process can be accelerated through reliable methods that detect desired traits early in the reproductive cycle. For instance, laser speckle technology distinguishes viable from nonviable seeds by analyzing quantitative and qualitative differences in speckle activity when seeds are exposed to laser light (Braga et al. 2003). Plant breeders can also focus on maintaining or increasing the levels of beneficial bioactive compounds while reducing harmful toxins. For instance, *Ginkgo biloba* L. extracts are standardized to include 24–27% flavonoid glycosides and 6–7% terpene lactones, with ginkgolic acid, a toxic component, kept below 5 ppm. Although such targets are typically reached during the extraction process, intentional selection of genotypes can help produce extracts that are naturally closer to the desired chemical profiles (Canter et al. 2005).

The benefits of domestication in Turkey can be illustrated by several examples. For instance, cultivating *Origanum majorana* L. increased essential-oil yield and generated greater profit compared to wild-collected material, owing to controlled production conditions. In both harvested and cultivated dried aerial parts, carvacrol was found to be the predominant essential-oil component (83.46% and 80.33%, respectively) (Bağcı et al. 2017). One study reported that both wild and cultivated *Salvia tomentosa* Mill. contain substantial levels of phenolic compounds, which are known for their strong antioxidant properties. Cultivation increased the total phenolic content, indicating that *S. tomentosa* can be effectively grown to support sustainable and standardized medicinal plant production (Dinçer et al. 2013). Gedikoğlu et al. (2024) conducted a comparative assessment of essential oils from lavender grown in a drought-impacted region of Turkey. In general, sages grow best in well-drained, light, limestone-based soils enriched with humus and grit, along with thorough weekly irrigation. For this reason, they are easily cultivated in dry garden conditions and are also recommended for use in rock gardens. In Turkey, sage species already under cultivation include *Salvia coccinea* Buc'hoz ex Etl., *S. farinacea* Benth., *S. microphylla* Sessé & Moc., *S. officinalis* L., *S. splendens* Sellow ex Nees, *S. superb* E. Vilm., and *S. transsylvanica* (Schur ex Griseb. & Schenk) Schur. (Karabacak et al. 2009). Research showed that applying cultivation practices to seeds of wild *Foeniculum vulgare* L. markedly increased both herb and seed yields, though it slightly reduced essential-oil content. However, the essential oils from cultivated plants demonstrated biological activity equal to or greater than that of wild plants against the tested phytopathogens. Thus, cultivation does not negatively influence essential-oil production or its biological effectiveness (Ok and Şanlı 2025).

Cultivating plants provides a sustainable approach to overcoming problems associated with herbal extracts, such as contamination, genetic variability, inconsistencies in extracts, misidentification, and phenotypic variation. By growing plants under controlled conditions, these challenges can be minimized while maintaining regulated levels of bioactive compounds and toxins. Traditional breeding improves agronomic and medicinal characteristics, while the use of molecular marker–assisted selection is becoming increasingly important. Progress in tissue culture and genetic transformation has also made it possible to alter major metabolic pathways. However, large-scale cultivation still encounters difficulties, including unpredictable market demand and a tendency for consumers to favor extracts sourced from the wild (Canter et al. 2005).

10.3.2 Loss of Genetic and Phytochemical Diversity in Medicinal Plants

The threats facing medicinal rare and endemic plants in Turkey have not been fully assessed. Key pressures include infrastructure expansion, intensive farming, industrial or mining operations, logging, overgrazing, overharvesting, pollution, and urbanization. For example, *Taxus baccata* L. forests have nearly vanished, and *Cedrus libani* A. Rich. survives only in limited areas of the Taurus Mountains (Akalın 2023). Despite improvements in cultivation and technology, numerous species in Turkey, such as *Allium* L. ssp., *Capparis Tourn.* ex L. ssp., *Ferula Tourn.* ex L. ssp., *Galanthus* L. ssp., *Glycyrrhiza glabra* L., *Gypsophila* L. ssp., *Orchis* Tourn. ex L. ssp., *Origanum* L. ssp., *Rosmarinus officinalis* L., *Salvia* L. ssp., *Satureja* L. ssp., and *Sideritis* L. ssp., are still predominantly collected from the wild (Özhatay et al. 1997; Akalın 2023).

For thousands of years, humans have strongly shaped domesticated crops through selection, producing phenotypes that differ greatly from their wild ancestors. While this has led to higher yields, a long-standing hypothesis suggests that domestication has also unintentionally reduced plants' natural defenses against herbivores (Whitehead et al. 2017). Cultivation is challenged by issues such as plant diseases and pests, contamination from agrochemical residues and heavy metals, as well as concerns over biodiversity conservation (Li et al. 2015). Many studies have shown that, over the course of long-term domestication, most crops have undergone genetic bottlenecks and a reduction in diversity compared with their wild relatives (Hyten et al. 2006; Yamasaki et al. 2007; Kovach and Mccouch 2008). Studying the genetic diversity of long-cultivated medicinal plants and clarifying their relationship to wild relatives is essential (Wang et al. 2020). The outcomes of plant domestication were influenced by agricultural conditions, human preferences, planting practices, and population genetic changes that occurred after the reduction in effective population size (Smýkal et al. 2018). Extensive evidence shows that domestication often reduces genetic diversity in many cultivated crops and medicinal plants (Hollingsworth et al. 2005; Lam et al. 2010; Gao et al. 2017; Zhang et al. 2017). Humans have intentionally and unintentionally favored traits like larger and more numerous seeds, reduced dispersal ability, and changes in secondary metabolites affecting phenology, taste, and seed dormancy (Meyer et al. 2012; Purugganan 2019; Gutierrez and Grillo 2022). A study examining domestication traits in 203 global food crops revealed that changes in secondary metabolites are one of the most frequent consequences of domestication (Meyer et al. 2012). Cultivated plants often show reduced performance compared with their wild ancestors in several ecological traits, such as drought tolerance, nutrient uptake, and pest and pathogen resistance (Koziol et al. 2012; Olsen and Wendel 2013; Whitehead et al. 2017; Gaillard et al. 2018; Isaac et al. 2021).

10.4 Applied Methods for Cultivation and Domestication of Medicinal Plants

Research on the genetic basis of domestication-related traits has mainly taken two approaches. The first uses classical quantitative genetics, analyzing the offspring of a cross between a wild and a cultivated parent to link genetic variation with the phenotypic trait being studied. This method has enabled the discovery of genes responsible for major morphological and physiological differences between domesticated plants and their wild ancestors. Recently, transcriptomic studies on species such as barley, cotton, maize, rice, and wheat have revealed that domestication causes a significant loss of allelic diversity. Current cultivated crops are estimated to retain only about 6% of the allelic variation found in their wild relatives (Alseekh et al. 2021).

Plant tissue culture plays a vital role in controlled vegetative propagation, particularly through micropropagation techniques such as organogenesis and somatic embryogenesis.

Organogenesis in plant tissue culture is a two-phase process that relies on plant growth regulators to initiate both root and shoot formation. In contrast, somatic embryogenesis is used to generate bipolar structures from callus tissue (Shahzad et al. 2017; Hesami et al. 2018).

Genome editing, especially through CRISPR/Cas9, serves as a highly efficient and versatile method for studying gene function and creating new germplasm, valued for its simplicity and broad applicability across organisms (Wang et al. 2022). CRISPR/Cas9 enables metabolic engineering in medicinal plants by allowing the targeted introduction of multiple genes at the same chromosomal site (Shabir 2021).

In-situ conservation is vital for safeguarding medicinal plants, as their therapeutic properties largely arise from secondary metabolites that are often triggered by specific environmental stimuli, which are difficult to reproduce under controlled cultivation (Coley et al. 2003; Figueiredo and Grelle 2009; Hasan et al. 2024). In-situ conservation, by protecting entire ecosystems within their natural habitats, ensures the preservation of indigenous plants while sustaining the integrity of complex ecological networks (Gepts 2006). Additionally, in-situ conservation supports greater biodiversity and strengthens the link between conserving resources and ensuring their sustainable use (Kumar et al. 2018).

Global in-situ conservation efforts have prioritized creating protected areas and adopting an ecosystem-based approach rather than a species-specific focus (Ma et al. 2012). The effectiveness of in-situ conservation depends on clear regulations, guidelines, and maintaining the natural growth conditions of medicinal plants within their native habitats (Volis and Blecher 2010; Hasan et al. 2024).

Ex-situ conservation, while not entirely separate from in-situ methods, serves as an effective complement, particularly for medicinal plants that are endangered, overexploited, scarce in number, slow-growing, or highly susceptible to replanting-related infections (Havens et al. 2006; Yu et al. 2010).

This strategy focuses on cultivating and adapting threatened species to secure their survival and, in some cases, to generate substantial amounts of plant material for medicinal use. It is often implemented as an urgent intervention to aid in the preservation of medicinal plants (Swarts and Dixon 2009; Sharma et al. 2022). Ex-situ conservation encompasses not only the cultivation of formerly wild medicinal plants in gardens outside their natural habitats but also the selection and storage of their reproductive materials in seed banks for future restoration. These efforts aim to support the survival of rare and endangered species by maintaining ecosystems under controlled conditions. While living collections often include only a limited number of individuals and therefore have restricted value for long-term genetic preservation, botanical gardens hold unique significance in these conservation initiatives (Shafi et al. 2021). Botanical gardens host a wide range of plant species grown under seminatural conditions, encompassing taxonomically and biologically diverse vegetation (Primack and Miller-Rushing 2009). Beyond serving as living repositories, they play an important role in medicinal plant conservation by improving propagation and cultivation methods and by supporting domestication and breeding programs (Maunder et al. 2001; Hasan et al. 2024).

Seed banks represent a more efficient strategy for conserving the genetic diversity of medicinal plants in ex-situ conditions than botanical gardens. They are strongly recommended for safeguarding the biological and genetic resources of wild plant species (Li and Pritchard 2009; Schoen and Brown 2001). Seed banks provide rapid access to plant samples for trait evaluation, generating valuable insights for conserving natural populations. They play a vital role in preserving seeds as genetic resources for medicinal plants and other crops (Hasan et al. 2024).

The framework of Good Agricultural Practices (GAP) for medicinal plants has been developed to regulate cultivation processes, safeguard product quality, and facilitate the standardization of herbal pharmaceuticals (Chan et al. 2012). This strategy seeks to ensure the production of high-quality, safe, and contaminant-free natural medicines by utilizing established knowledge to address diverse facets of cultivation and processing (Muchugi et al. 2008). GAP incorporate a wide range of components, including the cultivation and harvesting methods, environmental conditions of cultivation sites, germplasm management, as well as quality control measures such as assessment of metallic elements, chemical characterization of bioactive compounds, microscopic and macroscopic verification, and pesticide residue analysis (Makunga et al. 2008).

Organic farming has attracted growing interest due to its potential to establish economically feasible, environmentally sound, integrated, and sustainable production systems for diverse medicinal plants (Rigby and Cáceres 2001; Macilwain 2004). The primary objectives of organic cultivation of medicinal plants are to generate high-quality raw materials with greater efficiency while promoting the sustainable utilization and conservation of plant resources. A defining feature of this approach is the exclusion of synthetic fertilizers, pesticides, and herbicides.

Organic farming promotes environmental sustainability by utilizing farm-derived renewable resources to support the ecological functions of medicinal plants and preserve habitat balance. The application of organic fertilizers provides a continuous nutrient supply and improves soil structure, thereby enhancing the growth of medicinal plants and the biosynthesis of key metabolites. Empirical evidence indicates that plants cultivated with organic fertilizers exhibit greater biomass yield and elevated essential-oil content compared to those grown without such inputs (Hendawy and Khalid 2011; Hasan et al. 2024).

Recent advancements in breeding technologies, including gene-editing approaches such as artificial polyploidy, CRISPR-Cas9, and plant tissue culture, provide significant opportunities for improving both the productivity and pharmaceutical potential of medicinal plants (Hasan et al. 2024). Polyploid plants demonstrate improved morphological traits and greater tolerance to both biotic and abiotic environmental stresses relative to diploid species (Kaensaksiri et al. 2011). Moreover, genome duplication has been shown to enhance the biosynthesis of secondary metabolites in medicinally important plants, with polyploid species exhibiting elevated gene expression levels relative to their diploid counterparts (Majdi et al. 2011; Tavan et al. 2015).

Research indicates that crop improvement has introduced substantially more modifications than domestication alone. Many of these metabolic changes appear to be unintended, highlighting the value of metabolomic studies in evaluating the nutritional composition of crops and their wild relatives. The extensive metabolic reprogramming associated with both domestication and crop improvement should be considered when exploring de novo domestication of wild relatives or strategies aimed at engineering metabolically valuable crops (Fernie and Yan 2019).

The Genotype–Phenotype Database (www.gephebase.org) consolidates information on quantitative trait nucleotides in eukaryotes that drive evolutionary changes, offering a resource of candidate genes with validated genotype–phenotype associations for use in crop engineering or introgression. Research indicates that the majority of mutations favored during breeding in both plants and animals are loss-of-function, aligning with the notion that crop diversification has largely been driven by major-effect mutations advantageous under controlled cultivation but disadvantageous in natural ecosystems (Courtier-Orgogozo and Martin 2020; Alseekh et al. 2021).

Domestication and subsequent crop improvement have resulted in profound modifications in gene expression, phenotypic traits, and the production of secondary metabolites (Alseekh et al. 2021). Biotechnological approaches and the domestication of medicinal plant species represent viable strategies to mitigate the overexploitation of resources within medicinal plant value chains. Techniques such as genetic transformation, molecular marker-assisted selection, and tissue culture, which enable the modification of biosynthetic pathways for target metabolites, offer considerable potential for addressing limitations such as the inherently slow growth rates of wild medicinal plants (Canter et al. 2005). Challenges in agrotechnology hinder domestication pathways, particularly due to difficulties in identifying suitable cultivars, variability in the concentration of active ingredients in domesticated varieties, and issues related to diseases, low environmental tolerance or survival under cultivation, and pests (Kala et al. 2006). The domestication of most medicinal plants remains constrained by limited knowledge of management practices and insufficient basic research on their distribution, ecology, and genetic diversity. This limitation may explain the continued preference for wild plants in the medicinal value chain (Volenzo and Odiyo 2020; Padulosi et al. 2002).

Although domestication bottlenecks reduce genetic diversity, the overall census population size can increase when domesticated species are cultivated over wider areas (Smýkal et al. 2018). Moreover, when expanding domesticated populations encounter coexisting wild relatives in new regions, admixture and genetic introgression may occur through interbreeding. Such processes can introduce additional genetic components into the domesticated population, potentially mitigating the reduced genetic diversity associated with bottlenecks and natural selection (Larson et al. 2014; Hazzouri et al. 2015; Mathew et al. 2015). A genetic bottleneck typically arises during domestication or the initial founding events. The composition of gene pools may be further modified when domesticated species disperse away from their center of origin (Smýkal et al. 2018).

Understanding the natural genetic diversity and population structure within and among species is crucial for formulating effective conservation strategies aimed at safeguarding genetic resources and supporting genetic improvement efforts (Falk and Holsinger 1991; Wang et al. 2020).

The widespread adoption of high-yielding cultivars has led to the replacement and gradual loss of locally adapted landraces, particularly in tropical and subtropical regions, while the narrow genetic base of modern cultivars has further heightened their vulnerability to emerging races of diseases and insect pests. Consequently, breeders have increasingly turned to landraces and, in particular, crop wild relatives possessing beneficial traits such as resistance to pathogens and insect pests, as well as tolerance to drought, temperature, and salinity extremes, in order to introgress the corresponding genes or quantitative trait loci into modern cultivars to enhance their resilience (Bellard et al. 2012).

Production systems and the genetic resources that support them—including crop wild relatives found in cultivated areas, protected lands, and natural ecosystems like forests—are under serious threat. These threats stem from major land-use changes, overuse of resources, and both natural and human-caused disasters. Climate change is already having a significant impact on plant species, affecting their distribution, life cycles, and population sizes. At present, it is estimated that two out of every five plant species face the risk of extinction (Bellard et al. 2012; Antonelli et al. 2022). According to the International Union for Conservation of Nature (IUCN) Red List of Threatened Species, the primary threats to plants in their natural habitats include agriculture and aquaculture (32.8%), exploitation of biological resources (21.1%), alterations to natural systems (10.8%), urban and commercial development (10.5%), invasive species, pathogens, and genetic factors (6.5%), pollution (5.4%), and climate change (4.1%). To slow down ongoing genetic erosion and protect species at risk of extinction, it is essential to preserve genetic diversity, particularly within biodiversity hotspots that include both wild and domesticated species (Pironon et al. 2020; Ebert and Engels 2020).

Recent progress in sequencing technologies has revealed that plant responses to environmental conditions, including both abiotic and biotic stresses, are partly influenced by interactions with microbes and the microbiome. Furthermore, evidence suggests that cultivated crops may have weaker plant–microbiome interactions compared to their wild relatives (Friesen et al. 2011; Shi et al. 2019; Wagner 2021; Gutierrez and Grillo 2022). Gene-editing-based de novo domestication offers a promising strategy, particularly for enhancing plant–microbiome interactions, since the genetic mechanisms underlying microbiome assembly are still not well understood (Zsögön et al. 2018). Therefore, de novo domestication is particularly valuable in systems where the genetic basis of domestication is already well understood (Nasti and Voytas 2021; Gutierrez and Grillo 2022).

Understanding how medicinal and aromatic plants respond to NPK fertilization can help boost yields in nutrient-poor soils. Environmentally friendly, biologically compatible cultivation practices improve both productivity and bioactive compound levels while reducing reliance on chemical fertilizers. Integrated nutrient management—combining organic and inorganic inputs—effectively enhances yield and active components. In addition, mycorrhizal fungi promote plant growth and nutrient absorption, allowing reduced NPK use and minimizing environmental harm. Therefore, successful cultivation requires a balanced application of organic, inorganic, and biological fertilizers (Malik et al. 2011).

Preserving genetic diversity, safeguarding plant genetic resources, and ensuring their sustainable use should be a global priority. Effective monitoring of genetic erosion and the vulnerability of genetic diversity is essential for protecting rare and endangered plant species. Since no single conservation method is sufficient on its own, both in-situ and ex-situ approaches should be applied in a complementary way. Additionally, biotechnological tools play a vital role in characterizing plant genetic resources, conserving them, and supporting their use in breeding programs (Salgotra and Chauhan 2023). Research on plant domestication has greatly advanced our understanding of the genetic regulation of key and complex traits. In addition, modern omics technologies and gene-editing tools can provide deeper insights into the intricate developmental and biochemical pathways that shape these domesticated traits. An in-depth understanding of these pathways can help design effective strategies to control the phenotypic expression of unfavorable traits, thereby supporting crop improvement. Furthermore, detailed insights into developmental and biochemical pathways can enable the manipulation of gene expression or enzyme active sites, which in turn influence protein function and ultimately the phenotypic expression of domesticated traits (Singh and van der Knaap 2022).

In Turkey, in-situ conservation is carried out within Forest Recreation Areas, National Parks, Nature Conservation Areas, Nature Parks, Natural Monuments, and Wildlife Protection Zones managed by the Ministry of Forestry. These sites are administered by the General Directorate of National Parks and Hunting—Wildlife, and together support in-situ preservation efforts. However, many of these areas have traditionally been managed with a primary focus on recreation rather than strict conservation (Akalın 2023).

10.5 Conclusion

Safeguarding medicinal plants requires a delicate balance between conservation and innovation. On one hand, climate change, habitat degradation, and unsustainable use threaten their survival; on the other hand, cultivation and domestication—while enabling large-scale production—often reduce genetic diversity and compromise the natural richness of these species. Yet, modern molecular breeding, biotechnology, and advanced cultivation systems now provide tools to counter these challenges, ensuring stable phytochemical profiles and reliable medicinal value. By integrating traditional knowledge with scientific progress and fostering collaboration among communities, policymakers, and researchers, we can preserve both the cultural and biological legacy of medicinal plants. In this way, these species will not only remain vital reservoirs of biodiversity but also continue to serve as sustainable and standardized resources for healthcare worldwide. Ultimately, embracing this synergy between tradition and technology is the key to securing medicinal plants as enduring assets for nature, culture, and human well-being.

References

Aalinezhad S, Dabaghian F, Namdari A, Akaberi M, Emami SA (2025) Phytochemistry and pharmacology of alkaloids from *Papaver* spp.: a structure–activity based study. Phytochem Rev 24:585–657

Abbas M, Gururani MA, Ali A, Bajwa S, Hassan R, Batool SW, Imam M, Wei D (2024) Antimicrobial properties and therapeutic potential of bioactive compounds in *Nigella sativa*: a review. Molecules 29:4914

Abd-Elsamei AH, Mahgoub ES, Wafa HA, Al Aboud NM, Al-Khayri JM, El-Moneim DA, Fallatah TA, Alqurashi M, Fayad E, Alsenani A, Alsufyani H (2025) Genetic diversity, agronomic performance, and bioactive properties, of sweet and bitter white lupin (*Lupinus albus* L.) genotypes. Chil J Agric Res 85:661–676

Agarwal U, Kanupriya Tonk RK, Verma S (2025) A comprehensive review of supernatural coriander herb (*Coriandrum sativum*): phytochemical insights, pharmacological potential and future perspective. Phytochem Rev:1–47. https://doi.org/10.1007/s11101-025-10142-5

Akalın E, Özhatay FN, Güler N, Ersoy H, Başak N, Yeşil Y, Oral D, Demirci S (2013) The flora of Yıldız Mountains (Kırklareli) Biosphere Project area. T J Bot 37(2): 225–269

Akalın E (2023) Threatened medicinal and aromatic plants and conservation studies, in Turkey. In: Máthé Á, Turgut K (eds) Medicinal and aromatic plants of Turkey, Medicinal and aromatic plants of the world, vol 10. Springer, pp 121–130. https://doi.org/10.1007/978-3-031-43312-2_6

Akalın E, Gurdal B, Olcay B (2020) General overview on the conservation of medicinal plants in Turkey. Turk J Biodivers 3:86–94

Allaby RG, Ware RL, Kistler L (2019) A re-evaluation of the domestication bottleneck from archaeogenomic evidence. Evol Appl 12:29–37. https://doi.org/10.1111/eva.12680

Alseekh S, Scossa F, Wen W, Luo J, Yan J, Beleggia R, Klee HJ, Huang S, Papa R, Fernie AR (2021) Domestication of crop metabolomes: desired and unintended consequences. Trends Plant Sci 26(6):650–661

Al-Snafi AE (2024) Constituents and therapeutic activities of *Pimpinella anisum*: a review. Int J Biol Pharm Sci Arch 8:67–78

Altuntaş H, Öke-Altuntaş F, Sarıtan S, Çolak H (2025) Effect of zinc precursors on the antibacterial and photocatalytic applications of green synthesized ZnO nanostructures using *Calluna vulgaris*. Microsc Res Tech 88:2918–2933. https://doi.org/10.1002/jemt.70031

Anagnostou M, Tomou EM, Krigas N, Skaltsa H (2025) Phenolic constituents of Greek native *Salvia* species: a comprehensive review. Phytochem Rev 24:3443–3522

Antonelli A, Smith RJ, Fry C, Simmonds MSJ, Kersey PJ, Pritchard HW, Abbo MS, Acedo C, Adams J, Ainsworth AM et al (2022) State of the world's plants and fungi; research report. Royal Botanic Gardens; Sfumato Foundation, Kew; London

Ashry NA (2013) Plant biodiversity and biotechnology. In: Poltronieri P, Burbulis N, Fogher C (eds) Woodhead publishing series in biomedicine, from plant genomics to plant biotechnology. Woodhead Publishing, Sawston, pp 205–222

Babaei Rad S, Mumivand H, Mollaei S, Khadivi A (2025) Effect of drying methods on phenolic compounds and antioxidant activity of *Capparis spinosa* L. fruits. BMC Plant Biol 25:133

Bağcı Y, Kan Y, Doğu S, Çelik SA (2017) The essential oil compositions of *Origanum majorana* L. cultivated in Konya and collected from Mersin-Turkey. Indian J Pharm Educ Res 51:463–469

Bajaj YPS (1998) Biotechnology for the improvement of medicinal plants. Acta Hortic 457:37–45

Bellard C, Bertelsmeier C, Leadley P, Thuiller W, Courchamp F (2012) Impacts of climate change on the future of biodiversity. Ecol Lett 15:365–377

Bernath J (2002) Strategies and recent achievements in selection of medicinal and aromatic plants. Acta Hortic 576:115–128

Boudou F, Belakredar A, Keziz A, Alsaeedi H, Cornu D, Bechelany M, Barhoum A (2025) *Camellia sinensis* phytochemical profiling, drug-likeness, and antibacterial activity against gram-positive and gram-negative bacteria: in vitro and in silico insights. Front Chem 13:1555574

Braga RA, Dal Fabbro IM, Borem FM, Rabelo G, Arizaga R, Rabal HJ, Trivi M (2003) Assessment of seed viability by laser speckle technology. Biosyst Eng 86:287–294

Canter PH, Thomas H, Ernst E (2005) Bringing medicinal plants into cultivation: opportunities and challenges for biotechnology. Trends Biotechnol 23(4):180–185

Castañeda-Álvarez NP, Khoury CK, Achicanoy HA, Bernau V, Dempewolf H, Eastwood RJ, Guarino L, Harker RH, Jarvis A, Maxted N, Müller JV (2016) Global conservation priorities for crop wild relatives. Nat Plants 2:1–6. https://doi.org/10.1038/nplants.2016.22

Ceballosa G, Paul R, Ehrlichb L, Dirzo R (2017) Biological annihilation via the ongoing sixth mass extinction signaled by vertebrate population losses and declines. Proc Natl Acad Sci USA 114:E6089–E6096

Chan K, Shaw D, Simmonds MS, Leon CJ, Xu Q, Lu A, Sutherland I, Ignatova S, Zhu YP, Verpoorte R, Williamson EM (2012) Good practice in reviewing and publishing studies on herbal medicine, with special emphasis on traditional Chinese medicine and Chinese materia medica. J Ethnopharmacol 140(3):469–475. https://doi.org/10.1016/j.jep.2012.01.038

Charlwood BV, Pletsch M (2002) Manipulation of natural product accumulation in plants through genetic engineering. Int J Geogr Inf Syst 9:139–151

Choi Y, Jeong J, In J, Yang D (2003) Production of herbicide-resistant transgenic *Panax ginseng* through the introduction of the phosphinothricin acetyl transferase gene and successful soil transfer. Plant Cell Rep 21:563–568

Coley PD, Heller MV, Aizprua R, Araúz B, Flores N, Correa M, Gupta M, Solis PN, Ortega-Barría E, Romero LI, Gomez B (2003) Using ecological criteria to design plant collection strategies for drug discovery. Front Ecol Environ 1(8):421–428. https://doi.org/10.1890/1540-9295(2003)001[0421:UECTDP]

Costa MA, Zia ZQ, Davin LB, Lewis NG (1999) Toward engineering the metabolic pathways of cancer-preventing lignans in cereal grains and other crops. In: Romeo JT (ed) Phytochemicals in human health protection, nutrition, and plant defense, Recent advances in phytochemistry, vol 33. Springer, New York, pp 67–87

Courtier-Orgogozo V, Martin A (2020) The coding loci of evolution and domestication: current knowledge and implications for bio-inspired genome editing. J Exp Biol 223:jeb208934

Cowling WA, Buirchell BJ, Falk DEA (2009) Model for incorporating novel alleles from the primary gene pool into elite crop breeding programs while reselecting major genes for domestication or adaptation. Crop Pasture Sci 60:1009–1015

Coyne CJ, Shiv Kumar S, von Wettberg EB, Marques E, Berger JD, Redden RJ, Ellis TN, Brus J, Zablatzká L, Smýkal P (2020) Potential and limits of exploitation of crop wild relatives for pea, lentil and chickpea improvement. Legume Sci 2:e36. https://doi.org/10.1002/leg3.36

Dakhli N, López-Jiménez A, Cárdenas C, Hraoui M, Dhaouafi J, Bernal M, Sebai H, Medina MÁ (2025) *Urtica dioica* aqueous leaf extract: chemical composition and in vitro evaluation of biological activities. Int J Mol Sci 26:1220

Dalir M, Choobchian S, Abbasi E (2025) Identifying the relative importance of cultivating medicinal plants in improving rural livelihoods. Environ Dev 54:101159

Diamond J (2005) Collapse: how societies choose to fail or succeed. Penguin, London

Dinçer C, Tontul İ, Çam İB, Özdemir KS, Topuz A, Nadeem HŞ, Ay ST, Göktürk RS (2013) Phenolic composition and antioxidant activity of *Salvia tomentosa* Miller: effects of cultivation, harvesting year, and storage. Turk J Agric For 37:561–567

Djermane N, Brahmi M, Adli DE, Gliozzi M, Musolino V, Khellaf R, Erenler R, Arhab R, Garzoli S (2025) Chemical composition, antioxidant, antimicrobial, antidiabetic, and butyrylcholinesterase inhibitory activities in vitro of the essential oil and crude extracts of the aerial parts of *Thymus ciliatus*. Chem Biodivers 22:e202403318

Doebley JF, Gaut BS, Smith BD (2006) The molecular genetics of crop domestication. Cell 127(7):1309–1321. https://doi.org/10.1016/j.cell.2006.12.006

Dubey T, Guerra DJ (2002) Use of biotechnology for growing medicinal plants. Recent Prog Med Plants 5:47–61

Ebert AW, Engels JMM (2020) Plant biodiversity and genetic resources matter! Plants 9(12):1706

Ekşi G, Özkan AMG, Koyuncu M (2020) Garlic and onions: an eastern tale. J Ethnopharmacol 253:112675

El Oihabi M, Mehdi I, Mbarek O, Mdarhri Y, Ghanimi S, Allal LB, Lanjri AF, Ammari M (2025) Review of Moroccan *Cannabis sativa* L.: cultivation, phytochemistry, therapeutic uses, and biological activities in in-vitro, in-vivo and in-silico assays. Ethnobot Res Appl 31:1–25

Evenson RE, Gollin D (2003) Assessing the impact of the green revolution, 1960 to 2000. Science 300:758–762

Falk KE, Holsinger DF (1991) Genetics and conservation of rare plants. Oxford University Press, New York

Fernie AR, Yan J (2019) De novo domestication: an alternative route toward new crops for the future. Mol Plant 12:615–631

Ferreira JFS, Duke SO (1997) Approaches for maximizing biosynthesis of medicinal plant secondary metabolites. AgBiotech News Inf 9:309N–315N

Figueiredo MS, Grelle CEV (2009) Predicting global abundance of a threatened species from its occurrence: implications for conservation planning. Divers Distrib 15(1):117–121. https://doi.org/10.1111/j.1472-4642.2008.00525.x

Friesen ML, Porter SS, Stark SC, von Wettberg EJ, Sachs JL, Martinez-Romero E (2011) Microbially mediated plant functional traits. Annu Rev Ecol Evol Syst 42:23–46

Gaillard C, Langan L, Pfeiffer M, Kumar D, Martens C, Higgins SI, Scheiter S (2018) African shrub distribution emerges via a trade-off between height and sapwood conductivity. J Biogeogr 45:2815–2826

Gao Y, Yin S, Wu L, Dai D, Wang H, Liu C, Tang L (2017) Genetic diversity and structure of wild and cultivated *Amorphophallus paeoniifolius* populations in southwestern China as revealed by RAD-seq. Sci Rep 7:14183. https://doi.org/10.1038/s41598-017-14738-6

Gaut BS, Seymour DK, Liu Q, Zhou Y (2018) Demography and its effects on genomic variation in crop domestication. Nat Plants 4:512–520. https://doi.org/10.1038/s41477-018-0210-1

Gedikoğlu A, Öztürk Hİ, Özçoban A (2024) Analysis of the chemical composition, antimicrobial, and antioxidant qualities of microwave and supercritical CO_2-extracted *lavender* essential oils cultivated in a hyperarid region of Türkiye. Molecules 29:5605

Gepts P (2006) Plant genetic resources conservation and utilization: the accomplishments and future of a societal insurance policy. Crop Sci 46(5):2278–2292. https://doi.org/10.2135/cropsci2006.03.0169gas

Gibson EL, Wardel J, Watts CJ (1998) Fruit and vegetable consumption, nutritional knowledge and beliefs in mothers and children. Appetite 31:205–228

Gross BL, Olsen KM (2010) Genetic perspectives on crop domestication. Trends Plant Sci 15:529–537

Güner A, Aslan S, Ekim T, Vural M, Babaç MT (2012) Türkiye Bitkileri Listesi (Damarlı Bitkiler). Nezahat Gökyiğit Botanik Bahçesi ve Flora Araştırmaları Derneği Yayını, İstanbul

Gutierrez A, Grillo MA (2022) Effects of domestication on plant–microbiome interactions. Plant Cell Physiol 63(11):1654–1666

Hajjar R, Hodgkin T (2007) The use of wild relatives in crop improvement: a survey of developments over the last 20 years. Euphytica 156:1–13. https://doi.org/10.1007/s10681-007-9363-0

Hallauer AR (2011) Evolution of plant breeding. Crop Breed Appl Biotechnol 11:197–206

Hasan N, Laskar RA, Farooqui SA, Naaz N, Sharma N, Budakoti M, Joshi DC, Choudhary S, Bhinda MS (2024) Genetic improvement of medicinal and aromatic plant species: breeding techniques, conservative practices and future prospects. Crop Des 3(4):100080

Hasler CM, Blumberg JB (1999) Symposium on phytochemicals: biochemistry and physiology. J Nutr 129:756S–757S

Havens K, Vitt P, Maunder M, Guerrant EO, Dixon K (2006) Ex situ plant conservation and beyond. Bioscience 56(6):525–531. https://doi.org/10.1641/0006-3568(2006)56[525:ESPCAB]2.0.CO;2

Hazzouri KM, Flowers JM, Visser HJ, Khierallah HSM, Rosas U, Pham GM, Meyer RS, Johansen CK, Fresquez ZA, Masmoudi K, Haider N (2015) Whole genome re-sequencing of date palms yields insights into diversification of a fruit tree crop. Nat Commun 6:8824. https://doi.org/10.1038/ncomms9824

Hendawy SF, Khalid KA (2011) Effect of chemical and organic fertilizers on yield and essential oil of chamomile flower heads. Med Aromat Plant Sci Biotechnol 5(1):43–48

Hesami M, Daneshvar MH, Yoosefzadeh-Najafabadi M, Alizadeh M (2018) Effect of plant growth regulators on indirect shoot organogenesis of *Ficus religiosa* through seedling derived petiole segments. J Genet Eng Biotechnol 16(1):175–180. https://doi.org/10.1016/j.jgeb.2017.11.001

Hoban S, Campbell CD, da Silva JM, Ekblom R, Funk WC, Garner BA, Godoy JA, Kershaw F, MacDonald AJ, Mergeay J, Minter M (2021) Genetic diversity is considered important but interpreted narrowly in country reports to the convention on biological diversity: current actions and indicators are insufficient. Biol Conserv 261:109233

Hollingsworth PM, Dawson IK, Goodall-Copestake WP, Richardson JE, Weber JC, Montes CS, Pennington RT (2005) Do farmers reduce genetic diversity when they domesticate tropical trees? A case study from Amazonia. Mol Ecol 14:497–501. https://doi.org/10.1111/j.1365-294X.2005.02431.x

Hunie Tesfa K, Ewunetie AT, Gebeyehu CD (2025) Mechanism of antihypertensive and hypocholesterolemic effect of white lupine (*Lupinus albus* L.): a review. Pharmacol Res Nat Prod 7:100207

Hyten DL, Qijian S, Youlin Z, Ik-Young C, Nelson RL, Costa JM, Specht JE, Shoemaker RC, Cregan PB (2006) Impacts of genetic bottlenecks on soybean genome diversity. Proc Natl Acad Sci USA 103:16666–16671. https://doi.org/10.1073/pnas.0604379103

İlktaç M, Kutlu ZP, Çelebi Y, Çelik G (2024) Antimicrobial activity of *Pimpinella* – an overview. EMU J Pharm Sci 7(1):31–39

Isaac ME, Nimmo V, Gaudin ACM, Leptin A, Schmidt JE, Kallenbach CM, Martin A, Entz M, Carkner M, Rajcan I, Boyle TD (2021) Crop domestication, root trait syndromes, and soil nutrient acquisition in organic agroecosystems: a systematic review. Front Sustain Food Syst 5:716480

Jang H, Choi M, Jang KS (2024) Comprehensive phytochemical profiles and antioxidant activity of Korean local cultivars of red chili pepper (*Capsicum annuum* L.). Front Plant Sci 15:1333035

Kaensaksiri T, Soontornchainaksaeng P, Soonthornchareonnon N, Prathanturarug S (2011) In vitro induction of polyploidy in *Centella asiatica* (L.) Urban. Plant Cell Tissue Organ Cult 107:187–194. https://doi.org/10.1007/s11240-011-9969-8
Kala CP, Dhyani PP, Sajwan BS (2006) Developing the medicinal plants sector in northern India: challenges and opportunities. J Ethnobiol Ethnomed 2(32):15
Kar I, Das S, Mukherjee A, Patra AK (2025) *Mentha* herbs and their bioactive molecules on biological functions, production performance, health, and product quality in poultry and livestock. Trop Anim Health Prod 57:311
Karabacak E, Uysal İ, Doğan M (2009) Cultivated *Salvia* species in Turkey. Biol Divers Conserv 2:71–77
Karık Ü, Çinar O, Tunçtürk M, Sekeroglu N, Gezici S (2016) Essential oil composition of some sage (*Salvia* spp.) species cultivated in İzmir (Turkey) ecological conditions. Indian J Pharm Educ Res 52:S102–S107
Khoury CK, Amariles D, Soto JS, Diaz MV, Sotelo S, Sosa CC, Ramírez-Villegas J, Achicanoy HA, Velásquez-Tibatá J, Guarino L, León B (2019) Comprehensiveness of conservation of useful wild plants: an operational indicator for biodiversity and sustainable development targets. Ecol Indic 98:420–429
King A, Young G (1999) Characteristics and occurrence of phenolic phytochemicals. J Am Diet Assoc 24:213–218
Kingston DGI (2011) Modern natural products drug discovery and its relevance to biodiversity conservation. J Nat Prod 74:496–511. https://doi.org/10.1021/np100550t
Koçyiğit M, Demirci KS (2018) "Small protected areas" for conservation priorities in South Anatolia (Başkonuş Mountain-Kahramanmaraş). Eur J Biol 77:89–96. https://doi.org/10.26650/eurjbiol.2018.0013
Kovach MJ, Mccouch SR (2008) Leveraging natural diversity: Back through the bottleneck. Curr Opin Plant Biol 11:193–200. https://doi.org/10.1016/j.pbi.2007.12.006
Koziol L, Rieseberg LH, Kane N, Bever JD (2012) Reduced drought tolerance during domestication and the evolution of weediness results from tolerance-growth trade-offs: sunflower growth and drought tolerance trade-offs. Evolution 66:3803–3814
Kumar N, Hazra KK, Nath CP, Praharaj CS, Singh U (2018) Grain legumes for resource conservation and agricultural sustainability in South Asia. In: Narendra K, Hazra KK, Nath CP, Praharaj CS, Singh U (eds) Legumes for soil health and sustainable management. Springer, Singapore, pp 77–107. https://doi.org/10.1007/978-981-13-0253-4_3
Lalovski I, Krasteva M, Yordanov E, Hristov E, Dimitrov M, Parvova I (2025) Safety of active substances derived from *Rosa damascena* and their potential biological activity in humans: a systematic review. Pharmacia 72:1–10
Lam H, Xu M, Liu X, Chen X, Yang W, Wong G, Wong FL, Li MW, He W, Qin N, Wang B, Li J (2010) Resequencing of 31 wild and cultivated soybean genomes identifies patterns of genetic diversity and selection. Nat Genet 42:1053–1059. https://doi.org/10.1038/ng.715
Larson G, Piperno DR, Allaby RG, Purugganan MD, Andersson L, Arroyo-Kalin M, Barton L, Climer Vigueira C, Denham T, Dobney K, Doust AN (2014) Current perspectives and the future of domestication studies. Proc Natl Acad Sci USA 111:6139–6146. https://doi.org/10.1073/pnas.1323964111
Leigh DM, Hendry AP, Vázquez-Domínguez E, Friesen VL (2019) Estimated six per cent loss of genetic variation in wild populations since the industrial revolution. Evol Appl 12:1505–1512
Li DZ, Pritchard HW (2009) The science and economics of ex situ plant conservation. Trends Plant Sci 14(11):614–621. https://doi.org/10.1016/j.tplants.2009.09.005
Li X, Chen Y, Lai Y, Yang Q, Hu H, Wang Y (2015) Sustainable utilization of traditional Chinese medicine resources: systematic evaluation on different production modes. Evid Based Complement Alternat Med 2015:218901. https://doi.org/10.1155/2015/218901
Ma J, Rong K, Cheng K (2012) Research and practice on biodiversity in situ conservation in China: progress and prospect. Biodivers Sci 20(5):551–558
Macilwain C (2004) Organic: is it the future of farming? Nature 428(6985):792–794

Majdi M, Liu Q, Karimzadeh G, Malboobi MA, Beekwilder J, Cankar K, de Vos R, Todorovic S, Simonovic A, Bouwmeester H (2011) Biosynthesis and localization of parthenolide in glandular trichomes of feverfew (*Tanacetum parthenium* L. Schulz Bip.). Phytochemistry 72(14–15):1739–1750. https://doi.org/10.1016/j.phytochem.2011.04.021

Makunga NP, Philander LE, Smith M (2008) Current perspectives on an emerging formal natural products sector in South Africa. J Ethnopharmacol 119(3):365–375. https://doi.org/10.1016/j.jep.2008.07.020

Malik AA, Sanghmitra Suryapani SS, Javed Ahmad JA (2011) Chemical vs organic cultivation of medicinal and aromatic plants: the choice is clear. Int J Med Arom Plants 1:5–13

Manzoor S, Rakha A, Rasheed H, Bhat ZF, Khan MSA, Abdi G, Aadil RM (2025) A comprehensive review on anxiolytic effect of *Lavandula Angustifolia* Mill. in clinical studies. Food Sci Nutr 13:e70993

Masondo NA, Ndhlala AR, Aremu AO, Van Staden J, Finnie JF (2013) A comparison of the pharmacological properties of garden cultivated and muthi-market sold *Boweia volubilis*. S Afr J Bot 86:135–138. https://doi.org/10.1016/j.sajb.2013.02.169

Mathai K (2000) Nutrition in the adult years. In: Mahan LK, Escott-Stump S (eds) Krause's food, nutrition, and diet therapy, 10th edn. Saunders/Elsevier, St. Louis, pp 271, 274–275

Mathew LS, Seidel MA, George B, Mathew S, Spannagl M, Haberer G, Torres MF, Al-Dous EK, Al-Azwani EK, Diboun I, Krueger RR (2015) A genome-wide survey of date palm cultivars supports two major subpopulations in *Phoenix dactylifera*. G3 (Bethesda) 5:1429–1438. https://doi.org/10.1534/g3.115.018341

Maunder M, Higgens S, Culham A (2001) The effectiveness of botanic garden collections in supporting plant conservation: a European case study. Biodivers Conserv 10:383–401. https://doi.org/10.1023/A:1016666526878

Meagher E, Thomson C (1999) Vitamin and mineral therapy. In Morrison G, Hark L (eds) Medical Nutrition and Disease, 2nd ed., Malden 33–58

McChesney JD (1999) Quality of botanical preparations: environmental issues and methodology for detecting environmental contaminants. In Eskinazi D (ed) Botanical Medicine: Efficacy, quality assurance and regulation, Larchmont 127–131

Meyer RS, Purugganan MD (2013) Evolution of crop species: genetics of domestication and diversification. Nat Rev Genet 14:840–852

Meyer RS, Duval AE, Jensen HR (2012) Patterns and processes in crop domestication: an historical review and quantitative analysis of 203 global food crops. New Phytol 196:29–48. https://doi.org/10.1111/j.1469-8137.2012.04253.x

Mofokeng MM, Prinsloo G, Araya HT, Du Plooy CP, Sathekge NR, Amoo SO, Steyn JM (2020) Yield and metabolite production of *Pelargonium sidoides* DC. in response to irrigation and nitrogen management. Metabolites 10(6):219. https://doi.org/10.3390/metabo10060219

Mofokeng MM, Du Plooy CP, Araya HT, Amoo SO, Mokgehle SN, Pofu KM, Mashela PW (2022) Medicinal plant cultivation for sustainable use and commercialisation of high-value crops. S Afr J Bot Sci 118(7–8):1–7

Moumen BE, Bouzoubaa A, Drioiche A, Eddahmouny M, Al Kamaly O, Shahat AA, Touijer H, Hadi N, Kharchouf S, Cherrat A, Fadili K (2025) Unveiling the chemical composition, antioxidant, and antimicrobial potentials of *Foeniculum vulgare* Mill: a combined in vitro and in silico approach. Int J Mol Sci 26:4499

Moyo M, Aremu AO, Gruz J, Šubrtová M, Szüčová L, Doležal K, Van Staden J (2013) Conservation strategy for Pelargonium sidoides DC: phenolic profile and pharmacological activity of acclimatized plants derived from tissue culture. J Ethnopharmacol 149:557–561. https://doi.org/10.1016/j.jep.2013.07.021

Msegued Ayam I, Zinedine A, Lyagoubi A, Rocha JM, Raoui SM, Errachidi F (2025) Chemical composition, antioxidant, and antifungal activity of essential oil extracted from cumin seeds (*Cuminum cyminum* L.) planted in Morocco. Chem Biodivers 22:e02687

Muchugi A, Muluvi GM, Kindt R, Kadu CA, Simons AJ, Jamnadass RH (2008) Genetic structuring of important medicinal species of genus Warburgia as revealed by AFLP analysis. Tree Genet Genomes 4:787–795. https://doi.org/10.1007/s11295-008-0151-3

Nasti RA, Voytas DF (2021) Attaining the promise of plant gene editing at scale. Proc Natl Acad Sci USA 118:e2004846117

Nazari N, Sharifnia F, Salimpour F, Mehrnia M, Graan A, Zarafshar M (2025) Phytochemical richness, antioxidant activity, and molecular diversity in *Papaver* species from Iran's western and central regions. Front Plant Sci 16:1635867

Ncube NH, Gupta J (2025) *Foeniculum vulgare* (Fennel): a comprehensive review of its antidiabetic properties. Asian Pac J Trop Biomed 15:85–97

Niazian M (2019) Application of genetics and biotechnology for improving medicinal plants. Planta 249:953–973. https://doi.org/10.1007/s00425-019-03099-1

Ok FZ, Şanlı A (2025) Effects of domestication on essential oil yield, chemical profiling, and antifungal activities of bitter fennel (*Foeniculum vulgare* Mill.). Biochem Syst Ecol 120:104951

Olsen KM, Gross BL (2008) Detecting multiple origins of domesticated crops. Proc Natl Acad Sci USA 105:13701–13702. https://doi.org/10.1073/pnas.0807439105

Olsen KM, Wendel JF (2013) A bountiful harvest: genomic insights into crop domestication phenotypes. Annu Rev Plant Biol 64:47–70

Özhatay N (2006) Important plant areas along BTC pipeline in Turkey. BTC Sirketi, Istanbul

Özhatay N, Koyuncu M, Atay S, Byfield A (1997) The wild medicinal plant trade in Turkey. The Society for the Protection of Nature (DHKD), İstanbul

Özhatay N, Byfield A, Atay S (2010) Important plant areas in Turkey: 122 key Turkish botanical sites. WWF Turkey, Istanbul

Padulosi S, Leaman D, Quek P (2002) Challenges and opportunities in enhancing the conservation and use of medicinal and aromatic plants. Int J Geogr Inf Syst 9(4):243–267

Panis B, Nagel M, Van den Houwe I (2020) Challenges and prospects for the conservation of crop genetic resources in field genebanks, in in vitro collections and/or in liquid nitrogen. Plants 9:1634

Pironon S, Borrell JS, Ondo I, Douglas R, Phillips C, Khoury CK, Kantar MB, Fumia N, Soto Gomez M, Viruel J, Govaerts R (2020) Toward unifying global hotspots of wild and domesticated biodiversity. Plants 9:1128

Pokhrel B, Singh S, Pandey NK, Kumar B, Chaudhary AK, Das S, Mazumder R, Mazumder A, Sharma S, Joshi K, Patel D, Baghel DS (2026) Ethnopharmacological, phytochemical, and toxicological characteristics of *Dactylorhiza hatagirea*: a synoptic overview. In: Singh J, Vasudev H, Prakash C, Singh M (eds) Advanced manufacturing technologies in biomedical science. CRC Press, pp 477–486

Primack RB, Miller-Rushing AJ (2009) The role of botanical gardens in climate change research. New Phytol 182(2):303–313. https://doi.org/10.1111/j.1469-8137.2009.02800.x

Puffenberger EG (2021) Recessive diseases and founder genetics. In: Gonzaga-Jauregui C, Lupski JR (eds) Translational and applied genomics, genomics of rare diseases. Academic Press, Cambridge, pp 97–115

Purugganan MD (2019) Evolutionary insights into the nature of plant domestication. Curr Biol 29:R705–R714

Purugganan MD (2022) What is domestication? Trends Ecol Evol 37(8):663–671

Purugganan MD, Fuller DQ (2009) The nature of selection during plant domestication. Nature 457:843–848

Ray DK, Gerber JS, MacDonald GK, West PC (2015) Climate variation explains a third of global crop yield variability. Nat Commun 6:5989

Rigby D, Cáceres D (2001) Organic farming and the sustainability of agricultural systems. Agric Syst 68(1):21–40. https://doi.org/10.1016/S0308-521X(00)00060-3

Saito K, Yamazaki M, Anzai H, Yoneyama K, Murakoshi I (1992) Transgenic herbicide-resistant Atropa belladonna using an Ri binary vector and inheritance of the transgenic trait. Plant Cell Rep 11:219–224

Salgotra RK, Chauhan BS (2023) Genetic diversity, conservation, and utilization of plant genetic resources. Genes 14(1):174

Salman-Minkov A, Sabath N, Mayrose I (2016) Whole-genome duplication as a key factor in crop domestication. Nat Plants 2:16115

Saxena M, Saxena J, Nema R, Singh D, Gupta A (2013) Phytochemistry of medicinal plants. J Pharmacogn Phytochem 1(6):168–182

Schippmann U, Leaman DJ, Cunningham A (2002) Impact of cultivation and gathering of medicinal plants on biodiversity: global trends and issues. In: Biodiversity and the ecosystem approach in agriculture, forestry and fisheries. Food and Agriculture Organization, Rome, pp 142–167

Schoen DJ, Brown AH (2001) The conservation of wild plant species in seed banks: attention to both taxonomic coverage and population biology will improve the role of seed banks as conservation tools. BioScience 51(11):960–966. https://doi.org/10.1641/0006-3568(2001)051[0960:TCOWPS]2.0.CO;2

Sevindik M, Kilibarda S, Uysal I, Šavikin K, Banjac N, Popović-Đorđević J (2025) Overview of phytochemical composition and bioactivity of *Lavandula* species. Phytochem Rev:1–26. https://doi.org/10.1007/s11101-025-10180-z

Shabir PA (2021) CRISPR/Cas9-mediated genome editing in medicinal and aromatic plants: developments and applications. In: Aftab T, Hakeem KR (eds) Medicinal and aromatic plants. Academic Press, pp 209–221. https://doi.org/10.1016/B978-0-12-819590-1.00009-4

Shafi A, Hassan F, Zahoor I, Majeed U, Khanday FA (2021) Biodiversity, management and sustainable use of medicinal and aromatic plant resources. In: Tariq A, Khalid RH (eds) Medicinal aromatic plants: healthcare and industrial applications. Springer International Publishing, Cham, pp 85–111. https://doi.org/10.1007/978-3-030-58975-2_3

Shahzad A, Sharma S, Parveen S, Saeed T, Shaheen A, Akhtar R, Yadav V, Upadhyay A, Ahmad Z (2017) Historical perspective and basic principles of plant tissue culture. In: Abdin MZ, Kiran U, Kamaluddin, Ali A (eds) Plant biotechnology: principles and applications. Springer, Singapore, pp 1–36. https://doi.org/10.1007/978-981-10-2961-5_1

Sharma P, Roy M, Roy B (2022) A review on influence of floral biology, pollination efficiency and conservation strategies of endangered medicinal plant, *Rauvolfia serpentina* (L.) Benth. ex Kurz. Ann Phytomed 11(1):86–98

Shi S, Chang J, Tian L, Nasir F, Ji L, Li X, Chunjie T (2019) Comparative analysis of the rhizomicrobiome of the wild versus cultivated crop: insights from rice and soybean. Arch Microbiol 201:879–888

Shukla SK (2023) Conservation of medicinal plants: challenges and opportunities. J Med Bot 7:5–10

Silva GC, da Silva FP, Beirão GA, Severino JJ, Machado MD, Barbosa MP, Reyes GB, Rickli ME, Lopes AD, Jacomassi E, Nunes MG (2025) Phytochemical profile, antioxidant capacity, and photoprotective potential of Brazilian *Humulus lupulus*. Pharmaceuticals 18:1229

Singh J, van der Knaap E (2022) Unintended consequences of plant domestication. Plant Cell Physiol 63(11):1573–1583

Smýkal P, Nelson MN, Berger JD, von Wettberg EJ (2018) The impact of genetic changes during crop domestication. Agronomy 8:119. https://doi.org/10.3390/agronomy8030026

Swarts ND, Dixon KW (2009) Terrestrial orchid conservation in the age of extinction. Ann Bot 104(3):543–556. https://doi.org/10.1093/aob/mcp025

Tavan M, Mirjalili MH, Karimzadeh G (2015) In vitro polyploidy induction: changes in morphological, anatomical and phytochemical characteristics of *Thymus persicus* (Lamiaceae). Plant Cell Tissue Org Cult 122:573–583. https://doi.org/10.1007/s11240-015-0789-0

Tian Y, Cortés-Avendaño P, Yang B, Glorio-Paulet P, Repo-Carrasco-Valencia R, Suomela JP (2024) Flavonoid diversity in bitter and debittered seeds of *Andean lupin* (*Lupinus mutabilis* Sweet). Food Chem 442:138411. https://doi.org/10.1016/j.foodchem.2024.138411

TÜİK (2024) Türkiye İstatistik Kurumu. Published on the internet. http://www.tuik.gov.tr. Downloaded on 20 October 2025

Tunç T, Ozpinar H, Koçyiğit ÜM, Erkan S (2024) Exploring the antioxidant, antimicrobial, antidiabetic, and anticancer properties of *Capsicum annuum* L. (Samandağ Pepper): biochemical in silico insights. Chem Select 9(48):e202403760

Tuzlacı E (2016) Türkiye Bitkileri Genelekselİlaç Rehberi. İstanbul Tıp Kitabevleri, İstanbul, 1460p

Van Tassel DL, Tesdell O, Schlautman B, Rubin MJ, DeHaan LR, Crews TE, Streit Krug A (2020) New food crop domestication in the age of gene editing: genetic, agronomic and cultural

change remain co-evolutionarily entangled. Front Plant Sci 11:789. https://doi.org/10.3389/fpls.2020.00789
Vavílov NI, Vavylov MI, Vavílov NI, Dorofeev VF (1992) Origin and geography of cultivated plants. Cambridge University Press, Cambridge
Veronese P, Li X, Niu X, Weller SC, Bressan RA, Hasegawa PM (2001) Bioengineering mint crop improvement. Plant Cell Tissue Organ Cult 64:133–144
Volenzo T, Odiyo J (2020) Integrating endemic medicinal plants into the global value chains: the ecological degradation challenges and opportunities. Heliyon 6(9):e04970
Volis S, Blecher M (2010) Quasi in situ: a bridge between ex situ and in situ conservation of plants. Biodivers Conserv 19:2441–2454. https://doi.org/10.1007/s10531-010-9849-2
Von Wettberg EJB, Chang PL, Başdemir F, Carrasquila-Garcia N, Korbu LB, Moenga SM, Bedada G, Greenlon A, Moriuchi KS, Singh V, Cordeiro MA (2018) Ecology and genomics of an important crop wild relative as a prelude to agricultural innovation. Nat Commun 9:649
Von Wettberg E, Davis TM, Smýkal P (2020) Wild plants as source of new crops. Front Plant Sci 11:591554
Wagner MR (2021) Prioritizing host phenotype to understand microbiome heritability in plants. New Phytol 232:502–509
Wang X, Liu XQ, Ko YZ, Jin XL, Sun JH, Zhao ZY, Yuan QJ, Chiang YC, Huang LQ (2020) Genetic diversity and phylogeography of the important medical herb, cultivated huang-lian populations, and the wild relatives *Coptis* species in China. Front Genet 11:708
Wang W, Wang W, Pan Y, Tan C, Li H, Chen Y, Liu X, Wei J, Xu N, Han Y, Gu H (2022) A new gain-of-function OsGS2/GRF4 allele generated by CRISPR/Cas9 genome editing increases rice grain size and yield. Crop J 10(4):1207–1212. https://doi.org/10.1016/j.cj.2022.01.004
Warschefsky E, Penmetsa RV, Cook DR, Von Wettberg EJ (2014) Back to the wilds: tapping evolutionary adaptations for resilient crops through systematic hybridization with crop wild relatives. Am J Bot 101:1791–1800. https://doi.org/10.3732/ajb.1400116
Whitehead SR, Turcotte MM, Poveda K (2017) Domestication impacts on plant–herbivore interactions: a meta-analysis. Philos Trans R Soc B Biol Sci 372:20160034
WHO (1998) Regulatory situation of herbal medicines: a worldwide review; (No. WHO/TRM/98.1); Geneva
Willis KJ (2017a) Useful plants—medicines. In: State of the world's plants 2017: 100. Royal Botanic Gardens, Kew. ISBN: 978-1-84246-628-5
Willis KJ (ed) (2017b) State of the world's plants 2017. Report. Royal Botanic Gardens, Kew
Xego S, Kambizi L, Nchu F (2016) Threatened medicinal plants of South Africa: case of the family hyacinthaceae. Afr J Tradit Complement Altern Med 13:168–180. https://doi.org/10.4314/ajtcam.v13i3.20
Xiao PG (1991) The Chinese approach to medicinal plants. Their utilization and conservation. In: Akerele O, Heywood V, Synge H (eds) Conservation of medicinal plants. Cambridge University Press, Cambridge, pp 305–313
Xu Y (2010) Molecular plant breeding. CAB International, Wallingford, 734p
Yamasaki M, Wright SI, Mcmullen MD (2007) Genomic screening for artificial selection during domestication and improvement in maize. Ann Bot 100:967–973. https://doi.org/10.1093/aob/mcm173
Yu H, Xie C, Song J, Zhou Y, Chen S (2010) TCMGIS-II based prediction of medicinal plant distribution for conservation planning: a case study of *Rheum tanguticum*. Chin Med 5:1–9. https://doi.org/10.1186/1749-8546-5-31
Zafar N, Zafar A (2025) Pharmaceuticals and therapeutic potentialities of *Melissa officinalis* L. – a review. J Int Res Med Pharm Sci 20:8–18
Zhang J, Wang X, Yao J, Li Q, Liu F, Yotsukura N, Krupnova TN, Duan D (2017) Effect of domestication on the genetic diversity and structure of *Saccharina japonica* populations in China. Sci Rep 7:42158. https://doi.org/10.1038/srep42158
Zsögön A, Čermák T, Naves ER, Notini MM, Edel KH, Weinl S, Freschi L, Voytas DF, Kudla J, Peres LE (2018) De novo domestication of wild tomato using genome editing. Nat Biotechnol 36:1211–1216

Chapter 11
Conservation Strategies for Endangered Medicinal Plants in the Kurdistan Region of Iraq: Challenges and Opportunities

Rubar Hussein Muhammed-Salih

Abstract The Kurdistan Region of Iraq (KRG), a globally significant part of the Irano-Anatolian biodiversity hotspot, harbors a rich concentration of medicinal plants integral to both its unique ecosystems and millennia-old ethnobotanical heritage, yet these plants are critical for local healthcare, economic resilience, and cultural identity, their existence is under severe and compounding threat. This study examines the multifaceted challenges to medicinal plant conservation in the KRG, focusing on ecological pressures, socioeconomic barriers, and governance gaps. Key threats include accelerating climate change, unsustainable harvesting driven by economic hardship, habitat loss from urbanization, and weak institutional enforcement. These pressures, combined with a significant lack of comprehensive floristic data, threaten to cause irreversible loss of species, genetic erosion, and the disappearance of invaluable traditional knowledge. Addressing this dual ecological and ethnobotanical crisis requires a multi-pronged and interdisciplinary approach. We propose evidence-based conservation strategies, including the establishment of new protected areas, strengthening community-based conservation, and integrating ethnobotanical documentation with modern scientific research. Such an approach, which leverages both in situ and ex situ methods, is not only a biological necessity but also a cultural and national heritage priority.

Keywords Biodiversity hotspot · Climate change · Conservation strategies · Ethnobotany · Ex situ conservation · In situ conservation · Kurdistan Region of Iraq · Medicinal plants

---

R. H. Muhammed-Salih (✉)
Department of Biology, College of Education, University of Sulaimani, Sulaymaniyah, Kurdistan Region, Iraq
e-mail: rubar.salih@univsul.edu.iq

S. Jha et al. (eds.), *Genetic Diversity in Medicinal Plants*, Sustainable Development and Biodiversity 44, https://doi.org/10.1007/978-981-95-7281-6_11

11.1 Introduction

The Kurdistan Region is an autonomous area located in the north and northeast of Iraq. It shares borders with Syria to the west, Turkey to the north, and Iran to the east, with the rest of Iraq to its south. The region comprises four Kurdish-majority governorates: Erbil (its capital), Sulaymaniyah, Duhok, and Halabja (Fig. 11.1).

The Kurdistan Region of Iraq (KRG), situated in the mountainous northeast of the country, occupies a vital part of the Taurus-Zagros mountain range[1] and falls within the globally important Irano-Anatolian biodiversity hotspot, a region that contains approximately 6000 plant taxa, with around 2500 species endemic to its unique mountain systems (Al-Douri 2014). This ecological richness is underpinned by the diverse topography of the KRG, which spans a wide altitudinal gradient ranging from approximately 200 above sea level along the Sirwan River to 3,611 m at Cheekha Dar (Kurdish: Çîxî Derê, meaning "Black Tent"). Cheekha Dar, located along the Iraq-Iran border, is widely regarded as the highest peak in Iraq (CIA 2007). The second-highest mountain, Halgurd, reaches an elevation of approximately 3,607 m. collectively, this pronounced elevational variation gives rise to

Fig. 11.1 Administrative map of Kurdistan Region of Iraq

[1] The Zagros Mountains form an extensive mountain range stretching across Iran, northern Iraq, and southeastern Turkey. Extending for about 1600 km (990 miles), the range starts in northwestern Iran and follows along Iran's western border, reaching into parts of southeastern Turkey and northeastern Iraq. From there, it continues southeast until it reaches the Persian Gulf. The mountains cover the southern part of the Armenian Highlands and run across the western and southwestern Iranian Plateau, ending near the Strait of Hormuz (Doubek 2013).

a heterogenous landscape characterized by high mountain peaks, deep valleys, steep rocky slopes, and narrow gorges, which together contribute substantially to the region's ecological and floristic diversity. These features, coupled with precipitation levels ranging from 200 to 700 mm annually, result in a broad range of microclimates and ecological niches (Hajani et al. 2022). The region's vegetation is predominantly oak forest, which covers approximately 90% of its total forest area, further enhancing its ecological value (Amin et al. 2016).

Medicinal plants have been integral to traditional healthcare systems worldwide for millennia (Dolatkhahi et al. 2014), retaining significant economic and therapeutic value in global trade (Zhang et al. 2015). Kurdistan hosts exceptional plant diversity across its mountains, valleys, and plains, shaped by a varied climate and ecological niches (Amin et al. 2016; Mamdoh et al. 2023). The mountainous terrain of the KRG, with pronounced elevational gradients from arid lowlands to high-altitude pastures, marked variability in annual precepitation (approximately ~200–700 mm, Hajani et al. 2022), and diverse microclimates, contributes to the high diversity and abundance of medicinal plant species, underscoring the region's importance for both biodiversity and ethnobotanical research. Over 30 documented medicinal plant species have been recorded as endemic to the Kurdistan Region alone (Amin et al. 2016). Field studies in only four locations from the KRG have identified 206 wild medicinal plant species, with many of these species belonging to families with well-documented therapeutic uses (Hussain et al. 2024). This botanical richness is embedded in a long-standing tradition of plant-based healing, practiced for millennia across the broader Mesopotamian region. These plants are deeply interwoven into Kurdish cultural heritage, and are actively used in local therapies (Abdulwahid-Kurdi et al. 2023).

This region's ethnomedicinal heritage traces back to ancient Mesopotamian civilizations (Sumerian, Babylonian, Assyrian) (Al-Douri 2014), yet remains critically understudied. Traditional knowledge, termed GYA-DARMAN, transcends specialized healers; it is passed orally and preserved within families and local markets, such as Erbil's Qaysari Market (one of Iraq's oldest trading hubs, operational since the twelfth century) and Shaqlawa's herbal trade on Safeen Mountain, which highlight the central role of medicinal plants in both commerce and culture (Mati and de Boer 2011; Amin et al. 2016). Beyond health, plants carry symbolic and spiritual significance: for instance, *Prunus microcarpa* fruits are used both in remedies and in crafting prayer beads (Qadir and Khwarahm 2025). Seasonal practices such as drying plants for winter storage exemplify sophisticated resource management and community resilience strategies (Al-Douri 2014).

Traditional medicine serves as the primary healthcare resource for 70–80% of rural communities, especially in remote mountain villages, where access to modern medicines remains limited due to the accessibility and affordability of medicinal plants (WHO 2013; Shah et al. 2013). Remedies often involve decoctions and infusions of leaves, roots, and seeds (Abdulwahid-Kurdi et al. 2023). Women and elder healers play a central role in collecting, preparing, and trading plants, forming the backbone of local livelihoods. Thus, conserving these plants sustains not only biodiversity but also cultural identity, rural healthcare, and economic resilience.

The region's significance is further amplified by its role as a genetic reservoir for species with validated bioactivities, including antioxidant, antimicrobial, and anti-inflammatory properties confirmed in recent phytochemical studies (Mamdoh et al. 2024).

The purpose of this chapter is to provide a comprehensive synthesis of the conservation challenges and opportunities facing endangered medicinal plants in the KRG. It aims to (i) document the ecological, cultural, and economic significance of medicinal plants within this biodiversity-rich yet politically fragile region; (ii) identify the principal threats driving species decline, including climate change and over-harvesting; and (iii) critically evaluate current conservation measures while highlighting persistent policy, institutional, and socioeconomic gaps.

By integrating ethnobotanical heritage with ecological data and international conservation frameworks, the chapter seeks to outline evidence-based strategies that combine in situ and ex situ approaches, community participation, and emerging technologies. Ultimately, the chapter intends to link local realities with global biodiversity goals, offering a roadmap for safeguarding the KRG's medicinal plant heritage as both a biological necessity and a cultural imperative for future generations.

11.2 Threats to Medicinal Plant Biodiversity in the KRG

The medicinal flora heritage of the KRG is under severe threat, leading to genetic erosion and species loss. Climate change has intensified droughts, wildfires, and dust storms, while human pressures—including deforestation, unsustainable grazing, and rapid urbanization—accelerate habitat degradation (KBF 2020; Abdulwahid-Kurdi et al. 2023; Qadir and Khwarahm 2025). Political instability and inadequate governance structures further undermine the effectiveness of protection and conservation initiatives.

Agricultural expansion, deforestation, and urbanization in ecologically sensitive zones continue to degrade critical mountain ecosystems such as the Gara, Halgurd, Piramagrun, Bradost, and Korek ranges. In Soran, post-2003 urban sprawl has significantly affected natural plant sites (Abdulwahid-Kurdi et al. 2023). Spring habitats supporting species such as *Adiantum capillus-veneris* are particularly vulnerable (Naqishbandi 2014). Similarly, the diverse biogeography of areas like Safeen Mountain—characterized by *Crataegus* and *Quercus* trees, woody bushes, shrubs, and specialized grasses—faces growing development pressure (Abdullah et al. 2016). Traditional transhumance, while culturally significant, damages herbaceous layers and impedes the regeneration of alpine medicinal plants like *Primula vulgaris* and *Arnebia euchroma*. In addition, military activity and landmines restrict both conservation and research access, particularly in Qandil and Halgurd-Sakran national park. Although several areas have been designated as important plant areas (IPAs), enforcement remains weak. Ex situ conservation efforts are still in their infancy, with seed banking and botanical gardens at early stages of development.

The combined absence of robust data, underfunded institutions, and restricted access leaves many species unstudied and unmanaged.

Climate change poses additional stress. Reduced snowfall, shifting precipitation patterns, rising temperatures, and erratic rainfall disrupt phenology and pollination. Highland-adapted species such as *Colchicum kotschyi* and *Iris persica*, which depend on specific microclimates, are particularly at risk (Amin et al. 2016). On Safeen Mountain, where average temperatures range from 0 to 25 °C, Unique microclimates that sustain diverse flora from *Ranunculus* species to *Hypericum* varieties are highly vulnerable. MaxEnt (Maximum Entropy) models predict significant range contractions for *Thymus kotschyanus* and *Gundelia tournefortii* (Fig. 11.2), while high-altitude endemics face the risk of "mountaintop extinction" as their habitats shrink.

Overharvesting further threatens species survival. The increasing demand for medicinal plants, combined with unsustainable collection practices—especially of roots and bulbs—has put high pressure on species, such as *Ferula* spp., *Orchis* spp., *Allium* spp., *Nasturtium officinale*, *Dorema aucheri*, and *Lepidium persicum* and *Arum conophalloides* (Fig. 11.2). These are often collected during reproductive phases, accelerating population decline (Ahmad et al. 2023). Similarly, *Tulipa systole* faces unsustainable extraction, depleting populations faster than natural regeneration allows. Over 60% of medicinal plants sold in Erbil's markets are imported, signaling local depletion (Amin et al. 2016). In Koya, 36.7% of medicinal plants are wild-harvested, with decoction (57.1%) being the primary preparation method (Abdulrahman and Kayfi 2023).

Equally concerning is the erosion of traditional ecological knowledge. Elder healers—key custodians of ethnomedicinal wisdom—are aging (36.3% are between 51 and 60 years old in Koya), and only 27.3% of knowledge holders have a tertiary education. Among Qaysari vendors, just 2 of 18 learned from formal teachers, while others rely on familial or self-taught practices. Migration of younger generations, shifts toward urban employment, and reliance on modern healthcare further weaken knowledge transmission (Naqishbandi 2014; Abdulrahman and Kayfi 2023; Abdulrahman 2025).

The combined loss of medicinal plants and the oral knowledge system tied to them threatens both biodiversity and cultural heritage, leading to irreversible ecological and therapeutic losses. Alarmingly, few species from the KRG have been formally assessed for the IUCN Red List, underscoring the urgent need for comprehensive field surveys, ecological modeling, and community-based conservation initiatives.

11.3 Urgency of Conservation for Medicinal Plants

The conservation of medicinal plants in the KRG is not merely a biological necessity but also a cultural, social, and economic imperative. The ongoing crisis is both ecological and ethnobotanical: the loss of species threatens biodiversity and

Fig. 11.2 Some wild plants collected by local people and sold in the local market. (**a**, **b**) Young plants of *G. tournefortii*, (**c**) young plants of *A. conophalloides*, (**d**) young plants of *Peucedanum aucheri*, (**e**) fruits of *Pistacia eurycarpa*, fruit, (**f**) young plants of *A. eriophyllum*, and (**g**) rosette leaves and basal part of the stem of *Anchusa azurea*

ecosystem resilience, while the erosion of traditional knowledge undermines community identity and the local healthcare system.

The KRG harbors a significant share of Iraq's endemic flora. Of the country's 176 endemic species, 60% are already threatened, 9 are listed as Critically Endangered, 25 as Endangered, and 58 as Vulnerable (Youssef 2020). Several endemic species of the Kurdistan Region, such as *Allium eriophyllum* and *Astragalus carduchorum* (Fig. 11.2) (Amin et al. 2016), along with species documented on

Safeen Mountain, including *Pterocephalus nestorianus* and *Anemone biflora* (Abdullah et al. 2016), have yet to be investigated phytochemically despite their long-standing traditional importance. Without timely conservation action, these plants and the knowledge associated with them may vanish before their therapeutic potential is scientifically understood.

The decline of millennia-old Kurdish healing traditions further compounds this crisis. Once recorded in foundational texts such as Avicenna's *Canon of Medicine* but now fading (Naqishbandi 2014), these practices are both indigenous heritage and influenced by Arabic-Persian scholarly traditions and Islamic medicine. References to plants cited in the Qur′an, such as the olive and fig (Ghasemi et al. 2013; Shah et al. 2013), highlight the cultural and spiritual depth of this knowledge. Yet many applications remain undocumented, for instance, the reported use of *Syzygium aromaticum* for COVID-19 in Sulaymaniyah and Koya (Abdulrahman and Kayfi 2023).

Despite widespread local use, most endemic plants of the KRG remain phytochemically unexplored. Promising leads, like the anti-inflammatory embinin from *Iris persica* and 6-tuliposide A from *Tulipa systole* (Amin et al. 2016; Majeed and Salih 2023a), risk being lost before their value is fully realized. Ethnobotanical studies from Soran further underscore this potential, revealing high use value (HUV) and informant consensus factors (ICF) for species like *Olea europaea* and *Ficus carica*, plants that may hold significant promise for pharmacological development (Abdulwahid-Kurdi et al. 2023). Without conservation, this therapeutic and economic wealth may be irretrievably lost.

11.4 Endangered Medicinal Plants in the KRG

11.4.1 Key Species at Risk

The KRG hosts a remarkable diversity of medicinal plants, many of which are increasingly threatened by overharvesting, climate change, land-use change, and ecological specialization. These plant species are not only ecologically irreplaceable but also central to Kurdish cultural and medicinal traditions.

Crataegus spp.

Crataegus azarolus and *C. monogyna*, which are endemic to the Zagros oak forests, are valued for soil stabilization, biodiversity support, and cardiovascular treatments in Kurdish medicine (Radha and Khwarahm 2022; Hamad and Surucu 2024). Climate-induced stressors, wildfires, and unsustainable exploitation threaten their survival. Models project significant declines for *C. azarolus*, with one scenario estimating a 4.5–4.7% net habitat loss by 2100 under high emissions (Radha and Khwarahm 2022).

Ferula spp.
High-altitude perennials such as *Ferula asafetida* yield oleo-gum resin used for respiratory, digestive, and neurological disorders. Overharvesting, erosion-prone habitats, and land-use pressures place them at risk, despite their ecological roles in pollination and soil stabilization (Ebrahimian et al. 2013).

Orchis spp.
The unsustainable harvesting of tuberous orchids, specifically species such as *Orchis mascula,* is peimarily driven by the demand for salep. Salep, a flour derived from the tubers of these orchids, is a key ingredient in traditional Kurdish ice cream and sweet, creamy beverage. This practice leads to localized extirpation of orchid populations. Given their inheret sensitivity to environmental fluctuations and their restricted geographical distribution, these orchids serve as effective bioindicators (Wraith et al. 2020; Ahmad et al. 2023). However, persistent overcollection for cultural and culinary applications has critically endangered wild populations (Wraith et al. 2020).

Prunus arabica* and *P. microcarpa
Endemic to the eastern Zagros, these species are climate-sensitive. *P. arabica* is highly vulnerable to rising temperatures, while *P. microcarpa*, traditionally used as a mild laxative, is already listed as “Near Threatened” (Hamad and Surucu 2024; Qadir and Khwarahm 2025). Modeling forecasts a net habitat loss of over 2% for *P. microcarpa* by the end of the century, with new suitable habitats unlikely to offset these losses (Qadir and Khwarahm 2025).

Juniperus excelsa subsp. polycarpos
A keystone of Zagros juniper woodlands, showing dieback below 2400 m and poor regeneration under climate stress (Hamad and Surucu 2024).

Triglochin palustris (Juncaginaceae)
Representing the only member of the Juncaginaceae family in Iraq, this species is classified as “Critically Endangered.” It has been documented only twice near the Persian frontier. Major threats include agricultural expansion, pesticide use, urbanization, climate change, logging, drought, and overgrazing (Hama et al. 2024). The seeds have been used orally to treat ailments such as eczema.

Quercus aegilops
Dominant in the region’s oak forests, this species faces severe declines. Models project a 3.16–3.6% contraction of its range by 2070 under both low- and high-emission scenarios (Khwarahm 2020). As an ecosystem stabilizer, its decline has cascading effects on soil, water, and biodiversity.

Other notable endemic and threatened plants include *Echinops erinaceus*, *Pulicaria edmondsonii*, *Pteropyrum scoparium*, and *Stipa mandavillei*. Although not directly medicinal, they are critical to the biodiversity of the KRG (Hamad and Surucu 2024).

Medicinal plants, including genera such as *Crataegus*, *Prunus*, and *Quercus,* functions as ecological keystone species. These species play a critical role in stabilizing soils, supporting diverse wildlife populations, and regulating essential ecosystem processes, including hydrology and climate cycles. Their decline risks cascading ecological failures and the erosion of ecosystem services essential to human well-being.

The broader context is alarming. A 2021 review identified 176 endemic plant taxa in Iraq, of which 60% are threatened: 9 Critically Endangered, 25 Endangered, and 58 Vulnerable (Youssef 2020). Most of these taxa are concentrated in the Zagros Mountains within the Irano-Anatolian biodiversity hotspot, which is currently experiencing rapid biodiversity erosion. The lack of updated assessments leaves many listed as Data Deficient or Not Evaluated (Youssef 2020). Security challenges and limited access to remote areas further exacerbate the problem, leaving the true status of many populations unknown. Culturally significant plants such as *Mentha longifolia*, *Ocimum basilicum*, and *A. eriophyllum* (Fig. 11.2) remain unevaluated, emphasizing the need for localized IUCN Red List assessments.

11.4.2 Current Status: Distribution and Population Trends

Despite recent progress, the conservation status of endangered medicinal plants in the KRG remains poorly documented. Datasets are fragmented, and large knowledge gaps persist. The Kurdistan Botanical Foundation (KBF) has collected over 70,000 plant specimens, with the aim of preserving one million samples (KBF 2020). Earlier, Nature Iraq's Key Biodiversity Areas Project (2004–2008) surveyed more than 100 sites, some of which had not been revisited since 1979.

Globally, gaps remain stark; the IUCN Red List covers only a fraction of the world's estimated 350,000 vascular plants. Iraq is considered a "floristic knowledge coldspot," with fewer than one new species described per year over the last two decades, in contrast to dozens annually in Turkey and Iran (Baillie et al. 2004; Majeed and Salih 2023b; Bachman et al. 2019).

In the KRG, medicinal and endemic plants are concentrated in the Zagros Mountains, especially at elevations of 1000–1700 m, but their distribution extends further. Plains and steppes support species from families such as Asteraceae and Lamiaceae, which are widely utilized by local communities.

Riparian zones along rivers like the Great Zab and Sirwan also provide unique habitats. Localities such as Garmian, Rawandiz, and Amedi districts are particularly rich in both botanical diversity and ethnomedicinal knowledge. This broad distribution highlights the region's varied botanical heritage and the urgent need for wide-ranging conservation strategies.

11.5 Conservation Challenges and Opportunities

The KRG is facing a dual crisis concerning its medicinal plants; they are threatened by a mix of political, environmental, and cultural pressures, yet they also represent a crucial opportunity for both biological conservation and cultural preservation (Abdullah et al. 2022; Al-Obaidi et al. 2023). This region harbors more than 3300 plant species, many of which are concentrated in the Zagros Mountains, including a high proportion of endemic and medicinal taxa (Sulaiman et al. 2023). These plants are vital for traditional medicine, food security, and global biodiversity, but their future is at risk due to overharvesting, climate change, habitat loss, and the erosion of traditional ecological knowledge (Schippmann et al. 2002).

Recent ethnobotanical research highlights that wild food and medicinal plants such as *G. tournefortii*, *A. azurea*, *Malva sylvestris*, *Rheum ribes*, *Arum* spp., and *Dichoropetalum aromaticum* (Fig. 11.2) remain central to Kurdish diets and cultural practices, particularly in spring when they dominate local markets (Sulaiman et al. 2023). However, unsustainable harvesting methods, such as uprooting whole plants, are driving declines in local populations. Despite these threats, such species also represent significant opportunities for domestication and cultivation, which could strengthen food security and provide alternative livelihoods in conflict-affected and food-insecure areas (Sulaiman et al. 2023).

Iraq's broader biodiversity crisis further complicates the conservation context. Habitat degradation, deforestation, pollution, invasive species, and climate change have accelerated species loss, with Iraq classified among countries experiencing critically endangered ecoregions, particularly in the Mesopotamian marshes and the Zagros Mountain forest-steppe (Al-Obaidi et al. 2023). In the KRG, conservation medicine and One Health approaches are emerging as frameworks that link ecosystem health with human well-being, underscoring the need for integrated biodiversity protection (Al-Obaidi et al. 2023).

11.5.1 Political and Socioeconomic Barriers

The conservation of medicinal plants in the KRG is severely constrained by the region's history of conflict, weak governance, and socioeconomic instability. Ongoing political instability and enforcement gaps limit the effectiveness of environmental regulations, while poverty and food insecurity drive communities to overharvest and sell wild plants cheaply in local markets (Sulaiman et al. 2023). Moreover, younger generations are increasingly disconnected from traditional ecological knowledge, threatening the intergenerational transmission of sustainable foraging practices. Together, these barriers highlight the urgent need for coordinated conservation policies that integrate local communities, scientific research, and cross-border collaboration.

Chronic underfunding for environmental initiatives is a major barrier. Decades of war and political instability have weakened state institutions, leaving them with limited capacity for sustained biodiversity conservation projects (Al-Obaidi et al. 2023). Reforestation and anti-desertification campaigns have often collapsed due to a lack of investment, despite their urgency in combating soil erosion and land degradation. Restricted access to mined or militarized areas continues to impede floristic surveys, limiting baseline data for threatened species. Furthermore, political disputes—such as conflicts over oil revenues between the KRG and Baghdad—divert scarce resources away from conservation priorities. These systematic issues jeopardize the survival of culturally and economically important species such as *Q. aegilops*, which has long provided firewood, food (acorn), and traditional medicine but remains highly vulnerable to deforestation pressures (Sulaiman et al. 2023). By contrast, Turkey has invested in large-scale oak reforestation, and Iran has developed restoration programs in Zagros oak forests, underscoring the lack of comparable initiatives in the KRG.

Although Iraq and the KRG have established legal frameworks, including *Forest Law No. 10 of 2012*, enforcement is weak. Under-resourced institutions such as the "Forest Police" lack both capacity and political backing to curb illegal logging, overgrazing, and unsustainable harvesting practices (Al-Obaidi et al. 2023). Traditional customary rules that once regulated access to rangelands and wild plant resources have eroded, further enabling overexploitation. For instance, culturally important species like *G. tournefortii*, *A. azurea*, and *R. ribes* (Fig. 11.2) are now heavily uprooted for commercial sale, despite their ecological sensitivity, undermining natural regeneration (Sulaiman et al. 2023). Similarly, *Arum* spp., traditionally consumed after detoxification, is increasingly harvested unsustainably due to market demand. In neighboring Turkey, *G. tournefortii* harvesting is subject to stricter seasonal regulation, while in Iran, conservation agencies have piloted community-based management of *R. ribes* to prevent overharvesting. Weak institutional accountability and fragmented governance exacerbate these pressures, leaving biodiversity conservation largely dependent on local initiatives and sporadic NGO-led projects rather than consistent state enforcement.

Regional instability continues to hinder the establishment and effective management of protected areas. Military activities, unresolved security concerns, and landmine contamination prevent ecological monitoring in key biodiversity hotspots such as Qandil, Halgurd, Gara, and Bradost (Al-Obaidi et al. 2023). Even promising initiatives like Halgurd-Sakran National Park—the first formally declared national park in Iraq—progress slowly due to political fragmentation, lack of technical resources, and limited coordination between federal and regional authorities. As highlighted by recent ethnobotanical studies, the KRG remains a center of wild food and medicinal plant diversity, yet many culturally and economically important species are under-documented because vast areas remain inaccessible or unsurveyed (Sulaiman et al. 2023).

11.5.2 Economic Pressure and Overharvesting

Economic hardship continues to drive unsustainable harvesting of medicinal plants. During the 1990s embargo, widespread oak deforestation occurred as households turned to firewood for survival. Today, collectors frequently uproot whole plants, strip bark, and gather seeds to sell cheaply to intermediaries, repeating destructive practices that undermine regeneration. Recent incidents in Barzan illustrate the scale of the problem: illegal foragers were detained for uprooting wild spring herbs, with authorities forcing them to replant the 939 collected individuals (Hamad and Surucu 2024). Such poverty-driven exploitation creates a destructive cycle in which overharvesting depletes wild populations, reducing availability and pushing communities to rely even more heavily on extraction.

This pattern extends beyond medicinal plants to encompass the broader wild food plant economy that sustains many rural households. In the Kurdistan Region, young boys and girls, along with senior women and men, climb mountains to collect available wild food plants and sell them cheaply in local markets to shopkeepers (Sulaiman et al. 2023). The economic importance of this trade is substantial—some respondents in similar conflict-affected regions report that income generated from selling wild plants reaches up to 20% of their annual income, though security concerns often limit these practices (Sulaiman et al. 2023). Market prices reflect this economic dependency: species like *Arum* spp. sell for $3 USD/kg and *R. ribes* for $2.05 USD/kg in local markets, making them attractive income sources for vulnerable populations despite potential ecological costs.

The harvesting methods employed often prioritize immediate economic returns over sustainability. Species such as *A. eriophyllum* (Fig. 11.2) face particular pressure because "the whole plant is pulled with its bulb before reaching the flowering stage," preventing reproduction and seed set (Sulaiman et al. 2023). Similarly, *C. azarolus* shrubs are "largely cut to be used as firewood due to the fuel deficiency during the conflict," demonstrating how economic hardship drives multiple forms of plant exploitation beyond traditional harvesting.

This trend is not unique to the Kurdistan Region. Globally, wild harvesting remains the dominant supply source for most medicinal and aromatic plants, because it is cheaper than cultivation, despite its ecological costs. Globally, an estimated 52,000 plant species are used medicinally, and around 4160 of them—approximately 8%—are now threatened with extinction, largely due to overharvesting combined with habitat loss (Schippmann et al. 2002).

The regional conservation status reflects these pressures. Iraq's terrestrial ecosystem, which frames the country's biodiversity within major and minor Palearctic ecoregions, is increasingly classified as "Critical/Endangered" due to multiple threats including habitat transformation and unsustainable resource extraction (Al-Obaidi et al. 2023). In the Kurdistan Region, overharvesting of flagship medicinal species such as *R. ribes* (wild rhubarb), *G. tournefortii* (Kingr), and *M. sylvestris* (mallow) (Fig. 11.2) is common. *R ribes*, listed as Near Threatened on the IUCN Red List, is often uprooted before completing its life cycle, reducing regeneration capacity. The species appears in multiple regional ecoregions and is economically valuable in local markets, intensifying collection pressure. *G. tournefortii*, considered Data Deficient globally but locally declining, is dug out entirely during the spring harvest for market sale, removing reproductive individuals and halting seed production. In Syria, the species has reached "critically low" abundance with "almost absent" market presence, while in the Kurdistan Region, it remains widespread but faces increasing pressure (Sulaiman et al. 2023). Even *M. sylvestris*, categorized as Least Concern globally, faces heavy local extraction pressures that threaten wild stands in the KRG due to its popularity for food and medicine—the species maintains "high" estimated environmental sustainability and "widespread" market occurrence, but continued intensive harvesting could alter this status.

11.5.3 Ecological and Environmental Challenges

The KRG faces a convergence of ecological stressors—climate variability, land degradation, water scarcity, and anthropogenic pressures—that threaten biodiversity and accelerate the decline of medicinal plants. These forces act as "threat multipliers," compounding vulnerabilities in fragile mountain and steppe ecosystems, with particularly severe impacts on the biodiverse Zagros Mountain range, the region's ecological backbone.

11.5.3.1 Climate Change as a Threat Multiplier

Climate change is intensifying ecological risks across the KRG. Iraq has already experienced a rise of 1–2 °C in average temperatures since the 1970s, with projections of up to +2 °C by 2050 (World Bank 2021; Binder et al. 2022). Although northern Iraq receives more rainfall than central and southern regions, it is not immune: increasing drought frequency and erratic precipitation have altered ecological baselines (Abdulrahman 2025).

The Northern Zagros forests are especially vulnerable. Modeling studies predict additional warming of +3.7 °C by 2070 and reduced rainfall, leading to habitat losses of 4–5% for key species such as *P. microcarpa* and *Q. aegilops* (Khwarahm 2020; Qadir and Khwarahm 2025). Mountain ecosystems, home to many endemic plants with narrow altitudinal ranges, face disproportionate risks. Species near their climatic limits of survival, particularly those at higher elevations, are highly sensitive to warming, a trend reflected globally in tree line shifts and alpine flora declines (Kelly and Goulden 2008).

11.5.3.2 Drought and Water Scarcity

Drought is one of the most severe and immediate threats in the region. Major droughts in 1998–1999, 2007–2008, and 2009 devastated rain-fed agriculture and natural vegetation communities (Hameed et al. 2018; Price 2018). Within the broader Iraqi water crisis, the KRG suffers from both climatic drivers and anthropogenic pressures such as upstream damming and mismanagement, which threaten to push water availability below the poverty line of 1000 m^3 per person annually (Price 2018; Yousuf et al. 2018).

Prolonged dry seasons and irregular rainfall reduce soil moisture and groundwater recharge, impairing plant survival. In mountain ecosystems, steep slopes cause rapid runoff and poor retention, further limiting water-dependent medicinal species (Al-Kakey et al. 2024; Hamad and Surucu 2024). Forests dominated by *Quercus libani* and *Q. infectoria*, which sustain diverse medicinal understory plants, are increasingly stressed by reduced precipitation and higher evapotranspiration rates (Haidari and Rezaei 2013).

11.5.3.3 Soil Erosion and Land Degradation

Soil degradation is a critical ecological threat across Iraq, with up to 90% of land at risk of desertification and 45% of farmland already degraded (Abdulrahman 2025). In the KRG, steep mountainous terrain intensifies erosion risks. A MEDALUS-based assessment in Harrir (Erbil Governorate) shows about 68 km^2 under active degradation, with vast areas classified as fragile and critical (Hamad and Surucu 2024). Key drivers include overgrazing, deforestation, unsustainable land use, and intensified climate extremes.

Soil loss directly undermines medicinal plant habitats. Species richness in forest ecosystems is closely tied to soil quality and stability (Haidari and Rezaei 2013). Upland specialists such as *Ferula* spp. rely on developed soils for establishment; erosion reduces seedling recruitment and long-term population viability (Ebrahimian et al. 2013; Gaznayee et al. 2024). The herbaceous layer, home to the highest diversity of medicinal plants in Northern Zagros forests, is especially vulnerable to disturbance.

11.5.3.4 Invasive Species

Invasive alien plants are an emerging but understudied threat. Environmental disturbance, coupled with climate change, facilitates their spread into fragile mountain habitats. Aggressive invaders such as *Brassica tournefortii* and *Centaurea solstitialis* can displace native flora, while *Hypericum perforatum* increases fire risks in degraded areas (Ahmad et al. 2021; Radha and Khwarahm 2022). The risk is acute for endemic and rare medicinal species with narrow ecological niches. Invasive species, with their broad tolerances and strong competitive abilities, often thrive under conditions of disturbance and warming, further marginalizing vulnerable native taxa.

11.5.3.5 Pollution and Ecological Contamination

Pollution compounds ecological stress across the KRG. Iraq's history of conflict and industrial activity has left widespread contamination from oil residues, sulfur fires, heavy metals, and chemical pollutants (Zwijnenburg and Postma 2017; Abdulrahman 2025). Mining, urbanization, and intensified agriculture continue to add new pressures.

Pollutants not only reduce vegetation density and degrade habitats but also alter the phytochemical composition of medicinal plants, compromising both their ecological role and medicinal value. Contaminated soils and water bodies limit the ecological stability needed for sustaining rare and endemic species (Al-Kakey et al. 2024).

These ecological and environmental challenges are deeply interconnected. Climate warming accelerates soil erosion and reduces water availability, while pollution and invasive species exploit the instability that follows. For medicinal plants, this creates cascading vulnerabilities where multiple stressors overwhelm adaptive capacity.

The northern Zagros forests, which support the richest medicinal plant diversity in the region, are particularly at risk. Their ecological integrity depends on complex interactions among trees, shrubs, and herbaceous layers (Haidari and Rezaei 2013). The disruption of these relationships by compounding stressors threatens not only individual species but also the entire ecosystem, and with them, centuries of traditional medicinal knowledge.

11.5.4 Cultural and Knowledge Gaps

The conservation of endangered medicinal plants in the KRG faces threats not only from ecological pressures but also from cultural erosion, insufficient documentation, and the weak integration of traditional medicine into formal healthcare. While herbal remedies remain deeply rooted in Kurdish heritage, modernization and environmental change are steadily eroding both the practices and the knowledge base.

Historically, ethnobotanical knowledge in Kurdistan was transmitted orally across generations, embedded within family life and community practices rather than restricted to specialists (Amin et al. 2016). Yet today, urbanization, migration, and ecological degradation have fragmented this cultural continuity (Al-Douri 2014). Species such as *G. tournefortii*, valued both as food and medicine, illustrate the dual cultural and ecological losses caused by unsustainable harvesting. Naqishbandi's (2014) survey in Erbil recorded 32 medicinal species still in use, though with little standardization, while Amin et al. (2016) documented over 82 species sold in Erbil's Qaysari Market with more than 130 distinct uses. Much of this knowledge, however, remains undocumented and is vulnerable to disappearance. Without systematic recording and scientific validation, traditional therapeutic practices risk vanishing alongside rare plants.

Low public awareness compounds these threats by driving destructive practices such as uprooting entire plants, overharvesting, and unregulated trade. Conservation initiatives, such as those by KBF, have attempted to promote sustainable practices, but outreach remains limited, particularly among younger, urban populations. Commonly used species like *Matricaria chamomilla* (chamomile) for stomach disorders and *A. capillus-veneris* (maidenhair fern) for kidney stones continue to be heavily collected without sustainability measures (Amin et al. 2016). At the same time, many Iraqi herbalist shops rely on imported plants, often stored under poor conditions, which both displace native species in local markets and raise consumer safety concerns (Al-Douri 2014). Storage public education programs linking medicinal plant use to cultural heritage and sustainable practices are therefore urgently needed.

Traditional healers remain central to community health but face significant gaps in knowledge and training. Surveys show that nearly 70% of herbalists in Iraq have not completed primary education, relying mainly on inherited rather than scientific knowledge (Al-Douri 2014). This lack of formal training undermines safe dosage practices, increases risks of misidentification, and limits the effective integration of traditional medicine into modern healthcare. For example, while *Prangos peucedanifolia* is widely used for kidney disorders and *Hypericum oerforatum* for inflammation, preparation methods and dosages vary significantly between practitioners (Amin et al. 2016).

From a scientific perspective, the flora of Kurdistan remains underexplored. The genus *Iris* is well represented with 12 species, yet only *Iris persica* and *I. germanica* have been systematically studied for phytochemistry. Recent work on *Iris postii* has revealed antioxidant and anti-inflammatory compounds, lending support to its traditional uses (Amin et al. 2021). However, most endemic and rare species remain scientifically unstudied, widening the gap between traditional practices and modern validation. Weak regulation of herbal markets further exacerbates this problem, enabling adulteration, mislabeling, and unsafe long-term use (Al-Douri 2014).

Bridging these cultural and scientific gaps requires targeted interventions, including licensing systems, standardized training programs, and collaborative ethnobotanical–pharmacological research. Such measures would not only safeguard Kurdish ethnomedicinal heritage but also strengthen healthcare and biodiversity conservation outcomes across the region.

11.6 Conservation Strategies

The effective conservation of endangered medicinal plants in the KRG requires a comprehensive, multi-pronged strategy that integrates in situ and ex situ approaches, supported by robust policy frameworks, community participation, socioeconomic incentives, and emerging technologies. Evidence from international experience demonstrates that medicinal plants can serve as effective entry points for conservation initiatives. Their cultural and livelihood significance often motivates communities to protect habitats that might otherwise be overlooked (Hamilton 2004, 2006). Aligning KRG initiatives with the Global Strategy for Plant Conservation (GSPC) further strengthens the region's capacity to contribute to global biodiversity targets (Timpte et al. 2018). The GSPC provides a framework of 5 objectives and 16 targets, which have guided global plant conservation since 2002 and remain central to the post-2020 biodiversity framework (Royal Botanical Gardens, Kew 2020).

11.6.1 In Situ Conservation

In situ conservation, protecting plant species within their natural habitats, remains the cornerstone of biodiversity protection. It safeguards ecological processes, maintains genetic diversity, and ensures the long-term survival of wild medicinal plant populations. The GSPC's Target 7 emphasizes that at least 75% of known threatened species should be conserved in situ (Royal Botanical Gardens, Kew 2020), a benchmark that KRG conservation strategies should adopt.

11.6.1.1 Protected Areas and National Parks

Protected areas are central to in situ conservation. Iraq has initiated several efforts to create national parks and reserves, including the Halgurd-Sakran National Park in the Kurdistan Region, a biodiversity hotspot encompassing diverse mountainous habitats (Al-Douri 2014). The KBF has been instrumental in conducting floristic and vegetation surveys in mountainous regions such as Qaywan, Qaradagh, Azmar-Goyzha, Halgurd, and Sakran to provide the scientific basis for conservation planning.

Another milestone has been the designation of Key Biodiversity Areas (KBAs) in Iraq, including four priority sites in the KRG (Sulaymaniyah, Erbil, and Duhok governorates), which meet IUCN criteria based on species vulnerability and irreplaceability (Hamad and Surucu 2024). Recommendations from this initiative include granting these sites formal legal protection and establishing site-level support councils to oversee management and restoration.

Kew's Tropical Important Plant Areas (TIPA) program provides a valuable model, as it has enabled countries like Guinea and the British Virgin Islands to integrate plant-focused priorities into protected area management (Royal Botanical Gardens, Kew 2020). Adopting a similar framework in the KRG could help integrate threatened medicinal species into national park planning. Despite progress, challenges persist. Many protected areas face weak legal enforcement and political instability, limiting effective conservation (Hamad and Surucu 2024). Moreover, medicinal plants are rarely included as explicit conservation targets, echoing global findings that such species often fall outside formal management plans (Schippmann et al. 2002). To address this gap, medicinally important species such as *C. azarolus*,

Q. aegilops, and *P. microcarpa* must be explicitly incorporated into the protected area framework. This approach aligns with global priorities under the Global Strategy for Plant Conservation (Target 7), which calls for at least 75% of threatened plant species to be conserved in situ by 2020 (Timpte et al. 2018).

11.6.1.2 Community-Based and Local Healer-Led Conservation

Local communities, and especially traditional healers, are pivotal for in situ conservation. In the KRG, medicinal plants are deeply embedded in cultural and healthcare practices, with knowledge passed orally across generations (Abdullah et al. 2022). Recognizing and strengthening this knowledge is essential.

Community-led conservation offers culturally grounded and practical approaches. The KBAs initiative highlighted the importance of engaging local stakeholders, paralleling successes in projects like China's Great Green Wall, where community participation was critical (Hamad and Surucu 2024). By contrast, Iraq's failed greenbelt projects show that top-down approaches are unsustainable without grassroots buy-in.

The GSPC's Target 13 specifically calls for maintaining and increasing indigenous and local knowledge related to plant use (Royal Botanical Gardens, Kew 2020). Engaging traditional healers in codesigning conservation actions, such as identifying priority species, documenting traditional knowledge, promoting sustainable harvesting, and creating community-managed conservation zones, can transform potential exploiters into conservation stewards (Schippmann et al. 2002). These strategies are consistent with global lessons that link livelihood security with biodiversity protection, particularly where communities depend heavily on herbal medicine and plant-based incomes (Hamilton 2004, 2006).

International examples provide useful parallels. In Nepal, community-based harvesting systems improved sustainability; in Uganda's Bwindi Forest, adaptive governance allowing controlled access reduced conflict; and in WWF-Nepal, integrating sustainable harvesting with health services improved both conservation and local welfare (Schippmann et al. 2002). For the KRG, defining clear rights and responsibilities within conservation zones is vital. Without secure tenure and participation, regulations risk alienating communities and undermining objectives.

11.6.2 Ex Situ Conservation Strategies in the KRG

Ex situ conservation, preserving species outside their natural habitats, provides essential backup systems. In regions facing political instability, habitat degradation, and climate stress, it is an indispensable complement to in situ measures. Seed banks, botanical gardens, and research collections safeguard irreplaceable genetic resources and support restoration initiatives.

The Plant Conservation Report 2014 underscores that ex situ strategies are critical to achieving global targets; at least 75% of threatened plant species should be conserved in ex situ collections, preferably in the country of origin (Timpte et al. 2018; Royal Botanical Gardens, Kew 2020).

11.6.2.1 Seed/Gene Banks: Preserving Genetic Resources for the Future

Seed banks are among the most effective and cost-efficient tools for conserving plant diversity (Schippmann et al. 2002). In the KRG, a major step forward has been the establishment of the first regional Gene Bank in Sulaymaniyah (Hawari Shar Park), founded by the European Union and implemented by the Food and Agriculture Organization (FAO). This facility conserves wild relatives of crops, vegetables, and indigenous plants, building resilience in Iraq's agricultural and ecological systems (Food and Agriculture of the United Nations 2025).

The gene bank emphasizes traits such as drought tolerance, a crucial adaptation under climate change, and has the potential to repatriate genetic resources historically exported from Kurdistan (Hamad and Surucu 2024). In parallel, the KBF has set the ambitious goal of collecting one million specimens from each native species to protect biodiversity from extinction risks (KBF 2020).

Partnerships with initiatives such as the Millennium Seed Bank (Royal Botanical Gardens, Kew) link local conservation to global repositories (Royal Botanical Gardens, Kew 2020). However, medicinal plants are still underrepresented in seed collections worldwide (Hamilton 2004, 2006). By prioritizing medicinal and endemic flora, the Sulaymaniyah Genebank fills a critical global gap.

11.6.2.2 Botanical Gardens: Living Repositories and Educational Hubs

Botanical gardens are equally vital in ex situ conservation, serving as living collections for research, education, and restoration (Cunningham 2001). In the KRG, the University of Duhok Botanical Garden already manages ex situ collections of endangered species and maintains a seed bank for rare flora. Meanwhile, the KBF has approved plans for a botanical garden in Sulaymaniyah, designed to complement seed/gene banking and herbarium work (KBF 2020).

Kew's experience shows that botanical gardens can function as "living laboratories," offering controlled environments for propagation trials, public engagement, and awareness-building (Royal Botanical Gardens, Kew 2020). They also contribute to recovery programs and promote the integration of traditional knowledge into propagation practices (Hamilton 2006). Although gardens often fall short in capturing full genetic variation, they remain key tools for awareness-raising and conservation in biodiversity hotspots such as the KRG (Schippmann et al. 2002).

11.6.2.3 Strategic Importance and Future Directions

In the dynamic and often unstable context of the KRG, ex situ conservation has progressed from a complementary role to a central strategy for safeguarding plant diversity. Seed banks and botanical gardens serve as critical repositories of genetic material, ensuring the preservation of valuable plant resources for ecological restoration, scientific research, and sustainable development.

For long-term effectiveness, ex situ conservation initiatives in the KRG must place particular emphasis on medicinal species within seed and gene banks as well as botanical gardens, thereby addressing their persistent underrepresentation at the Millennium Seed Bank. These initiatives also require secure, sustained funding and the integration of ex-situ facilities into national biodiversity policies.

By aligning local initiatives with global biodiversity goals and ensuring meaningful community involvement, the KRG can safeguard its unique medicinal plant heritage for future generations.

11.6.3 Policy and Legal Frameworks: Aligning National Priorities with International Commitments

The KRG has enacted several legislative measures aimed at environmental protection. These include the Forest Law (Law No. 10 of 2021), which prohibits tree cutting and outlines penalties for violations, and the law on Agricultural Quarantine, which regulates the import and circulation of plant materials to prevent pest spread. In addition, the Ministry of Agriculture Law (Law No. 1 of 2007) articulates broader goals for sustainable agricultural and natural resource management.

However, enforcement remains deeply problematic. The Forest Law, for instance, is frequently violated with limited legal repercussions. Enforcement bodies such as the “Forest Police” are severely under-resourced and face systematic challenges, including frequent reports of legal logging and large-scale attacks on tree populations (Hamad and Surucu 2024). Furthermore, national legislation often fails to account for the subsistence-level use and economic importance of medicinal plants, leading to insufficient protection frameworks. The lack of formal recognition and regulation for traditional medicinal plant use further exacerbates the issue, with many wild species being harvested unsustainably in the absence of legal safeguards (Schippmann et al. 2002).

There are some positive developments. Initiatives such as the ongoing formulation of a national strategy for establishing protected areas and the National Biodiversity Strategy and Action Plan (NBSAP 2015–2020) signal progress. Yet their success depends on political will, sustainable financing, and institutional capacity, all of which remain limited.

At the international level, Iraq is a signatory to key multilateral environmental agreements, including the Convention on Biological Diversity (CBD), the Convention on International Trade in Endangered Species of Wild Fauna and Flora (CITES), and the UN Convention to Combat Desertification (UNCCD). These treaties provide comprehensive frameworks for plant conservation, with specific instruments such as the CBD's Global Strategy for Plant Conservation and SDG Target 15, both of which aim to halt biodiversity loss and promote sustainable land management.

Nevertheless, translating these commitments into effective action within the KRG has been severely constrained. Weak institutions, pervasive corruption, limited government collaboration with conservation organizations, such as the KBF, and broader political instability in post-conflict Iraq have all contributed to a pronounced implementation gap (Hamad and Surucu 2024). The legacies of war and terrorism-related environmental degradation have further eroded conservation infrastructure and institutional capacity (Al-Douri 2014).

The post-2020 Biodiversity Framework underscores the need for plant-specific milestones and indicators to ensure that plant conservation is central to biodiversity strategies (Royal Botanical Garden, Kew 2020). For the KRG, aligning domestic legal frameworks with these global standards is vital to closing the gap between international commitments and regional realities.

The persistent disconnect between well-designed policies and their practical enforcement highlights the need for a paradigm shift in conservation governance in the KRG. Policy efforts must extend beyond ratification toward robust and accountable implementation. This requires strengthening enforcement mechanisms, ensuring adequate funding for institutions, combating corruption, and embedding transparency in governance. It also entails recognizing and integrating traditional ecological knowledge and customary laws into statutory frameworks, while formalizing the role of traditional healers' community-based conservation within legal structures.

A hybrid governance approach, combining statutory law with customary practices and international mandates, offers a culturally grounded yet legally enforceable pathway forward. International partners, including multilateral agencies and NGOs, should prioritize capacity building for enforcement agencies, legal institutions, and civil society actors in order to address the persistent implementation gap (Fig. 11.3).

Without such integrated and pragmatic measures, even the most ambitious conservation laws will remain aspirational, leaving the KRG's rich medicinal plant heritage increasingly vulnerable to escalating threats.

11.6.4 Emerging Technologies for Biodiversity Monitoring and Plant Conservation in the KRG

Emerging technologies hold transformative potential for biodiversity conservation, especially in ecologically sensitive and data-deficient regions such as the KRG. These tools complement traditional methods, enabling rapid, cost-effective, and large-scale ecological monitoring, species identification, and strategic conservation planning.

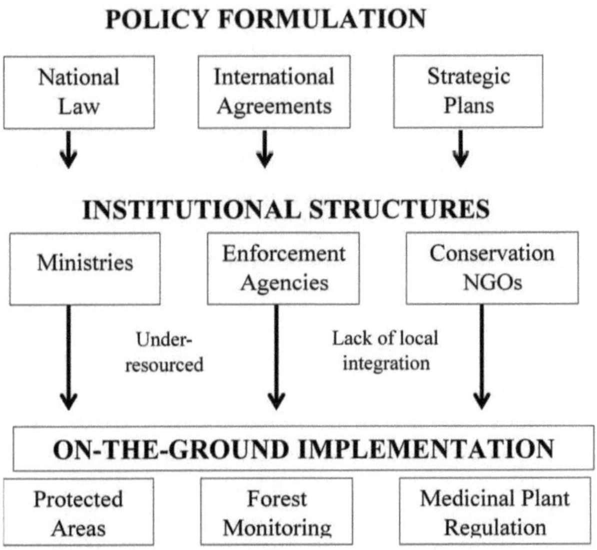

Fig. 11.3 Connection between conservation policy formulation and implementation in the KRG

11.6.4.1 DNA Barcoding and Molecular Approaches

DNA barcoding has revolutionized biodiversity assessment by providing rapid, accurate, and cost-effective species identification, even for cryptic taxa or fragmented plant material. In plants, widely used loci include the ITS region, *mat*K, and *rbc*L (Salih et al. 2017; Hama et al. 2024). This approach is particularly valuable in the KRG, where traditional taxonomic expertise is limited, and political instability and rugged terrain have left many species undocumented. In a region often referred to as a "floristic knowledge cold-spot," this method accelerates species cataloging and provides a critical foundation for conservation planning (Farooq et al. 2020).

Recent advances highlight that DNA barcoding is not limited to species identification but extends to broader biodiversity applications. The development of reference DNA barcode libraries (e.g., BOLD, GenBank) and the rise of metabarcoding and environmental DNA (eDNA) allow researchers to characterize entire communities, monitor invasive species, and even reconstruct past ecosystems from soil or permafrost samples (Pompanon and Shehzad 2025; Koblmüller 2023). This is particularly relevant for the KRG, where field surveys are constrained, and cryptic taxa may otherwise be overlooked.

Moreover, DNA barcoding has proven crucial in herbal medicine authentication, with alarming global evidence of high adulteration rates in herbal powders and plant-based products (Koblmüller 2023). In the KRG, where medicinal plant markets play a vital role in livelihoods and healthcare, barcoding supports monitoring of endangered species, detection of substitution or adulteration, and improved traceability and quality assurance. By creating economic incentives for sustainable harvesting and discouraging fraudulent practices, DNA barcoding not only strengthens conservation goals but also safeguards public health and traditional knowledge systems.

11.6.4.2 Remote Sensing and GIS Applications

Remote sensing (RS), particularly when integrated with geographic information systems (GIS), provides unparalleled capacity to track ecological and land-use changes across large, inaccessible areas. In the KRG, where surveys are often limited by rugged terrain, landmines, and sociopolitical constraints, RS-GIS technologies are indispensable (Hussein et al. 2025).

Key applications include mapping vegetation cover with high-resolution satellite imagery such as Landsat OLI and RapidEye. Studies conducted in Duhok (2014) and across Erbil, Sulaymaniyah, Halabja, and Garmian (2015) demonstrated that vegetation covers only 8–12% of the KRG's total area, with the densest coverage occurring between 500 and 1500 m elevation (Ministry of Agriculture and Water Resources, Kurdistan Region, Iraq 2015). These datasets establish critical baselines for conservation planning while underscoring the vulnerability of lowland and urbanized districts. Their ability to be updated at decadal intervals also offers a cost-effective alternative to conventional ground-based surveys.

Beyond ecology, remote sensing has also been applied to archaeological and cultural heritage monitoring, further highlighting its interdisciplinary utility for regional planning, environmental governance, and sustainable development.

11.6.4.3 Species Distribution Models (SDMs)

Species distribution models (SDMs), such as Maxent, when combined with GIS, provide robust tools for predicting both current and future species ranges under climate change scenarios. These models assess habitat suitability based on environmental variables, thereby identifying key areas of species distribution and priority areas for conservation action (Hamad and Surucu 2024).

In the KRG, Maxent models applied to species such as *C. azarolus* and *C. monogyna* have revealed potential range contractions and altitudinal shifts linked to climate change. Integrating RS-GIS datasets with SDMs offers practical insights for designing climate-resilient strategies. For instance, findings on vegetation cover and elevation-specific distribution can guide the designation of protected areas, prioritization of forest management zones, and establishment of ecological corridors that facilitate species migration under shifting climate envelopes.

11.7 Opportunities and Future Directions

Although the conservation of medicinal plants in the KRG faces serious challenges, a growing range of opportunities offers promising pathways toward sustainable, science-based, and culturally grounded strategies. The future of medicinal plant conservation in the region will depend on strengthening partnerships, integrating traditional knowledge with modern science, adopting climate-resilient approaches, and harnessing technological innovations.

11.7.1 Strengthening Local and Global Partnerships

Successful conservation requires inclusive collaboration among diverse stakeholders, including traditional healers, academic institutions, NGOs, local communities, and international organizations. Local NGOs such as Nature Iraq and the KBF are at the forefront, conducting biological surveys, raising environmental awareness, developing plant databases, and leading gene banking and botanical garden initiatives. Universities such as the University of Sulaimani, the University of Duhok, and the American University of Sulaymaniyah contribute through research, capacity building, and joint projects with international partners, including National Geographic.

Cross-border cooperation is particularly critical due to the transboundary nature of biodiversity in the Irano-Anatolian hotspot and the Zagros Mountains. Shared endemic species and common threats, such as desertification and climate change, necessitate joint monitoring, habitat protection, and coordinated conservation measures with neighboring countries including Iran, Turkey, and Syria. Models such as the *Hawizeh Marsh Ramsar Site* and partnerships with global institutions like the *Millennium Seed Bank* and the *Global Partnership for Plant Conservation* highlight the benefits of coordinated conservation and knowledge exchange.

11.7.2 Integrating Traditional Knowledge with Modern Science

The KRG's ethnobotanical heritage holds centuries of empirically refined knowledge on medicinal plant use, sustainable harvesting, and ecological stewardship. Yet, rapid urbanization, cultural change, and generational shifts threaten the survival of this oral tradition.

Systematic ethnobotanical documentation is essential, encompassing surveys with traditional healers and quantitative analyses such as Use Value and Informant Consensus Factor. Initiatives such as the Iraq Medicinal Plant Database represent key steps toward preserving this knowledge and embedding it within conservation planning. Combining traditional knowledge with scientific approaches, such as species distribution modeling, genetic analysis, and phytochemical studies, can strengthen conservation strategies while ensuring cultural relevance and local ownership.

Community engagement in afforestation, habitat restoration, and cultivation programs is also vital. Collaborative research between scientists and local ethno-scientists can guide experimental cultivation and sustainable management, ensuring that conservation efforts remain socially equitable, culturally appropriate, and ecologically resilient.

11.8 Conclusion

The conservation of endangered medicinal plants in KRG represents both an urgent ecological challenge and a cultural imperative. As part of the Irano-Anatolian biodiversity hotspot, the KRG harbors exceptional botanical diversity and an ethnomedicinal heritage deeply rooted in Mesopotamian civilization and Kurdish tradition. Yet this wealth is increasingly imperiled by climate change, habitat degradation, overharvesting, political instability, and the erosion of traditional ecological knowledge.

The evidence presented throughout this chapter underscores a dual crisis: biodiversity loss threatens the survival of endemic and ecologically significant plant species, while cultural erosion undermines traditional healing practices and local healthcare systems. Weak institutional enforcement, fragmented governance, and underfunded conservation initiatives exacerbate these risks, leaving many species unprotected and knowledge systems undocumented.

Despite these challenges, promising opportunities exist. In situ strategies, such as strengthening protected areas, integrating community-based management, and explicitly prioritizing medicinal plants in conservation planning, offer culturally grounded and ecologically effective solutions. Ex situ measures, including seed and gene banking as well as botanical gardens, are becoming increasingly vital in the KRG's volatile context, safeguarding genetic resources and enabling future restoration. Emerging technologies, particularly DNA barcoding, remote sensing, and species distribution modeling, provide powerful tools to bridge critical data gaps and inform climate-resilient strategies.

Equally important is the alignment of KRG conservation efforts with international frameworks such as the Global Strategy for Plant Conservation, CITES, and the post-2020 Global Biodiversity Framework. Achieving this requires a paradigm shift from policy formulation to robust and transparent implementation, coupled with meaningful integration of traditional knowledge systems and active community engagement.

Ultimately, conserving medicinal plants in the KRG is not only about protecting biodiversity but also about preserving cultural identity, sustaining rural livelihoods, and safeguarding public health. By linking local stewardship with global commitments, investing in institutional capacity, and fostering cross-border collaboration, the region can transform its endangered medicinal flora from a symbol of vulnerability into a foundation for resilience and sustainable development. Without such integrative and pragmatic action, however, the KRG risks irreversible ecological and ethnobotanical losses with profound consequences for future generations.

References

Abdullah FO, Hussain FH, Sardar AS, Vita-Finzi P, Vidari G (2016) Phytochemistry and ethnopharmacology of medicinal plants used on Safeen Mountain in the Kurdistan Region of Iraq. Nat Prod Commun 11(12):1934578X1601101236. https://doi.org/10.1177/1934578X1601101236

Abdullah FO, Hussain FH, Sardar AS, Gilardoni G, Thu ZM, Vidari G (2022) Bio-active compounds from Teucrium plants used in the traditional medicine of Kurdistan Region, Iraq. Molecules 27(10):3116. https://doi.org/10.3390/molecules27103116

Abdulrahman SA (2025) Iraq's desertification: the complication of environmental security issues. Nat Built Soc Environ Health 1(1):81–98. https://doi.org/10.63095/NBSEH.25.720252

Abdulrahman MD, Kayfi S (2023) Medicinal plants traditionally used in Koya, Kurdistan Region of Iraq. AIP Conf Proc 2554(1):090004. https://doi.org/10.1063/5.0103752

Abdulwahid-Kurdi SJ, Abdulwahid MJ, Magaji U, Aghwan ZA, Atan R, Hamadamin KA (2023) Medicinal plants used in Soran district Kurdistan Region of Iraq, an ethnobotanicals study. J Pharm Pharmacogn Res 11(1):1–32. https://doi.org/10.56499/jppres22.1484_11.1.1

Ahmad SA, Ahmad NS, Salih S (2021) Invasive alien species in Iraq. In: Pullaiah T, Lelmini MR (eds) Invasive alien species: observations and issues from around the world. Wiley, pp 126–150. https://doi.org/10.1002/9781119607045.ch18

Ahmad SA, Salih RM, Hama SY (2023) Morphological characterization and conservation status for the herbarium specimens of Orchidaceae in Kurdistan Iraq. Ibn AL-Haitham J Pure Appl Sci 36(2):55–84. https://doi.org/10.30526/36.2.3031

Al-Douri NA (2014) Some important medicinal plants in Iraq. Int J Adv Herb Altern Med 2(1):10–20

Al-Kakey O, Dunger V, Al-Mukhtar M, Gaznayee HA (2024) Evaluating meteorological drought and its impacts on vegetation cover and surface water in the Headwater of Little Zab River Basin. Revista de Ciências Agroveterinárias 23(4):739–750. https://doi.org/10.5965/223811712342024739

Al-Obaidi RM, Hussein RH, Shaker AS, Ameen QA, Ararat K, Khidhir ZK, Arif SK, Al-Sada IHA (2023) Conservation medicine and biodiversity in Iraq. In: Aguilar-Marcelino L, Abbas RZ, Khan A, Younus M, Saeed NM (eds) One health triad, vol I. Unique Scientific Publishers, Faisalabad, pp 209–219. https://doi.org/10.47278/book.oht/2023.31

Amin HI, Ibrahim MF, Hussain FH, Sardar AS, Vidari G (2016) Phytochemistry and ethnopharmacology of some medicinal plants used in the Kurdistan Region of Iraq. Nat Prod Commun 11(3):1934578X1601100306. https://doi.org/10.1177/1934578X1601100306

Amin HI, Hussain FH, Najmaldin SK, Thu ZM, Ibrahim MF, Gilardoni G, Vidari G (2021) Phytochemistry and biological activities of Iris species growing in Iraqi Kurdistan and phenolic constituents of the traditional plant Iris postii. Molecules 26(2):264

Bachman SP, Field R, Reader T, Raimondo D, Donaldson J, Schatz GE, Nic Lughadha E (2019) Progress, challenges and opportunities for Red Listing. Biol Conserv 234:45–55. https://doi.org/10.1016/j.biocon.2019.03.002

Baillie J, Hilton-Taylor C, Stuart SN (eds) (2004) IUCN Red List of Threatened Species: a global species assessment. IUCN, Gland

Binder L, Sedova B, Rüttinger L, Tomalka J, Gleixner S, Hauf Y, Lange S, Viehoff A, Hornsby M, Dieffenbacher J, Pereira LC (2022) Climate risk profile Iraq. https://publications.pik-potsdam.de/rest/items/item_27727_3/component/file_27728/content

Central Intelligence Agency (CIA) (2007) The World Factbook: Iraq. https://www.cia.gov

Cunningham AB (2001) Applied ethnobotany: people, wild plant use and conservation. Routledge. https://doi.org/10.4324/9781849776073

Dolatkhahi M, Dolatkhahi A, Nejad JB (2014) Ethnobotanical study of medicinal plants used in Arjan–Parishan protected area in Fars Province of Iran. Avicenna J Phytomed 4(6):402–412. PMID: 25386404; PMCID: PMC4224954

Doubek J (2013) Zagros folded zone (map). Wikimedia Commons. Available at: https://commons.wikimedia.org/wiki/File:Zagros_Folded_Zone.jpg. Accessed 20 Sept 2025

Ebrahimian V, Bagheri R, Mohseni M, Poormirzaei A (2013) Ecological and phenological study on Ferulago angulate in the Hezar Mountains and Bondar Henza, Kerman, Iran. J Rangeland Sci 3(4):277–285. https://sid.ir/paper/623467/en

Farooq Q, Shakir M, Ejaz F, Zafar T, Durrani K, Ullah A (2020) Role of DNA barcoding in plant biodiversity conservation. Sch Int J Biochem 3(3):48–52. https://doi.org/10.36348/sijb.2020.v03i03.002

Food and Agriculture Organization of the United Nations (2025) First gene/seed bank in Kurdistan region to preserve Iraq's agricultural heritage. FAO. https://www.fao.org/neareast/news/details/first-gene-seed-bank-in-kurdistan-region-to-preserve-iraq-agricultural-heritage/en

Gaznayee HAA, Sabr HA, Galalaey AMK, Shakir SB, Omer FO, Kareem SHS, Ababakr SH, Mohammad RK, Ahmad FM (2024) Integrate terrain variables and rapid eye satellite in vegetation indices, for identifying forest cover area and density: a case study in mountainous Iraqi Kurdistan Region (IKR). Passer J Basic Appl Sci 6(1):107–114. https://doi.org/10.24271/psr.2024.192393

Ghasemi PA, Momeni M, Bahmani M (2013) Ethnobotanical study of medicinal plants used by Kurd tribe in Dehloran and Abdanan districts, Ilam province, Iran. Afr J Tradit Complement Altern Med 10(2):368–385. https://doi.org/10.4314/ajtcam.v10i2.24

Haidari M, Rezaei D (2013) Study of plant diversity in the Northern Zagros forest (case study: Marivan region). Int J Adv Biol Biomed Res 1(1):1–0. http://ijabbr.com/upload/IJABBR-2013-1001.pdf

Hajani E, Shajee K, Kaleel F, Abdulhaq H (2022) Characteristics of changes in rainfall data in the Kurdistan Region, Iraq. Arab J Geosci 15(6):509. https://doi.org/10.1007/s12517-022-09809-6

Hama SY, Salih RHM, Ahmad SA (2024) Morphological study and conservation status of Juncagenaceae in Kurdistan Region, Iraq. Ibn AL-Haitham J Pure Appl Sci 37(3):81–87. https://doi.org/10.30526/37.3.3688

Hamad KO, Surucu A (2024) Land degradation sensitivity and desertification risk in Harrir region, northern Iraq. Heliyon 10(5):e27123. https://doi.org/10.1016/j.heliyon.2024.e27123

Hameed M, Ahmadalipour A, Moradkhani H (2018) Apprehensive drought characteristics over Iraq: results of a multidecadal spatiotemporal assessment. Geosciences 8(2):58. https://doi.org/10.3390/geosciences8020058

Hamilton AC (2004) Medicinal plants, conservation and livelihoods. Biodiversity and Conservation 13(8): 1477–1517. https://doi.org/10.1023/B:BIOC.0000021333.23413.42

Hamilton A (2006) Plant conservation: an ecosystem approach, 1st edn. Routledge. https://doi.org/10.4324/9781849772181

Hussain FHS, Majedi S, Abdulsattar Faraj T, Ozdemir M, Ahamad J, Karim TH, Dilawer Issa K, Q Mustafa M (2024) Documentation of ethnomedicinal uses of wild plants growing in Kodo Mountain by Kurdish Tribe of Iraq. J Chem Soc Pak 46(4):168–184. https://doi.org/10.52568/001432/JCSP/46.02.2024

Hussein ZE, Alhamadani OZ, Hussein WA (2025) Utilizing artificial neural networks for 2D coordinate transformation: a case study on the reference system of Iraq. Iraqi J Sci 66:955–969. https://doi.org/10.24996/ijs.2025.66.2.30

Kelly AE, Goulden ML (2008) Rapid shifts in plant distribution with recent climate change. Proceedings of the National Academy of Sciences (PNAS) 105(33): 11823–11826. https://doi.org/10.1073/pnas.0802891105

Khwarahm NR (2020) Mapping current and potential future distributions of the oak tree (*Quercus aegilops*) in the Kurdistan Region, Iraq. Ecol Process 9(1):1–16. https://doi.org/10.1186/s13717-020-00259-0

Koblmüller S (2023) DNA barcodes for evolution and biodiversity. Diversity 15(9):1003. https://doi.org/10.3390/d15091003

Kurdistan Botanical Foundation (KBF) (2020) Kurdistan Botanical Foundation. Kurdipedia. https://www.kurdipedia.org/default.aspx?q=20200723143508386090&lng=8

Majeed H, Salih R (2023a) Two new subspecies of *Crocus biflorus* L. (Iridaceae) from the Iraqi Kurdistan Region using morphological and molecular methods. Italus Hortus 30(1):52–67. https://doi.org/10.26353/j.itahort/2023.1.5267

Majeed HO, Salih RHM (2023b) Iris zagrica subsp. penjwenica; Iris zagrica subsp. warteica; two new subspecies from the Iraqi Kurdistan Region belonging to Iris section Zagrica. Passer 5(2):404–409. https://doi.org/10.24271/PSR.2023.409295.1357

Mamdoh A, Jamil LA, Hussain FHS (2023) Systematic review on traditional use, phytochemical and pharmacological properties of some medicinal plants growing in Kurdistan Region–Iraq. Eurasian J Sci Eng 9(2):4. https://doi.org/10.23918/eajse.v9i2p4

Mamdoh AN, Jamil LA, Hussain FHS (2024) Comparative analysis of medicinal plants from Kurdistan Region-Iraq for chemical constituents and bioactivities. Plant Sci Today 11(2):861–868. https://doi.org/10.14719/pst.3582

Mati E, de Boer H (2011) Ethnobotany and trade of medicinal plants in the Qaysari Market, Kurdish Autonomous Region, Iraq. J Ethnopharmacol 133(2):490–510. https://doi.org/10.1016/j.jep.2010.10.023

Ministry of Agriculture and Water Resources, Kurdistan Region, Iraq (2015) Mapping and estimating vegetation coverage in Iraqi Kurdistan Region using remote sensing and GIS. Report, Ministry of Agriculture and Water Resources, Kurdistan Region, Iraq

Naqishbandi A (2014) Plants used in Iraqi traditional medicine in Erbil-Kurdistan Region. Zanco J Med Sci 18(3):811–815. https://doi.org/10.15218/zjms.2014.0038

Pompanon F, Shehzad W (2025) DNA barcodes to characterize biodiversity. Encyclopédie de l'Environnement. http://www.encyclopedie-environnement.org/?p=693

Price RA (2018) Environmental risks in Iraq. K4D Helpdesk Report. Institute of Development Studies, Brighton

Qadir RY, Khwarahm NR (2025) Current and projected future spatial distribution patterns of *Prunus microcarpa* in the Kurdistan Region of Iraq. Biology 14(4):358. https://doi.org/10.3390/biology14040358

Radha KO, Khwarahm NR (2022) An integrated approach to map the impact of climate change on the distributions of *Crataegus azarolus* and *Crataegus monogyna* in Kurdistan Region, Iraq. Sustainability 14(21):14621. https://doi.org/10.3390/su142114621

Royal Botanic Gardens, Kew (2020) Kew and the global strategy for plant conservation. Retrieved from http://kew.org/sites/default/files/2020-01/Kew%20and%20the%20Global%20Strategy%20for%20Plant%20Conservation.pdf

Salih RHM, Majeský Ľ, Schwarzacher T, Gornall R, Heslop-Harrison P (2017) Complete chloroplast genomes from apomictic *Taraxacum* (Asteraceae): identity and variation between three microspecies. PLoS One 12(2):e0168008. https://doi.org/10.1371/journal.pone.0168008

Schippmann U, Leaman DJ, Cunningham AB (2002) Impact of cultivation and gathering of medicinal plants on biodiversity: global trends and issues. In: FAO (ed) Biodiversity and the ecosystem approach in agriculture, forestry and fisheries. FAO, pp 142–167

Shah A, Marwat SK, Gohar F, Khan A, Bhatti KH, Amin M, Din NU, Ahmad M, Zafar M (2013) Ethnobotanical study of medicinal plants of semi-tribal area of Makerwal & Gulla Khel (lying between Khyber Pakhtunkhwa and Punjab provinces), Pakistan. Am J Plant Sci 4:98–116. https://doi.org/10.4236/ajps.2013.41015

Sulaiman N, Aziz MA, Stryamets N, Mattalia G, Zocchi DM, Ahmed HM, Manduzai AK, Shah AA, Faiz A, Sõukand R, Polesny Z (2023) The importance of becoming tamed: wild food plants as possible novel crops in selected food-insecure regions. Horticulturae 9(2):171. https://doi.org/10.3390/horticulturae9020171

Timpte M, Marquard E, Paulsch C (2018) Analysis of the strategic plan 2011–2020 of the Convention on Biological Diversity (CBD) and first discussions of resulting recommendations for a post-2020 CBD framework – full study report. Institute of Biodiversity – Network (IBN). https://doi.org/10.13140/RG.2.2.27623.09124

World Bank (2021) World development indicators. https://databank.worldbank.org/source/world-development-indicators

World Health Organization (WHO) (2013) WHO traditional medicine strategy: 2014–2023. World Health Organization, Geneva. https://www.who.int/publications/i/item/9789241506096

Wraith J, Norman P, Pickering C (2020) Orchid conservation and research: an analysis of gaps and priorities for globally Red Listed species. Ambio 49(10):1601–1611. https://doi.org/10.1007/s13280-019-01306-7

Youssef S (2020) Endemic plant species of Iraq: from floristic diversity to critical analysis review. J Duhok Univ 23(2):90–105. https://doi.org/10.26682/ajuod.2020.23.2.12

Yousuf MA, Rapantova N, Younis JH (2018) Sustainable water management in Iraq (Kurdistan) as a challenge for governmental responsibility. Water 10(11):1651. https://doi.org/10.3390/w10111651

Zhang S, Wang R, Zeng W, Zhu W, Zhang X, Wu C, Song J, Zheng Y, Chen P (2015) Resource investigation of traditional medicinal plant *Panax japonicus* (T. Nees) CA Mey and its varieties in China. J Ethnopharmacol 166:79–85. https://doi.org/10.1016/j.jep.2015.02.051

Zwijnenburg W, Postma F (2017) Living under a black sky: conflict pollution and environmental health concerns in Iraq. Pax, Utrecht, pp 1–36

Chapter 12
Conservation by Use and Genetic Diversity of Medicinal Plants: Case Studies in Southern Brazil Atlantic Forest

Tiago Montagna, Adelar Mantovani, Alexandre Mariot, Alexandre Siminski, and Maurício Sedrez dos Reis

Abstract The Atlantic Forest harbors one of the planet's greatest biodiversities, characterized by a high degree of plant endemism. However, historical and contemporary processes of land use and occupation have resulted in the loss of about 75%

T. Montagna (✉)
Núcleo de Pesquisas em Florestas Tropicais, Universidade Federal de Santa Catarina (UFSC), Florianópolis, Santa Catarina, Brazil

Pós-Graduação em Recursos Genéticos Vegetais, UFSC, Florianópolis, Santa Catarina, Brazil
e-mail: t.montagna@ufsc.br

A. Mantovani
Núcleo de Pesquisas em Florestas Tropicais, Universidade Federal de Santa Catarina (UFSC), Florianópolis, Santa Catarina, Brazil

Laboratório de Engenharia Florestal, Pós-Graduação em Engenharia Florestal, Pós-Graduação em Produção Vegetal, Universidade do Estado de Santa Catarina, Lages, Santa Catarina, Brazil
e-mail: adelar.mantovani@udesc.br

A. Mariot
Núcleo de Pesquisas em Florestas Tropicais, Universidade Federal de Santa Catarina (UFSC), Florianópolis, Santa Catarina, Brazil

Drimys Agroambiental LTDA, Florianópolis, Santa Catarina, Brazil

A. Siminski
Núcleo de Pesquisas em Florestas Tropicais, Universidade Federal de Santa Catarina (UFSC), Florianópolis, Santa Catarina, Brazil

Centro Reforma de Restauração Ecológica, UFSC, Curitibanos, Santa Catarina, Brazil

Núcleo de Estudos em SocioAgroBiodiversidade, Pós-Graduação em Ecossistemas Agrícolas e Naturais, UFSC, Curitibanos, Santa Catarina, Brazil

M. S. dos Reis
Núcleo de Pesquisas em Florestas Tropicais, Universidade Federal de Santa Catarina (UFSC), Florianópolis, Santa Catarina, Brazil

Núcleo de Estudos em SocioAgroBiodiversidade, Pós-Graduação em Ecossistemas Agrícolas e Naturais, UFSC, Curitibanos, Santa Catarina, Brazil

S. Jha et al. (eds.), *Genetic Diversity in Medicinal Plants*, Sustainable Development and Biodiversity 44, https://doi.org/10.1007/978-981-95-7281-6_12

of its original cover. Despite this, human populations have inhabited the biome for millennia, using a wide variety of native plants for medicinal purposes. This chapter aims to discuss how the use of native medicinal species can serve as a tool for conserving genetic diversity, biodiversity, and the traditional knowledge associated with them. Three case studies are presented: *Varronia curassavica*, *Ilex paraguariensis*, and *Drimys brasiliensis*. For each species, aspects related to medicinal use, ecology, geographic distribution, raw material acquisition, and genetic diversity are addressed. The integrated analysis of these studies shows that managing these species can simultaneously contribute to maintaining genetic variability and generating income for extractivist communities. Furthermore, the potential of medicinal plants in the Atlantic Forest goes beyond economic and therapeutic aspects, as their use is embedded in traditional landscape management systems that promote the conservation of forest cover, plant diversity, and ecological interactions. The expansion and strengthening of these systems depend on integrated actions and scientific support, which are essential for characterizing management practices and valuing the traditional knowledge associated with the conservation of the Atlantic Forest's biological diversity.

Keywords *Drimys brasiliensis* · *Ilex paraguariensis* · *Varronia curassavica* · Non-timber forest products

12.1 Southern Brazil Atlantic Forest

The Southern Region of Brazil is characterized by a remarkable diversity of ecosystems, predominantly within the Atlantic Forest and the Pampa (Temperate Grassland), as well as in core areas of the Cerrado in the state of Paraná. Among these, the Atlantic Forest, the primary focus of this chapter, covers around 394,922 km^2, accounting for roughly 68% of the southern Brazilian territory. This biome encompasses nearly the entire state of Paraná (98%), all of Santa Catarina (100%), and a portion of Rio Grande do Sul (37%) (IBGE 2024). Renowned for its exceptional biodiversity, the Atlantic Forest hosts over 20,000 species of angiosperms and numerous endemic species, ranking it among the five global regions with the highest levels of endemism among vascular plants (Myers et al. 2000).

However, this biome is critically threatened due to the extensive reduction and fragmentation of its original vegetation cover, a consequence of Brazil's historical land use and occupation processes. The region encompasses the country's largest industrial, agricultural, and silvicultural centers, and is home to approximately 70% of the Brazilian population (MapBiomas 2023). Consequently, the Atlantic Forest and its associated ecosystems have been reduced to about 25% of their original extent, with most remaining fragments consisting of secondary forest formations (Fundação SOS Mata Atlântica 2023; MapBiomas 2023). This habitat loss and fragmentation have led to the classification of the Atlantic Forest as one of the 36 global biodiversity hotspots, regions recognized for their exceptional biological richness and high levels of threat (Myers et al. 2000).

Historically, human populations have inhabited the territory of southern Brazil for at least 5000 years (Dias and Jacobus 2003; Bueno and Dias 2015). Throughout this period, they have engaged in various interactions with local ecosystems and their species, particularly plants. According to Clement et al. (2021), approximately 2913 plant species in the Atlantic Forest are considered useful, such as for medicinal, nutritional, cultural, or industrial purposes, reflecting a long-standing tradition of human use. Although the utilization of native species has declined with increased occupation and land-use intensification—especially from the nineteenth century onward—many of these species continue to hold significant cultural importance, contributing to food security and income generation (Coradin et al. 2011). It is important to note that many of these relationships are linked to self-consumption, subsistence practices, religious rituals, or extractivism, which often render these resources "invisible" or undervalued in official statistics (Siminski et al. 2016).

The number of native species utilized and their significance for local and traditional communities, including family farmers, are continually increasing and being reassessed. Research has recognized these social actors as key agents in the protection of their surrounding natural environments (Pereira and Diegues 2010). In this context, conservation strategies that incorporate biocultural approaches can expand their scope by promoting the preservation of both biodiversity and local cultures, thereby fostering integrated socio-ecological systems that benefit both nature and communities (Gavin et al. 2015). Within this framework, Levis et al. (2024), building on concepts from the United Nations Environment Programme (UNEP), introduce examples of "Hope Spots" within terrestrial socio-ecological systems. These are defined as areas capable of significantly enhancing socio-ecological resilience, where local communities and public engagement can be strategically combined with advanced science, engineering, and technology—including geospatial tools—to maximize efforts in conserving vulnerable biodiversity (Levis et al. 2024).

From this perspective, biodiversity conservation through sustainable use is a scientific approach that posits that local and traditional populations, including family farmers, utilize resources from natural ecosystems in a sustainable manner (Diegues et al. 2001; Bensusan 2006; Vieira-da-Silva et al. 2009; Reis et al. 2018; Tagliari et al. 2021). Conservation through use can support the maintenance of ecological processes within ecosystems while also enabling families to utilize their resources for subsistence and income generation.

In southern Brazil, research emphasizes the importance of management systems and extractive practices, such as the harvesting of yerba mate (*Ilex paraguariensis* A.St.-Hil.) beneath Brazilian pine (*Araucaria angustifolia* (Bertol.) Kuntze) in areas of *caívas* and *faxinais* (Reis et al. 2018; Marques et al. 2019), as well as extractive systems for *Pinhões*—pine nuts (Zechini et al. 2012, 2018; Magnanti et al. 2017; Tagliari et al. 2023). Yerba mate and pine nuts are among the most significant non-timber forest products (NTFPs) in terms of economic value and production volume in southern Brazil (IBGE 2023). They are integral components of agroforestry systems organized by local communities, which value and conserve these species due to their long history of use, nutritional importance, and sociocultural significance.

Considering legislation and social demand for resources, there is a need to understand the systems in use and to study aspects of the autoecology of the species in demand. This approach aims to evaluate systems that promote the conservation of forest landscapes on properties while simultaneously supporting in situ (on-farm) conservation of these resources. Therefore, understanding the dynamics of landscape change, population structure, and genetic diversity of a given species—particularly those targeted for resource extraction—enables the characterization of the effects of local strategies and practices that favor conservation through sustainable use (Reis et al. 2018; Milanesi et al. 2013, 2021).

In this chapter, three case studies of native Atlantic Forest species with medicinal uses will be highlighted: *erva baleeira* (*Varronia curassavica* Jacq.), yerba mate (*Ilex paraguariensis*), and *cataia* (*Drimys brasiliensis* Miers) (Fig. 12.1). These species have well-documented traditional medicinal applications, supported by scientific studies that confirm the efficacy of their active compounds. An additional important aspect of convergence is that the social groups associated with the use of these plants represent potential partners in the conservation of the Atlantic Forest. This partnership occurs both through traditional knowledge related to the management of these species and ecosystems, and through the presence of significant remnant populations of these plants on private lands, which underscores the importance of involving local communities in biodiversity conservation.

12.2 Case Studies

*12.2.1 Erva Baleeira (*Varronia curassavica *Jacq.)*

12.2.1.1 Medicinal Use

Varronia curassavica, popularly known in Brazil as *erva baleeira*, is a shrub species native to the Atlantic Forest with medicinal properties. The species was historically used by whale hunters to treat wounds and bruises, hence its popular name *erva baleeira* (Kinupp 2007), roughly translated as "whaleboat herb." The plant is valued for its anti-inflammatory, analgesic, and wound-healing effects (Falkenberg 2011; Sertié et al. 2005), as well as its antibacterial properties (Michielin et al. 2009). Antirheumatic and hemostatic effects have also been attributed to *V. curassavica* (Akisue et al. 1983). The species is also used in traditional preparations, such as infusions in alcohol (*garrafadas*) and teas, and is employed by local healers and folk practitioners, reflecting a rich body of associated knowledge and practices (Penteado 2020). The compound α-humulene, found particularly in the leaves, is considered the primary agent responsible for the plant's anti-inflammatory properties (Medeiros et al. 2007). Recent research shows that the species produces

Fig. 12.1 Studied species. (**a**) Flowering individual of *Drimys brasiliensis*, (**b**) bark exploitation on *D. brasiliensis*, (**c**) flowering individual of *Varronia curassavica*, (**d**) detailed view of *V. curassavica* fruits, (**e**) individual of *Ilex paraguariensis*, (**f**) detailed view of floral buds of *I. paraguariensis*. (Two photographs of *Varronia curassavica* are reused in this figure with permission from Marcia Patricia Hoeltgebarum)

"cordialin A," a compound with potential for cervical cancer treatment (Baveloni et al. 2024).

This scientific validation contributed to the development of phytomedicines, such as Acheflan®, a pharmaceutical product based on *V. curassavica* extract developed by a consortium of Brazilian universities in partnership with Aché Laboratórios Farmacêuticos S.A., with an investment exceeding R$ 15 million over 7 years of research (Ereno 2005). The patented herbal medicine derived from the plant generates significant revenue: in 2012, annual sales surpassed US$ 9.4 million, according to data from Aché (Hoeltgebaum and Reis 2017). Given its recognized medicinal potential, the species was included in the National List of Medicinal Plants of Interest to the Brazilian Unified Health System (RENISUS) (Ministério da Saúde do Brasil 2009). Today, various products derived from the species, such as teas, oils, extracts, and creams, can be found in both formal and informal markets.

12.2.1.2 Occurrence Area and Ecological Aspects

Varronia curassavica exhibits a broad distribution range, occurring from Central America to southern Brazil. In Brazil, it spans from Ceará to Rio Grande do Sul, forming dense clusters in humid coastal areas (Smith 1970; Lorenzi and Matos 2008). The species also inhabits *Cerrado* (savanna) and *Caatinga* (dry woodland) ecosystems (Rodal and Nascimento 2002; Mendes et al. 2015), as well as *campo rupestre* (rocky grassland) vegetation (Gastauer et al. 2012).

The species is hermaphroditic and presents self-incompatibility mechanisms such as heterostyly and protogyny, promoting outcrossing (Opler et al. 1975; Ganders 1979; Hoeltgebaum et al. 2018). It displays a generalist pollination syndrome, with flowers visited by insects from 6 orders and 24 families, predominantly Hymenoptera (50% of visits), Diptera (29%), and Lepidoptera (16%) (Hoeltgebaum et al. 2018). The floral morphology facilitates access to short-proboscid visitors, particularly favoring pollination by flies, small bees, and ants (Hoeltgebaum et al. 2018). Seed dispersal occurs mainly by birds and ants. Omnivorous birds (e.g., *Turdus amaurochalinus*, *Mimus saturninus*) swallow fruits whole, enabling long-distance dispersal, while the ant *Acromyrmex striatus* disperses seeds over short distances, often to nutrient-rich microsites, enhancing genetic connectivity and seedling establishment across different environments (Hoeltgebaum et al. 2018).

Varronia curassavica exhibits positive population growth in natural environments, with recruitment rates exceeding mortality and a reverse "J-shaped" size distribution (Hoeltgebaum et al. 2020). Individuals present a clumped spatial distribution, primarily in dry lowlands and dune transition zones, yet lack significant spatial genetic structure, indicating random genotype distribution due to outcrossing systems and high gene flow mediated by generalist pollinators (Hoeltgebaum et al. 2020). These traits demonstrate how environmental heterogeneity shapes both demographic dynamics and genetic diversity in populations from *restinga*, highlighting the species' high adaptive capacity in such variable ecosystems (Hoeltgebaum et al. 2020).

12.2.1.3 Extractivism, Management, and Raw Material Provenance

The use of *V. curassavica* in folk medicine and its pharmaceutical importance have made the plant an increasingly exploited resource. Extractivism remains an important method for obtaining its raw plant material, based on traditional knowledge, especially among local communities that use the plant in their traditional health systems (Penteado 2020). However, although extractivism can become a source of income for fishing and farming families, unmanaged intensification may lead to overexploitation, reducing populations, diminishing genetic variability, and jeopardizing the species' survival (Sebbenn et al. 2008).

The industrial use of *V. curassavica*, on the other hand, requires raw material with greater uniformity to enhance the efficiency and consistency of industrial processes. In this regard, characterizing propagation, cultivation, harvesting, and storage practices for the species is crucial to support large-scale production. To date, studies have described its in vitro propagation (Vieira et al. 2013) and stem cuttings (Mendes et al. 2014), but it can also be propagated via seeds. Different cultivation practices have also been studied, including varying light intensities (Lapa 2006; Feijó et al. 2014) and planting densities (Montanari and Magalhaes 2004), as well as the influence of chemical fertilization on biomass production (Magalhães et al. 2004).

Other studies aim to characterize the influence of environmental, ecological, and genetic factors on essential oil production and medicinal compounds. For example, the essential oil yield of *V. curassavica* increases with higher light irradiance, and the oil composition varies in compound quantities and concentrations when extracted from plants grown under different light intensities (Feijó et al. 2014). Additionally, while its essential oil content does not change depending on leaf collection time, two monoterpenes were detected at specific hours: sabinene was detected only in collections at 6:00 a.m., while aristolene was not detected at 3:00 p.m. (Queiroz et al. 2016). The production of *V. curassavica* in agroforestry systems with cacao and rubber trees has proven viable, with the species showing the highest dry biomass and essential oil yields when harvested at 18 months of cultivation (Barazetti et al. 2022). Finally, genetic variability has proven to influence its essential oil composition (Marques et al. 2019; Queiroz et al. 2019).

Collectively, these findings indicate multiple options for obtaining raw material, suited for either more artisanal or more industrial processes, as well as an important variation in the yield and composition of *V. curassavica* essential oil depending on environmental, management, and genetic aspects. This variability is highly significant, as it enables product diversification and expands the potential for cultivation across diverse environments.

12.2.1.4 Genetic Diversity and Conservation

Studies using different molecular markers (isozymes and microsatellites) indicate that *V. curassavica* displays moderate to high genetic diversity, with heterozygosity levels consistent with widely distributed outcrossing species (Hoeltgebaum et al. 2015; Hoeltgebaum and Reis 2017). Hoeltgebaum et al. (2015) found an average genetic diversity (H_E) of 0.111 among three populations from Santa Catarina Island, with fixation indices (F_{IS}) ranging from −0.273 to 0.222, revealing heterozygote excess in two populations. Genetic divergence between populations was moderate (F_{ST} = 0.079), indicating some degree of subdivision. In another study of populations from three distinct municipalities, Hoeltgebaum and Reis (2017) demonstrated that the species is autotetraploid and exhibits high genetic variability (H_E = 0.665–0.748), with low differentiation between populations and high gene flow (Nm = 4.87), consistent with its outcrossing system and efficient dispersal. No significant spatial genetic structure was detected, suggesting substantial intrapopulation gene flow, despite the clumped distribution.

The high genetic variability depicted implies that these populations could be prioritized for conservation and could serve as genetic sources for future breeding programs. For instance, the essential oil from different accessions of *V. curassavica* presented different activity when tested to control white spot disease in fish (Nizio et al. 2018). This result shows how the genetic and phenotypic variability of the species can be used to develop new medicines. It is also important to emphasize that the species is polyploid, a trait sometimes associated with enhanced productivity and industrial performance in plant species (Sattler et al. 2016; Trojak-Goluch et al. 2021). However, habitat fragmentation and overharvesting threaten these critical genetic reservoirs. Therefore, while protected areas are vital to preserve existing diversity, implementing sustainable management strategies, including integration into agroecosystems, is equally necessary, as well as exploring the variability of the species.

*12.2.2 Yerba Mate (*Ilex paraguariensis *A. St. Hil.)*

12.2.2.1 Medicinal Use

Ilex paraguariensis, popularly known as yerba mate, has been used as a tonic and stimulant beverage since the times of Indigenous peoples of South America (Edwin and Reitz 1967). When the Spanish arrived in what is now the state of Paraná in 1554, they encountered the Guarani people regularly consuming a beverage called caá-i, prepared with dried and ground leaves mixed with water and sipped through a bamboo straw (Aranha 1967). Its use dates back to even earlier times, with yerba mate leaves found in Inca tombs in Peru (Edwin and Reitz 1967).

The medicinal potential of yerba mate goes beyond its stimulant activity, as a wide diversity of nutritional and bioactive compounds is found in its leaves. Multiple vitamins—A, B1, B2, B3, B5, B6, C, and E—as well as minerals, particularly calcium, magnesium, and phosphorus; polyphenols, alkaloids (notably caffeine), and saponins are present (Song et al. 2025). The same review indicates that the various compounds found in yerba mate exhibit confirmed biological activities: antioxidant, anti-inflammatory, anticancer, and anti-obesity.

Yerba mate infusion intake increased tumor latency and reduced tumor volume in rats (Santiano et al. 2023). The same study also reported that the infusion reduced viability, proliferation, and adhesion of three tumor cell lines in vitro, compared to the control group (water). Another preclinical study demonstrated the potential of aqueous yerba mate extract in obesity treatment, due to its ability to regulate genes associated with inflammation, white adipose and liver metabolism, and hepatic damage and fibrosis (Tolouei et al. 2025). Aqueous yerba mate extract also promotes satiety and reduces body weight in rats (Hussein et al. 2011). Most of the studies cited involve in vitro or preclinical trials using rodents, underscoring the need for further research in humans.

12.2.2.2 Occurrence Area and Ecological Aspects

Yerba mate has an occurrence range of approximately 540,000 km^2, distributed across Brazil, Bolivia, Paraguay, Uruguay, and Argentina (Edwin and Reitz 1967; Oliveira and Rotta 1985). Brazil encompasses approximately 80% of its distribution, with natural populations mainly found in the states of Santa Catarina, Paraná, Rio Grande do Sul, and Mato Grosso do Sul, and to a lesser extent in São Paulo and Minas Gerais (Edwin and Reitz 1967). It is a small tree, typically reaching 3–5 m in height under cultivation, though it can grow up to 30 m (Carvalho 2003). A typical understory species of the Araucaria Forest, yerba mate thrives in shaded environments (Edwin and Reitz 1967), although it can also be cultivated in full light, as in commercial plantations (Penteado Jr and Goulart 2019).

Yerba mate is a dioecious species, with male and female flowers occurring on separate individuals. Its reproductive phenology is relatively variable, with flowering occurring from September to December and fruiting between December and April (Edwin and Reitz 1967; Sousa et al. 2003; Pires et al. 2014; Mattos 2015). This species interacts with multiple faunal components. Its flowers are visited and pollinated by insects from the orders Coleoptera, Diptera, Hymenoptera, and Lepidoptera, representing at least 18 families (Pires et al. 2014; Mattos 2015). Yerba mate fruits are consumed by at least 14 bird species (Colussi and Prestes 2011; Mattos 2015), which are the main seed-dispersal agents for the species.

12.2.2.3 Extractivism, Management, and Raw Material Provenance

Yerba mate is currently Brazil's leading non-timber forest product in terms of annual production and extraction volume (425,800 tons) and ranks second in annual production value (R$ 589.6 million) (IBGE 2023). Most of this production is concentrated in southern Brazil, primarily originating from traditional extractive systems like *caívas* (Marques et al. 2018, 2019), managed by family farmers, as well as monoculture plantations operated by both farmers and industry companies (Penteado Jr and Goulart 2019).

The production process involves both farmers and companies specialized in yerba mate commercialization, forming a supply chain that flows predominantly from producers to companies. Systematic pruning of branches and leaves is conducted to obtain raw material, typically every 2–3 years (Marques 2014; Mattos 2015; Dortzbach et al. 2018; Marques et al. 2019). Shorter pruning cycles are employed in monoculture plantations (Penteado Jr and Goulart 2019).

The traditional management/extractive system called *caíva* involves exploitation in forested landscapes dominated by *A. angustifolia* (in the canopy) with various Myrtaceae species present, along with cattle integrated into the production system (Marques et al. 2014, 2019; Mattos 2015; Dortzbach et al. 2018). This production system holds a protected designation of origin, "*Erva-Mate Planalto Norte Catarinense*," due to its production method under the shade of Araucaria and other native trees (Dortzbach et al. 2018). Additionally, it commands a higher market value (Marques 2014; Mattos 2015).

The traditional yerba mate extraction process functions as an agroforestry system, enabling the reconciliation of forest remnant conservation with production while promoting conservation through the sustainable use of multiple Atlantic Forest biome species (Reis et al. 2018).

12.2.2.4 Genetic Diversity and Conservation

Studies using molecular markers show that yerba mate populations generally exhibit high levels of genetic diversity (Gauer and Cavalli-Molina 2000; Gottlieb et al. 2011; Cascales et al. 2014; Reis et al. 2018; Paiva et al. 2020; Bittencourt et al. 2022). For example, yerba mate populations located in forested landscapes managed for leaf production showed diversity indices of the same magnitude (mean $H_E = 0.233$ – isoenzymes) as a population located within a conservation unit (Reis et al. 2018).

High levels of diversity were also estimated for other populations located in protected areas. Populations from Chapecó/SC (National Forest) and Teixeira Soares/RS (municipal park), of different ages, exhibited genetic diversity ranging from 0.584 to 0.660 (SSR) (Bittencourt et al. 2022). Moreover, low genetic divergence was observed between individuals from populations of different ages, indicating sufficient internal gene flow to maintain genetic diversity across generations (Bittencourt et al. 2022). Conversely, populations managed in monocultural landscapes also serve as important repositories of genetic diversity. This is the case for four cultivated populations from the provinces of Misiones and Corrientes (Argentina), which showed high genetic diversity (mean $H_E = 0.569$ – ISSR).

The studies presented demonstrate that yerba mate's genetic diversity is being conserved through various means: protected areas, traditional systems, and conventional cultivation practices. The role of traditional management systems in conserving the species' diversity stands out. The system described by Reis et al. (2018) is characterized by Araucaria landscapes in which yerba mate is the primary production focus. Nevertheless, yerba mate is frequently present in other traditional systems, such as those centered on *pinhão* production (Rover and Magnanti 2021; Rodrigues 2024), even when not cultivated as the main product. This broadens the possibilities for conserving its genetic diversity, even when the species is not the productive focus.

*12.2.3 Cataia (*Drimys brasiliensis *Miers)*

12.2.3.1 Medicinal Use

Drimys brasiliensis, known as *cataia* or *casca-de-anta*, is a tree native to the Atlantic Forest, and its bark has been used medicinally through the exploitation of plants in natural populations. The bark of Drimys species was commercialized as "Winter's bark" or Cortex Winteranus in the Universal Pharmacopoeia (Ecorce de Winter in French, "Winter's bark" in the Anglo-American tradition). It was discovered incidentally by Captain Winter, one of Sir Francis Drake's lieutenants, when a 1577 storm forced him to take refuge in the Strait of Magellan to treat his crew's scurvy (Trinta and Santos 1997).

The Araucanian aborigines celebrated rituals under the canopy of one species of this genus (Pio Corrêa 1931). The vernacular name *casca-de-anta* ("tapir's bark") refers to observations of sick tapirs (*Tapirus americanus*) consuming their bark. In Santa Catarina, the two occurring species (*D. brasiliensis* and *D. angustifolia*) are used in the *Planalto Catarinense* region (SC/Brazil) as a meat condiment, with their bark ground into powder and employed as a substitute for black pepper (Trinta and Santos 1997).

In folk medicine, *D. brasiliensis* is used internally as a stimulant, antispasmodic, aromatic, antidiarrheal, and antipyretic, for uterine hemorrhage, and in certain digestive tract disorders (Simões et al. 1986). The literature on this species reports several chemical compounds: sesquiterpenes, terpenoids, and lignans isolated from the bark; terpenoids and flavonoids with antitumoral action from the leaves; and cardioactive substances detected in young leaves. In pharmaceutical applications, it is marketed as a tincture and elixir, prescribed for digestive tract disorders and as an antipyretic (Simões et al. 1986).

Drimys species contain numerous bioactive compounds, particularly in bark and leaves. These include drimane-type sesquiterpenes (polygodial, methoxy-coumaroyl-oxy-polygodial, drimanial), flavonoids, tannins, pectin, resins, and aromatic essential oils (Cechinel-Filho et al. 1998; Jansen and De Groot 2004; Fratoni et al. 2016). Essential oil analysis reveals a predominance of monoterpenes (93%), with α-pinene (55.47%) as the primary constituent, followed by β-pinene (18.96%), limonene (4.16%), and 4-terpineol (4.28%), plus minor components like myrcene, sabinene, and camphene (Radomski and Marques 2005). Leaves contain significantly higher essential oil yields (1.09–1.39%) than bark (0.14–0.15%), though with similar compositions (Maggioni et al. 2018).

The compounds demonstrate multiple bioactivities. Hydroalcoholic bark extracts show notable anti-inflammatory and analgesic effects, primarily attributed to polygodial (Cechinel-Filho et al. 1998; Malheiros et al. 2005). Nonpolar fractions contain alkaloids and exhibit strong antimicrobial activity against gram-positive bacteria (*Bacillus cereus* and *Staphylococcus* strains) and *Candida albicans* (Sobottka et al. 2020). Meanwhile, polar fractions rich in tannins and flavonoids display distinct phytochemical profiles (Sobottka et al. 2020). Notably, drimanes from *D. brasiliensis* show cytotoxicity against leukemia cell lines (K562, Nalm6), suggesting their anticancer potential (Fratoni et al. 2016).

12.2.3.2 Occurrence Area and Ecological Aspects

Drimys brasiliensis occurs in Brazil in the states of Bahia, Minas Gerais, Espírito Santo, Rio de Janeiro, São Paulo, Paraná, Santa Catarina, and Rio Grande do Sul, mostly associated with *Araucaria angustifolia* in Mixed Ombrophilous Forest (Trinta and Santos 1997). The species flowering was reported from July to April (Trinta and Santos 1997) and December to March (Mariot et al. 2014a) in Santa Catarina, while observed to occur from June to November in São Paulo (Gottsberger et al. 1980).

The main floral visitors and potential pollinators are Coleoptera, Diptera, and Hymenoptera, with occasional visits by Thysanoptera, Hemiptera, and Lepidoptera, attracted by scent to feed on stigmatic exudates and pollen (Gottsberger et al. 1980; Mariot et al. 2014a). Its fruits are berry-like, dark-streaked when ripe, with reniform, glossy-black seeds showing dormancy due to embryonic immaturity (Abreu et al. 2005), fruiting from October in Santa Catarina (Trinta and Santos 1997) and between September and January in Botucatu (São Paulo) (Gottsberger et al. 1980).

Fruiting occurs year-round in Caçador (Santa Catarina), with overlapping production across years (Mariot et al. 2014a). The fruits are consumed by birds that assist in seed dispersal. In São Paulo, recorded dispersers include the Rufous-headed Tanager (*Schistochlamys ruficapillus*), Great Kiskadee (*Pitangus sulphuratus*), and Diademed Tanager (*Stephanophorus diadematus*) (Gottsberger et al. 1980), while in Santa Catarina, the Olivaceous Elaenia (*Elaenia mesoleuca*) was observed (Mariot et al. 2014a).

12.2.3.3 Extractivism, Management, and Raw Material Provenance

No commercial plantations exist yet, though the vegetative propagation capacity of *D. brasiliensis* was evaluated using leaf explants and axillary or apical buds for clonal regeneration (Martins and Astarita 2005), and stem cuttings (Machado et al. 2011). Although vegetative propagation is viable, seed propagation remains the best propagation method for the species, despite embryonic immaturity.

Since raw material currently comes from natural populations, a study evaluated bark regrowth capacity under different bark-harvesting methods in natural environments, as well as their influence on population dynamics and reproductive biology (Mariot et al. 2014b). Although none of the treatments modified the diameter at breast height increment rates, height, or phenological behavior. Nevertheless, wider bark strips showed lower regeneration rates and higher pest and disease incidence, indicating that optimal strip dimensions should be 2 cm wide × 2 m height, spaced 4 cm apart.

This study suggests that *D. brasiliensis* has high potential for sustainable management in natural populations, demonstrating possibilities to generate income for landowners while contributing to the Atlantic Forest use and conservation.

12.2.3.4 Genetic Diversity and Conservation

The genetic diversity and distribution of eight natural populations of *D. brasiliensis* in southern Brazil were studied by Mariot et al. (2019). The research found low interpopulation diversity ($F_{ST} = 0.05$), suggesting that populations share significant portions of the species' total genetic diversity, while the population fixation index (F_{IS}) signals potential genetic diversity loss.

The significantly lower inbreeding coefficient observed in seedlings versus adult trees in the EPAG population suggests that *D. brasiliensis* can produce offspring in Hardy–Weinberg equilibrium (Mariot et al. 2019). In contrast, maintained inbreeding values for the other two studied populations also suggest occurrences of biparental inbreeding and even self-fertilization (Mariot et al. 2019). These results imply that inbreeding levels may be linked to pollinator availability and diversity, meaning that maintaining genetic diversity depends on conserving the fauna associated with *D. brasiliensis* pollination.

The species shows a disjunct distribution in Santa Catarina State, with isolated populations occurring on hilltops in Dense Ombrophilous Forest (Trinta and Santos 1997). Although some gene flow between isolated and continuous populations is possible, it likely occurs at low frequencies, particularly given the Atlantic Forest biome fragmentation (Ribeiro et al. 2009; Vibrans et al. 2012).

Nonrandom genotypic distribution was found for one population of *D. brasiliensis* (Mariot et al. 2019). At this site, the findings suggest that seeds should be collected from plants spaced at least 40 m apart to maximize genetic diversity capture. In the absence of direct estimations, this distance could be used as criterion for seed collection in other populations.

12.3 Genetic Diversity: A Conservation Tool for Multispecies Management Systems

Genetic diversity refers to the heritable biological variation accumulated throughout a species' evolutionary processes. This diversity can be examined at different levels, primarily interspecific diversity (genetic differences between species) and intraspecific diversity (genetic variability among individuals of the same species) (Frankham et al. 2010; Freeland et al. 2011). Intraspecific variation manifests through genetic polymorphisms, primarily originating from accumulated spontaneous mutations in DNA nucleotide sequences across generations (Hartl and Clark 2007).

These mutations can be neutral, deleterious, or beneficial, with their frequency in populations influenced by evolutionary processes like natural selection, genetic drift, gene flow, and mutation. Maintaining genetic diversity is crucial for population adaptability, enabling responses to environmental changes and selective pressures, making it essential for biodiversity conservation (Allendorf et al. 2013).

Intraspecific genetic variation can be assessed and quantified using molecular markers (e.g., isozymes, microsatellites, SNPs). For intraspecific diversity, researchers can investigate how genetic variation is distributed across a species' range. This involves analyzing diversity within local populations and between different populations or geographic regions. These distribution patterns reflect population genetic structure, often shaped by geographic barriers, evolutionary history, dispersal patterns, and reproductive strategies.

Spatial genetic structure studies enable the identification of conservation management units, reveal historical colonization and migration processes, and inform policies for genetic diversity preservation and resource use (Mariot et al. 2002; Hoeltgebaum et al. 2015). Understanding genetic diversity at different levels and distribution patterns is therefore essential for not only evolutionary biology and population genetics, but for applied practices like endangered species management, ecological restoration, and natural population monitoring (Frankham 2005; Allendorf et al. 2013). Spatial genetic structure can also inform distances for seed collection, aiming to retain genetic diversity, which is an important outcome of these studies (Mariot et al. 2019).

For species dependent on biotic interactions for reproduction (e.g., pollination), environmental changes may significantly affect reproductive system behavior (e.g., increasing the proportion of related crosses). These changes may negatively impact the species' adaptive capacity (inbreeding). Therefore, understanding reproductive systems and monitoring them across reproductive cycles is essential for preventing and minimizing impacts from reduced genetic diversity.

Similarly, gene flow can be inferred by assessing genetic diversity in individuals. This allows for quantifying gene flow between demographic cohorts of a species, verifying diversity maintenance across different developmental stages (Costa et al. 2020; Bittencourt et al. 2022). This information is crucial for guiding and/or correcting resource use strategies to ensure genetic diversity maintenance across demographic cohorts, enabling species to maintain their evolutionary and adaptive trajectory in the landscape.

Currently, genetic diversity assessment tools are becoming increasingly accessible and viable for evaluating and monitoring plant resource use. However, in practice, such assessment remains predominantly restricted to research on populations of currently used species. The challenge lies in making these monitoring tools routine for high-demand species that may consequently experience population declines and lose evolutionary capacity, increasing diversity loss risks.

12.4 Concluding Remarks and Perspectives

The case studies presented here demonstrate that the potential of medicinal plants in the Atlantic Forest extends beyond strictly economic or therapeutic aspects. Their use is often embedded in traditional landscape management systems, which are structured around maintaining forest cover, plant species diversity, and interactions with wildlife, ensuring the continuity of ecosystem services associated with environmental conservation.

These management systems are deeply rooted in traditional knowledge and hold significant cultural value for local communities. Thus, the production of medicinal plants in these contexts not only contributes to biodiversity conservation but also promotes the appreciation and transmission of traditional knowledge, reinforcing the cultural identity of the involved populations.

It is worth noting that these systems host various other medicinal species, such as *Monteverdia truncate* (Nees) Biral, *Bromelia antiacantha* Bertol., *Baccharis* spp., *Schinus terebinthifolius* Raddi, *Mikania* spp., and several Myrtaceae species (Coradin et al. 2011). The presence of these species reinforces the multifunctionality of traditional systems, enabling the reconciliation of sustainable use of multiple medicinal plants, biodiversity conservation, and the maintenance of associated ecosystem services.

Although promising, expanding the use of medicinal plants in traditional management systems requires integrated actions across multiple spheres. Scientific research plays a fundamental role in this process by contributing to the characterization of management techniques, landscape structures, and compositions that favor the multiple use of resources, as well as documenting and valuing associated traditional knowledge. Furthermore, it can demonstrate how these traditional production systems effectively contribute to conserving the remarkable taxonomic, genetic, and functional diversity present in the Atlantic Forest.

References

Abreu DCAD, Kuniyoshi YS, Medeiros ACSD, Nogueira AC (2005) Morphological characterization of fruits and seeds of *Drimys brasiliensis* Miers. (Winteraceae). Rev Bras Sementes 27:67–74

Akisue MK, Oliveira F, Moraes MS, Akisue G, Mancini B (1983) Pharmacognostic characterization of *Cordia verbenacea* DC. (Boraginaceae) drug and tincture. Rev Cienc Farm 5:69–82

Allendorf FW, Luikart G, Aitken SN (2013) Genetics and the conservation of populations, 2nd edn. Wiley-Blackwell, Oxford

Aranha LFS (1967) O mercado ervateiro. Boletim n. 48. Faculdade de Ciências Econômicas e Administração USP, São Paulo

Barazetti VM, Lima TM, Sodré GA, Gross E (2022) Management perspectives to maximize secondary metabolite production from medicinal plants in agroforestry systems. Agrofor Syst 96:681–695. https://doi.org/10.1007/s10457-022-00731-0

Baveloni FG, Scardelato JA, Carvalho FA, Paino IMM, Medina-Alarcón KP, Pereira-da-Silva MA, Chorilli M, Santos AG, Soares CP (2024) Cordialin A isolatedfrom *Varronia curassavica* Jacq. loaded in nanoemulsion as potential cytotoxic agent on human cervical tumor cells. Chem Pap 78:6141–6158. https://doi.org/10.1007/s11696-024-03531-1

Bensusan N (2006) Conservação da biodiversidade em áreas protegidas. Editora FGV, Rio de Janeiro

Bittencourt L, Silva DT, Nascimento B, Saiki FA, Stedille LIB, Costa NCF, Mantovani A (2022) Genetic diversity in successive age cohorts of *Ilex paraguariensis* in Southern Brazil. For Sci 68:291–298. https://doi.org/10.1093/forsci/fxac012

Bueno L, Dias ASP (2015) Initial settlement of South America: contributions from the Brazilian context. Estud Av 29:119–147. https://doi.org/10.1590/S0103-40142015000100009

Carvalho PER (2003) Espécies arbóreas Brasileiras. Embrapa Florestas, Colombo

Cascales J, Bracco M, Poggio L, Gottlieb AM (2014) Genetic diversity of wild germplasm of *Ilex paraguariensis* St. Hil. from Uruguay. Genetica 142:563–573. https://doi.org/10.1007/s10709-014-9804-3

Cechinel-Filho V, Schlemper V, Santos ARS, Pinheiro TR, Yunes RA, Mendes GL, Calixto JB, Delle Monache F (1998) Isolation and identification of active compounds from *Drimys winteri* barks. J Ethnopharmacol 62:223–227. https://doi.org/10.1016/S0378-8741(98)00069-5

Clement CR, Casas A, Parra-Rondinel FA, Levis C, Peroni N, Hanazaki N, Cortés-Zárraga L, Rangel-Landa S, Alves RP, Ferreira MJ, Cassino MF, Coelho SD, Cruz-Soriano A, Pancorbo-Olivera M, Blancas J, Martínez-Ballesté A, Lemes G, Lotero-Velásquez E, Bertin VM, Mazzochini GG (2021) Disentangling domestication from food production systems in the Neotropics. Quaternary 4:4. https://doi.org/10.3390/quat4010004

Colussi J, Prestes NP (2011) Frugivory by birds in *Myrciaria trunciflora* (Mart) O. Berg. (Myrtaceae), *Eugenia uniflora* L. (Myrtaceae), and *Ilex paraguariensis* St. Hil. in northern Rio Grande do Sul. Rev Bras Ornitol 19:48–55

Coradin L, Siminski A, Reis A (2011) Espécies da Flora Nativa de Valor Econômico Atual ou Potencial. Série Biodiversidade 40

Costa NCF, Stedille LIB, Lauterjung MB, Montagna T, Candido-Ribeiro R, Bernardi AP, Mantovani A, Reis MS, Nodari RO (2020) Spatiotemporal variation in mating system and genetic diversity of *Araucaria angustifolia*: implications for conservation and seed collection. For Ecol Manag 481:118716. https://doi.org/10.1016/j.foreco.2020.118716

Dias AS, Jacobus AL (2003) How ancient is the settlement of Southern Brazil? Rev CEPA 27:39–67

Diegues ACS, Sant'Ana RV, Arruda E (2001) Saberes tradicionais e biodiversidade no Brasil. Ministério do Meio Ambiente, Brasília

Dortzbach D, Neppel G, Trabaquini K, Vieira VF (2018) Indicação Geográfica Erva-mate do Planalto Norte Catarinense: produto. Epagri, Florianópolis

Edwin G, Reitz R (1967) Aquifoliaceae. In: Reitz R (ed) Flora Ilustrada Catarinense, parte I, fasc. Aqui. Herbário Barbosa Rodrigues, Itajaí

Ereno D (2005) Da natureza para a Farmácia. Pesqui FAPESP. http://revistapesquisa.fapesp.br

Falkenberg MB (2011) *Varronia curassavica*. In: Coradin L, Siminski A, Reis A (eds) Espécies Nativas da Flora Brasileira de Valor Econômico Atual ou Potencial. MMA, Brasília

Feijó EVDS, Oliveira RAD, Costa LCDB (2014) Light affects *Varronia curassavica* essential oil yield by increasing trichome frequency. Rev Bras Farmacogn 24:516–523. https://doi.org/10.1016/j.bjp.2014.10.005

Frankham R (2005) Genetics and extinction. Biol Conserv 126:131–140. https://doi.org/10.1016/j.biocon.2005.05.002

Frankham R, Ballou JD, Briscoe DA (2010) Introduction to conservation genetics, 2nd edn. Cambridge University Press

Fratoni E, Claudino VD, Yunes RA, Franchi GC Jr, Nowill AE, Cechinel-Filho V, Delle Monache F, Malheiros A (2016) Further drimane sesquiterpenes from *Drimys brasiliensis* stem barks with cytotoxic potential. Naunyn-Schmiedebergs Arch Pharmacol 389:791–797. https://doi.org/10.1007/s00210-016-1241-7

Freeland JR, Kirk H, Petersen SD (2011) Molecular ecology, 2nd edn. Wiley-Blackwell
Fundação SOS Mata Atlântica (2023) Atlas dos Remanescentes Florestais. https://cms.sosma.org.br
Ganders FR (1979) The biology of heterostyly. N Z J Bot 17:607–635. https://doi.org/10.1080/0028825X.1979.10432574
Gastauer M, Messias MCTB, Neto JAAM (2012) Floristic composition, species richness and diversity of *campo rupestre* vegetation from the Itacolomi State Park, Minas Gerais, Brazil. Environ Nat Resour Res 2:115–130
Gauer L, Cavalli-Molina S (2000) Genetic variation in natural populations of *Ilex paraguariensis* A. St.-Hil. (Aquifoliaceae) using RAPD markers. Heredity 84:647–656. https://doi.org/10.1046/j.1365-2540.2000.00687.x
Gavin MC, McCarter J, Mead A, Berkes F, Stepp JR, Peterson D, Tang R (2015) Defining biocultural approaches to conservation. Trends Ecol Evol 30:140–145. https://doi.org/10.1016/j.tree.2014.12.005
Gottlieb AM, Giberti GC, Poggio L (2011) Evaluacióndel germoplasma de *Ilex paraguariensis* e *Ilexdumosa* (Aquifoliaceae). Bol Soc Argent Bot 46:113–123
Gottsberger G, Silberbauer-Gottsberger I, Ehrendorfer F (1980) Reproductive biology in the primitive relic angiosperm *Drimys brasiliensis* (Winteraceae). Plant Syst Evol 135:11–39. https://doi.org/10.1007/BF00983005
Hartl DL, Clark AG (2007) Principles of population genetics, 4th edn. Sinauer Associates, Sunderland
Hoeltgebaum MP, Reis MS (2017) Genetic diversity and population structure of *Varronia curassavica*: a medicinal polyploid species in a threatened ecosystem. J Hered 108:415–423. https://doi.org/10.1093/jhered/esx010
Hoeltgebaum MP, Bernardi A, Montagna T, Reis MS (2015) Genetic diversity and structure of *Varronia curassavica* Jacq. populations in resting as of Santa Catarina Island. Rev Bras Plantas Med 17:1083–1090
Hoeltgebaum MP, Montagna T, Lando AP, Puttkammer C, Orth AI, Guerra MP, Reis MS (2018) Reproductive biology of *Varronia curassavica* Jacq. (Boraginaceae). An Acad Bras Cienc 90:87–102. https://doi.org/10.1590/0001-3765201820160273
Hoeltgebaum MP, Lauterjung MB, Montagna T, Candido-Ribeiro R, Vieira W, Bernardi AP, Cristofolini C, Reis MS (2020) Genetic and demographic aspects of *Varronia curassavica* Jacq. in a heterogeneous coastal ecosystem. An Acad Bras Ciênc 92:e20180532. https://doi.org/10.1590/0001-3765202020180532
Hussein GME, Matsuda H, Nakamura S, Hamao M, Akiyama T, Tamura K, Yoshikawa M (2011) Mate tea (*Ilex paraguariensis*) promotes satiety and body weight lowering in mice: involvement of glucagon-like peptide-1. Biol Pharm Bull 34:1849–1855. https://doi.org/10.1248/bpb.34.1849
IBGE (2023) Produção da extração vegetal e da silvicultura 2023. https://www.ibge.gov.br
IBGE (2024) Bioma predominante por município para fins estatísticos. https://biblioteca.ibge.gov.br
Jansen BJM, De Groot A (2004) Occurrence, biological activity and synthesis of drimane sesquiterpenoids. Nat Prod Rep 21:449–477. https://doi.org/10.1039/B311170A
Kinupp VF (2007) Riqueza de plantas alimentícias não-convencionais na região metropolitana de Porto Alegre. PhD thesis, Universidade Federal do Rio Grande do Sul
Lapa FS (2006) Cordiacurassavica (Jacq.) Roem. & Schult.: Influência de fatores ambientais no crescimento e na produção de metabólitos. MSc thesis, Universidade Federal de Santa Catarina
Levis C, Flores BM, Campos-Silva JV, Peroni N, Staal A, Padgurschi MCG, Dorshow W, Moraes B, Schmidt M, Kuikuro TW, Kuikuro H, Wauja K, Kuikuro K, Kuikuro A, Fausto C, Franchetto B, Watling J, Lima H, Heckenberger M, Clement CR (2024) Contributions of human cultures to biodiversity and ecosystem conservation. Nat Ecol Evol 8:866–879. https://doi.org/10.1038/s41559-024-02356-1
Lorenzi H, Matos FJA (2008) Plantas medicinais no Brasil. Instituto Plantarum, Nova Odessa. 544

Machado TM, Pieri C, Lima GP, Ono EO (2011) Effect of plant regulators on vegetative propagation of *Drimys brasiliensis* Miers. (Winteraceae). Naturalia 34:87–94

Magalhães PM, Figueira GM, Montanari Júnior I (2004) Influence of NPK fertilization on *Cordia verbenacea* DC. biomass yield. In: International symposium breeding research on medicinal and aromatic plants, Campinas

Maggioni RA, Netto ER, Radomski MI, Oliveira EB, Deschamps C, Zuffellato-Ribas KC (2018) Produção de Óleo essencial de folhas e cascas de Drimys brasiliensis Miers (casca-d'anta). Rev Cubana Plantas Med 23(4). https://revplantasmedicinales.sld.cu/index.php/pla/article/view/714

Magnanti NJ, Rover OJ, Siminski A, Santos KL, Siddique I (2017) Conservação pelo uso de espécies ameaçadas: Manejo da Araucária em Sistemas Agroflorestais Agroecológicos. In: Siddique I et al (eds) Construindo Conhecimentos Sobre Agroflorestas em Rede. UFSC, Florianópolis

Malheiros A, Cechinel Filho V, Schmitt CB, Yunes RA, Escalante A, Svetaz L, Zacchino S, Delle Monache F (2005) Antifungal activity of drimane sesquiterpenes from *Drimys brasiliensis* using bioassay-guided fractionation. J Pharm Pharm Sci 8:335–339

MapBiomas (2023) Mapeamento anual de cobertura e uso da terra no Brasil entre 1985 a 2022. https://brasil.mapbiomas.org

Mariot A, Di Stasi LC, Reis MS (2002) Genetic diversity in natural populations of *Piper cernuum*. J Hered 93:365–369. https://doi.org/10.1093/jhered/93.5.365

Mariot A, Mantovani A, Bittencourt R, Reis MS (2014a) Aspects of reproductive biology of *Drimys brasiliensis* Miers (Winteraceae) in Floresta Ombrófila Mista, South of Brazil. Ciênc Florest 24:877–888. https://doi.org/10.1590/1980-509820142404008

Mariot A, Mantovani A, Reis MS (2014b) Bark harvesting systems of *Drimys brasiliensis* Miersin the Brazilian Atlantic Rainforest. An Acad Bras Cienc 86(3):1315–1326

Mariot A, Montagna T, Reis MS (2019) Genetic diversity and structure of *Drimys brasiliensis* in southern Brazil: insights for conservation. J For Res 31:1325–1332. https://doi.org/10.1007/s11676-019-00934-9

Marques AC (2014) As paisagens do mate e a conservação socioambiental: um estudo junto aos agricultores familiares do Planalto Norte Catarinense. PhD thesis, Universidade Federal do Paraná

Marques A, Denardin VF, Reis MS, Wisniewski C (2014) As paisagens do mate no planalto norte catarinense. In: Desenvolvimento territorial: políticas públicas brasileiras, experiências internacionais e a indicação geográfica como referência. Liber Ars, São Paulo, pp 33–46

Marques A, Reis MS, Denardin VFA (2018) A araucária e a erva-mate: constituindo paisagens de uso, conflito e conservação da biodiversidade. In: Fronteiras fluídas: florestas com araucárias na América Meridional. Oikos, São Leopoldo, pp 250–263

Marques AC, Reis MS, Denardin VF (2019) Yerba mate landscapes: forest use and socio-environmental conservation. Ambient Soc 22:e02822. https://doi.org/10.1590/1809-4422asoc201702822vu2019L3AO

Martins FS, Astarita LV (2005) Estabelecimento in vitro de *Drimys brasiliensis* Miers. In: Resumos do 56° Congresso Nacional de Botânica

Mattos AG (2015) Conservação pelo uso de populações de *Ilex paraguariensis* A. St. Hil, em sistemas extrativistas no planalto norte catarinense. PhD thesis, Universidade Federal de Santa Catarina

Medeiros DPW, Lopes AV, Zickel CS (2007) Phenology of woody species in tropical coastal vegetation, northeastern Brazil. Flora 202:513–520. https://doi.org/10.1016/j.flora.2006.11.002

Mendes ADR, Lacerda THS, Rocha SMG, Martins ER (2014) Reguladores vegetais e substratos no enraizamento de estacas de erva-baleeira (*Varronia curassavica* Jacq.). Rev Bras Plantas Med 16:262–270. https://doi.org/10.1590/S1516-05722014000200015

Mendes ADR, Nascimento CR, Queiroz TB, Pinto VB, Martins ER (2015) Ecogeografia de populações de erva-baleeira (*Varronia curassavica*) no Norte e Vale do Jequitinhonha em Minas Gerais. Ciênc Rural 45:418–424. https://doi.org/10.1590/0103-8478cr20140196

Michielin EMZ, Salvador AA, Riehl CAS, Smânia A Jr, Smânia EFA, Ferreira RSR (2009) Chemical composition and antibacterial activity of *Cordia verbenacea* extracts obtained by different methods. Bioresour Technol 100:6615–6623. https://doi.org/10.1016/j.biortech.2009.07.061

Milanesi LS, Peroni N, Reis MS (2013) Use of the palm *Euterpe edulis* Martius in landscapes managed by German-origin migrants in Southern Brazil. J Ethnobiol Ethnomed 9:47. https://doi.org/10.1186/1746-4269-9-47

Milanesi LS, Montagna T, Reis MS, Peroni N (2021) Population biology of *Euterpe edulis* Martius (Arecaceae) in managed landscapes in Southern Brazil. Econ Bot 75:144–157. https://doi.org/10.1007/s12231-021-09519-2

Ministério da Saúde do Brasil (2009) Relação de Plantas Medicinais de Interesse ao SUS. http://portalsaude.saude.gov.br

Montanari I, Magalhaes PM (2004) Production of *Cordia verbenaceae* DC according to plantation density. In: III International symposium breeding research on medicinal and aromatic plants, Campinas

Myers N, Mittermeier RA, Mittermeier CG, Fonseca GAB, Kent J (2000) Biodiversity hotspots for conservation priorities. Nature 403:853–858. https://doi.org/10.1038/35002501

Nizio DADC, Blank AF, Sampaio TS, Brito FDA, Andrade TM, Arrigoni-Blank MDF, Maria AN (2018) Distillation methods affect the chemical composition of *Varronia curassavica* Jacq. essential oil? Biosci J 34:629–639

Oliveira YMM, Rotta E (1985) Natural distributionareaof *Ilex paraguariensis* St. Hil. In: Seminário sobre atualidades e perspectivas florestais, Curitiba

Opler PA, Baker WL, Frankie GW (1975) Reproductive biology of some Costa Rican *Cordia* species (Boraginaceae). Biotropica 7:234–247

Paiva DI, Cascales J, Rosetti MEN, Scherer RA, Gauchat ME, Gottlieb AM (2020) Unraveling the genetic complexity of a cultivated breeding population of *Ilex paraguariensis* St. Hil. An Acad Bras Ciênc 92:e20190113. https://doi.org/10.1590/0001-3765202020190113

Penteado DGB (2020) Uso tradicional de *Varronia curassavica* Jacq. (Boraginacea) e sua contribuição para o aperfeiçoamento da lei de acesso e repartição de benefícios (n° 13.123/2015). MSc thesis, Universidade Federal de Santa Catarina

Penteado JF Jr, Goulart ICGR (2019) Erva 20: Sistema de produção para erva-mate. Embrapa, Brasília

Pereira BE, Diegues AC (2010) Conhecimento de populações tradicionais como possibilidade de conservação da natureza: uma reflexão sobre a perspectiva da etnoconservação. Desenvolv Meio Ambiente 22:37–50. https://doi.org/10.5380/dma.v22i0.16054

Pio Corrêa M (1931) Dicionário das plantas úteis do Brasil e das exóticas cultivadas. Imprensa Nacional, Rio de Janeiro

Pires EZ, Stedille LIB, Machado S, Mantovani A, Bortoluzzi RL (2014) Reproductive biology of erva-mate (*Ilex paraguariensis* A. St. Hil) in a remnant of High Montane Mixed Ombrophilous Forest. Rev Ciênc Agrovet 13:171–180

Queiroz TB, Mendes ADR, Silva JCRL, Fonseca FSA, Martins ER (2016) Essential oil content and chemical composition of *Varronia curassavica* Jaqc. according to collection time. Rev Bras Plantas Med 18:356–362. https://doi.org/10.1590/1983-084X/15_116

Queiroz TB, Fonseca FSA, Mendes ADR, Azevedo AM, Martins ER (2019) Chemical diversity of accessions of the *in vivo* germplasm bank of *Varronia curassavica* (Jacq.). Acta Sci Agron 42:e42726. https://doi.org/10.4025/actasciagron.v42i1.42726

Radomski MI, Marques MOM (2005) Composição do óleo essencial da cataia (*Drimys brasiliensis*). In: Simpósio brasileiro de óleos essenciais, 3. Resumos. Instituto Agronômico, Campinas

Reis MS, Montagna T, Mattos AG, Filippon S, Ladio AH, Marques ADC, Zechini AA, Peroni N, Mantovani A (2018) Domesticated landscapes in Araucaria Forests, Southern Brazil: a multispecies local conservation-by-use system. Front Ecol Evol 6:11. https://doi.org/10.3389/fevo.2018.00011

Ribeiro MC, Metzger JP, Martensen AC, Ponzoni FJ, Hirota MM (2009) The Brazilian Atlantic Forest: how much is left, and how is the remaining forest distributed? Implications for conservation. Biol Conserv 142:1141–1153. https://doi.org/10.1016/j.biocon.2009.02.021

Rodal MJR, Nascimento LM (2002) Levantamento florístico da floresta serrana da reserva biológica de Serra Negra, microrregião de Itaparica, Pernambuco, Brasil. Acta Bot Bras 16:481–500. https://doi.org/10.1590/S0102-33062002000400009
Rodrigues NC (2024) Caracterização do sistema tradicional de produção de pinhão no Planalto Serrano Catarinense. MSc thesis, Universidade Federal de Santa Catarina
Rover OJ, Magnanti NJ (2021) Extrativismo do Pinhão: soberania e segurança alimentar, agroecologia e gênero. In: Mota DM et al (eds) Soberania alimentar: biodiversidade, cultura e relações de gênero. Embrapa, Brasília
Santiano FE, Fernández MLA, Espino M, Zyla LE, Rey L, Gómez SE, Bruna FA, Pistone-Creydt V, Pietrobon E, Pérez Elizalde R, Silva MF, Carón RW, López Fontana CM (2023) Protectiveeffectsofyerba mate (*Ilex paraguariensis*) on prostate cancer development. Nutrition 108:111957. https://doi.org/10.1016/j.nut.2022.111957
Sattler MC, Carvalho CR, Clarindo WR (2016) The polyploidy and its key role in plant breeding. Planta 243:281–296. https://doi.org/10.1007/s00425-015-2450-x
Sebbenn AM, Degen B, Azevedo VCR, Silva MB, de Lacerda AEB, Ciampi AY, Kanashiro M, Carneiro FS, Thompson I, Loveless MD (2008) Modelling the long-term impacts of selective logging on genetic diversity and demographic structure of four tropical tree species in the Amazon forest. For Ecol Manag 254:335–349. https://doi.org/10.1016/j.foreco.2007.08.009
Sertié JAA, Woisky RG, Wiezel G, Rodrigues M (2005) Pharmacological assay of Cordia verbenacea V: oral and topical anti-inflammatory activity, analgesic effect and fetus toxicity of a crude leaf extract. Phytomedicine 12(5):338–344. https://doi.org/10.1016/j.phymed.2003.09.013
Siminski A, Santos KL, Wendt JGN (2016) Rescuing agroforestry as strategy for agriculture in Southern Brazil. J For Res 27:739–746. https://doi.org/10.1007/s11676-016-0232-3
Simões CMO, Mentz LA, Schenkel EP, Irgang BE, Stehmann JR (1986) Plantas da medicina popular no Rio Grande do Sul. Ed. da Universidade/UFRGS, Porto Alegre
Smith LB (1970) Boraginaceae. In: Flora Ilustrada de Santa Catarina. Herbário Barbosa Rodrigues, Itajaí
Sobottka AM, Luz CBD, Tonial F, Rossato-Grando LG, Bertol CD (2020) Potencial antimicrobiano e caracterização fitoquímica de extratos da casca do caule de *Drimys brasiliensis* Miers (Winteraceae). Rev Ciênc Méd Biol 19:66–71
Song Y, Cao J, Cao F, Su E (2025) A systematic review on *Ilex paraguariensis* A. St. Hil. J Food Compos Anal 142:107466. https://doi.org/10.1016/j.jfca.2025.107466
Sousa VA, Daros TL, Sturion JA (2003) Reproductive phenology of *Ilex paraguariensis* St. Hil. Embrapa Florestas, Brasília
Tagliari MM, Levis C, Flores BM, Blanco GD, Freitas CT, Bogoni JA, Vieilledent G, Peroni N (2021) Collaborative management as a way to enhance Araucaria Forest resilience. Perspect Ecol Conserv 19:131–142. https://doi.org/10.1016/j.pecon.2021.03.002
Tagliari MSM, Bogoni JA, Blanco GD, Cruz AP, Peroni N (2023) Disrupting a socio-ecological system: could traditional ecological knowledge be the key to preserving the Araucaria Forest in Brazil under climate change? Clim Chang 176:20–40. https://doi.org/10.1007/s10584-022-03477-x
Tolouei SEL, Marcon R, Vilela FC, Freitas CS, Heller M, Andrade EL, Macedo Júnior SJ, dos Santos AA, Rocha RF, Fadanni GP, Marques N, Siqueira Júnior JM, Calixto JB (2025) Preclinical development of a standardized extract of *Ilex paraguariensis* A. St.-Hil. for the treatment of obesity and metabolic syndrome. Pharmacol Res 213:107607. https://doi.org/10.1016/j.phrs.2025.107607
Trinta EF, Santos E (1997) Flora Ilustrada Catarinense: Winteraceae. Herbário Barbosa Rodrigues, Itajaí
Trojak-Goluch A, Kawka-Lipińska M, Wielgusz K, Praczyk M (2021) Polyploidy in industrial crops: applications and perspectives in plant breeding. Agronomy 11:2574. https://doi.org/10.3390/agronomy11122574
Vibrans AC, McRoberts RE, Lingner DV, Nicoletti AL, Moser P (2012) Extensão original e atual da cobertura florestal de Santa Catarina. In: Vibrans AC et al (eds) Inventário florístico florestal de Santa Catarina, vol 1. Edifurb, Blumenau

Vieira DD, Conceição AS, Melo IM, Stapf MNS (2013) A família Boraginaceae sensu lato na APA Serra Branca/Raso da Catarina, Bahia, Brasil. Rodriguésia 64:151–168. https://doi.org/10.1590/S2175-78602013000100013

Vieira-da-Silva C, Miguel LA, Reis MS (2009) A comercialização do pinhão de *Araucaria angustifolia* no Distrito de Taquara Verde, município de Caçador-SC. Rev Bras Agroecol 4:1–5

Zechini AA, Schussler G, Silva JZ, Mattos AG, Peroni N, Mantovani A, Reis MS (2012) Produção, comercialização e identificação de variedades de pinhão no entorno da Floresta Nacional de Três Barras—SC. Biodivers Bras 2:74–82

Zechini AA, Lauterjung MB, Candido-Ribeiro R, Montagna T, Bernardi AP, Hoeltgebaum MP, Mantovani A, Reis MS (2018) Genetic conservation of *Araucaria angustifolia* through traditional land use. Econ Bot 72:166–179. https://doi.org/10.1007/s12231-018-9414-6

Chapter 13
Impacts of Climate Change on the Genetic and Phytochemical Diversity of Medicinal Plants in Biodiversity Hotspot Regions

Uddipta Borthakur, Nibedita Sarma, Beauty Rani Basumatary, Sulagna Gautom Phukan, Kunal Boro, Sanghamitra Gogoi, Nandini Sikdar, Bikash Kumar Kundu, Bikash Kalita, and Bhaben Tanti

Abstract Biodiversity hotspots, harbouring a significant proportion of the world's endemic species, are repositories of immense medicinal plant wealth. Medicinal plants are of utmost importance in both traditional and modern pharmacotherapies, serving as primary sources of bioactive compounds in the development of drugs. However, accelerating global climate change poses serious threats to their survival, genetic variability, and phytochemical stability. Changes in temperature regimes, precipitation patterns, and the increasing frequency of extreme climatic events are reshaping habitats, disrupting reproductive biology and interactions with pollinators, and resulting in fragmented populations and genetic erosion. Simultaneously, climate-induced abiotic stress alters metabolic pathways, leading to qualitative and quantitative changes in major secondary metabolites, thereby affecting the medicinal potency and safety. This chapter highlights the ecological and genetic responses of medicinal plants to climate stress, phytochemical profile alterations, and changes in therapeutic attributes, based on evidence derived from biodiversity hotspots around the world, such as the Himalayas, the Amazon Basin, and the Mediterranean region. Updated conservation approaches discussed in this chapter include classical and molecular tools, metabolomics, ecological modelling, and community-based strategies for the preservation of medicinal plant resources. The chapter provides critical insights into emerging challenges and future perspectives for protecting the genetic and phytochemical diversity of medicinal plants under the changing climatic regimes.

U. Borthakur · N. Sarma · B. R. Basumatary · S. G. Phukan · K. Boro · S. Gogoi · N. Sikdar · B. K. Kundu · B. Tanti (✉)
Department of Botany, Gauhati University, Guwahati, Assam, India
e-mail: uddipta@gauhati.ac.in; nibedita@gauhati.ac.in; sgphukan@gauhati.ac.in; kunal@gauhati.ac.in; sanghamitragogoi@gauhati.ac.in; btanti@gauhati.ac.in

B. Kalita
Department of Botany, Assam Don Bosco University, Tapesia Garden, Kamarkuchi, Sonapur, Assam, India

S. Jha et al. (eds.), *Genetic Diversity in Medicinal Plants*, Sustainable Development and Biodiversity 44, https://doi.org/10.1007/978-981-95-7281-6_13

Keywords Medicinal plants · Climate change · Genetic diversity · Biodiversity hotspots · Phytochemicals

13.1 Introduction

Understanding the impacts of climate change on medicinal plants first requires acknowledging the fundamental role of climate itself. Climate represents the long-term patterns of temperature, precipitation, humidity, and seasonal cycles that shape ecosystems and govern the distribution, growth, and survival of plant species. Any sustained shift in these climatic parameters, whether gradual or abrupt, can influence ecological balances. Climate change refers to these long-term alterations in the Earth's climate system, driven by both natural variability and, more prominently in recent decades, human-induced factors. Anthropogenic greenhouse gas emissions, large-scale deforestation, industrial expansion, unsustainable agricultural practices, and rapid urbanization collectively accelerate global warming and disrupt atmospheric and hydrological cycles. These changes not only intensify extreme weather events but also modify ecological niches at both macro- and micro-levels (Sofowora et al. 2013; Pan et al. 2014).

Biodiversity hotspot regions, which harbour exceptional concentrations of endemic species and are simultaneously experiencing high rates of habitat loss, are particularly vulnerable to these climatic shifts. These hotspots, such as the Indo-Burma region, Western Ghats–Sri Lanka, and Eastern Himalayas, serve as reservoirs of ancient lineages and rich genetic diversity. Medicinal plants within these landscapes have evolved intricate genetic, biochemical, and physiological adaptations that are aligned with localized climatic conditions. However, the accelerating pace of climate change threatens to destabilize these finely tuned systems. Altered temperature regimes, unpredictable rainfall, increased frequency of droughts, and shifting pest–pathogen dynamics can exert significant pressure on plant populations, driving changes in genetic composition, population structure, and biosynthetic pathways (Gurib-Fakim 2006).

As phytochemical profiles of medicinal plants are closely regulated by environmental cues, even subtle climatic deviations can influence the production, concentration, and diversity of secondary metabolites. These compounds, like alkaloids, phenolics, terpenoids, flavonoids, and others, form the biochemical foundation for the therapeutic value of medicinal species. Consequently, climate-driven alterations may enhance, suppress, or qualitatively transform phytochemical outputs, with profound implications for traditional medicine, pharmacological research, and conservation biology (Lindsey 2020).

Positioning the discussion within the broader context of climate dynamics and hotspot vulnerability allows a more integrated understanding of how climate change affects both the genetic integrity and phytochemical resilience of medicinal plants. This perspective not only strengthens the theme of the chapter but also highlights the urgent need for adaptive conservation strategies, sustainable utilization frameworks, and interdisciplinary research to safeguard the medicinal plant wealth residing in global biodiversity hotspots (Ummenhofer and Meehl 2017).

Medicinal plants have been central to human health for centuries and remain a primary healthcare resource for billions worldwide. They underpin modern pharmacology and continue to play a key role in traditional systems like Ayurveda, Siddha, Traditional Chinese Medicine, and various indigenous practices. Beyond their therapeutic value, these species are closely tied to cultural and socioeconomic traditions, reflecting deep community–nature relationships. Many major drugs, including aspirin, digoxin, and morphine, originate from plants, underscoring their enduring relevance. The rising global interest in herbal medicine and phytotherapy further highlights their importance and the urgent need to conserve these valuable resources for future generations (Hamilton 2004; Alamgir 2017; Wang et al. 2022).

However, this global heritage now faces unprecedented challenges due to climate change. Human-induced greenhouse gas emissions, particularly carbon dioxide (CO_2), have risen dramatically from ~280 ppm in pre-industrial times to over 410 ppm in recent decades (IPCC 2019) and are projected to continue increasing. These emissions drive shifts in global temperature, altered precipitation regimes, glacial retreat, and more frequent extreme weather events, leading to widespread ecosystem instability (Reddy et al. 2010). Such environmental disruptions pose direct threats to medicinal plants, which are highly sensitive to variations in temperature, soil moisture, and atmospheric CO_2.

The interaction of medicinal plants with their environment reflects the intricate balance of the natural world. These species often evolve in specific ecological niches, developing unique phytochemical profiles as adaptive responses to local stresses, such as pests, drought, and temperature extremes (Lambert et al. 1997). These same bioactive compounds, like alkaloids, terpenoids, flavonoids, and phenolics, are the basis of their medicinal efficacy. Yet climate change disrupts these finely tuned interactions, leading to quantitative and qualitative alterations in phytochemical composition (Ahmed et al. 2019; Sharma et al. 2020). Elevated CO_2 may enhance biomass but dilute metabolite concentrations, while drought and heat stress can unpredictably alter secondary metabolism, reduce efficacy, or even enhance toxic constituents (Yang et al. 2018). Such variability challenges both pharmaceutical standardization and the stability of traditional medical practices.

Equally concerning are the impacts of climate change on the distribution and genetic diversity of medicinal plants. Species are increasingly forced to migrate to higher altitudes or latitudes, yet many face barriers to dispersal or adaptation (Sykes and Prentice 1996). The Himalayan region, a global biodiversity hotspot, illustrates this vulnerability, with endemic medicinal species like *Taxus*, *Picrorhiza*, and *Nardostachys* already experiencing range shifts and population declines (Dhar et al. 2002). The loss of such species would have ecological, medical, and economic consequences, eroding local livelihoods and limiting opportunities for new drug discovery.

Economically, medicinal plants contribute substantially to global industries, from pharmaceuticals and dietary supplements to cosmetics. This sector supports rural livelihoods in developing nations and contributes to international trade and research (Mustafa et al. 2017). Yet unsustainable harvesting, coupled with climate-driven habitat degradation, threatens both biodiversity and economic stability. Conservation through sustainable cultivation, habitat protection, and integration of traditional knowledge is therefore not just an ecological necessity but also a socio-economic imperative (Mondal and Palit 2022). The purpose of this chapter is to analyse the impacts of climate change on medicinal plants and to highlight adaptation and mitigation measures crucial for safeguarding their diversity and medicinal properties for posterity.

13.2 Impact of Climate Change on Genetic Diversity of Medicinal Plants

13.2.1 Genetic Impacts of Climate Change

Climate change poses a significant risk to the genetic diversity of medicinal plants. It alters temperatures, precipitation patterns, and ecosystem stability, especially in biodiversity hotspot areas. Medicinal plants with specific ecological needs, such as *Nardostachys jatamansi* and *Picrorhiza kurroa* in the Himalayas, are facing serious range reductions due to higher temperatures and reduced snowfall. By 2050, habitat loss could reach 30–60% (Gairola et al. 2010; Chandra et al. 2022). These shifts cause habitat fragmentation, limit gene flow, and lead to inbreeding and loss of alleles (Shrestha et al. 2012).

13.2.2 Climate Change and Genetic Structure/ Population Dynamics

Insect-pollinated plants like *Valeriana jatamansi* and *Swertia chirayita* are seeing less successful pollination because climate change disrupts the timing of flowering and pollinator activity (Chakraborty et al. 2015). Similarly, tropical medicinal plants, such as *Withania somnifera*, *Gymnema sylvestre*, and *Rauvolfia serpentina*, have distinct genetic populations across changing ecological zones, showing interrupted genetic connections (Udayakumar et al. 2013; Khan and Shah 2016). In areas like the Western Ghats, erratic rainfall has worsened genetic divergence. Overharvesting drought-resistant plants like *Aloe vera* and replacing wild populations with uniform varieties have sped up genetic loss and sameness (Kumar et al. 2022). Conservation strategies need to include population genetics, ecological modelling, and in situ/ex situ methods like seed banks and cryopreservation to safeguard genetic resources as climates change (FAO 2010).

13.2.3 Genetic Erosion, Fragmentation, and Population Bottlenecks

Climate change, habitat loss, and overexploitation are leading to major genetic loss in medicinal plants. Rising temperatures, inconsistent rainfall, and diminished snow cover are fragmenting habitats, particularly for alpine species like *N. jatamansi*, *Rheum emodi*, *P. kurroa*, and *S. chirayita*, with habitat losses predicted to reach 60% by 2050 (Joshi and Dhawan 2007). Fragmentation limits gene flow, decreases genetic diversity, and encourages inbreeding and genetic drift. A similar trend is visible in tropical species, such as *W. somnifera*, *R. serpentina*, and *G. sylvestre*, where loss of alleles further decreases diversity. Extreme weather events like prolonged droughts, heatwaves, and floods make matters worse by causing population crashes, sharply cutting their numbers, and leading to the loss of rare traits. Populations of *P. kurroa*, *Valeriana jatamansi*, and *Rhodiola rosea* show less genetic variation when faced with combined climatic and harvesting pressures (Mykhailenko et al. 2025). Likewise, *Harpagophytum procumbens* in southern Africa experiences genetic bottlenecks due to shifting ranges and unsustainable harvesting (Muzila et al. 2014). Such losses threaten evolutionary potential and chemical diversity. In *Ocimum sanctum*, restriction linked to stress decreased the variety of chemical traits (Rastogi et al. 2019). It is important to note that populations may look healthy while harbouring hidden genetic weaknesses that threaten their resilience. Thus, preserving genetic diversity is crucial for ecological adaptation and medicinal use. Conservation efforts should focus on maintaining landscape connections, restoring habitats, and using advanced molecular markers to identify and protect populations with rich genetic diversity in environmentally sensitive areas.

13.2.4 Disruption of Pollination and Gene Flow

Climate change is affecting flowering times, pollinator activity, and plant distributions, which disrupts gene flow in medicinal plants. *Angelica glauca*, *V. jatamansi*, and *S. chirayita* often bloom at times that do not align with their pollinators, leading to fewer seeds and poor reproduction (Renner and Zohner 2018). Many medicinal plants depend on specific insect pollinators, which are either moving to new locations or declining due to changes in temperature and rainfall. For example, the high-altitude Himalayan poppy *Meconopsis integrifolia* suffers from reduced pollination at higher elevations, where pollen availability doubles at 4452 m compared to lower altitudes (Wu et al. 2015). Pollinator scarcity and habitat fragmentation together limit cross-pollination and genetic exchange. In *Panax ginseng*, studies show decreased gene flow between separate populations, often resulting from mismatched flowering times and uneven visits by pollinators. This causes inbreeding and lower genetic diversity (Zhuravlev et al. 2010). In species where seeds are dispersed by wind or animals, changes in the environment and local climate can obstruct these natural movement paths. To protect genetic diversity in climate-sensitive medicinal plants, it is vital to restore pollinator habitats and maintain natural corridors that facilitate reproduction and distribution.

13.3 Climate-Induced Changes in Phytochemical Profiles of Medicinal Plants

Climate change is increasingly influencing the quality, quantity, and overall integrity of phytochemicals in medicinal plants. Variations in temperature, rainfall, atmospheric CO_2, and the frequency of extreme weather events directly affect the biosynthesis, accumulation, and composition of secondary metabolites. These climatic stressors not only modify phytochemical concentrations but can also alter structural diversity and pharmacodynamic effectiveness (Hasnu et al. 2025).

Different climate factors trigger distinct metabolic responses. Elevated CO_2 generally enhances the production of carbon-based compounds, such as flavonoids and tannins, due to increased photosynthetic carbon flow (Jo et al. 2022). In contrast, drought, salinity, and temperature stress stimulate the synthesis of ROS-scavenging metabolites, including alkaloids and terpenoids. These stresses can suppress key bioactive constituents or lead to the emergence of novel phytochemical profiles. For example, rising temperatures and drought reduce flavonoid and phenolic content in *Vitis vinifera* and *Lavandula angustifolia*, while salinity stress increases aloin production in *A. vera* (Sharma 2024; Alum 2024). Similarly, drought enhances thymol and carvacrol levels in *Thymus daenensis*, strengthening its antimicrobial activity (Bistgani et al. 2023).

Quantitatively, climate-induced fluctuations in alkaloids, terpenoids, and glycosides depend on the intensity and duration of stress exposure. Increased soil salinity and heat have been shown to lower phenolic content and alter antioxidant potential in several medicinal species (Bistgani et al. 2019). Phytochemical variability also occurs across ecological gradients; for instance, *A. vera* from tropical and semi-arid zones exhibits distinct phenolic profiles and antioxidant capacities (Kumar et al. 2017). These variations reflect deeper transcriptional and epigenetic adjustments that regulate biosynthetic pathways under environmental stress (Jan et al. 2021).

Plant secondary metabolites, though not essential for primary metabolic processes like growth and reproduction, contribute to survival by acting as pigments, antioxidants, and enzyme inhibitors, and by defending against UV radiation, pathogens, and oxidative damage (Hu et al. 2018). The biosynthesis of secondary metabolites in plants is interconnected with multiple pathways that are initiated by primary metabolites. As climate change advances, the increasing frequency and severity of stresses, such as higher atmospheric CO_2 levels, extreme temperatures, drought, salinity, and heavy metal contamination, are collectively reshaping plant metabolism and altering the diversity and abundance of bioactive compounds they synthesize (Jangpangi et al. 2025).

Elevated CO_2 and temperature can either enhance or suppress secondary metabolite production, depending on species and stress duration (Pant et al. 2021; Jamloki et al. 2021). Abiotic stresses activate specific genes and signalling pathways involved in the synthesis of bioactive compounds, antioxidants, and phytohormones (Akula and Ravishankar 2011; Mahajan et al. 2020). Epigenetic modifications, particularly DNA/histone methylation, modulate gene expression under stress, influencing metabolite biosynthesis (Perrone and Martinelli 2020). Environmental cues, such as drought, salinity, and temperature extremes, activate stress-responsive transcription factors, such as WRKY, MYB, NAC, bZIP, and AP2/ERF, which function as a gene regulatory network governing the shikimate, mevalonate (MVA), and methylerythritol phosphate (MEP) pathways (Sharma and Puri 2017). These pathways are responsible for the biosynthesis of phenolics, flavonoids, alkaloids, and terpenoids that are critical compounds not only for plant defence but also for their therapeutic applications (Kumar et al. 2021; Cadavid et al. 2023).

Studies on *Polygonatum kingianum* (Qian et al. 2021) and *Helianthus tuberosus* revealed that drought stress alters the expression of genes in the phenylpropanoid and flavonoid biosynthesis pathways, impacting the levels of compounds like flavonoids, lignin, anthocyanins, and gingerol. Key biosynthetic genes, such as phenylalanine ammonia lyase (PAL), chalcone synthase (CHS), flavonol synthase (FLS), and anthocyanidin synthase (ANS), are often upregulated under drought, while some metabolite pathways may be downregulated depending on the plant species and recovery phase (Ahmed et al. 2021).

13.4 Impacts of Climate Change on Pharmacological Efficacy and Medical Safety

The therapeutic potential of medicinal plants is intricately linked to the concentration, ratio, and structural integrity of their bioactive constituents. Climate change is considered a critical factor influencing the pharmacological reliability and safety of phytochemicals derived from medicinal plants. For instance, Berberine demonstrates improved anti-inflammatory and antidiabetic activity under stress-enhanced accumulation; however, this may be accompanied by hepatotoxic risks depending on dosage and pharmacokinetics (Alum 2024). Likewise, *W. somnifera* (Ashwagandha or Indian ginseng), widely used for its adaptogenic and anti-inflammatory properties, exhibits enhanced withanolide content under heat stress, compromising its efficacy in stress-related disorders (Sharma and Puri 2017). Similarly, *P. quinquefolius* (American ginseng) shows altered ginsenoside profiles when cultivated under elevated CO_2 and temperature regimes, affecting its pharmacodynamics and bioavailability (Jochum et al. 2007).

Beyond efficacy, climate-induced metabolomic shifts raise safety concerns. Stress conditions may trigger the accumulation of novel or unintended phytochemicals, some of which may be toxic or interact adversely with other compounds. For example, *Rhodiola rosea*, a popular adaptogen, has shown increased levels of salidroside and rosavin under UV-B stress, but also elevated levels of unknown phenolic derivatives that have not been fully characterized for safety (El Gendy et al. 2023). Such variability complicates standardization and dosage accuracy, especially in traditional medicine systems where quality control mechanisms may be limited. The impacts of climate change on medicinal plant diversity at different biodiversity-rich regions are summarized in Table 13.1.

13.5 Strategies to Minimize the Impacts of Climate Change on Medicinal Plants

Minimizing the impacts of climate change on the genetic and phytochemical diversity of medicinal plants requires integrative strategies that blend ecological, technical, social, and policy-driven efforts. It explores critical conservation strategies, including in situ and ex situ methods, emphasizes the role of traditional knowledge and community engagement, examines the potential of climate-resilient genetic approaches and assisted migration, and reviews supportive policy frameworks for effective implementation.

13.5.1 Role of Climate-Resilient Genotypes and Assisted Migration to Minimize the Impact of Climate Change on Medicinal Plants

The selection of climate-resilient genotypes involves using advanced genomic approaches, such as genome-wide association studies and landscape-level genomic analysis, to identify genes responsible for traits like drought tolerance, temperature resilience, and phytochemical stability (Omondi et al. 2024). These adaptive genotypes can be integrated into both in situ and ex situ conservation strategies to strengthen the survival and productivity of medicinal plants under changing environmental conditions (Van Daele et al. 2024). Assisted migration, which involves deliberately relocating species or genetically diverse populations to regions expected to remain suitable under future climate conditions, is another key strategy. Approaches such as choosing seeds from areas with similar future climates, using seeds from different regions, and selecting planting materials based on future conditions can help plants adapt better to climate change. Effective application requires careful ecological assessment, risk evaluation, long-term monitoring, and adherence to ecological and ethical guidelines (Prober et al. 2015).

Table 13.1 Climate change–induced threats and impacts on medicinal plant diversity at various biodiversity hotspot regions

Scientific name	Climate change induced threats	Effects	Regions	References
Rhodiola rosea	Glacial retreat, altered snowmelt ("elevation squeeze"), overharvest	Loss/habitat contraction, genetic bottleneck, reduced regeneration, risk of extinction	Himalayas, Altai Mountains, Qinghai-Tibetan Plateau	You et al. (2018); Mykhailenko et al. (2025)
Stryphnodendron pulcherrimum	Habitat loss, warming, drought	Habitat loss, local extinction risk	Amazonia, Cerrado, Pantanal	Gomes et al. (2021a, b)
Crocus sativus	Rising temperatures, erratic rain	Drought/ waterlogging, affected corm growth, unstable phytochemical profile	Mediterranean Basin, Southern Europe	Husaini (2014); Pirasteh-Anosheh et al. (2023)
Boswellia sacra	Warmer, less rain, overexploitation	Habitat/reproductive loss, population decline, resin drop	Oman, Yemen, Ethiopia, Somalia	Johnson et al. (2025)
Nardostachys jatamansi	Rising temperatures, shifting climate, overharvesting	Chlorophyll loss, reduced biomass, oxidative stress, habitat loss, genetic bottlenecks, population collapse	Himalayas, Qinghai, Tibet	Chauhan et al. (2021)
Pilocarpus microphyllus	Diminishing suitable habitat, water stress	Declining alkaloid concentrations, poor regeneration, supply chain risk	Brazil, Amazonia	Amaral et al. (2022)
Origanum vulgare	Droughts, heat waves	Risk of extinction, active compound shifts	Mediterranean Basin	Mansinhos et al. (2024)
Commiphora wightii	Erratic rainfall, aridity, salinization	Lower resin, poor regeneration	India, North Africa, Middle East	La Montagna et al. (2023)
Aconitum heterophyllum	Rising temperatures, reduced snow, drought	Shrinking habitat, lower alkaloids, poor seed survival	Himalayas	Gaira et al. (2011)
Cinchona spp.	Drought, extreme weather	Reduced bark/ quinine concentration, seedling decline	Amazonia	Vergara et al. (2023)
Stachys inflata	Climate variability, drying	Growth decline, metabolite changes	Mediterranean Basin	Shaban et al. (2023)
Withania somnifera	Heat waves, short seasons	Reduced alkaloids, poor flowering/seed set	India, Central Asia, Middle East	Senkoro et al. (2024)

(continued)

Table 13.1 (continued)

Scientific name	Climate change induced threats	Effects	Regions	References
Saussurea laniceps	Elevated temperature, overcollection	Shifting/squeezed habitat, mountaintop extinction, decreased plant height	Himalayas	Peng et al. (2019); Zhao et al. (2024)
Croton lechleri	Increasing droughts, fire	Reduced sap, population decline	Amazonia	Feng et al. (2021)
Laurus nobilis	Temperature extremes, precipitation pattern shifts	Habitat loss, range contraction	Mediterranean Basin	Akyol et al. (2023)
Warburgia salutaris	Drought, higher temperatures, overharvest, land use change	Habitat loss, population decline	Southern and Eastern Africa	Senkoro et al. (2020, 2024)
Tabernaemontana spp.	Warmer climate, habitat fragmentation	Fragmented habitat, limited genetic exchange	Amazonia	Prosser and Brain (2024)
Aloe vera	Drought, soil degradation, high UV	Stunted growth, gel composition change, root stress	India, Africa, Middle East, Americas	Kumar et al. (2017); Karmaoui and Moussafir (2024)
Podophyllum hexandrum	Soil moisture decline, elevated temperature	Lower podophyllotoxin, poor seed dispersal	Himalayas	Gulzar et al. (2024)
Psychotria viridis	Droughts, lower humidity	Lower alkaloids, reduced bark/leaves' active contents	Amazonia	Cavalcante et al. (2018)
Mentha pulegium	Long dry seasons, erratic rainfall	Biomass loss, phytochemical reduction, population loss	Mediterranean Basin	Laftouhi et al. (2023)
Ziziphusspina-christi	Water table loss, warming, drought	Lower fruit yield, more pests	North Africa, Middle East, South Asia	Zait et al. (2020)
Valeriana jatamansi	Drought during flowering	Root essential oil loss, poor flowering	Himalayas	Charmakar et al. (2021)
Copaifera spp.	Reduced rainfall, fire, humidity	Resin yield decline, loss of older trees	Amazonia	Souza et al. (2019); da Costa et al. (2025)
Artemisia herba-alba	Less rain, drought	Lower biomass, habitat squeeze	North Africa, Middle East, South Asia	Verma and Verma (2023)
Sideritis spp.	Warming, invasive plants, greening, dry weather	Habitat loss, genetic diversity decrease, local extinction risk	Mediterranean Basin, Balkans	Theodoridis et al. (2024)

(continued)

Table 13.1 (continued)

Scientific name	Climate change induced threats	Effects	Regions	References
Capparis decidua	Erratic monsoons, soil erosion	Poor seedling, fruiting cycle, shrinking habitat	India, North Africa, Middle East	Ahmad et al. (2022); Iqbal et al. (2025)
Bergenia ciliata	Higher temperature, reduced snow	Lower biomass, local extinction	Himalayas	Kaliyathan et al. (2016)
Cayaponia tayuya	Drought, land-use change	Habitat/regenerative loss	Amazonia	Amaral et al. (2023)
Aristolochia paucinervis	Less wet habitats, drying trends	Wild shrinking populations, genetic loss	Mediterranean Basin	Berjano et al. (2006)
Arnica montana	Summer aridity, temperature and precipitation fluctuation	Lower survival, reduced growth and reproduction rate	European montane grasslands	Vikane et al. (2019); Stanik et al. (2020, 2021)

13.5.2 Role of Traditional Knowledge and Community Participation

Community seed banks offer frontline defence for the conservation of climate-resilient plant genetic resources. These are locally governed institutions that conserve, exchange, and regenerate seeds adapted to specific agroecological conditions. A study conducted in the Hindu Kush Himalaya region of Nepal documented the conservation of 117 traditional crop varieties in a community-managed seed bank. This initiative not only helped preserve locally adapted genotypes but also significantly reduced seed input costs for farmers (Vernooy et al. 2013). In Uganda, the Kiziba Community Seed Bank has contributed to the enhancement of on-farm genetic diversity while also serving as a living repository of indigenous agricultural knowledge and traditional seed management practices (Vernooy et al. 2017). Globally, community seed banks in Ethiopia, India, and South Africa are increasingly recognized in strategic climate adaptation plans. Community seed banks fulfil multiple roles in conservation: preserving rare genotypes, enhancing seed access, strengthening food sovereignty, and supporting local knowledge (Vernooy et al. 2015, 2017). They also function as platforms for sharing seeds and knowledge, documented in biodiversity registers and guided by secure benefit-sharing mechanisms (Vernooy et al. 2017).

13.5.3 Policy Frameworks and Global Initiatives

International-level frameworks such as the Convention on Biological Diversity, the Nagoya Protocol on Access and Benefit Sharing, the Sustainable Development Goals, and the Global Plan of Action for Plant Genetic Resources for Food and

Agriculture promote the conservation of genetic diversity, fair sharing of benefits, and the equitable use of biological resources. For example, Indonesia's national conservation assessments recommend combining in situ and ex situ strategies, guided by these global instruments. At the regional level, policies play a vital role in supporting community seed banks, establishing small protected areas known as micro-reserves, encouraging participatory plant breeding, and integrating traditional knowledge systems into conservation planning. To make conservation efforts more responsive to climate change, adaptive planning should include the use of geographic information system-based gap analysis to identify areas that act as natural refuges and regions with high genetic diversity. Best practices for implementation include developing transparent methods for valuing community contributions, strengthening local genetic resource centres through capacity-building, offering financial support for community-led conservation activities, and ensuring legal frameworks that protect farmers' seed rights and promote fair benefit-sharing mechanisms.

13.5.4 Sustainable Harvesting and Bioprospecting

Sustainable harvesting ensures long-term availability of medicinal and economically important plants (Hamilton 2004; Kala 2015; Hasnu et al. 2025). Bioprospecting, the systematic search for novel bioactive compounds, has advanced with the integration of ethnobotanical knowledge, molecular profiling, and phytochemical screening (Newman and Cragg 2020). High-throughput in vitro assays and advanced chromatographic techniques, such as high-performance liquid chromatography (HPLC), high-performance thin-layer chromatography (HPTLC), and liquid chromatography-mass spectrometry (LC-MS), enable efficient screening and identification of active plant constituents (Halder et al. 2024). Molecular markers, DNA barcoding, and metabolomics now provide new avenues for understanding genetic diversity, traceability, preventing overexploitation, and ensuring quality control (Chen et al. 2010, 2023; Ganie et al. 2015). Case studies highlight the benefits of participatory conservation. In the Eastern Himalayas, community-driven bioprospecting improved conservation of the endangered *S. chirayita* (Joshi and Dhawan 2007). Advances in plant tissue and cell culture further ensure reliable production of valuable compounds, bypassing pressures on wild populations (Pithiya et al. 2022; Kirakosyan et al. 2004). Bioprospecting frameworks must balance innovation with equity. The Nagoya Protocol emphasizes fair access and benefit-sharing (Kamau et al. 2010); protecting indigenous rights while encouraging sustainable use. Traditional knowledge remains central to these efforts, providing insights into plant-based medicine and guiding conservation priorities (Hasnu et al. 2025). Together, sustainable harvesting and bioprospecting demonstrate how modern science and traditional wisdom can converge to secure biodiversity and global health solutions (Fig. 13.1).

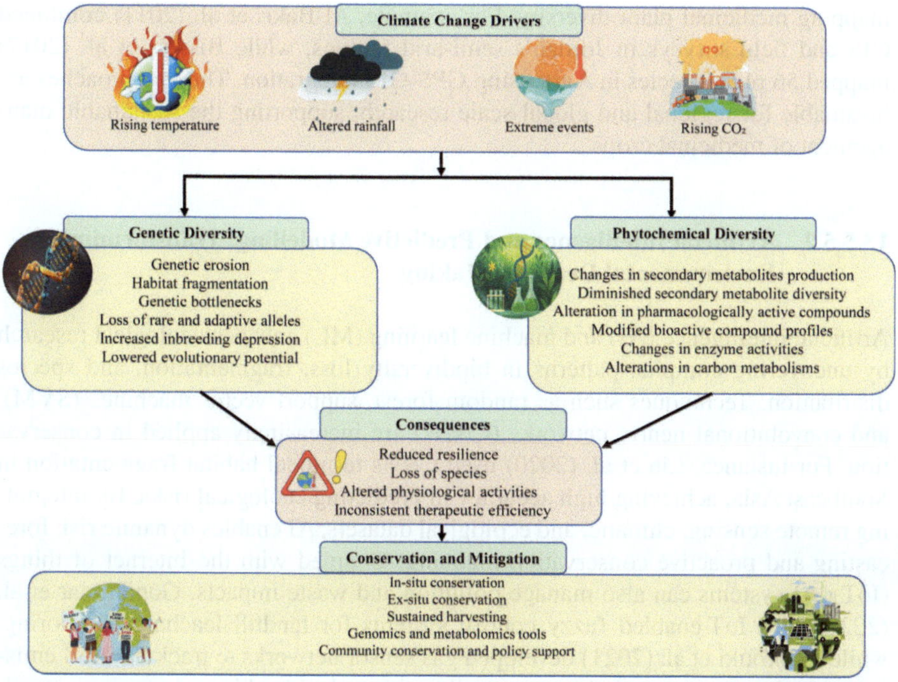

Fig. 13.1 Conceptual framework showing how climate change drivers affect genetic and phytochemical diversity of medicinal plants, leading to species loss, reduced resilience, and the need for conservation strategies

13.5.5 Emerging Tools and Technologies for Monitoring and Minimizing Impacts of Climatic Changes

13.5.5.1 Remote Sensing and Geospatial Mapping: Tracking Environmental Change in Real Time

Remote sensing technologies integrated with geographic information system (GIS) provide essential tools for monitoring land-use change, habitat loss, and climate-driven shifts in ecosystems. Satellite platforms such as MODIS, Landsat, and Sentinel generate high-resolution temporal and spatial data critical for tracking deforestation, desertification, and phenological changes. Kumar et al. (2021a, b) demonstrated this potential by detecting the upward migration of alpine medicinal plants in the Western Himalayas, highlighting ecosystem sensitivity to warming. Geospatial analytics also enable species distribution modelling (SDM) and habitat suitability assessments, supporting the identification of conservation hotspots and climate refugia. UAVs (drones) further enhance localized monitoring of vegetation health and land-use dynamics. Specialized software such as GeoMedia, integrated land and water information system (ILWIS), and earth resources (ER) Mapper are widely used for crop modelling and spatial analysis. Practical applications include mapping medicinal plant diversity. For example, Al-Bakri et al. (2011) combined GIS and field surveys in Jordan's semi-arid regions, while Biswas et al. (2017) mapped 56 plant species in Agra using GPS-GIS integration. These approaches are invaluable for regional and global-scale research, supporting the sustainable management of medicinal crops.

13.5.5.2 Artificial Intelligence and Predictive Modelling: Transforming Environmental Decision-Making

Artificial intelligence (AI) and machine learning (ML) enhance ecological research by uncovering complex patterns in biodiversity loss, fragmentation, and species distribution. Techniques such as random forest, support vector machines (SVM), and convolutional neural networks (CNNs) are increasingly applied in conservation. For instance, Liu et al. (2020) used CNNs to model habitat fragmentation in Southeast Asia, achieving high accuracy in predicting ecological risks. By integrating remote sensing, climatic, and ecological datasets, AI enables dynamic risk forecasting and proactive conservation planning. Coupled with the Internet of things (IoT), AI systems can also manage pollution and waste impacts. Gopikumar et al. (2021) used IoT-enabled fuzzy control systems for landfill leachate monitoring, while Mabrouki et al. (2021) developed gas sensor networks to track harmful emissions. Such technologies help protect soil fertility, plant health, and ecosystem resilience. AI-driven systems thus provide policymakers with powerful tools for energy efficiency, emission reduction, and cost-effective climate adaptation (Guo et al. 2022).

13.5.5.3 Genomics: Decoding Biodiversity for Effective Conservation and Climate Adaptation

Genomic tools are increasingly vital for guiding conservation strategies aimed at safeguarding medicinal plant diversity under changing climates. In situ conservation through protected areas, micro-reserves, and community-managed forests allows species to retain evolutionary potential while maintaining interactions with their native ecosystems (Xia et al. 2022). Climate projections suggest that more than 50% of key medicinal plants in regions like Indonesia may lose suitable habitats by 2050 under various representative concentration pathways (RCP) scenarios, emphasizing the urgency of such measures (Cahyaningsih et al. 2021). Ex situ approaches, including seed banks, cryopreservation, and botanical garden collections, provide secure genetic repositories supported by biotechnological tools, such as tissue culture, micropropagation, and synthetic seeds (Chen et al. 2016). Integrating genomic data with these in situ and ex situ strategies such as using germplasm reintroduction or habitat restoration based on genetic diversity patterns, strengthens adaptive capacity and ensures long-term resilience of medicinal plant resources in the face of climate change (Xia et al. 2022).

Genomics has revolutionized the study of biodiversity by revealing the genetic mechanisms underpinning adaptation, resilience, and evolutionary processes. Tools, such as whole-genome sequencing, RNA-Seq, and environmental DNA (eDNA), enable researchers to identify genetic variants linked to drought tolerance, salinity resistance, and phenotypic plasticity in medicinal and wild plants (Taberlet et al. 2018). High-throughput sequencing platforms facilitate the development of molecular markers like SSRs and SNPs, which are crucial for monitoring genetic diversity and breeding climate-resilient crops (Varshney et al. 2021). Genotyping-by-sequencing (GBS) has expanded the pool of molecular markers, enabling genome-wide association studies (GWAS) to validate trait-linked haplotypes (Elshire et al. 2011; Poland and Rife 2012). Similarly, MutMap helped identify the *OsRR22* gene conferring salt tolerance in rice (Abe et al. 2012; Zhang et al. 2019). Transcriptomics further supports climate adaptation research. RNA-Seq has been used to profile stress-induced gene expression (Wang et al. 2009), while pan-genomics and graph-based assemblies capture genomic diversity at the population level (Yan et al. 2023a, b). High-quality genome assemblies, such as those developed for drought-tolerant maize germplasm CIMBL55, highlight the power of third-generation sequencing and mapping technologies (Hu et al. 2018; Yang et al. 2019). Beyond crop improvement, genomic data aid phylogenetic reconstruction, species delimitation, and conservation prioritization in biodiversity hotspots (Rieseberg and Willis 2007; Savolainen et al. 2013). These advances support ex situ conservation, seed banks, and in situ monitoring, aligning with global frameworks, such as the Convention on Biological Diversity and FAO's Global Plan of Action. In sum, genomics provides a foundation for both climate adaptation and biodiversity conservation of medicinal plants.

13.6 Research Gaps and Limitations

Despite growing interest, several gaps remain in understanding how climate change affects medicinal plants. One major issue is the lack of sufficient studies on advanced tools like genomics and metabolomics, especially for lesser-known species. Most studies rely on basic traits, which restrict insight into how plants respond at the molecular level (Napier et al. 2023). Species distribution models often ignore fine-scale environmental factors, reducing their accuracy. Fragmented and remote habitats are less represented in research. Habitat fragmentation reduces gene flow and increases selective pressures, as shown in *Primula elatior* populations (Van Daele et al. 2023). The social dimensions of climate resilience are also neglected. Traditional knowledge and local practices are rarely integrated into research and policy, despite their importance for sustainable development. Similarly, the effects of climate factors like CO_2, ozone, and heat on bioactive compound production are still poorly understood (Jangpangi et al. 2025). Finally, extinction rates have increased drastically, driven by climate change and habitat loss. This threatens medicinal species before they are studied or conserved (Pimm et al. 2014). Linking extinction risk to conservation planning remains a key research need.

13.7 Strategic Recommendations

The survival and effectiveness of medicinal plants are under serious threat from human-induced stressors and climate change. It is advised to use a comprehensive, scientifically based approach with a focus on conservation, technology development, governmental support, and community involvement to guarantee their continuous availability and efficacy. Integrated in situ and ex situ conservation is essential. In situ conservation maintains natural evolutionary processes, while ex situ approaches, such as seed banking, cryopreservation, and tissue culture, offer security against habitat loss (Rayalu et al. 2024; Gaur 2024). Additionally, Sharma et al. (2025) studied biotechnological interventions, including tissue culture, micropropagation, and CRISPR/Cas9 genome editing, which enabled mass propagation and targeted enhancement of medicinal traits. These innovations reduce wild harvesting pressure and improve phytochemical stability. Omics technologies such as genomics, transcriptomics, proteomics, and metabolomics can be used, which offer insights into plant responses to climate stress and support the identification of resilient genotypes for breeding and conservation (Kumar et al. 2021a, b; Chaturvedi 2022). Integrating traditional ecological knowledge and community-based conservation enhances resource stewardship and promotes sustainable use. According to Gupta and Koontz (2019), local participation in harvesting and benefit-sharing strengthens conservation outcomes. Genetic diversity in cultivation should be prioritized over monoculture to enhance adaptive capacity and maintain consistent medicinal quality, according to Van Daele et al. (2023).

13.8 Conclusion

Climate change poses a dual challenge to medicinal plants, affecting both their genetic and phytochemical stability. While existing studies highlight these risks, there remain significant gaps that need to be addressed. This includes integrating omics-based research into conservation efforts, involving local communities in monitoring and adopting cultivation practices that preserve both genetic and chemical diversity. By implementing climate-smart in situ and ex situ conservation, improving landscape connectivity, and fostering collaboration between science and traditional knowledge, we can protect the biological and therapeutic foundations of medicinal plants for future generations.

References

Abe A, Kosugi S, Yoshida K, Natsume S, Takagi H, Kanzaki H, Terauchi R (2012) Genome sequencing reveals agronomically important loci in rice using MutMap. Nat Biotechnol 30(2):174–178. https://doi.org/10.1038/nbt.2095

Ahmad M, Iqbal Z, Bushra Hussain A, Abdullah M, Alataway A, Mattar MA (2022) Seasonal changes modulate the rhizosphere of desert plant species. Agronomy 13(1):57. https://doi.org/10.3390/agronomy13010057

Ahmed S, Stepp J, Orians C, Griffin T, Matyas C, Robbat A, Cash SB (2019) Effects of extreme drought on tea (*Camellia sinensis*) functional quality validate indigenous farmer knowledge and sensory preferences in tropical China. PLoS One 14(5):e0217011. https://doi.org/10.1371/journal.pone.0217011

Ahmed U, Rao MJ, Qi C, Xie Q, Noushahi HA, Yaseen M, Zheng B (2021) Expression profiling of flavonoid biosynthesis genes and secondary metabolites accumulation in Populus under drought stress. Molecules 26(18):5546. https://doi.org/10.3390/molecules26185546

Akula R, Ravishankar GA (2011) Influence of abiotic stress signals on secondary metabolites in plants. Plant Signal Behav 6(11):1720–1731. https://doi.org/10.4161/psb.6.11.17613

Akyol A, Örücü ÖK, Arslan ES, Sarıkaya AG (2023) Predicting of the current and future geographical distribution of *Laurus nobilis* L. under the effects of climate change. Environ Monit Assess 195(4):459. https://doi.org/10.1007/s10661-023-11086-z

Alamgir ANM (2017) Therapeutic use of medicinal plants and their extracts: volume 1. Springer, Cham. https://doi.org/10.1007/978-3-319-63862-1

Al-Bakri J, Al-Eisawi D, Damhoureyeh S, Oran S (2011) GIS-based analysis of spatial distribution of medicinal and herbal plants in arid and semi-arid zones in the Northwest of Jordan. Ann Arid Zone 50(2):99–115. https://doi.org/10.56093/aaz.v50i2.63697

Alum EU (2024) Climate change and its impact on the bioactive compound profile of medicinal plants: implications for global health. Plant Signal Behav 19(1):2419683. https://doi.org/10.1080/15592324.2024.2419683

Amaral GC, Pezzopane JE, de Souza Nóia Júnior R, Martínez MF, Fonseca MD, Gibson EL, Toledo JV, Pezzopane JR, Klippel VH, Xavier TM (2022) *Pilocarpus microphyllus* seedling growth threatened by climate change: an ecophysiological approach. Theor Appl Climatol 147(1):347–361. https://doi.org/10.1007/s00704-021-03831-6

Amaral GC, Pezzopane JE, de Souza Noia Junior R, Fonseca MD, Martínez MF, de Oliveira Gomes V, Toledo JV, Pezzopane JR, Martín RT (2023) Climate change and the growth of Amazonian species seedlings: an ecophysiological approach to *Euterpe oleracea*. New For 54(2):269–287. https://doi.org/10.1007/s11056-022-09921-1

Berjano R, de Vega C, Arista M, Ortiz PL, Talavera S (2006) A multi-year study of factors affecting fruit production in *Aristolochia paucinervis* (Aristolochiaceae). Am J Bot 93(4):599–606. https://doi.org/10.3732/ajb.93.4.599

Bistgani ZE, Hashemi M, DaCosta M, Craker L, Maggi F, Morshedloo MR (2019) Effect of salinity stress on the physiological characteristics, phenolic compounds and antioxidant activity of *Thymus vulgaris* L. and *Thymus daenensis* Celak. Ind Crop Prod 135:311–320. https://doi.org/10.1016/j.indcrop.2019.04.055

Bistgani ZE, Hashemi M, Akbari P, Mumivand H (2023) Influence of drought stress and foliar application of chitosan on nutrient accumulation and phenolic composition of *Thymus daenensis* Celak. Crop Sci 63(2):921–935. https://doi.org/10.1002/csc2.20901

Biswas B, Walker S, Varun M (2017) Web GIS based identification and mapping of medicinal plants: a case study of Agra (U.P.), India. Plant Arch 17(1):8–20

Cadavid IC, Balbinott N, Margis R (2023) Beyond transcription factors: more regulatory layers affecting soybean gene expression under abiotic stress. Genet Mol Biol 46(1 Suppl 1):e20220166. https://doi.org/10.1590/1678-4685-GMB-2022-0166

Cahyaningsih R, Phillips J, Brehm JM, Gaisberger H, Maxted N (2021) Climate change impact on medicinal plants in Indonesia. Glob Ecol Conserv 30:e01752. https://doi.org/10.1016/j.gecco.2021.e01752

Cavalcante AD, Cardoso GA, Oliveira FLD, Bearzoti E, Okuma AA, Duarte LP, Vieira-Filho SA (2018) Influence of environmental factors and cultural methods on the content of N,N-dimethyltryptamine in *Psychotria viridis* (Rubiaceae). J Braz Chem Soc 29(6):1245–1255. https://doi.org/10.21577/0103-5053.20170221

Chakraborty S, Mukherjee D, Baskey S (2015) Floral homeostasis breakdown in endangered plant *Valeriana jatamansi* Jones (Valerianaceae) in North Eastern Himalayan region. Am J Plant Sci 6(19):3119. https://doi.org/10.4236/ajps.2015.619304

Chandra N, Singh G, Lingwal S, Bisht MPS, Tewari LM, Joshi VC (2022) Ecological status of alpine medicinal and aromatic plants of Western Himalaya. J Herbs Spices Med Plants 28(1):73–88. https://doi.org/10.1080/10496475.2021.1963903

Charmakar S, Kunwar RM, Sharma HP, Rimal B, Baral S, Joshi N, Gauli K, Acharya RP, Oli BM (2021) Production, distribution, use and trade of *Valeriana jatamansi* Jones in Nepal. Glob Ecol Conserv 30:e01792. https://doi.org/10.1016/j.gecco.2021.e01792

Chaturvedi T (2022) Maximizing genetic gain in Kalmegh. Front Plant Sci 13:1042222. https://doi.org/10.3389/fpls.2022.1042222

Chauhan HK, Oli S, Bisht AK, Meredith C, Leaman D (2021) Review of the biology, uses and conservation of the critically endangered endemic Himalayan species *Nardostachys jatamansi* (Caprifoliaceae). Biodivers Conserv 30(12):3315–3333. https://link.springer.com/article/10.1007%2Fs10531-021-02269-6

Chen S, Yao H, Han J, Liu C, Song J, Shi L, Leon C (2010) Validation of the ITS2 region as a novel DNA barcode for identifying medicinal plant species. PLoS One 5(1):e8613. https://doi.org/10.1371/journal.pone.0008613

Chen SL, Yu H, Luo HM, Wu Q, Li CF, Steinmetz A (2016) Conservation and sustainable use of medicinal plants: problems, progress, and prospects. Chin Med 11(1):37. https://doi.org/10.1186/s13020-016-0108-7

Chen S, Dang D, Liu Y, Ji S, Zheng H, Zhao C, Ruan Y (2023) Genome-wide association study presents insights into the genetic architecture of drought tolerance in maize seedlings under field water-deficit conditions. Front Plant Sci 14:1165582. https://doi.org/10.3389/fpls.2023.1165582

da Costa P, de Castilho CV, Cito AC, Barbosa RI, Kaminski PE, Martins K, de Oliveira Wadt LH (2025) Impact of oleoresin harvesting on the reproductive phenology of *Copaifera pubiflora* Benth. (Leguminosae) in a precipitation gradient in Northern Brazilian Amazon. For Ecol Manag 578:122462. https://doi.org/10.1016/j.foreco.2024.122462

Dhar U, Rawal RS, Upreti J (2002) Setting priorities for conservation of medicinal plants: a case study in the Indian Himalaya. Biol Conserv 95(1):57–65. https://doi.org/10.1016/S0006-3207(00)00199-9

El Gendy ANG, Fouad R, Omer EA, Cock IE (2023) Effects of climate change on medicinal plants and their active constituents. In: Hasanuzzaman M (ed) Climate-resilient agriculture, vol 1: crop responses and agroecological perspectives. Springer International Publishing, Cham, pp 125–156. https://doi.org/10.1007/978-3-031-37424-1_6

Elshire RJ, Glaubitz JC, Sun Q, Poland JA, Kawamoto K, Buckler ES, Mitchell SE (2011) A robust, simple genotyping-by-sequencing (GBS) approach for high diversity species. PLoS One 6(5):e19379. https://doi.org/10.1371/journal.pone.0019379

Feng X, Merow C, Liu Z, Park DS, Roehrdanz PR, Maitner B, Enquist BJ (2021) How deregulation, drought and increasing fire impact Amazonian biodiversity. Nature 597(7877):516–521. https://doi.org/10.1038/s41586-021-03876-7

Food and Agriculture Organization of the United Nations (2010) The second report on the state of the world's plant genetic resources for food and agriculture. https://www.researchgate.net/publication/259973338

Gaira KS, Dhar U, Belwal OK (2011) Potential of herbarium records to sequence phenological pattern: A case study of *Aconitum heterophyllum* in the Himalaya. Biodiv Cons 20(10):2201–2210. https://doi.org/10.1007/s10531-011-0082-4

Gairola S, Shariff NM, Bhatt A, Kala CP (2010) Influence of climate change on production of secondary chemicals in high altitude medicinal plants: issues needs immediate attention. J Med Plant Res 4(18):1825–1829. https://doi.org/10.5897/JMPR10.354

Ganie SH, Upadhyay P, Das S, Sharma MP (2015) Authentication of medicinal plants by DNA markers. Plant Gene 4:83–99. https://doi.org/10.1016/j.plgene.2015.10.002

Gaur R (2024) Assessing the impact of climate change on medicinal plants: strategies for sustainable management. Afr J Biomed Res 27(3S):2381–2390. https://doi.org/10.53555/AJBR.v27i3S.2647

Gomes PW, Pamplona TC, Navegantes-Lima KC, Quadros LB, Oliveira AL, Santos SM, da Silva MN (2021a) Chemical composition and antibacterial action of *Stryphnodendron pulcherrimum* bark extract, "barbatimão" species: evaluation of its use as a topical agent. Arab J Chem 14(6):103183. https://doi.org/10.1016/j.arabjc.2021.103183

Gomes P, Quirós-Guerrero L, Silva C, Pamplona S, Boutin JA, Eberlin M, Wolfender JL, Silva M (2021b) Feature-based molecular network-guided dereplication of natural bioactive products from leaves of *Stryphnodendron pulcherrimum* (Willd.) Hochr. Metabolites 11(5):281. https://doi.org/10.3390/metabo11050281

Gopikumar S, Raja S, Robinson YH, Shanmuganathan V, Chang H, Rho S (2021) A method of landfill leachate management using internet of things for sustainable smart city development. Sustain Cities Soc 66:102521. https://doi.org/10.1016/j.scs.2020.102521

Gulzar H, Sofi P, Pala NA, Peerzada IA, Shah IA, Nazir S, Kaif M (2024) Effect of spacing and shading intensity on growth performance of *Podophyllum hexandrum* Royle: a critically endangered plant species of Western Himalaya. Ecol Front 44:105–111. https://doi.org/10.1016/j.chnaes.2023.07.007

Guo Q, Ren M, Wu S, Sun Y, Wang J, Wang Q, Chen Y (2022) Applications of artificial intelligence in the field of air pollution: a bibliometric analysis. Front Public Health 10:933665. https://doi.org/10.3389/fpubh.2022.933665.2972

Gupta D, Koontz TM (2019) Working together? Synergies in government and NGO roles for community forestry in the Indian Himalayas. World Dev 114:326–340. https://doi.org/10.1016/j.worlddev.2018.09.016

Gurib-Fakim A (2006) Medicinal plants: traditions of yesterday and drugs of tomorrow. Mol Asp Med 27(1):1–93. https://doi.org/10.1016/j.mam.2005.07.008

Halder M, Kundu A, Jha S (2024) Secondary metabolites identification techniques of the current era. In: Mérillon JM, Ramawat KG (eds) Plant specialized metabolites: phytochemistry, ecology and biotechnology. Springer Nature Switzerland, Cham, pp 1–41

Hamilton AC (2004) Medicinal plants, conservation and livelihoods. Biodivers Conserv 13(8):1477–1517. https://doi.org/10.1023/B:BIOC.0000021333.23413.42

Hasnu S, Kundu BK, Basumatary BR, Goswami D, Som S, Rehman M, Tanti B (2025) Phytochemical profiling and antioxidant potential in the beans of *Vanilla borneensis* Rolfe – a rare, endemic, and threatened Orchid of Assam, India. Pharmacol Res 9:100372. https://doi.org/10.1016/j.prenap.2025.100372

Hu L, Robert CA, Cadot S, Zhang XI, Ye M, Li B, Erb M (2018) Root exudate metabolites drive plant-soil feedbacks on growth and defense by shaping the rhizosphere microbiota. Nat Commun 9(1):2738. https://doi.org/10.1038/s41467-018-05122-7

Husaini AM (2014) Challenges of climate change: omics-based biology of saffron plants and organic agricultural biotechnology for sustainable saffron production. GM Crops Food 5(2):97–105. https://doi.org/10.4161/gmcr.29436

IPCC (2019) Climate change and land: an IPCC special report. Intergovernmental Panel on Climate Change. https://www.ipcc.ch/srccl/

Iqbal U, Sana A, Abid S, Rafiq S, Sharif M, Wahab A, Ahmad KS (2025) Resistance and resilience of desert-dwelling species *Capparis decidua* (Forssk.) Edgew to desert ecosystem. Int J Environ Sci Technol 22:14497–14518. https://doi.org/10.1007/s13762-025-06579-y

Jamloki A, Bhattacharyya M, Nautiyal MC, Patni B (2021) Elucidating the relevance of high temperature and elevated CO_2 in plant secondary metabolites (PSMs) production. Heliyon 7(8):e07709. https://doi.org/10.1016/j.heliyon.2021.e07709

Jan R, Asaf S, Numan M, Lubna, Kim KM (2021) Plant secondary metabolite biosynthesis and transcriptional regulation in response to biotic and abiotic stress conditions. Agronomy 11(5):968. https://doi.org/10.3390/agronomy11050968

Jangpangi D, Patni B, Chandola V, Chandra S (2025) Medicinal plants in a changing climate: understanding the links between environmental stress and secondary metabolite synthesis. Front Plant Sci 16:1587337. https://doi.org/10.3389/fpls.2025.1587337

Jo NY, Lee J, Byeon JE, Park HJ, Ryoo JW, Hwang SG (2022) Elevated CO_2 concentration induces changes in plant growth, transcriptome, and antioxidant activity in fennel (*Foeniculum vulgare* Mill.). Front Plant Sci 13:1067713. https://doi.org/10.3389/fpls.2022.1067713

Jochum GM, Mudge KW, Thomas RB (2007) Elevated temperatures increase leaf senescence and root secondary metabolite concentrations in the understory herb *Panax quinquefolius* (Araliaceae). Am J Bot 94(5):819–826. https://doi.org/10.3732/ajb.94.5.819

Johnson S, Said AB, Vahalík P, Karas L, Willson MS, Bongers F (2025) Rapid conservation assessment of *Boswellia sacra* in Oman reveals complex threat and population patterns. J Arid Environ 229:105368. https://doi.org/10.1016/j.jaridenv.2025.105368

Joshi P, Dhawan V (2007) Analysis of genetic diversity among *Swertia chirayita* genotypes. Biol Plant 51(4):764–768. https://doi.org/10.1007/s10535-007-0156-z

Kala CP (2015) Indigenous knowledge on medicinal plants: lessons learned from communities in India. J Ethnobiol Ethnomed 11(1):20. https://doi.org/10.1186/s13002-015-0006-9

Kaliyathan NM, Sharma AK, Verma AK, Singh A (2016) Species distribution modeling of *Bergenia ciliata* in Mussoorie range of lesser Himalayas. Ann For 24(2):149–158

Kamau EC, Fedder B, Winter G (2010) The Nagoya Protocol on access to genetic resources and benefit sharing: what is new and what are the implications for provider and user countries and the scientific community? Law Environ Dev J 6(3):246–262. https://papers.ssrn.com/sol3/papers.cfm?abstract_id=2178975

Karmaoui A, Moussafir Z (2024) *Aloe vera* plant: characterizing the components and activities and exploring some stress factors associated with climate change. In: Water-soil-plant-animal nexus in the era of climate change. IGI Global Scientific Publishing, pp 1–27. https://doi.org/10.4018/978-1-6684-9838-5.ch001

Khan S, Shah RA (2016) Assessment of genetic diversity among India Ginseng, *Withania somnifera* (L) Dunal using RAPD and ISSR markers. Res Biotechnol 7:1–10. https://doi.org/10.19071/rib.2016.v7.2818

Kirakosyan A, Sirvent TM, Gibson DM, Kaufman PB (2004) The production of hypericins and hyperforin by in vitro cultures of St. John's wort (*Hypericum perforatum*). Biotechnol Appl Biochem 39(1):71–81. https://doi.org/10.1042/BA20030144

Kumar A (2021) Omics technologies for sustainable agriculture. Spring. https://doi.org/10.1007/978-981-16-2956-3

Kumar S, Yadav A, Yadav M, Yadav JP (2017) Effect of climate change on phytochemical diversity, total phenolic content and in vitro antioxidant activity of *Aloe vera* (L.) Burm.f. BMC Res Notes 10(1):60. https://doi.org/10.1186/s13104-017-2385-3

Kumar R, Das S, Mishra M, Choudhury DR, Sharma K, Kumari A, Singh R (2021a) Emerging roles of NAC transcription factor in medicinal plants: progress and prospects. 3 Biotech 11(10):425. https://doi.org/10.1007/s13205-021-02970-x

Kumar V, Singh P, Singh PK, Rahim MS, Sharma V, Roy J, Sharma H (2021b) *Trillium govanianum*. In: Himalayan medicinal plants: advances in botany, production & research. Academic Press, pp 243–257. https://doi.org/10.1016/B978-0-12-823151-7.00006-4

Kumar R, Salar RK, Naik PK, Yadav M, Kumar A, Kumar A, Chhokar V (2022) Elucidation of genetic diversity and population structure of sixty genotypes of *Aloe vera* using AFLP markers. S Afr J Bot 147:1146–1155. https://doi.org/10.1016/j.sajb.2021.06.008

La Montagna D, Attorre F, Hamdiah S, Maděra P, Malatesta L, Vahalík P, De Sanctis M (2023) Climate change effects on the potential distribution of the endemic *Commiphora species* (Burseraceae) on the island of Socotra. Front For Glob Change 6:1183858. https://doi.org/10.3389/ffgc.2023.1183858

Laftouhi A, Eloutassi N, Ech-Chihbi E, Rais Z, Taleb A, Assouguem A, Ullah R, Kara M, Fidan H, Beniken M, Taleb M (2023) Impact of climate change on the chemical compositions and antioxidant activity of *Mentha pulegium* L. ACS Omega 8(49):46598–46607. https://doi.org/10.1021/acsomega.3c05564

Lambert J, Srivastava J, Vietmeyer N (1997) Medicinal plants: rescuing a global heritage, World Bank Technical Paper No. 355. World Bank, Washington, DC. http://documents.worldbank.org/curated/en/538541468242090002

Lindsey R (2020) Climate change: atmospheric carbon dioxide. NOAA. Climate.gov. Retrieved from https://www.climate.gov

Liu D, Jiang W, Mu L, Wang S (2020) Streamflow prediction using deep learning neural network: case study of Yangtze River. IEEE Access 8:90069–90086

Mabrouki J, Azrour M, Fattah G, Dhiba D, El Hajjaji S (2021) Intelligent monitoring system for biogas detection based on the Internet of Things: Mohammedia, Morocco city landfill case. Big Data Min Anal 4(1):10–17. https://doi.org/10.26599/BDMA.2020.9020017

Mahajan M, Kuiry R, Pal PK (2020) Understanding the consequence of environmental stress for accumulation of secondary metabolites in medicinal and aromatic plants. J Appl Res Med Aromat Plants 18:100255. https://doi.org/10.1016/j.jarmap.2020.100255

Mansinhos I, Goncalves S, Romano A (2024) How climate change-related abiotic factors affect the production of industrial valuable compounds in Lamiaceae plant species: a review. Front Plant Sci 15:1370810. https://doi.org/10.3389/fpls.2024.1370810

Mondal TK, Palit P (2022) Impact of climate change on medicinal plants: a review. Genet Resour Crop Evol 69:1613–1632. https://doi.org/10.1007/s10722-021-01284-9

Mustafa G, Arif R, Atta A, Sharif S, Jamil A (2017) Bioactive compounds from medicinal plants and their importance in drug discovery in Pakistan. Matrix Sci Pharma 1(1):17–26. https://doi.org/10.26480/msp.01.2017.17.26

Muzila M, Werlemark G, Ortiz R, Sehic J, Fatih M, Setshogo M, Nybom H (2014) Assessment of diversity in *Harpagophytum* with RAPD and ISSR markers provides evidence of introgression. Hereditas 151(4–5):91–101. https://doi.org/10.1111/hrd2.00052

Mykhailenko O, Jalil B, McGaw LJ, Echeverría J, Takubessi M, Heinrich M (2025) Climate change and the sustainable use of medicinal plants: a call for "new" research strategies. Front Pharmacol 15:1496792. https://doi.org/10.3389/fphar.2024.1496792

Napier JD, Heckman RW, Juenger TE (2023) Gene-by-environment interactions in plants: molecular mechanisms, environmental drivers, and adaptive plasticity. Plant Cell 35(1):109–124. https://doi.org/10.1093/plcell/koac322

Newman DJ, Cragg GM (2020) Natural products as sources of new drugs over the nearly four decades from 01/1981 to 09/2019. J Nat Prod 83(3):770–803. https://doi.org/10.1021/acs.jnatprod.9b01285

Omondi EO, Lin CY, Huang SM, Liao CA, Lin YP, Oliva R, van Zonneveld M (2024) Landscape genomics reveals genetic signals of environmental adaptation of African wild eggplants. Ecol Evol 14(7):e11662. https://doi.org/10.1002/ece3.11662

Pan SY, Litscher G, Gao SH, Zhou SF, Yu ZL, Chen HQ, Ko KM (2014) Historical perspective of traditional indigenous medical practices: the current renaissance and conservation of herbal resources. Evid Based Complement Alternat Med 2014:525340. https://doi.org/10.1155/2014/525340

Pant P, Pandey S, Dall'Acqua S (2021) The influence of environmental conditions on secondary metabolites in medicinal plants: a literature review. Chem Biodivers 18(11):e2100345. https://doi.org/10.1002/cbdv.202100345

Peng D, Sun L, Pritchard HW, Yang J, Sun H, Li Z (2019) Species distribution modelling and seed germination of four threatened snow lotus (*Saussurea*), and their implication for conservation. Glob Ecol Conserv 17:e00565. https://doi.org/10.1016/j.gecco.2019.e00565

Perrone A, Martinelli F (2020) Plant stress biology in epigenomic era. Plant Sci 294:110376. https://doi.org/10.1016/j.plantsci.2019.110376
Pimm SL, Jenkins CN, Abell R, Brooks TM (2014) The biodiversity of species and their rates of extinction, distribution. Science 344(6187):1246752. https://doi.org/10.1126/science.1246752
Pirasteh-Anosheh H, Babaie-Zarch MJ, Nasrabadi M, Parnian A, Alavi-Siney SM, Beyrami H, Race M (2023) Climate and management factors influence saffron yield in different environments. Agrosyst Geosci Environ 6(3):e20418. https://doi.org/10.1002/agg2.20418
Pithiya MB, Sharma SK, Sharma SK, Sharma M, Kotwal N (2022) Advancements and challenges in plant tissue culture: a comprehensive overview. J Plant Biota 1(1):12–16. https://doi.org/10.51470/JPB.2022.1.1.12
Poland JA, Rife TW (2012) Genotyping-by-sequencing for plant breeding and genetics. Plant Genome 5(3):92–102. https://doi.org/10.3835/plantgenome2012.05.0005
Prober SM, Byrne M, McLean EH, Steane DA, Potts BM, Vaillancourt RE, Stock WD (2015) Climate-adjusted provenancing: a strategy for climate-resilient ecological restoration. Front Ecol Evol 3:65. https://doi.org/10.3389/fevo.2015.00065
Prosser RS, Brain RA (2024) Where have all the flowers gone? A systematic evaluation of factors driving native terrestrial plant decline in North America. Environ Sci Pollut Res 31(35):48460–48483. https://doi.org/10.1007/s11356-024-34349-9
Qian H, Xu Z, Cong K, Zhu X, Zhang L, Wang J, Ji P (2021) Transcriptomic responses to drought stress in *Polygonatum kingianum* tuber. BMC Plant Biol 21(1):537. https://doi.org/10.1186/s12870-021-03297-8
Rastogi S, Shah S, Kumar R, Vashisth D, Akhtar MQ, Kumar A, Shasany AK (2019) *Ocimum* metabolomics in response to abiotic stresses: cold, flood, drought and salinity. PLoS One 14(2):1–26. https://doi.org/10.1371/journal.pone.0210903
Rayalu MB, Siddulu N, Raju KG (2024) Exploring the genetic diversity of medicinal plants: conservation and biotechnological approaches. J Med Plants Stud 12(4):396–401. https://doi.org/10.22271/plants.2024.v12.i4e.1732
Reddy AR, Chaitanya KV, Vivekanandan M (2010) Drought-induced responses of photosynthesis and antioxidant metabolism in higher plants. J Plant Physiol 161(11):1189–1202. https://doi.org/10.1016/j.jplph.2004.01.013
Renner SS, Zohner CM (2018) Climate change and phenological mismatch in trophic interactions among plants, insects and vertebrates. Ann Rev Ecol Evol Syst 49(1):165–182. https://doi.org/10.1146/annurev-ecolsys-110617-062535
Rieseberg LH, Willis JH (2007) Plant speciation. Nat Rev Genet 8(5):285–295. https://doi.org/10.1038/nrg2042
Savolainen O, Lascoux M, Merilä J (2013) Ecological genomics of local adaptation. Nat Rev Genet 14(11):807–820. https://doi.org/10.1038/nrg3522
Senkoro AM, Talhinhas P, Simões F, Loureiro J, Chambel L, Gonçalves S, Eloy L, Cordeiro J, Oliveira MM, Vasconcelos MW (2020) The genetic legacy of fragmentation and overexploitation in the threatened medicinal African pepper-bark tree, *Warburgia salutaris*. Sci Rep 10(1):19725. https://doi.org/10.1038/s41598-020-76654-6
Senkoro AM, Munt DD, Shackleton CM, Ribeiro-Barros AI, Voeks RA (2024) The case of a threatened medicinal tree with optimistic prospects under climate change. Glob Ecol Conserv 54:e03126. https://doi.org/10.1016/j.gecco.2024.e03126
Shaban M, GhehsarehArdestani E, Ebrahimi A, Borhani M (2023) Climate change impacts on optimal habitat of *Stachys inflata* medicinal plant in central Iran. Sci Rep 13(1):6580. https://doi.org/10.1038/s41598-023-33660-8
Sharma MK (2024) Exploring the biochemical profiles of medicinal plants cultivated under stressful environmental conditions. Curr Agric Res J 12(1):81–103. https://doi.org/10.12944/CARJ.12.1.07
Sharma M, Puri S (2017) Temperature stress mediated oxidative and antioxidant defense in *Withania somnifera* (L.) Dunal. J Chem Pharm Sci 10(5):892–899. https://www.jchps.com/issues/Volume%2010_Issue%202/0361216.pdf

Sharma A, Kumar V, Shahzad B, Tanveer M, Sidhu GPS, Handa N, Bali AS (2020) Worldwide pesticide usage and its impacts on ecosystem. SN Appl Sci 2:1446. https://doi.org/10.1007/s42452-020-03245-3

Sharma N, Thakur K, Zinta R (2025) Genome editing research initiatives and regulatory landscape of genome edited crops in India. Transgenic Res 34:13. https://doi.org/10.1007/s11248-025-00432-1

Shrestha UB, Gautam S, Bawa KS (2012) Widespread climate change in the Himalayas and associated changes in local ecosystems. PLoS One 7(5):36541. https://doi.org/10.1371/journal.pone.0036741

Sofowora A, Ogunbodede E, Onayade A (2013) The role and place of medicinal plants in the strategies for disease prevention. Afr J Tradit Complement Alternat Med 10(5):210–229. https://doi.org/10.4314/ajtcam.v10i5.2

Souza ML, Lovato MB, Fagundes M, Valladares F, Lemos-Filho JP (2019) Soil fertility and rainfall during specific phenological phases affect seed trait variation in a widely distributed Neotropical tree, *Copaifera langsdorffii*. Am J Bot 106(8):1096–1105. https://doi.org/10.1002/ajb2.1333

Stanik N, Lampei C, Rosenthal G (2020) Summer aridity rather than management shapes fitness-related functional traits of the threatened mountain plant *Arnica montana*. Ecol Evol 10(11):5069–5078. https://doi.org/10.1002/ece3.6259

Stanik N, Lampei C, Rosenthal G (2021) Drought stress triggers differential survival and functional trait responses in the establishment of *Arnica montana* seedlings. Plant Biol 23(6):1086–1096. https://doi.org/10.1111/plb.13306

Sykes MT, Prentice IC (1996) Climate change, tree species distributions and forest dynamics: a case study in the mixed conifer/hardwoods zone of northern Europe. Clim Chang 34:161–177

Taberlet P, Bonin A, Zinger L, Coissac E (2018) Environmental DNA: for biodiversity research and monitoring. Oxford University Press. https://doi.org/10.1093/oso/9780198767220.001.0001

Theodoridis S, Hickler T, Thines M (2024) Mountain greening and rising temperatures erode habitats of ironwort (*Sideritis*), an important natural medicinal resource. Plants People Planet 6(4):862–874. https://doi.org/10.1002/ppp3.10497

Udayakumar R, Kasthurirengan S, Mariashibu TS, Sudhakar B, Ganapathi A, Kim EJ, Kim SC (2013) Analysis of genetic variation among populations of *Withania somnifera* (L.) in South India based on RAPD markers. Eur J Med Plants 3(2):266–280

Ummenhofer CC, Meehl GA (2017) Extreme weather and climate events with ecological relevance: a review. Philos Trans R Soc B Biol Sci 372(1723):20160135. https://doi.org/10.1098/rstb.2016.0135

Van Daele F, Honnay O, DeKort H (2023) Habitat fragmentation reshapes genomic footprints of selection in a forest herb. arXiv preprint arXiv:2310.16004. https://doi.org/10.48550/arXiv.2310.16004

Van Daele F, Honnay O, Janssens S, DeKort H (2024) Habitat fragmentation affects climate adaptation in a forest herb. J Ecol 112(2):246–264. https://doi.org/10.1111/1365-2745.14225

Varshney RK, Sinha P, Singh VK, Kumar A, Zhang Q, Bennetzen JL (2021) 5Gs for crop genetic improvement. Curr Opin Plant Biol 60:101986. https://doi.org/10.1016/j.pbi.2020.101986

Vergara AJ, Cieza-Tarrillo D, Ocaña C, Quiñonez L, Idrogo-Vasquez G, Muñoz-Astecker LD, Auquiñivin-Silva EA, Cruzalegui RJ, Arbizu CI (2023) Current and future spatial distribution of the genus *Cinchona* in Peru: opportunities for conservation in the face of climate change. Sustainability 15(19):14109. https://doi.org/10.3390/su151914109

Verma PK, Verma S (2023) Biosynthesis of essential oils in *Artemisia* species and conservation through in vitro propagation. In: Mishra MK, Kumari N (eds) Plants for immunity and conservation strategies. Springer Nature, Singapore, pp 133–152. https://doi.org/10.1007/978-981-99-2824-8_9

Vernooy R, Shrestha P, Chaudhary P (2013) Community seed banks in Nepal past, present, future. In: Proceedings of a National Workshop, 14–15 June 2012, Pokhara, Nepal. https://hdl.handle.net/10568/68933

Vernooy R, Shrestha P, Sthapit B (2015) Community seed banks: origins, evolutions and prospects. Bioversity International. https://doi.org/10.4324/9781315886329

Vernooy R, Sthapit B, Otieno G, Shrestha P, Gupta A (2017) The roles of community seed banks in climate change adaption. Dev Pract 27(3):316–327. https://doi.org/10.1080/09614524.2017.1294653

Vikane JH, Rydgren K, Jongejans E, Vandvik V (2019) Rainfall and temperature change drive *Arnica montana* population dynamics at the Northern distribution edge. Oecologia 191(3):565–578. https://doi.org/10.1007/s00442-019-04519-5

Wang Z, Gerstein M, Snyder M (2009) RNA-Seq: a revolutionary tool for transcriptomics. Nat Rev Genet 10(1):57–63. https://doi.org/10.1038/nrg2484

Wang J, Li Y, Luo H, Zhang X, Deng Z, Chen S (2022) Advances in medicinal plant research in the post-genomic era. Hortic Res 9:uhac003. https://doi.org/10.1093/hr/uhac003

Wu Y, Liu YR, Peng H, Yang Y, Liu GL, Cao GX, Zhang Q (2015) Pollination ecology of alpine herb *Meconopsis integrifolia* at different altitudes. Chin J Plant Ecol 39(1):1. https://doi.org/10.17521/cjpe.2015.0001

Xia C, Huang Y, Qi Y, Yang X, Xue T, Hu R, Yu S (2022) Developing long-term conservation priority planning for medicinal plants in China by combining conservation status with diversity hotspot analyses and climate change prediction. BMC Biol 20(1):89. https://doi.org/10.1186/s12915-022-01285-4

Yan H, Sun M, Zhang Z, Jin Y, Zhang A, Lin C, Huang L (2023a) Pangenomic analysis identifies structural variation associated with heat tolerance in pearl millet. Nat Genet 55(3):507–518. https://doi.org/10.1038/s41588-023-01302-4

Yan J, Wang P, Zhu L, Zhang Y, Yang S, Gao Q, Zhao J (2023b) A graph-based pan-genome reveals structural variation, expression differences and heat-resistance genes in pearl millet. Nat Genet 55(4):602–612. https://doi.org/10.1038/s41588-023-01283-1

Yang L, Wen KS, Ruan X, Zhao YX, Wei F, Wang Q (2018) Response of plant secondary metabolites to environmental factors. Molecules 23(4):762. https://doi.org/10.3390/molecules23040762

Yang N, Xu XW, Wang RR, Peng WL, Cai L, Song JM, Li ZM (2019) Contributions of *Zea mays* subspecies mexicana haplotypes to modern maize. Plant Biotechnol J 17(4):688–698. https://doi.org/10.1111/pbi.12958

You J, Qin X, Ranjitkar S, Lougheed SC, Wang M, Zhou W, Ouyang D, Zhou Y, Xu J, Zhang W, Wang Y, Yang J, Song Z (2018) Response to climate change of montane herbaceous plants in the genus *Rhodiola* predicted by ecological niche modelling. Sci Rep 8(1):5879. https://doi.org/10.1038/s41598-018-24360-9

Zait Y, Konsens I, Schwartz A (2020) Elucidating the limiting factors for regeneration and successful establishment of the thermophilic tree *Ziziphusspina-christi* under a changing climate. Sci Rep 10(1):14335. https://doi.org/10.1038/s41598-020-71276-4

Zhang D, Chen L, Li D, Lv B, Chen Y, Chen J, Chen Y (2019) OsRR22 regulates salt tolerance in rice. Plant Physiol 180(4):2225–2239. https://doi.org/10.1104/pp.19.00137

Zhao R, Wang S, Chen S (2024) Predicting the potential habitat suitability of *Saussurea* species in China under future climate scenarios using the optimized maximum entropy (MaxEnt) model. J Clean Prod 474:14352. https://doi.org/10.1016/j.jclepro.2024.143552

Zhuravlev YN, Reunova GD, Kats IL, Muzarok TI, Bondar AA (2010) Genetic variability and population structure of endangered *Panax ginseng* in the Russian Primorye. Chin Med 5(1):21. https://doi.org/10.1186/1749-8546-5-21

Chapter 14
Genetic Diversity, Threats, and Conservation Management of Indonesian Medicinal Plants with Special Emphasis on Sundaland Hotspot

Kusumadewi S. Yulita, Didi Usmadi, and Henti H. Rachmat

Abstract Medicinal plants are an important component of global biodiversity, as they play a significant role in public health and cultural practices and generate substantial income for producing countries. It is estimated that over 50,000 plant species possess medicinal properties closely linked to their genetic diversity. Genetic variation present in medicinal plants is often considered a key factor in determining species or population resilience to environmental change and their phytochemical content, such as terpenoids, alkaloids, and phenolics. In many countries, the production of herbal and modern medicines still relies on wild populations of medicinal plants. At the same time, most wild medicinal plant species remain inadequately explored for their genetic and phytochemical properties. Medicinal plants, particularly the wild populations, are now facing threats from habitat degradation, overharvesting, and climate change. Such threats will ultimately lead to population declines and the loss of crucial genetic variability. Serious threats are posed to endemic species with limited genetic information. In this context, conservation genetics approaches are essential for protecting medicinal plant genetic resources. Molecular marker approaches such as microsatellites, SNPs, and NGS technologies (e.g., MIG-seq) enable more accurate assessments of population structure, variation, and dynamics and help identify priority populations for conservation. This chapter provides an overview of the genetic diversity, threats, and conservation management

---

K. S. Yulita (✉) · H. H. Rachmat
Research Center for Ecology, National Research and Innovation Agency (BRIN), Cibinong, Bogor, Indonesia
e-mail: kusu002@brin.go.id; hent003@brin.go.id

D. Usmadi
Research Center for Biota Systems, National Research and Innovation Agency (BRIN), Cibinong, Bogor, Indonesia
e-mail: didi020@brin.go.id

S. Jha et al. (eds.), *Genetic Diversity in Medicinal Plants*, Sustainable Development and Biodiversity 44, https://doi.org/10.1007/978-981-95-7281-6_14

strategies of medicinal plants, with a particular emphasis on the Sundaland biodiversity hotspot.

Keywords Genetic diversity · Medicinal plants · Conservation · Biotechnology · Climate change · Sundaland hotspot · Biological resources

14.1 Introduction

Medicinal plants are an important component of the global biodiversity as they provide benefits to public health, cultural practices, and economic aspects. For centuries, medicinal plants have been used as natural resources for treating diseases and maintaining human health. Medicinal plants provide key bioactive compounds that are important to traditional herbal medicine and the modern pharmaceutical industry. Medicinal plants are employed to treat various acute and chronic illnesses, as well as a range of physical and psychological disorders (Halder and Jha 2023).

Globally, more than 50,000 plant species are known to possess medicinal properties. The unique value of medicinal plants comes from genetic diversity that is not yet fully explored, particularly the wild populations. Genetic diversity that is reflected through allelic richness, genotypic variation, and population structure is a crucial foundation for a species' resilience to internal and external stresses (Maxted et al. 2020). Genetic variation also influences the production of phytochemical compounds, such as terpenoids, alkaloids, and phenolics, which determine their therapeutic potential (Viccini et al. 2014; Abd-Elgawad and Alotaibi 2019). When genetic diversity declines—for example, due to overharvesting and habitat degradation—the resilience of medicinal plant populations can decline as well. Thus, it eventually affects their ability to produce important secondary metabolites (Li et al. 2025).

The genetic and biochemical potential of most medicinal plants, especially in the Sundaland region, has not been thoroughly studied. Many species are still collected directly from the wild without sustainable cultivation systems; thus, their survival depends on the preservation of their natural habitat. In the past two decades, habitat destruction, overexploitation, and climate change have become serious threats to the sustainability of medicinal plant populations. Forest fragmentation, land conversion, and environmental degradation have led to declines in wild populations and the loss of valuable genetic variation. This situation is worsened by the lack of effort to domesticate and preserve wild populations, and by limited information on genetic structure, intraspecific diversity, and reproductive systems, especially in endemic species. The continuing decline of wild populations can lead to species extinction, ultimately eliminating the potential for the discovery of new drugs derived from natural resources (Schippmann et al. 2006).

The genetic and biochemical potential of medicinal plants, particularly in Sundaland, remains largely unknown. Many medicinal plant species traded or used directly depend on wild harvesting due to the lack of sustainable cultivation systems (except for a few common, widely cultivated species). Their survival is mainly dependent on the condition of their natural habitats. In the past two decades, habitat destruction, excessive harvesting, and climate change have become serious threats to medicinal plants. The population declines in their natural habitats can lead to the loss of genetic variation that is crucial for long-term adaptation and survival. This situation is further worsened by minimal domestication efforts, low investment in conservation, and limited information on the genetic structure, intraspecific diversity, and reproductive systems of various species, particularly of endemic species. If this trend of population decline continues, some populations or medicinal plant species may become extinct. The loss of these species not only reduces biodiversity but also prevents the discovery of new bioactive compounds valuable for future drug development.

In this context, conservation genetics-based approaches are important to protect medicinal plant genetic resources. The use of molecular markers such as microsatellites, SNPs, and NGS-based methods (e.g., MIG-seq) allows a reliable assessment of genetic variation within and among populations. These data can be used to identify priority conservation populations, detect genetic erosion, and formulate science-based management strategies. Integrating genetic approaches with ecological, ethnobotanical, and biopharmacological information is a strategic step in understanding the links between genetic diversity, ecosystem function, and human benefits.

In this chapter, an overview is provided of the genetic diversity, threats, and conservation management of medicinal plants, with special emphasis on the Sundaland hotspot region, including geographic distribution, genetic structure, and population dynamics, as well as management strategies for the current and future protection of medicinal plants. It is expected that this chapter will provide useful information for scientific and practical reference to researchers, policymakers, and natural resource managers in understanding and preserving the genetic diversity of medicinal plants—especially in the Sundaland region.

14.2 Present Threats to Medicinal Plants in a Global Perspective

The number of medicinal plant species in the world ranges between 50,000 and 80,000, of which 15,000 are threatened with extinction (Chen et al. 2016; Halder and Jha 2023; IUCN 2025). The largest number of threatened plants comes from the Fabaceae family, followed by Rubiaceae, Zingiberaceae, Asphodelaceae, and Lauraceae. Species of Fabaceae are the highest in Angiosperms; thus, the number of medicinal plants is in accordance with this evidence. The highest numbers of threatened genera are *Aloe* L (Asphodelaceae), followed by *Psychotria* (Rubiaceae) and *Etlingera* (Zingiberaceae). These are plant genera that dominate the pharmaceutical and cosmetic industries. The genus *Aloe*, the largest group in Asphodelaceae, is confined to the Old World. *Aloe* species, especially *A. vera*, have a long history of use as medicinal remedies and are widely applied in the food, healthcare, and pharmaceutical sectors (Salehi et al. 2018). 121 *Aloe* species are threatened with extinction, but this does not include the famous *Aloe vera*. The second-largest genus is *Psychotria* (Rubiaceae), which is known for its high alkaloid content (de Andrade et al. 2025). The alkaloid compounds have pharmacological effects, including pain relief, reduced inflammation, antimicrobial, and anticancer activity, thereby enhancing their value for medicinal applications. Fifty-four species of this genus are now threatened with extinction and therefore require further attention to their harvest and trade management. The third largest threatened genus is *Etlingera* (Zingiberaceae), which has a long history of usage for culinary and traditional medicine mainly due to its aromatic rhizomes, leaves, and inflorescences, which are rich in essential oils and bioactive compounds with antimicrobial, anti-inflammatory, and antioxidant properties (Poulsen 2006; Chan et al. 2011).

Most of these threatened medicinal plants are distributed in the Indomalayan region (33.2%), followed by the Afrotropical (26.5%) and Neotropical (18.7%) regions, consistent with the region's rich medicinal plant resources and their extensive use. Chen et al. (2016) reported that China and India have the highest usage of medicinal plants, accounting for 11,146 and 7500, respectively. Threats to extinction are even more impactful on the endemic species. Of the recorded endemic threatened medicinal plants, eight species have already become extinct in the neotropical regions. *Angostura ossana* (Rutaceae) was reported to be extinct in 2015, having been rarely found since 1965 due to overexploitation of its wood (Gonzalez-Oliva 2020). The bark of the tree is used to substitute cinchona. Seven species of *Brugmansia* (Solanaceae) from Neotropics—i.e., *B. arborea, B. aurea, B. insignis, B. sanguinea, B. suaveolens, B. versicolor,* and *B. vulcanicola*—are also reported to be extinct in the wild. de Feo (2004) reported a wide range of uses for these species, including therapeutic, divinatory, phytotherapeutic, initiation rites, illness, and black magic. Meanwhile, the critically endangered species are primarily recorded in the Indomalayan regions, specifically in Sundaland, the Eastern Himalayas, and Madagascar. These biogeographic hotspot regions are home to many endemic plants, but have also been areas where significant land conversions have occurred.

Many medicinal plant species worldwide are currently facing various threats, from anthropogenic activities and other environmental changes that affect their survival and utilization. According to assessments published in the IUCN Red List of Threatened Species, agriculture and aquaculture (27.4%) pose the greatest threats to medicinal plants. The next threat comes from residential and commercial development, which contributes 15.0% of the natural habitat conversion. Such conversion of natural habitats significantly alters the forest landscape, leading to habitat fragmentation that disrupts gene flow between populations and increases the occurrence of inbreeding (Aguilar et al. 2006). Inbreeding often results in the loss of adaptive alleles that may contribute to a species' resilience to environmental change. This was exemplified by a study on *Panax quinquefolius* (American ginseng, Araliaceae) (Cruse-Sanders and Hamrick 2004). The results of the study showed that fragmented populations exhibit reduced heterozygosity and increased inbreeding depression, resulting in a decline in pharmacological quality and increased susceptibility to pathogens. Similar patterns have also been observed in *Mentha cervina* and *Mentha pulegium* (Lamiaceae) in Portugal, where the destruction of natural habitats, the spread of invasive species, and overharvesting driven by human activities such as stream modification, vegetation clearance, and grazing pressure have accelerated genetic erosion in both wild and cultivated populations (Póvoa et al. 2023). Such examples illustrate that once habitats are fragmented and populations decline, genetic drift occurs, thereby increasing the risk of extinction (Ticktin 2004).

Data from the IUCN Red List for Threatened species revealed that the use of biological resources contributed 26.7% of the threat to medicinal plants. The demand for herbal medicines and botanical ingredients in the global market is increasing rapidly, and to some extent has led to the excessive harvesting of wild populations, especially species with high economic value. According to Halder and Jha (2023), approximately 65–85% of the population in developing countries depend on medicinal plants as a primary source of healthcare. Meanwhile, more than 25% of prescription drugs in developed countries are derived from wild species or populations (Chen et al. 2016). The collection of wild medicinal plants is often carried out through destructive methods, such as debarking, uprooting, or extracting reproductive tissues. For example, *Eurycoma longifolia* (Simaroubaceae) plants are mainly uprooted, while *Cinnamomum verum* (Lauraceae) is extracted by debarking. When harvesting techniques are not well managed, extraction pressure not only reduces population size but also selectively removes the most valuable genotypes, thus reducing levels of genetic diversity. Another example from Brazil, where 86% of medicinal plants supplying local and industrial markets are still collected from the wild, often through destructive harvesting practices (De Melo et al. 2009). Other examples include *Prunus africana* (African cherry, Rosaceae), which is exploited for its phytosterol-rich bark, and *Aquilaria* spp., which is widely harvested for the production of agarwood (Thymelaeaceae) and other agarwood-producing species, targeted for resinous heartwood; and *Taxus* spp. (yew, Taxaceae), which are heavily harvested for the production of paclitaxel, an anticancer drug (Schiff et al. 1979; Weaver 2014). These species are now listed in CITES Appendix, indicating international concern about sustainable harvest and trade (Cunningham et al. 2018). Without careful, sustainable management, this will lead to a significant decline in their populations (Bisht et al. 2016; Chen et al. 2019). The consequence is not only a decline in wild populations but also the loss of rare genotypes, which are critical for long-term adaptation and survival.

Other threat categories with lower percentages include climate change and extreme weather (4.7%), energy production and mining (4.6%), transportation and service corridors (4.5%), and invasive and other problematic species, genes, and diseases (4.1%). Climate change can indirectly affect the growth and development of medicinal plants by altering their reproductive biology (Mykhailenko et al. 2025). Changes of reproductive behavior affect the pollination agents of medicinal plants. The loss or decline in pollinator populations may reduce cross-pollination and gene flow, thereby increasing selfing rates and further narrowing genetic variation (Potts et al. 2016). Species with low genetic diversity tend to exhibit more uniform phenological responses to increasing temperatures, making them more vulnerable to environmental changes including climate change (Cahyaningsih et al. 2021a; Doi et al. 2010). Several studies (Wan et al. 2024; Laftouhi et al. 2025; Liu et al. 2025) have shown that changes in temperature, CO_2 levels, rainfall, and drought affect the biosynthesis and accumulation of secondary metabolites that determine the phytochemical quality of medicinal plants. Li et al. (2021) further reported that increasing temperature was positively correlated with increases in total phenolic, flavonoid, chlorogenic acid, and hyperoside contents in *Houttuynia cordata* (Saururaceae).

Changes in temperature and precipitation, combined with more frequent droughts, heat waves, and frosts, cause shifts or reductions in species' ranges, particularly for plants with restricted distribution, such as narrow-endemic species. A study by Xu et al. (2024) revealed that Chinese medicinal plant species are gradually expanding into higher-altitude and higher-latitude regions, and that their natural habitats are expected to decline in the coming decades. Another study, employing models to predict further range reduction under future warming, has already documented habitat fragmentation in *Gentiana manshurica* (Gentianaceae), an endangered medicinal species (Zou et al. 2023). It is therefore crucial to incorporate genetic data into climate adaptation strategies, particularly when designing protected areas to ensure coverage of genetic variation in areas experiencing niche shifts due to climate change (Hoban et al. 2020).

Other threats with the smallest percentages are pollution (0.7%), geological events (0.7%), and other categories (0.5%), including invasive species. Pollution resulting from modern agricultural practices and other industrial activities, such as drilling, steelmaking, and the burning of fossil fuels, further affects the growth and metabolism of medicinal plants (Ditta et al. 2022; Rani et al. 2022). Soil or water pollution may directly affect populations of medicinal plants by toxicity, particularly to plants with limited distributions. Geological events such as landslides or volcanic eruptions can destroy habitats on a local scale or even on a large scale if they involve massive landslides. Invasive species, pests, and pathogens further disrupt ecological interactions, causing rapid declines particularly in genetically homogeneous populations (Pimentel et al. 2005). The presence of invasive species can displace native medicinal plants through competition, leading to a decline in their population sizes (Shabbir and Bajwa 2006).

Finally, the erosion of traditional knowledge and biocultural diversity represents a less tangible but equally critical threat. Currently, the younger generation is less interested in studying and practicing their ancestors' traditional knowledge, as preferences have shifted toward modern medicine, particularly given the ease of obtaining medication, which is now widely available on e-commerce platforms. In fact, local communities have historically maintained the genetic diversity of local landraces, which are adapted to their local environments (Bussmann 2013). As cultural practices shift or are replaced by modern, commercial businesses that prioritize practicality and ease of access to medicinal ingredients, these reservoirs of genetic and phytochemical diversity risk being lost (Khoury et al. 2013).

Overall, these data confirm that the primary threats to medicinal plants encompass a range of anthropogenic activities and climate change. Addressing these threats requires integrated conservation approaches that balance in situ and ex situ strategies, recognize customary rights, ensure sustainable use and trade, and acknowledge genetic diversity as a fundamental component of medicinal plant conservation.

14.3 Genetic Diversity, Threats, and Conservation Management of Medicinal Plants in the Sundaland Hotspot

Sundaland is a massive landmass connecting the present-day Malay Peninsula, Sumatra, Borneo, Java, and numerous smaller islands in between. It is considered one of the 36 global biodiversity hotspots (Myers et al. 2000; Noss et al. 2015). Geologically, this region represents a vast continental shelf that was repeatedly exposed and submerged during the Pleistocene glacial cycles. Such fluctuations in sea levels allowed biotic exchange across the present-day islands, followed by recurrent isolation, processes that together fostered extensive speciation and high levels of endemism. Today, Sundaland's landscapes are dominated by diverse forests, ranging from mangroves to montane forests, with several unique ecosystems, including karst, heath, and peat swamp habitats. Each of these ecological zones supports distinct plant assemblages, contributing to the region's extraordinary floristic richness.

Sundaland is estimated to comprise over 25,000 vascular plant species (Myers et al. 2000), many of which are endemic and possess medicinal properties of cultural and pharmacological importance. Families of large trees that form the main tropical forest formations in Sundaland, such as Annonaceae, Dipterocarpaceae, Euphorbiaceae, Fagaceae, Lauraceae, and Pinaceae, provide ethnomedicinal value for their resins, barks, flowers, and even leaves. Herbaceous plants such as Araceae, Orchidaceae, and Zingiberaceae are valued for their rhizomes, stems, and leaves, as well as their chemically diverse metabolites. Additionally, shrubs, small trees, and understory plants such as kayubajakah (*Spatholobus littoralis*, Fabaceae), akarkuning (*Arcangelisia flava*, Menispermaceae), and pasakbumi (*Eurycoma longifolia*, Simaroubaceae) are important medicinal plant resources in the Sundaland region, harvested for their wood and roots.

Many medicinal plants from Sundaland have been studied using modern biomedical approaches informed by traditional knowledge. Some examples include camptothecin, derived from *Camptotheca acuminata* (Nyssaceae) as a precursor to anticancer therapies, and calanolide A from *Calophyllum lanigerum* var. *austrocoriaceum* (Calophyllaceae) in Borneo, which shows potent anti-HIV activity (Teoh 2016). Active biocompounds such as quassinoids and eurypeptides derived from *E. longifolia* (Simaroubaceae) and cryptocaryone from *Cryptocarya* spp. (Lauraceaceae) are also promising bioprospects of medicinal plants from Borneo. A tree species native to the island of Borneo, Kratom (*Mitragyna speciosa*, Rubiaceae), which has recently gained popularity due to its psychoactive effects, is also considered a highly promising source of medicinal compounds.

The medicinal properties of plants can be determined by their genetic makeup that influences the concentration and composition of secondary metabolites, such as alkaloids, terpenoids, and phenolics, which directly impact therapeutic efficacy. The genetic diversity of some medicinal plants in Sundaland has been studied using various approaches, including RAPD, ISSR, AFLP, and next-generation sequencing for genomic analyses. For example, *Baeckea frutescens* (Myrtaceae), an aromatic medicinal shrub widely used in traditional therapies, was assessed with NGS-derived SSR markers in Peninsular Malaysia. Results of this study showed a high level of observed heterozygosity (H_o = 0.900), moderate expected heterozygosity (H_e = 0.484), and evidence of clonal populations indicated by high genetic similarity, thus indicating the need to include more samples representing broader population levels to fully capture the genetic representation (Alias et al. 2021). *Etlingera elatior* (Zingiberaceae), a widely used medicinal and ornamental ginger species, shows considerable genetic variation among 57 accessions collected from seven locations across Peninsular Malaysia (Ismail et al. 2019). *Labisia pumila* (Primulaceae) and *Eurycoma longifolia* (Simaroubaceae) are examples of medicinal plants from the Sundaland region that have been extensively studied for their genetic diversity. *L. pimula*, locally known as kacip fatimah and used in women's health remedies, has been studied for its genetic diversity using ISSR markers (Karimi et al. 2014), SSR (Tnah et al. 2014), DNA barcodes, and HPLC (Azli et al. 2024). Meanwhile, *E. longifolia* is a medicinal plant widely used as an aphrodisiac for men and has been studied using RFLP markers (Azli et al. 2024); SSR (Lee et al. 2018a); SSR and ISSR (Alttaher et al. 2021); and DNA barcodes (Yulita et al. 2022).

Many medicinal plants possess variation in their phytochemical content that may reflect genetic differences among populations. An example of a popular medicinal plant is *Andrographis paniculata* (Acanthaceae), which exhibits significant variation in andrographolide content across Southeast Asia, and these differences are related to genetic structure among populations. A similar finding was observed in *A. paniculata* from Indonesia, as revealed by studies using ISSR and RAPD, which revealed low genetic diversity but high genetic differentiation among locations (Arif et al. 2020). Therefore, maintaining broad genetic diversity is crucial for maintaining phytochemical variation.

14.4 Threats, Conservation, and Management Strategy

The conservation of medicinal plant genetic diversity is an important issue that extends beyond species preservation, and can also protect the diversity of bioactive compounds that determine their therapeutic efficacy and capacity to adapt to environmental stress. At a global context, the issue of conservation and management of medicinal plants has already been stated in the global commitments of the UN-SDG target and the Kunming-Montreal Global Biodiversity Framework (KMGBF), i.e., target 9 (Sustainably manage wild species to provide social, economic, and environmental benefits for people), target 4 (urgent management actions and conservation, maintain and restore the genetic diversity within and between populations of native, wild and domesticated species), target 5 (sustainable use and harvesting and trade of wild species), and target 9 (sustainable management and use of wild species) (document of CoP 15 2022).

At present, the Sundaland region faces substantial environmental pressures that threaten the survival of native medicinal plants. A major threat in this region is land conversion. Most of the natural habitats of medicinal plants have been heavily converted, mainly to agricultural plantations such as oil palm and rubber, resulting in many wild populations being fragmented. Even so, there are still many local communities that depend on medicinal plants in the forest. For millennia, indigenous peoples such as the Orang Asli of Peninsular Malaysia and the Dayak of Borneo have developed sophisticated systems of traditional medicine based on this botanical richness. Ethnobotanical surveys indicate that 2000–3000 species are commonly used in healthcare across the region (de Padua et al. 1999; de Boer et al. 2014). Generally, local races of a type of plant possess superior characteristics that are not found in cultivated plant species. By continuing to utilize medicinal plant resources in the forest, the genetic diversity of local medicinal plants is indirectly conserved.

From a sociocultural aspect, changes have also occurred, such as erosion of traditional ecological knowledge among younger generations (Samuel et al. 2010). As younger generations increasingly embrace more modern, urban, and global cultures, ethnomedical knowledge, traditionally passed down orally, is being lost in many communities. This loss of knowledge distances younger generations from their ancestral cultural heritage (Samuel et al. 2010).

The trend of consuming herbal medicines through e-commerce platforms is growing rapidly in Sundaland countries, particularly for high-demand species such as *Zingiber officinale* (ginger). This high market demand has often resulted in smuggling, illegal trade, and excessive harvesting, for example, *Aquilaria malaccensis* (Agarwood, Thymelaeaceae; Soehartono and Newton 2001) and *Eurycoma longifolia* (Simaroubaceae; Kartikawati et al. 2014). Currently, herbal products and their derivatives from those species are widely traded on e-commerce platforms and in conventional stores. Overharvesting of heavily demanded medicinal plants still poses a serious challenge, threatening the sustainability of their natural populations.

An additional challenge in the genetic conservation of medicinal plants in the Sundaland region is the lack of baseline data on genetic diversity. For the vast majority of Sundaland's medicinal plants, the extent and distribution of their genetic diversity remain mostly unknown, making prioritization and planning difficult. Cahyaningsih et al. (2021b) conducted a gap analysis of Indonesian medicinal plants and highlighted gaps in characterization and documentation. In addition, the FAO country report (Indonesia) (FAO 2011) also lists lack of characterization/evaluation and documentation as a major constraint for PGRFA management.

A special strategy for the conservation and sustainable management of medicinal plants in the Sundaland region is necessary because species in this region have distributions that cross borders. Four countries share geographic regions within Sundaland: Malaysia, Singapore, which lies on the Malay Peninsula, and Borneo Island, shared between Indonesia, Malaysia, and Brunei Darussalam. Therefore, conservation efforts for the medicinal plants in the Sundaland region require an integrated approach that incorporates ecological, genetic, and cultural dimensions. In particular, the socioeconomic dimension plays an equally important role to the biological aspect, as successful conservation depends heavily on local community support, participatory mechanisms, and equitable benefit sharing from the use of genetic resources. Integrated law enforcement across countries is necessary to prevent illegal trade and smuggling.

Effective conservation and management of medicinal plants requires an integrated strategy that combines in situ and ex situ approaches and ensures the active participation of local communities. In situ conservation still remains as the main base because it allows natural evolution and adaptation of species within intact ecosystems. In situ conservation practice should involve the protection of natural habitats and the establishment of special conservation areas that ensure the continuity of gene flow and the long-term sustainability of wild populations. In Indonesia, Cahyaningsih et al. (2021b) proposed several potential locations for in situ conservation of medicinal plants that could serve as a basis for national conservation planning.

These efforts need to be complemented by ex situ strategies that focus on the preservation and maintenance of genetic resources outside their natural habitats. Ex situ conservation actions should include the establishment of a seed bank, particularly for orthodox seeds, development of in vitro culture and vegetative propagation techniques for recalcitrant seeds, nursery, and planting living collections in botanical gardens and arboretums (Chen et al. 2016). The synergy between in situ and ex situ strengthens the genetic and ecological resilience of species, while ensuring the ability of essential genetic material for research, breeding, and population recovery.

In addition to technical approaches, empowering local communities is a key aspect of successful conservation. Indigenous peoples and local communities are valuable resources of ethnobotanical knowledge and traditional wisdom. Their participation should be highly acknowledged and actively involved in the management practices. Interdisciplinary research combining botany, genetics, pharmacology, and social sciences will provide a comprehensive framework for addressing the region's complex conservation challenges.

At the policy level, implementing the Nagoya Protocol on Access and Benefit Sharing (ABS) is crucial to ensure that the use of genetic resources provides tangible benefits to local communities and contributes to conservation funding (Robinson et al. 2021). Furthermore, a robust policy framework and regional cooperation are also essential. Enforcement of CITES regulations for high-value species such as *Aquilaria*, the integration of priority medicinal plants into National Biodiversity Action Plans (NBSAPs), and cross-border collaboration between Indonesia, Malaysia, Brunei Darussalam, and Thailand are strategic steps to support sustainable medicinal plant conservation in the Sundaland region.

14.5 Genetic Structure, Threats, and Management Strategy of Indonesian Medicinal Trees: *Taxus sumatrana, Eurycoma longifolia, Styrax paralleloneurus*, and Agarwood-Producing Species

Indonesia, situated in the biodiversity hotspot of Sundaland, is home to 5490 medicinal plants (Cahyaningsih et al. 2021c). These plants not only support local livelihoods and traditional knowledge systems but also provide compounds of global pharmaceutical importance. Research on medicinal plants in Indonesia is under the auspices of the National Research and Innovation Agency, in collaboration with the Ministry of Health. Indonesia once had a national research project on medicinal plants and herbal medicines conducted from 2012 to 2017. This activity collected 28,000 medicinal plant specimens from 405 ethnic groups in 34 provinces in Indonesia. For Indonesians, herbal medicine, known as jamu, is a staple of daily life, particularly that derived from the Zingiberaceae family. Studies on genetic diversity in Indonesia's medicinal flora have been extensively conducted, particularly on popular species in the Zingiberaceae, Piperaceae, palms, and cinnamon families. Research conducted using a molecular genetic approach is generally used to solve problems of identification and infraspecific diversity.

However, many native medicinal plants are now facing serious threats from excessive harvesting, habitat degradation, and climate change. Assessing their ecological status and genetic diversity is therefore essential to guide conservation planning, ensure population resilience, and maintain the evolutionary potential of these species. We present three native medicinal plants from Indonesia: *Taxus sumatranas l* or *Taxus wallichiana, Eurycoma longifolia, Styrax paralleloneurus,* and *Aquilaria* spp.

14.5.1 Taxus sumatranas.l

Taxus sumatrana (Miq.) de Laub or *Taxus sumatranas.l*, commonly known as Sumatran Yew (Fig. 14.1), is naturally distributed across Taiwan, Sulawesi (Indonesia), and extends to its southernmost range in mainland Sumatra in Indonesia (De Laubenfels 1988). The species is unique within the *Taxus* genus as it reaches the southernmost limit of the genus's distribution, occurring naturally along the Bukit Barisan Selatan Mountain belt in Sumatra's forests (Rachmat et al. 2016). It is critically important for its taxol content, a compound used in cancer treatment (Zhou et al. 2010; Weaver 2014). However, in Indonesia, its restricted distribution, slow growth, dioecious nature, and habitat destruction pose significant threats (Sunandar et al. 2019).

Understanding population structure and genetic diversity is critical for conservation planning, enabling the identification of unique lineages, assessment of environmental vulnerability, and development of targeted seed collection strategies (Hoban

Fig. 14.1 *Taxus sumatrana* planted in Cibodas Botanical Garden, West Java-Indonesia (**a**), the stem and bark (**b**), and leaves (**c**). (Photo: H.H. Rachmat 2007)

et al. 2020). Such knowledge is essential for habitat management and restoration to preserve evolutionary potential (Willi et al. 2022). In *Taxus sumatrana*, RAPD analysis of the Kerinci population (Jambi Province, Sumatra) revealed moderate genetic diversity (53.89% polymorphic loci), with the highest heterozygosity observed in the Mount Tujuh subpopulation (H_e = 0.2044) (Frianto et al. 2021). In contrast, chloroplast DNA (cpDNA) sequences (*trn*L-*trn*F, *psbC-trnS*, *rbcL*) showed no variation among individuals from Mount Kerinci (Jambi) and Mount Dempo (South Sumatra), despite their separation by >350 km along the Bukit Barisan Selatan Mountain belt (Rachmat et al. 2016). This lack of cpDNA divergence suggests a recent evolutionary bottleneck in Indonesian populations. Phylogenetic analysis incorporating a third population from Toba (North Sumatra) identified only one indel across all Sumatran populations, with no single-nucleotide polymorphisms (Kamal et al. 2023). These findings highlight the importance of integrating nuclear and chloroplast markers to elucidate both contemporary gene flow and historical demographic processes in conservation planning.

Taxus sumatrana populations exhibit irregular age distributions with significantly reduced seedling and sapling recruitment. However, ecological diversity remains stable as indicated by high importance value index (IVI) scores (Sunandar et al. 2019). The distinct floristic composition at each site highlights its individual ecological importance, requiring conservation approaches that combine habitat preservation with enrichment planting to address demographic imbalances while maintaining current vegetation patterns. In Sumatra's protected forests where this species grows, habitat disturbances violate environmental regulations, yet ongoing pressures from land encroachment and illegal timber harvesting continue to threaten populations (Rachmat et al. 2016). Conservation efforts must enhance forest monitoring and protection across the Bukit Barisan Mountain belt, supported by stricter enforcement of illegal logging laws and community engagement initiatives. Propagation through shoot cuttings has shown promising results, with 50–65% rooting success using auxin treatment in soil–husk growing media (Susilowati et al. 2019a). At the same time, enrichment planting offers potential for habitat restoration (Sunandar et al. 2019). Further research into seed dormancy mechanisms remains essential to improve natural regeneration outcomes for this ecologically and medically valuable species.

14.5.2 Eurycoma longifolia

Eurycoma longifolia Jack (pasak bumi or tongkatali) (Fig. 14.2) is a highly valued medicinal plant native to Southeast Asian rainforests, including Indonesia. Vegetation analyses across several study sites reveal concerning population structures, characterized by disproportionate age-class distributions. In Batang Lubu

Fig. 14.2 *Eurycoma longifolia* woody shrub grown in natural habitat (**a**) and their compound leaves (**b**). (Photo: H.H. Rachmat 2021)

Sutam Forest (North Sumatra), the importance value index (IV I) demonstrates a sharp decline from seedlings (21.26%) to complete absence at pole and tree stages, indicating recruitment failure (Susilowati et al. b, c). Similar patterns emerge at the Muka Kuning Nature Tourism Park (Riau Islands), where IVI values drop from 21.9% at the sapling stage to 0% at the tree stage. The absence of the tree stage is likely to reflect the species' natural growth habit as a small tree or woody shrub (Susilowati et al. 2023). Despite these demographic imbalances, forest ecosystems hosting *E. longifolia* maintain high biodiversity, with Shannon–Wiener indices ranging from 2.27 to 3.22 and evenness indices exceeding 0.85 across sites, suggesting resilient plant communities capable of withstanding anthropogenic pressures (Susilowati et al. 2019b, 2021, 2023). Horizontal stand structures typically exhibit inverted J patterns, while vertical stratification remains intact, dominated by intermediate canopy layers (stratum C), indicating structurally mature forests that could support natural regeneration if harvesting pressures are mitigated (Susilowati et al. 2021).

Molecular analyses of Sumatran *E. longifolia* populations reveal limited genetic variation in the chloroplast trnL-F region (960 bp), with 99.9% sequence similarity to reference genomes (MH751519, GU593014). However, five unique nucleotide variations were identified in the trnL intron region, distinguishing Indonesian specimens from other *Eurycoma* species (Yulita et al. 2022). Phylogenetic analyses confirm the taxonomic identity of Sumatran populations, clustering them with other *E. longifolia* accessions while maintaining subtle genetic distinctiveness. The A/T--rich composition (A: 30.2%, T: 34.5%) of the trnL-F region aligns with the typical chloroplast genome architecture, providing reliable DNA barcodes for species authentication (Yulita et al. 2022). These genetic markers are particularly valuable for tracing the origin of illegally traded materials, given the species' prevalence in international markets.

The species faces severe anthropogenic threats, primarily driven by demand for its medicinal roots, which contain bioactive compounds with aphrodisiac, antimalarial, and anticancer properties. In Batang Lubu Sutam Forest—a major source region—illegal harvesting has caused local population collapses, with annual exports to Malaysia and neighboring countries reaching commercial scales (Susilowati et al. 2019b). The absence of mature reproductive individuals (pole and tree stages) across multiple study sites demonstrates the unsustainable nature of current extraction practices (Susilowati et al. 2019b, 2023). While some forests maintain good overall regeneration capacity (62.5% of species show healthy recruitment), *E. longifolia itself* exhibits poor regeneration status in overexploited areas, classified as "no regeneration" in 8.3% of studied species assemblages (Susilowati et al. 2021).

Effective conservation requires multi-pronged approaches addressing both ecological and socioeconomic factors. Site-specific protection measures are essential, particularly in production forests where strengthened surveillance and law enforcement could curb illegal harvesting (Susilowati et al. 2019b). The high genetic uniformity suggests that even small protected populations could preserve representative genetic diversity, making targeted area protection feasible. In Muka Kuning in the

Riau Islands, stricter enforcement of nature tourism park regulations is recommended to prevent biodiversity loss (Susilowati et al. 2023). DNA barcoding techniques using the *trn*L-F marker can support these efforts by enabling traceability of traded materials (Yulita et al. 2022).

Active restoration is critical where natural regeneration fails, while sustainable harvesting protocols must account for the species' slow maturation, potentially implementing rotational collection systems in production forests. Community-based conservation programs could align local livelihoods with protection goals, particularly in source regions like Batang Lubu Sutam (Susilowati et al. 2019b, 2021). Long-term monitoring should track IVI trends across growth stages as key performance indicators for conservation success. Critical knowledge gaps remain regarding the species' reproductive ecology and population genetics. While chloroplast markers show limited variation, nuclear DNA studies are needed to assess outcrossing rates and genetic connectivity between fragmented populations. Research on seed dormancy and propagation techniques could improve ex situ conservation efforts. Multisite studies comparing protected versus harvested populations would clarify minimum viable population thresholds. Such research should inform adaptive management frameworks, ensuring this ecologically and economically valuable species persists across its native range.

14.5.3 Styrax paralleloneurus

Styrax paralleloneurus (formerly classified as *S. sumatrana*) (Fig. 14.3) is an endemic canopy tree species native to North Sumatra, Indonesia, primarily distributed across the districts of Pakpak Bharat, Humbang Hasundutan, and Tapanuli Utara along the Bukit Barisan Mountain range (Rachmat et al. 2017; Susilowati et al. 2019c). The species is ecologically and economically significant for its production of benzoin resin, which contains high levels of rosin and cinnamic acid. This resin serves as a primary source of income for local communities through traditional agroforestry systems (García-Fernández et al. 2003; Saputra et al. 2020). Traditional benzoin gardens maintain a seminatural forest structure through low-intensity management, in which selective thinning of competing vegetation allows the persistence of coppicing species while promoting the natural regeneration of forest flora (García-Fernández et al. 2003). This system creates an intermediate disturbance regime that preserves tree diversity comparable to primary forests (Shannon index >3.0), though with reduced representation of mature-habitat specialists and endemic species (García-Fernández et al. 2003). The cultivation practices demonstrate a sustainable balance between resin extraction and biodiversity conservation, with abandoned gardens showing rapid successional recovery toward primary forest composition.

Molecular analyses reveal significant genetic structuring among *S. paralleloneurus* populations, with chloroplast *trn*L-*trn*F sequences (941 bp) identifying four distinct haplotypes across North Sumatra (Rachmat et al. 2017). Strong population

Fig. 14.3 *Styrax paralleloneurus* in community garden (**a**), the tree (**b**), tapped trunk (**c**), crown branching (**d**), and the leaves and fruit (**e**). (Photo: H.H. Rachmat 2025)

differentiation (F_{ST} = 0.80952 between Pakpak Bharat and Tapanuli Utara) reflects isolation by the rugged topography of the Bukit Barisan mountains, which act as barriers to gene flow (Rachmat et al. 2017). Nucleotide diversity remains low (π_{silent} = 0–3.33 × 10^{-3}; π_{nonsyn} = 0–5.9 × 10^{-4}), suggesting historical bottlenecks or limited interpopulation migration (Rachmat et al. 2017). Phylogenetically, Sumatran populations cluster closely with Chinese relatives (*S. suberifolius*, *S. chinensis*; 63% bootstrap support), indicating trans-regional evolutionary connections (Susilowati et al. 2019c). However, standard DNA barcodes (*matK* + *rbcL*; 1479 bp combined) fail to resolve species-level distinctions within Sumatran *Styrax*, highlighting the need for more variable nuclear markers (Susilowati et al. 2017). Findings from the genetic analysis indicate that every extant population contributes irreplaceable allelic diversity shaped by local microclimatic pressures, reinforcing the necessity of protecting all populations to sustain long-term evolutionary resilience.

Climate change and land conversion pose existential threats to *S. paralleloneurus*. Species distribution modeling (MaxEnt) projects an 81–95% reduction in suitable habitats by 2070 under RCP4.5 and RCP8.5 scenarios, driven primarily by rising temperatures (mean coldest quarter >19 °C) and deforestation (Saputra and

Lee 2021). Currently, only 8.91% (663,222 ha) of North Sumatra provides optimal conditions, with shrubs, gardens, and secondary forests being key land cover types (Saputra et al. 2020). The species' reliance on specific thermal regimes (13–19 °C) and well-drained soils makes it particularly vulnerable to warming trends (Saputra and Lee 2021). Additionally, unsustainable land-use practices exacerbate habitat degradation, with soil analyses identifying root limitations, nutrient depletion, and erosion as major constraints to cultivation (Rahmawaty et al. 2023). These abiotic stresses compound the biological vulnerability inherent in the species' fragmented metapopulation structure and low genetic diversity.

Protecting the remaining genetic diversity requires landscape-scale approaches. In situ conservation should prioritize elevational corridors along Bukit Barisan to facilitate climate-driven range shifts, focusing on moderately suitable lands (Rahmawaty et al. 2023; Saputra and Lee 2021). Existing benzoin gardens warrant formal recognition as "domesticated conservation units," given their proven capacity to maintain forest structure and facilitate natural regeneration (García-Fernández et al. 2003). Ex situ collections should capture all four haplotypes, with particular emphasis on the divergent Pakpak Bharat population (Rachmat et al. 2017). Propagation efforts must address soil limitations through organic amendments and terracing to improve water retention and nutrient availability (Rahmawaty et al. 2023).

Community-based management remains central to conservation success. Traditional garden systems should be enhanced through: (1) enrichment planting using genetically diverse stock, (2) phased harvesting to ensure adequate reproductive output, and (3) value-chain development to incentivize sustainable practices (García-Fernández et al. 2003; Saputra et al. 2020). Land-use planning must integrate climate projections, restricting conversions in areas predicted to remain suitable under future scenarios (Saputra and Lee 2021). Developing more sensitive molecular tools (e.g., SNP panels) will enable monitoring of genetic erosion and illegal trade (Susilowati et al. 2017, 2019c).

Critical unanswered questions include: (1) pollen/seed dispersal mechanisms governing gene flow, (2) adaptive potential to environmental changes, and (3) symbiotic relationships affecting resin production. Resolving taxonomic uncertainties through genomic approaches will clarify conservation units, while assisted migration trials could test population resilience (Susilowati et al. 2019c). Longitudinal studies tracking genetic diversity in managed versus wild populations would inform breeding strategies. By addressing these gaps while implementing landscape-scale protection, Indonesia can safeguard this ecologically and culturally significant species against the increasing pressures of anthropogenic activities.

14.5.4 Agarwood-Producing Trees

Agarwood, a valuable oleoresin produced primarily by species of the Thymelaeaceae family (particularly *Aquilaria* spp. and *Gyrinops* spp., Fig. 14.4), represents one of Indonesia's most economically important non-timber forest products (Lukman et al.

Fig. 14.4 The natural regeneration of *Aquilaria hirta* (**a**), *Aquilaria malaccensis* (**b**), and the illegal hunting of agarwood from *Aquilaria malaccensis* trees. (Photo: H.H. Rachmat 2024, 2025)

2022; Pratama et al. 2025). As the principal source of gaharu in international trade, these species face significant conservation challenges due to overexploitation, with populations primarily concentrated in Sumatra and eastern Kalimantan (Lee et al. 2018a, b; Soehartono and Newton 2000). The resin, formed in response to wounding or fungal infection, has been traditionally used for medicinal purposes (including sedation and detoxification) and in incense production (Tanaka and Ito 2020a). Currently, 26 agarwood-producing species across three plant families (Thymelaeaceae, Euphorbiaceae, and Fabaceae) have been documented in Indonesia, though many are declining and were listed in CITES Appendix II in 2005 (Tanaka and Ito 2020b; Lukman et al. 2022). Their vulnerable IUCN status reflects threats from overharvesting and habitat loss (Soehartono and Newton 2000). Conservation requires protecting natural stands while establishing sustainable cultivation.

Several studies on the genetic diversity and phylogeny of agarwood-producing trees have been conducted on naturally growing populations in Indonesia. Phylogenetic analysis has been conducted on 28 Indonesian agarwood-producing species originating from ex situ and in situ populations using ITS and *trn*L-*trn*F markers. This study revealed paraphyly in these genera when *A. cumingiana* was excluded. Notably, these molecular markers effectively distinguished species distribution patterns across Wallace's Line, separating East and West Indonesian populations. The study also highlights ongoing taxonomic challenges in identifying agarwood species and underscores the urgent need for enhanced conservation strategies (Lee et al. 2018a, b). From 66 agarwood accessions using *matK*, ITS, and *trn*L-*trn*F to specimens from Aceh (Indonesia), the study revealed taxonomic discordance: morphological clustering identified *Aquilaria beccariana* (10.61%),

A. malaccensis (10.61%), and *A. microcarpa* (78.79%). At the same time, molecular data showed conflicting species assignments across markers (Lukman et al. 2022). The above results were consistent with the phylogenetic analysis, which revealed paraphyly in *Aquilaria* and *Gyrinops* (Eurlings and Gravendeel 2005), and have been observed across different countries, demonstrating incongruence between traditional morphological classification (stamen number) and molecular clades. Calyx morphology, leaf venation, and indumentum show stronger phylogenetic correlation, supporting the taxonomic merger of *Gyrinops* into *Aquilaria*.

Another study on DNA barcoding of agarwood collected from four major islands—Sumbawa, Lombok, Sulawesi, and Kalimantan, including commercial incense sticks, revealed that the species of the traded agarwood trees are not only *Aquilaria* spp. but also confirmed the *Gyrinops* species—currently unlisted in Japanese non-pharmacopoeial standards (2018) as approved agarwood sources (Tanaka and Ito 2020a). Thus, this opens the door to DNA barcoding for species identification in the traceability of agarwood products. Research on evaluating *Aquilaria microcarpa* populations in South and West Kalimantan was conducted using RAPD markers (Qiptiyah et al. 2021). Analysis of 44 samples revealed low genetic diversity (uHE: 0.152–0.249), with 49 polymorphic loci from nine stable primers. Populations exhibited significant differentiation (genetic distance: 0.1733), evidenced by private alleles at distinct loci (G18/540 in Gumbil, G18/550 in Sanggau) and by AMOVA-confirmed interpopulation variation, highlighting implications for seed-source management in plantation programs. Analysis of 41 *A. malaccensis* samples from Sumatra mainland, West Kalimantan, and small islands adjacent to the two mainlands (Bintan, Lingga, and Bunguran Island) using *trn*L-*trn*F and *trn*H-*psb*A markers identified six haplotypes with distinct phytoregional distribution. Bunguran Island exhibited unique haplotypes and shared one with haplotypes from Sumatra's adjacent islands, suggesting historical connections to Sundaland. Despite low nucleotide diversity ($\pi = 0$–0.00215), significant genetic differentiation was observed (max $F_{ST} = 0.83686$ between Bunguran and West Kalimantan), highlighting the importance of haplotype-specific conservation strategies for ex situ programs of *A. malaccensis* (Rachmat et al. 2024).

In conclusion, the case studies presented above suggest that medicinal plant genetic resources require integrated approaches. *T. sumatrana*, which is highly valued for its taxol content, suffers from limited distribution, slow growth, and poor natural regeneration. Similarly, *Eurycoma longifolia* (pasak bumi or tongkat ali) is threatened by unsustainable root harvesting practices. Endemic species such as *Styrax paralleloneurus* further highlight the interplay between ecological and cultural significance. Cultivated in traditional benzoin gardens, this resin-producing tree supports community livelihoods while maintaining seminatural forest structures. Agarwood-producing trees (*Aquilaria* spp.) face similar pressures, with international trade driving population depletion. Strengthening in situ protection, developing ex situ collections, applying molecular tools for species identification and traceability, and empowering community-based management are all critical steps. Such strategies will not only safeguard ecological resilience and genetic diversity but also secure the long-term sustainability of species that hold both cultural significance and global medicinal value.

14.6 The Impact of Cultivation and Domestication on the Genetic and Phytochemical Aspects of Zingiberaceae Species

Zingiberaceae, or the ginger family, is a monocotyledonous flowering plant that has long been utilized in many cultures for its rich aromatic compounds and bioactive metabolites. This plant has been highly valued for centuries for its role in traditional healing systems, culinary traditions, and cultural rituals. Currently, the family comprises 1884 species across 50 genera (POWO 2025), a notable increase from 1600 species previously reported by Christenhusz and Byng (2016). It has a pantropical distribution, with the highest species diversity occurring in Southeast Asia (Zou et al. 2022). High morphological variation among ginger species (Wicaksana et al. 2011) and specialized pollination systems (Gao et al. 2005) observed in ginger plants make Zingiberaceae well adapted to various ecosystems, including karst (Nor-Syahaiza et al. 2023). Plant organs derived from Zingiberaceae—mainly rhizomes—have been widely utilized in both traditional and modern medicine since many species contain diverse medicinal properties such as antioxidant, anti-inflammatory, anticancer, neurotrophic, cosmeceutical, antifungal, and antimicrobial bioactivities (Chen et al. 2008; Soares et al. 2018; Kravchenko et al. 2019; Han et al. 2021; Alolga et al. 2022; Destryana et al. 2024; Sumi et al. 2024). *Zingiber officinale* (common ginger) is the most popular species, widely cultivated for use in the food, cosmetic, and pharmaceutical industries. Domestic utilization and global market demand for ginger have been rapidly growing since it was first domesticated and its popularity spread worldwide. Most livelihoods in rural areas of Southeast Asia have this species in their yards. The leading producers of ginger in the International market are India (32.75%), China (21.41%), Nigeria (12.40%), and Bangladesh (10.80%). The main importing countries include Japan, the USA, the UK, and Saudi Arabia (Rankethkumbura 2024). Global production of ginger in 2022 reached 4.8 million metric tons from cultivation areas of 456,747 ha worldwide. The global market demand for high-quality ginger requires cultivation practices that produce market-preferred qualities, such as high volatile oil content, low fiber content, and sufficient nonvolatile ether extract (Oladosu et al. 2024). Cultivation practices using advanced technologies have been widely implemented in Zingiberaceae, ranging from basic agricultural practices to omics technologies. However, it is crucial to understand how these cultivation practices have affected the genetic structure and phytochemical composition of cultivated species. This section discusses the impact of cultivation and domestication on the genetic and phytochemical properties of Zingiberaceae, as well as the conservation efforts that should be undertaken.

14.6.1 History of Domestication of Zingiberaceae

Domestication and cultivation are often seen as alternative approaches to wild harvesting. The history of ginger domestication dates back approximately 3000 years, and is closely associated with the Austronesian migration routes (Sasikumar 2005). Archaeobotanical and linguistic evidence suggests that early human populations in Southeast Asia utilized rhizomatous plants such as turmeric (*Curcuma longa*), ginger (*Zingiber officinale*), and galangal (*Alpinia galanga*) for food and medicine. Gingers are then widely spread across the globe as a result of intensive use and subsequent dispersal through prehistoric Austronesian migration routes, followed by maritime trade networks linking South and Southeast Asia with the Middle East and the Mediterranean (Kizhakkayil and Sasikumar 2011). Human selection during early cultivation favored morphological and phytochemical traits such as rhizome size, aroma, pigment intensity, and bioactive compound content.

Domestication pathways in Zingiberaceae were primarily influenced by vegetative propagation, a practice that facilitated the preservation of desirable clones but limited sexual recombination. This form of propagation led to the fixation of favorable genotypes and also restricted gene flow, resulting in a narrow genetic base (Oladosu et al. 2024). Consequently, cultivated populations often exhibit high uniformity and reduced fertility compared to their wild relatives. Such genetic bottlenecks are evident across various genera of Zingiberaceae, including *Zingiber, Curcuma,* and *Amomum* (Ashraf et al. 2014; Balada et al. 2021; Gogoi et al. 2023). Li et al. (2022) demonstrated that domestication is strongly associated with increased production of secondary metabolites and a reduction in allelic diversity.

14.6.2 Genetic and Phytochemical Changes Under Domestication

A long history of selective breeding has reduced genetic diversity in several species of Zingiberaceae. Vegetative propagation through rhizomes has been the predominant mode of reproduction in this family. In addition, agricultural practices during the domestication of the Zingiberaceae family may have affected reproductive systems, thereby impacting levels of genetic diversity. This was exemplified by the study of Huang et al. (2021) on *Zingiber corallinum* and *Z. nudicarpum. Zingiber corallinum* is a self-pollinating species that maintains its diversity through local adaptation and genetic drift. This species exhibits lower genetic diversity among populations than the cross-pollinating species of *Z. nudicarpum* (Huang et al. 2019, 2021). Higher genetic diversity is observed *in Z. montanum*, which reproduces generatively, compared to common ginger (*Z. officinale*), which has lower genetic diversity due to intensive vegetative propagation (Thomas et al. 2016). *Zingiber montanum* (cassumunar ginger) is reported by Bua-in and Paisooksantivatana (2010) to retain high genetic diversity, as shown by RAPD analyses, despite being clonally propagated. On the other hand, *Z. officinale* often exhibits lower genetic variation in cultivation due to intensive vegetative propagation and limited sexual reproduction (Kizhakkayil and Sasikumar 2011). Another example is *Curcuma longa*, which shows lower allelic diversity in cultivated varieties than in its wild relatives. This decline in alleles is due to clonal propagation practices and intensive selection for superior traits, such as rhizome yield and curcumin content (Sasikumar 2005). Another study of genetic diversity in *Amomum villosum* (cardamom) using ISSR markers, however, showed considerable genetic variation even in established cultivation areas. Vegetative propagation combined with local selection has led to some clustering and reduced gene flow among populations (Xu et al. 2022). These examples revealed that domestication and habitat fragmentation in many medicinal and crop plants can lead to genetic bottlenecks, reduced heterozygosity, and inbreeding, especially where propagation is largely asexual (Huang et al. 2021). Furthermore, reproductive strategies are considered to influence patterns of genetic diversity. Cross-pollinating species, such as *Zingiber nudicarpum*, tend to maintain higher levels of diversity compared to self-pollinating or clonally reproducing species, such as *Z. corallinum* (Huang et al. 2019, 2021). Thus, the genetic structure of Zingiberaceae populations reflects a complex interaction between reproductive biology and the intensity of human cultivation practices.

Domestication and cultivation practices not only affect the genetic diversity of Zingiberaceae plants but can also modify their phytochemical composition. For example, a long period of intensive domestication has led to significant variation in bioactive compounds, such as ethyl-trans-p-methoxycinnamate, among *Kaempferia galanga* accessions (Raina et al. 2015). Similar findings of increased levels of secondary metabolites, i.e., phenolics, flavonoids, and alkaloids, are also recorded in some domesticated ginger plants, such as *Zingiber rubens* and *C. longa*. The plants exhibit higher levels of antioxidant and antibacterial activities, accompanied by a more complex phytochemical profile (Zaman and Kayang 2025; Windarsih et al. 2025). In addition, selective breeding has also been reported to increase the levels of important bioactive compounds, such as gingerols and curcuminoids, in cultivated ginger and turmeric (Gaur et al. 2016; Sandeep et al. 2016).

The phytochemical profile of Zingiberaceae is also strongly influenced by environmental factors and cultivation conditions, such as soil composition, temperature, light intensity, and altitude, which can significantly affect the accumulation of secondary metabolites. Mahawer et al. (2023) reported that *Zingiber roseum* exhibits significant variations in bioactive compound content across locations at different altitudes. Conversely, cultivated *Hedychium spicatum* exhibits changes in metal and fatty acid levels compared with its wild relatives (Lohani et al. 2015).

14.7 Modern Cultivation Practices and Phytochemical Enhancement

Advances in modern agriculture have introduced innovative cultivation techniques to improve the productivity, quality, and phytochemical profiles of Zingiberaceae plants. The application of fertigation systems and the use of biofertilizers and other fertilizers are standard cultivation practices. These systems have been shown to modify phytochemical expression and affect the accumulation of secondary metabolites such as gingerol, shogaol, curcumin, and other essential oils (Nur Diyana et al. 2024; Akbari et al. 2022), which can ultimately increase the production of secondary metabolites (Satya et al. 2024). Similar findings were reported by Gaur et al. (2016) and Sandeep et al. (2016), who showed that differences in elevation and soil moisture significantly affect the essential oil profiles of several Zingiberaceae species.

Tissue culture techniques are also a common approach in modern Zingiberaceae production because they enable the mass propagation of genetically uniform, pathogen-free plants with stable metabolic performance. It also provides somaclonal variation in breeding to produce superior genotypes with higher secondary metabolite content (Ferreira et al. 2023). This approach provides highly controlled environmental conditions, allowing the biosynthesis of high-value compounds such as curcumin, zingiberene, and various essential oil components to be optimized by manipulating media, light, and metabolic precursors (Windarsih et al. 2025). Micropropagation and in vitro cultivation have also been shown to increase phenolic and flavonoid levels in *Curcuma zedoaria* and *Kaempferia angustifolia* (Subramanian et al. 2025; Saensouk et al. 2023).

The development of genomics, metabolomics, and phenomics has opened up new opportunities to uncover bioactive compounds in ginger plants. Mass spectrometry (MS)-based metabolomics approach can be used to characterize secondary metabolite profiles, assess interspecies variation, and study the effects of environmental influences. Integrating metabolomics-MS techniques with multivariate analysis has been shown to improve the accuracy of assessing *Z. officinale* quality, including the quantification of gingerol and other bioactive components (Salem et al. 2021). Furthermore, volatilomic and metabolomic studies of various Zingiberaceae species have successfully revealed the diversity of their metabolite profiles. These studies demonstrate how differences in the biosynthetic pathways of volatile (fragrant) and nonvolatile compounds determine their pharmacological properties and economic value (Zhang et al. 2023). A study on *Alpinia hainanensis* integrating phenomic, metabolomic, and flavoromic approaches successfully mapped how various quality indicators—such as key bioactive compounds and aroma components—are formed and regulated during fruit development. These findings provide a deeper understanding of the dynamics of high-value metabolite production in the Zingiberaceae family (Chen et al. 2025). Thus, the integration of these various 'omics' approaches provides a strong scientific base for developing plant breeding strategies, improving phytochemical quality, and creating superior cultivars with greater metabolite production stability under various environmental conditions.

14.8 Conservation Strategies and Breeding Opportunities

The decline in genetic diversity observed in cultivated Zingiberaceae species poses a significant challenge to conservation efforts and breeding programs, as successful conservation and breeding of germplasm depend heavily on the availability of phenotypic variation to improve plant resilience and productivity. Species that have been intensively domesticated tend to have low genetic variation or even genetic homogeneity in clonal populations. Cultivated populations of *Scrophularia ningpoensis* in China exhibit a substantial reduction in genetic diversity, with only three haplotypes compared to 21 in wild populations. Intensive selection for desirable medicinal traits has been a significant factor in this decline, producing marked genetic differences between cultivated and natural populations (Chen et al. 2014).

Conservation of natural populations and wild relatives of Zingiberaceae species is critical to preserve genetic resources, as they may possess important alleles for resilience and adaptability to environmental change. Genetic resources of these wild populations and species can be preserved through ex situ conservation strategies, including the establishment of germplasm collections and tissue culture repositories. Such collections are also crucial for breeding initiatives. Utilizing wild germplasm can recover lost allelic variation and improve resistance to environmental and biological challenges.

Despite its high economic value, the Zingiberaceae family faces serious challenges in maintaining its sustainability. To address this challenge, a crucial first step is to understand the state of its genetic diversity. A study of 10 Zingiberaceae species from East India revealed an interesting pattern: genetic diversity is more widely distributed across species than concentrated within a single species (Mohanty et al. 2014). This means that each species and population harbors unique genetic potential. If one population becomes extinct, we could lose a valuable gene pool that cannot be found elsewhere.

Another study on *Curcuma wenyujin* reported that although wild populations still possess considerable genetic diversity, cultivated populations are more vulnerable to reduced genetic variation (Zheng et al. 2015). This emphasizes that protecting wild populations in their natural habitat is just as important as developing cultivated varieties. This is where ex situ conservation techniques, particularly through tissue culture and cryopreservation, play a crucial role. This method allows us to preserve the genetic material of endangered Zingiberaceae species for the long term, away from habitat disturbance and overexploitation (Chakraborty et al. 2023). This secure germplasm collection is like a "gene bank," ready for use at any time, for example, to research new bioactive compounds with anti-allergy properties (Das et al. 2024). However, conservation alone is not enough. To ensure that Zingiberaceae can meet future needs, intelligent breeding programs must be designed. Research on two Zingiber species revealed that cross-pollinated species tend to have populations with highly divergent genetic traits (Huang et al. 2021). This fact provides valuable guidance for breeders that to achieve superior traits, we need to explore and cross genetic material from different populations. Therefore, conservation and breeding should be carried out together and integrated. Preserved germplasm collections provide genetic "raw material," while modern breeding is to create new varieties of Zingiberaceae that are more disease resistant, adaptable to climate change, and rich in bioactive compounds. Integrating these two approaches is key to preserving Zingiberaceae while using them sustainably.

In conclusion, domestication through cultivation practices in Zingiberaceae has successfully increased the economic value and pharmacological benefits of these plants globally, but also poses the risk of genetic erosion and changes in phytochemical composition. On the other hand, selective breeding and modern cultivation practices can increase the production of bioactive compounds in some species. Hence, conservation of natural populations of Zingiberaceae remains crucial to maintain genetic diversity and ensure their long-term resilience, adaptability, and phytochemical stability. Future research should involve genomic and metabolomic approaches to discover active biocompounds and optimize the production of high-value phytochemical compounds while conserving genetic diversity. A balance between sustainable utilization and conservation will ensure that Zingiberaceae species continue to provide ecological, medicinal, and economic benefits for future generations.

References

Abd-Elgawad ME, Alotaibi MO (2019) Genetic diversity among Saudi *Peganum harmala* and *Rhazya stricta* populations using chemical and ISSR markers. Curr Pharm Biotechnol 20(13):1134–1146. https://doi.org/10.2174/1389201020666190619105249

Aguilar R, Ashworth L, Galetto L, Aizen MA (2006) Plant reproductive susceptibility to habitat fragmentation: review and synthesis through a meta-analysis. Ecol Lett 9(8):968–980. https://doi.org/10.1111/j.1461-0248.2006.00927.x

Akbari GA, Miri SA, Zeinali H (2022) Investigating the effects of different fertilizers and cultivation media on the yield and active ingredients of *Zingiber officinale* Rosc. J Med Plants 21(84):75–86. https://doi.org/10.52547/jmp.21.84.75

Alias NN, Lee CT, Zakaria NF, Lee SL, Ng KKS, Ng CH, Tnah LH, Basherudin N (2021) Development of genic SSR markers from transcriptome sequences of *Baeckea frutescens* (cucur atap) for genetic diversity assessment. J Trop Plant Physiol 13(1):11. https://doi.org/10.56999/jtpp.2021.13.1.11

Alolga RN, Wang F, Zhang X, Li J, Tran L-SP, Yin X (2022) Bioactive compounds from the Zingiberaceae family with known antioxidant activities for possible therapeutic uses. Antioxidants 11(7):1281. https://doi.org/10.3390/antiox11071281

Alttaher AGA, Yusof ZNB, Shaharuddin NA (2021) Direct shoot organogenesis and clonal fidelity confirmation of Tongkat Ali (*Eurycoma longifolia*) using molecular markers. Int J Agric Biol 26:9–16. https://doi.org/10.17957/IJAB/15.1802

Arif MF, Subositi D, Sari AN, Aristya GR, Lesmana I, Sri Kasiamdari R (2020) Genetic diversity of green chireta (*Andrographis paniculata* (Burm.f.) Wall. ex Nees) from Indonesia based on ISSR and RAPD markers. Malay Appl Biol 49(1):61–68. https://doi.org/10.55230/mabjournal.v49i1.1655

Ashraf K, Ahmad A, Chaudhary A, Mujeeb M, Ahmad S, Amir M, Mallick N (2014) Genetic diversity analysis of *Zingiber officinale* Roscoe by RAPD collected from subcontinent of India. Saudi J Biol Sci 21(2):159–165

Azli SNAM, Bakar AA, Shahimi S et al (2024) Molecular approach for identification and validation of *Eurycoma longifolia, Labisia pumila,* and *Orthosiphon stamineus* in herbal products. Beni-Suef Univ J Basic Appl Sci 13:23. https://doi.org/10.1186/s43088-024-00481-1

Balada C, Castro M, Fassio C, Zamora A, Marchant MJ, Acevedo W, Guzmán L (2021) Genetic diversity and biological activity of *Curcuma longa* ecotypes from Rapa Nui using molecular markers. Saudi J Biol Sci 28(1):707–716

Bisht VK, Negi JS, Bhandari AK (2016) Check on extinction of medicinal herbs in Uttarakhand: no need to uproot. Nat Acad Sci Lett 39(3):233–235. https://doi.org/10.1007/s40009-016-0437-1

Bua-in S, Paisooksantivatana Y (2010) Study of clonally propagated cassumunar ginger (*Zingiber montanum* (Koenig) Link ex Dietr.) and its relation to wild Zingiber species from Thailand revealed by RAPD markers. Genet Resour Crop Evol 57(3):405–414

Bussmann RW (2013) The globalization of traditional medicine in northern Peru: from shamanism to molecules. Evid Based Complement Alternat Med 2013:291903. https://doi.org/10.1155/2013/291903

Cahyaningsih R, Phillips J, Magos Brehm J, Gaisberger H, Maxted N (2021a) Climate change impact on medicinal plants in Indonesia. Glob Ecol Conserv 30:e01752. https://doi.org/10.1016/j.gecco.2021.e01752

Cahyaningsih R, Brehm JM, Maxted N (2021b) Gap analysis of Indonesian priority medicinal plant species as part of their conservation planning. Glob Ecol Conserv 26:e01459. https://doi.org/10.1016/j.gecco.2021.e01459

Cahyaningsih R, Brehm JM, Maxted N (2021c) Setting the priority medicinal plants for conservation in Indonesia. Genet Resour Crop Evol 68(8):2019–2050. https://doi.org/10.1007/s10722-021-01115-6

Chakraborty A, Santra I, Haque SM, Ghosh B (2023) In vitro conservation of commercial and threatened members of Zingiberaceae: an Indian scenario. Biodivers Conserv 32(7):2155–2195

Chan EWC, Wong SK, Chan HT (2011) Phytochemistry and pharmacological properties of Etlingera species: a review. Pharm J 3(22):6–10. https://doi.org/10.5530/pj.2011.22.2

Chen IN, Chang CC, Ng CC, Wang CY, Shyu YT, Chang TL (2008) Antioxidant and antimicrobial activity of Zingiberaceae plants in Taiwan. Plant Foods Hum Nutr 63(1):15–20. https://doi.org/10.1007/s11130-007-0063-7

Chen C, Li P, Wang R-H, Schaal BA, Fu C-X (2014) The population genetics of cultivation: domestication of a traditional Chinese medicine, *Scrophularia ningpoensis* Hemsl. (Scrophulariaceae). PLoS One 9(8):e105064. https://doi.org/10.1371/journal.pone.0105064

Chen S-L, Yu H, Luo H-M, Wu Q, Li C-F, Steinmetz A (2016) Conservation and sustainable use of medicinal plants: problems, progress, and prospects. Chin Med 11(1):37. https://doi.org/10.1186/s13020-016-0108-7

Chen G, Sun W, Wang X et al (2019) Conserving threatened widespread species: a case study using a traditional medicinal plant in Asia. Biodivers Conserv 28:213–227. https://doi.org/10.1007/s10531-018-1648-1

Chen X, Tian Q, Chen Z, Jiang Q, Hu X, Wang D, Wu T, Luo X, Yuan L, Yu F (2025) Integrative phenomics, metabolomics, and flavoromics reveal key quality indicators during the formation of flavor and bioactive compounds in *Alpinia hainanensis* (Zingiberaceae) fruit. Food Chem 486:144602

Christenhusz MJM, Byng JW (2016) The number of known plant species in the world and its annual increase. Phytotaxa 261(3):201–217. https://doi.org/10.11646/phytotaxa.261.3.1

Cruse-Sanders JM, Hamrick JL (2004) Genetic diversity in harvested and protected populations of wild American ginseng, *Panax quinquefolius* L. Am J Bot 91(4):540–548. https://doi.org/10.3732/ajb.91.4.540

Cunningham AB, Brinckmann JA, Schippmann U (2018) Non-detriment findings for timber, medicinal and aromatic plant species listed in CITES Appendices II and III. CITES

de Andrade CML, Rohlfs IAC, Caetano TTV et al (2025) The redesigned Psychotria genus: medicinal potential of alkaloids and biotechnological strategies for metabolite improvement. Phytochem Rev. https://doi.org/10.1007/s11101-025-10154-1

de Padua LS et al (eds) (1999) Plant resources of South-East Asia (PROSEA) — volume on medicinal plants. E-depot. https://edepot.wur.nl/411255

de Boer HJ, Cotingting C (2014) Medicinal plants for women's healthcare in southeast Asia: a meta-analysis of their traditional use, chemical constituents, and pharmacology. J Ethnopharmacol 151(2):747–767

Das J, Sanglyne MW, Kurbah LA, Das MC (2024) Progress and prospect in ex situ conservation of Zingiberaceae harboring antiallergic compounds. In: Biotechnology of medicinal plants with antiallergy properties: research trends and prospects. Springer Nature, Singapore, pp 567–610

De Feo V (2004) The ritual use of Brugmansia species in traditional Andean medicine in northern Peru. Econ Bot 58(Supplement):S221–S229. https://doi.org/10.1663/0013-0001(2004)58[S221:TRUOBS]2.0.CO;2

De Laubenfels D (1988) Flora Malesiana. Coniferales, 3rd edn., vol 10, issue 1. Kluwer Academic Publishers

De Melo JG, De Amorim ELC, De Albuquerque UP (2009) Native medicinal plants commercialized in Brazil: priorities for conservation. Environ Monit Assess 156:567–580. https://doi.org/10.1007/s10661-008-0506-0

Destryana RA, Estiasih T, Sukardi, Pranowo D (2024) Zingiberaceae rhizome essential oil: a review of chemical composition, biological activity, and application in food industry. IOP Conf Ser Earth Environ Sci 1299:012010. https://doi.org/10.1088/1755-1315/1299/1/012010

Ditta A, Ullah N, Li X, Soomro GS, Imtiaz M, Mehmood S, Jan AU, Rizwan MS, Rizwan M, Ahmad I (2022) Heavy metal pollution and medicinal plants: an overview. In: Husen A (ed) Environmental pollution and medicinal plants. CRC Press, pp 145–164. https://doi.org/10.1201/9781003178866-8

Doi H, Takahashi M, Katano I (2010) Genetic diversity increases regional variation in phenological dates in response to climate change. Glob Chang Biol 16(1):373–379. https://doi.org/10.1111/j.1365-2486.2009.01993.x

Eurlings MCM, Gravendeel B (2005) trnL–trnF sequence data imply paraphyly of *Aquilaria* and *Gyrinops* (Thymelaeaceae) and provide new perspectives for agarwood identification. Plant Syst Evol 254(1):1–12. https://doi.org/10.1007/s00606-005-0312-x

Ferreira MDS, Rocha ADJ, Nascimento FDS, Oliveira WDDS, Soares JMDS, Rebouças TA, Morais Lino LS, Haddad F, Ferreira CF, Santos-Serejo JAD, Fernández JS (2023) The role of somaclonal variation in plant genetic improvement: a systematic review. Agronomy 13(3):730

Frianto D, Prihartini I, Novriyanti E, Widyatmoko CAYPB (2021) The genetic potential of *Taxus sumatrana* medicinal plants in Kerinci Regency. IOP Conf Ser Earth Environ Sci 914(1):012076. https://doi.org/10.1088/1755-1315/914/1/012076

Food and Agriculture Organization of the United Nations (2011) Country Report on the State of Plant Genetic Resources for Food and Agriculture — Indonesia. FAO

Gao JY, Ren PY, Li QJ (2005) Advances in the study of breeding system and pollination biology of gingers (Zingiberaceae and Costaceae). J Syst Evol 43(6):574–585. https://doi.org/10.1360/aps040074

García-Fernández C, Casado MA, Ruiz Pérez M (2003) Benzoin gardens in North Sumatra, Indonesia: effects of management on tree diversity. Conserv Biol 17(3):829–836. https://doi.org/10.1046/j.1523-1739.2003.01487.x

Gaur M, Das A, Sahoo RK, Mohanty S, Joshi RK, Subudhi E (2016) Comparative transcriptome analysis of ginger variety Suprabha from two different agro-climatic zones of Odisha. Genom Data 9:42–43. https://doi.org/10.1016/j.gdata.2016.06.014

Gogoi A, Munda S, Paw M, Begum T, Siddiqui MH, Gaafar ARZ, Kesawat MS, Lal M (2023) Molecular genetic divergence analysis amongst high curcumin lines of golden crop (*Curcuma longa* L.) using SSR marker and use in trait-specific breeding. Sci Rep 13(1):19690

Gonzalez-Oliva L (2020) Angostura ossana. In: The IUCN red list of threatened species 2020, p e.T31675A149822436. https://doi.org/10.2305/IUCN.UK.2020-3.RLTS.T31675A149822436.en

Halder M, Jha S (2023) The current status of population extinction and biodiversity crisis of medicinal plants. In: Jha S, Halder M (eds) Medicinal plants: biodiversity, biotechnology and conservation, Sustainable development and biodiversity series. Springer Nature Singapore, Singapore

Han AR, Kim H, Piao D, Jung CH, Seo EK (2021) Phytochemicals and bioactivities of *Zingiber cassumunar* Roxb. Molecules 26(8):2377. https://doi.org/10.3390/molecules26082377

Hoban S, Bruford M, D'Urban Jackson J, Lopes-Fernandes M, Heuertz M, Hohenlohe PA, Paz-Vinas I, Sjögren-Gulve P, Segelbacher G, Vernesi C, Aitken S, Bertola LD, Bloomer P, Breed M, Rodríguez-Correa H, Funk WC, Grueber CE, Hunter ME, Jaffe R et al (2020) Genetic diversity targets and indicators in the CBD Post-2020 Global Biodiversity Framework must be improved. Biol Conserv 248:108654. https://doi.org/10.1016/j.biocon.2020.108654

Huang R, Chu QH, Lu GH, Wang YQ (2019) Comparative studies on population genetic structure of two closely related selfing and outcrossing *Zingiber* species in Hainan Island. Sci Rep 9:17997. https://doi.org/10.1038/s41598-019-54526-y

Huang R, Zhang ZD, Wang Y, Wang YQ (2021) Genetic variation and genetic structure within metapopulations of two closely related selfing and outcrossing *Zingiber* species (Zingiberaceae). AoB Plants 13(1):plaa065. https://doi.org/10.1093/aobpla/plaa065

Ismail NA, Rafii MY, Mahmud TMM, Hanafi MM, Miah G (2019) Genetic diversity of torch ginger (*Etlingera elatior*) germplasm revealed by ISSR and SSR markers. BioMed Res Int 2019:5904804. https://doi.org/10.1155/2019/5904804

IUCN (2025) The IUCN red list of threatened species. Version 2023-3. https://www.iucnredlist.org

Kamal I, Arrofaha N, Susila, Yulita KS, Rachmat HH, Susilowati A (2023) Genetic variation and phylogenetic position of *Taxus sumatrana* (Miq.) de Laub. (Taxaceae) in Sumatra, Indonesia based on trnL–trnF region. Biodivers 24(9):5155–5166. https://doi.org/10.13057/biodiv/d240960

Karimi E, Haji-Javadi M, Saad MS, Khoo BY (2014) Genetic relationship among *Labisia pumila* (Myrsinaceae) species based on ISSR-PCR. Plant Syst Evol 300:193–201. https://doi.org/10.4238/2014.April.29.8

Kartikawati SM, Amir E, Zuhud M, Hikmat A, Kartodihardjo H (2014) Habitat preferences, distribution pattern, and root weight estimation of pasakbumi (*Eurycoma longifolia* Jack.). J Man Hut Trop XX(1):43–50. https://doi.org/10.7226/jtfm.20.1.43

Khoury CK, Greene S, Wiersema J, Maxted N, Jarvis A, Struik PC (2013) An inventory of crop wild relatives of the United States. Crop Sci 53:1496–1508. https://doi.org/10.2135/cropsci2012.10.0585

Kizhakkayil J, Sasikumar B (2011) Diversity, characterization and utilization of ginger: a review. Plant Genet Resour 9(3):464–477. https://doi.org/10.1017/S1479262111000670

Kravchenko I, Eberle L, Nesterkina M, Kobernik A (2019) Anti-inflammatory and analgesic activity of ointment based on dense ginger extract (*Zingiber officinale*). J Herbmed Pharmacol 8(2):126–132. https://doi.org/10.15171/jhp.2019.20

Laftouhi A, Elrherabi A, Farihi A, Nasr FA, Hmamou A, Mahraz MA, Bouhrim M, Al-zharani M, Qurtam AA, Eloutassi N, Taleb A, Taleb M (2025) Impact of climate change on the composition and bioactivity of *Inula viscosa* essential oils: in vitro, in silico, and ADME analysis. Food Sci Nutr 13:e70845. https://doi.org/10.1002/fsn3.70845

Lee CT, Norlia B, Tnah LH, Lee SL, Ng CH, Ng KKS et al (2018a) Isolation and characterisation of SSR markers in tongkatali (*Eurycoma longifolia*) using next-generation sequencing approach. J Trop For Sci 30(3):279–291

Lee SY, Turjaman M, Mohamed R (2018b) Phylogenetic relatedness of several agarwood-producing taxa (Thymelaeaceae) from Indonesia. Trop Life Sci Res 29(2):13–28. https://doi.org/10.21315/tlsr2018.29.2.2

Li S, Lin K-H, Wu C-W, Chang Y-S (2021) Effect of temperature on growth, physiological, and antioxidant characteristics in *Houttuynia cordata*. Not Bot Horti Agrobot Cluj Napoca 49(4):12536. https://doi.org/10.15835/nbha49412536

Li WX, Li JL, He JJ, Zhang HL, Luo P, Wei Y, Zhao MT (2022) Genetic diversity and population structure of germplasm resources of *Amomum villosum* based on SSR markers. China J Chin Mater Med 47(17):4618–4626

Li Y, Zhao W, Zuo X et al (2025) Habitat changes due to cultivation alter the genetic diversity and secondary metabolic changes of medicinal plants. Sci Rep. https://doi.org/10.1038/s41598-025-20010-z

Liu Z, Zhou L, Yan B, Liu Z (2025) Comprehensive assessment of climate-driven distribution dynamics and anthocyanin variation in *Lycium ruthenicum* using MaxEnt, HPLC, and chemometric approaches. Ind Crop Prod 233:121359. https://doi.org/10.1016/j.indcrop.2025.121359

Lohani N, Tewari G, Joshi GC, Kishor K (2015) Comparative phytochemical analysis of wild and cultivated rhizomes of *Hedychium spicatum* Buch.-Ham. of north west Himalaya. J Indian Chem Soc 92(6):105–109

Lukman, Dinarti D, Siregar UJ, Turjaman M, Sudarsono (2022) Characterization and identification of agarwood-producing plants (*Aquilaria* spp.) from North Aceh, Indonesia, based on morphological and molecular markers. Biodivers 23(9):4861–4871. https://doi.org/10.13057/biodiv/d230955

Mahawer SK, Kumar R, Karakoti H, Prakash O, Kumar S, Rawat DS, Maslowiecka J, Isidorov VA, de Oliveira MS (2023) Altitudinal variability in chemical profile and pesticidal properties of *Zingiber roseum* (Roxb.) Roscoe extracts. Biochem Syst Ecol 110:104708. https://doi.org/10.1016/j.bse.2023.104708

Maxted N, Hunter D, Ortiz Rios R (2020) Plant genetic conservation. Cambridge University Press. UK

Mohanty S, Panda MK, Acharya L, Nayak S (2014) Genetic diversity and gene differentiation among ten species of Zingiberaceae from Eastern India. 3 Biotech 4(4):383–390

Myers N, Mittermeier RA, Mittermeier CG, Da Fonseca GAB, Kent J (2000) Biodiversity hotspots for conservation priorities. Nature 403(6772):853–858. https://doi.org/10.1038/35002501

Mykhailenko O, Jalil B, McGaw LJ, Echeverría J, Takubessi M, Heinrich M (2025) Climate change and the sustainable use of medicinal plants: a call for "new" research strategies. Front Pharmacol 15:1496792. https://doi.org/10.3389/fphar.2024.1496792

Nor-Syahaiza AZ, Appalasamy S, Suhana Z, Arumugam N, Nurdiyanah-Atikah MR, Mahani MC (2023) Anatomical study of limestone Zingiberaceae species in Jeli District, Kelantan. J Trop For Sci 35(1):28–38. https://doi.org/10.26525/jtfs2023.35.1.28

Noss RF, Platt WJ, Sorrie BA, Weakley AS, Means DB, Costanza J, Peet RK (2015) How global biodiversity hotspots may go unrecognized: lessons from the North American Coastal Plain. Divers Distrib 21(2):236–244. https://doi.org/10.1111/ddi.12278

Nur Diyana A, Nor Azlan SN, Mohd Lazim MI, Hussin A, Mansor A, Ibrahim AH (2024) Greener approaches to improve phenolic acid composition and antioxidant activity of different cultivation methods of *Zingiber officinale* Roscoe extracts. Food Res 6(Suppl. 3):67–73. https://doi.org/10.26656/fr.2017.6(S3).1

Oladosu Y, Rafii MY, Arolu F, Murugesu S, Chukwu SC, Salisu MA, Kadar AI (2024) Genetic diversity and utilization of ginger (*Zingiber officinale*) for varietal improvement: a review. AIMS Agric Food 9(1):183–208

Pimentel D, Zuniga R, Morrison D (2005) Update on the environmental and economic costs associated with alien-invasive species in the United States. Ecol Econ 52(3):273–288. https://doi.org/10.1016/j.ecolecon.2004.10.002

Potts SG, Biesmeijer JC, Kremen C, Neumann P, Schweiger O, Kunin WE (2016) Global pollinator declines: trends, impacts and drivers. Trends Ecol Evol 25(6):345–353. https://doi.org/10.1016/j.tree.2010.01.007

Poulsen AD (2006) Etlingera of Borneo. Natural History Publications, Borneo

Póvoa O, Lopes V, Barata AM, Farinha N (2023) Monitoring genetic erosion of aromatic and medicinal plant species in Alentejo (South Portugal). Plants 12(14):2588. https://doi.org/10.3390/plants12142588

POWO-Plants of the World Online (2025) Zingiberaceae. Royal Botanic Gardens, Kew. https://powo.science.kew.org/taxon/urn:lsid:ipni.org:names:30004541-2

Pratama BA, Atikah TD, Susanti R, Jakalalana S, Ferdinandus SP, Maail RS (2025) Stand structure and biomass estimation for cultivation of Agarwood-producing species on Buru Island, Maluku Province. Indonesia Media Konservasi 30(1):63–72. https://doi.org/10.29244/medkon.30.1.63

Qiptiyah M, Widyatmoko CAYPB, Nurtjahjaningsih ILG, Prihatini I (2021) Genetic diversity of *Aquilaria microcarpa* Baill. in Kalimantan using RAPD markers. IOP Conf Ser Earth Environ Sci 914:012039. https://doi.org/10.1088/1755-1315/914/1/012039

Rachmat HH, Subiakto A, Kamiya K (2016) Short communication: genetic diversity and conservation strategy considerations for highly valuable medicinal tree of *Taxus sumatrana* in Indonesia. Biodivers 17(2):487–491. https://doi.org/10.13057/biodiv/d170213

Rachmat HH, Susilowati A, Elfiati D, Hartini KS, Faradillah WN (2017) Strong genetic differentiation of the endemic rosin-producing tree *Styrax sumatrana* (Styracaceae) in North Sumatra, Indonesia. Biodivers 18(4):1331–1335. https://doi.org/10.13057/biodiv/d180407

Rachmat HH, Kamiya K, Yulita KS, Hidayat A, Susilowati A (2024) Genetic structuring of the wild *Aquilaria malaccensis* population from Indonesia inferred by trnL-trnF and trnH-psbA. Iran J Sci 48(3):541–552. https://doi.org/10.1007/s40995-024-01623-x

Rahmawaty, Ismail MH, Rauf A, Batubara R, Elena Sitorus EW, Simamora Z, Ginting EF (2023) Land characteristics and land suitability assessment for *Styrax* sp. in Humbang Hasundutan Regency, North Sumatra, Indonesia. Heliyon 9(7):e16936. https://doi.org/10.1016/j.heliyon.2023.e16936

Raina AP, Abraham Z, Sivaraj N (2015) Diversity analysis of *Kaempferia galanga* L. germplasm from South India using DIVA-GIS approach. Indus Crops Prod 69:433–439. https://doi.org/10.1016/j.indcrop.2015.02.052

Rani N, Joshi M, Sagar A, Sharma HR (2022) Potential impacts of environmental pollution on the growth and metabolism of medicinal plants: an overview. In: Husen A (ed) Environmental pollution and medicinal plants. CRC Press. https://doi.org/10.1201/9781003178866-1

Robinson DF, Abdel-Latif A, Richerzhagen C (2021) The Nagoya Protocol and its implications for research on medicinal plants. In: Phytomedicine. Springer, Cham, pp 1–17

Rankethkumbura KW (2024) A comprehensive analysis of the current landscape of ginger cultivation and market dynamics in Sri Lanka World 20(2):734–746

Saensouk S, Yaowachai W, Chumroenphat T, Nonthalee S, Saensouk P (2023) In vitro regeneration, transplantation and phytochemical profiles of *Kaempferia angustifolia* Roscoe. Not Bot Horti Agrobot Cluj Napoca 51(4):13190. https://doi.org/10.15835/nbha51413190

Salehi B, Albayrak S, Antolak H, Kręgiel D, Pawlikowska E, Sharifi-Rad M, Uprety Y, Tsouh Fokou PV, Yousef Z, Amiruddin Zakaria Z, Varoni EM, Sharopov F, Martins N, Iriti M, Sharifi-Rad J (2018) *Aloe* genus plants: from farm to food applications and phytopharmacotherapy. Int J Mol Sci 19(9):2843. https://doi.org/10.3390/ijms19092843

Salem MA, Zayed A, Alseekh S, Fernie AR, Giavalisco P (2021) The integration of MS-based metabolomics and multivariate data analysis allows for improved quality assessment of *Zingiber officinale* Roscoe. Phytochemistry 190:112843

Samuel AJ, Kalusalingam A, Chellappan DK, George A, Kumar V (2010) Ethnomedical survey of plants used by the Orang Asli in Kampung Bawong, Perak, West Malaysia. J Ethnobiol Ethnomed 6(1):5. https://doi.org/10.1186/1746-4269-6-5

Sandeep IS, Kuanar A, Akbar A, Kar B, Dasa S, Mishra A, Sial P, Naik PK, Nayak S, Mohanty S (2016) Agroclimatic zone based metabolic profiling of turmeric (*Curcuma longa* L.) for phytochemical yield optimization. Ind Crop Prod 85:229–240. https://doi.org/10.1016/j.indcrop.2016.03.007

Saputra MH, Sagala SAH, Lee HS (2020) Species distribution of *Styrax sumatrana* in North Sumatra using Maxent modelling approach. Forum Geogr 33(2):196–208. https://doi.org/10.23917/forgeo.v33i2.9056

Sasikumar B (2005) Genetic resources of Curcuma: diversity, characterization and utilization. Plant Genet Resour Newslett 144:27–31

Satya, Khan T, Hashmi K, Raza S, Gupta S, Joshi S (2024) Plant response to gold nanoparticles in terms of growth, development, production, and protection: an overview. In: Plant response to gold nanoparticles: plant growth, development, production, and protection, pp 1–35. https://doi.org/10.1007/978-981-97-7174-5_1

Schiff PB, Fant J, Horwitz SB (1979) Promotion of microtubule assembly in vitro by taxol. Nature 277:665–667. https://doi.org/10.1038/277665a0

Schippmann U, Leaman DJ, Cunningham AB (2006) A comparison of cultivation and wild collection of medicinal and aromatic plants under sustainability aspects. In: Bogers RJ, Craker LE, Lange D (eds) Medicinal and aromatic plants. Springer, Dordrecht, pp 75–95

Shabbir A, Bajwa R (2006) Distribution of parthenium weed (*Parthenium hysterophorus* L.), an alien invasive weed species threatening the biodiversity of Islamabad. Weed Biol Manag 6(2):89–95. https://doi.org/10.1111/j.1445-6664.2006.00202.x

Soares AA, Jacomassi E, da Mata R, Lopes KFC, Borges JL, Pereira UP, Germano RM, Otutumi LK, Martins LA, Gonçalves DD (2018) Antimicrobial activity of *Zingiber officinale* Roscoe and *Alpinia purpurata* (Vieill.) K. Schum. (Zingiberaceae): a review. Semin Ciênc Agrár 39(4):1849–1862. https://doi.org/10.5433/1679-0359.2018v39n4p1849

Soehartono T, Newton AC (2000) Conservation and sustainable use of tropical trees in the genus *Aquilaria* I. Status and distribution in Indonesia. Biol Conserv 96(1):83–94. https://doi.org/10.1016/S0006-3207(00)00055-0

Soehartono T, Newton AC (2001) Conservation and sustainable use of tropical trees in the genus *Aquilaria* II. The impact of gaharu harvesting in Indonesia. Biol Conserv 97(1):29–41. https://doi.org/10.1016/S0006-3207(00)00089-6

Subramanian M, Gantait S, Jaafar JN, Ismail MF, Sinniah UR (2025) Micropropagation of white turmeric (*Curcuma zedoaria* (Christm.) Roscoe) and establishment of adventitious root culture for the production of phytochemicals. Ind Crops Prod 223:120101. https://doi.org/10.1016/j.indcrop.2024.120101

Sumi MJ, Zaman SB, Imran S, Rhaman MS (2024) A review on the ethnopharmacological importance and biochemical composition of medicinal plants within the Zingiberaceae family. Plant Sci Today 11(Suppl. 1):275–286. https://doi.org/10.14719/pst.3514

Sunandar AD, Barus SP, Ksuwanda W, Saputra MH (2019) Vegetation diversity and conservation implications on habitat of Taxus (*Taxus sumatrana* Miq. de Laub) in Northern Sumatra:365–371. https://doi.org/10.5220/0008554603650371

Susilowati A, Hendalastuti H, Kholibrina CUTR, Ramadhani R (2017) Short communication: weak delineation of *Styrax* species growing in North Sumatra, Indonesia by matK + rbcL gene. Biodivers J Biol Divers 18(3):1270–1274. https://doi.org/10.13057/biodiv/d180353

Susilowati A, Rachmat HH, Kholibrina CR, Hartini KS, Rambe HA (2019a) Propagation for conserving endangered taxol producing tree *Taxus sumatrana* through shoot cuttings technique. J Phys Conf Ser 1282(1):0–5. https://doi.org/10.1088/1742-6596/1282/1/012110

Susilowati A, Rachmat HH, Elfiati D, Hasibuan MH (2019b) The composition and diversity of plant species in pasakbumi's (*Eurycoma longifolia*) habitat in Batang Lubu Sutam forest, North Sumatra, Indonesia. Biodivers 20(2):413–418. https://doi.org/10.13057/biodiv/d200215

Susilowati A, Rachmat HH, Fadilah WN, Kusuma YS (2019c) Phylogeny of Kemenyan Toba (*Styrax sumatrana*) inferred from trnL-trnF chloroplast DNA sequence. In: Proceedings of the international conference on natural resources and technology (ICONART 2019), vol 1, pp 26–29. https://doi.org/10.5220/0008387300260029

Susilowati A, Hidayat A, Zaitunah A, Riniarti M, Afandi R (2021) Woody species stand structure and regeneration status in long jack (*Eurycoma longifolia* Jack) habitat in Batang Lubu Sutam Forest, Padang Lawas, North Sumatra. J Sylva Lestari 9(3):466–474. https://doi.org/10.23960/jsl.v9i3.545

Susilowati A, Rachmat HH, Yulita KS, Wijaya K (2023) Floristic composition and structure of *Eurycoma longifolia* habitat in Muka Kuning Nature Tourism Park, Riau Islands, Indonesia. Biodivers 24(5):2836–2842. https://doi.org/10.13057/biodiv/d240537

Saputra M, Lee HS (2021) Evaluation of climate change impacts on the potential distribution of Styrax sumatrana in North Sumatra, Indonesia. Sustainability 13(2):462

Tanaka S, Ito M (2020a) DNA barcoding for identification of agarwood source species using trnL-trnF and matK DNA sequences. J Nat Med 74(1):42–50. https://doi.org/10.1007/s11418-019-01338-z

Tanaka S, Ito M (2020b) Species identification of Indonesian agarwood using a DNA-barcoding method. J Nat Med 74(1):323–330. https://doi.org/10.1007/s11418-019-01362-z

Teoh ES (2016) Medicinal orchids of Asia. Springer. https://doi.org/10.1007/978-3-319-24274-3

Thomas GE, Geetha KA, Augustine L, Mamiyil S, Thomas G (2016) Analyses between reproductive behavior, genetic diversity and Pythium responsiveness in *Zingiber* spp. reveal an adaptive significance for hemiclonality. Front Plant Sci 7:1913. https://doi.org/10.3389/fpls.2016.01913

Ticktin T (2004) The ecological implications of harvesting non-timber forest products. J Appl Ecol 41(1):11–21. https://doi.org/10.1111/j.1365-2664.2004.00859.x

Tnah LH, Lee CT, Lee SL, Ng CH, Ng KKS (2014) Development of microsatellites in *Labisia pumila* (Myrsinaceae), an economically important Malaysian herb. Appl Plant Sci 2(6):1400019. https://doi.org/10.3732/apps.1400019

Viccini LF, Silveira RS, do Vale AA, de Campos JMS, Reis AC, Santos MO, Campos VR, Carpanez AG, Grazul RM (2014) Citral and linalool content has been correlated to DNA content in *Lippia alba* (Mill.) N.E. Brown (Verbenaceae). Ind Crop Prod 59:14–19. https://doi.org/10.1016/j.indcrop.2014.04.028

Wan GZ, Li QQ, Jin L, Chen J (2024) Integrated approach to predicting habitat suitability and evaluating quality variations of *Notopterygium franchetii* under climate change. Sci Rep 14:26927. https://doi.org/10.1038/s41598-024-77824-6

Weaver BA (2014) How Taxol/paclitaxel kills cancer cells. Mol Biol Cell 25(18):2677–2681. https://doi.org/10.1091/mbc.E14-04-0916

Wicaksana N, Gilani SA, Ahmad D, Kikuchi A, Watanabe KN (2011) Morphological and molecular characterization of underutilized medicinal wild ginger (*Zingiber barbatum* Wall.) from Myanmar. Plant Genet Resour 9(4):531–542. https://doi.org/10.1017/S1479262111000840

Willi Y, Kristensen TN, Sgro CM, Weeks AR, Ørsted M, Hoffmann AA (2022) Conservation genetics as a management tool: the five best-supported paradigms to assist the management of threatened species. Proc Natl Acad Sci USA 119(1):1–10. https://doi.org/10.1073/pnas.2105076119

Windarsih A, Indrianingsih AW, Suryani R, Wiyono T, Noviana E, Darsih C, Rahayu E, Pangestuti R, Asari SM, Pratiwi SI, Apriyana W, Rosyida VT, Sulistyowaty MI et al (2025) Evaluation of the antioxidant, antidiabetic, and antibacterial activity of *Curcuma domestica, Zingiber officinale, Alpinia galanga, Curcuma xanthorrhiza,* and *Kaempferia galanga* concurrent with its metabolite's constituent analysis. Chem Biodivers:e202500645. https://doi.org/10.1002/cbdv.202500645

Xu J, Yang B, Li M, Li Z, Tu Y, Tang L, He G (2022) Research on germplasm diversity of *Amomum villosum* Lour in genuine producing area. PLoS One 17(8):e0268246

Xu J, Wu Y, Wang S, Xu Y, Du C, Rohani ER, Han R (2024) Predicting the potential distribution of *Arisaema heterophyllum* in China under current and future climate change based on ArcGIS and MaxEnt model. Plant Biosyst 158(6):1326–1334. https://doi.org/10.1080/11263504.2024.2407813

Yulita KS, Susilowati A, Rachmat HH, Susila, Hidayat A, Dwiyanti FG (2022) Molecular identification of *Eurycoma longifolia* Jack from Sumatra, Indonesia using trnL-F region. Biodivers 23(3):1375–1383. https://doi.org/10.13057/biodiv/d230323

Zaman SS, Kayang H (2025) Research on nutritional, phytochemical and antioxidant properties of *Zingiber rubens* Roxb. Proc Indian Natl Sci Acad. https://doi.org/10.1007/s43538-025-00463-6

Zhang Y, Su R, Yuan H, Zhou H, Jiangfang Y, Liu X, Luo J (2023) Widely targeted volatilomics and metabolomics analysis reveal the metabolic composition and diversity of Zingiberaceae plants. Meta 13(6):700

Zheng WH, Zhuo Y, Liang L, Ding WY, Liang LY, Wang XF (2015) Conservation and population genetic diversity of *Curcuma wenyujin* (Zingiberaceae), a multifunctional medicinal herb. Genet Mol Res 4(3):10422–10432

Zhou X, Zhu H, Liu L, Lin J, Tang K (2010) A review: recent advances and future prospects of taxol-producing endophytic fungi. Appl Microbiol Biotechnol 86(6):1707–1717. https://doi.org/10.1007/s00253-010-2546-y

Zou P, Newman M, Liao JP (2022) Systematics of Zingiberaceae. Grana 61(6):448–470

Zou H, Chen B, Zhang B, Zhou X, Zhang X, Zhang X, Wang J (2023) Conservation planning for the endemic and endangered medicinal plants under climate change and human disturbance: a case study of *Gentiana manshurica* in China. Front Plant Sci 14:1184556. https://doi.org/10.3389/fpls.2023.1184556

Chapter 15
An Overview of Threats of Genetic Diversity in Medicinal Plants

Gowri Rajkumar and Samarathunga M. P. G. P. B. Abeyrathna

Abstract Medicinal plants have been essential to human health and well-being for ages, forming the basis of traditional medicine and contemporary pharmaceuticals. Over 80% of the global population depends on plant-based medications for basic healthcare, which proves that preserving the genetic diversity of these plants is paramount. The genetic diversity of medicinal plants is critical to their adaptability, resistance, and medical efficacy, allowing them to thrive in a variety of environmental conditions while also producing bioactive chemicals with therapeutic effects. However, this genetic diversity is under tremendous threat from a variety of human-caused and natural sources. Several natural factors that threaten genetic diversity, such as climate change, natural disasters such as wildfires, earthquakes, etc., pests and diseases, habitat destruction, invasive species, genetic drift, natural selection, and geographical isolation, affect the ability of medicinal plants to adapt to environmental changes as well as their disease resistance. Deforestation, overharvesting, agricultural expansion and monoculture, environmental pollution, the introduction of genetically modified plants, several industrial activities, and urbanization can be considered some of the human-influenced factors that threaten genetic diversity. A thorough and sustainable approach is necessary to counteract these threats. In-situ and ex-situ conservation, encouraging sustainable harvesting methods, investing in scientific research on these problems and promoting awareness of the need to preserve genetic diversity, can all help conservation efforts. This initiative will benefit both current and future generations while supporting the preservation of the delicate balance between the natural environment and human health.

Keywords Genetic diversity · Natural threats · Anthropogenic threats · Prevention · Medicinal plants

G. Rajkumar (✉) · Samarathunga M. P. G. P. B. Abeyrathna
Department of Botany, Faculty of Science, University of Jaffna, Jaffna, Sri Lanka
e-mail: gowrir@univ.jfn.ac.lk

S. Jha et al. (eds.), *Genetic Diversity in Medicinal Plants*, Sustainable Development and Biodiversity 44, https://doi.org/10.1007/978-981-95-7281-6_15

15.1 Introduction

Medicinal herbs have historically played an important role in human health and well-being, acting as the foundation for both traditional healthcare systems and modern medications. Ancient civilisations such as Egypt, China, and India relied heavily on medicinal flora, and this legacy continues today, with more than 80% of the global population relying on plant-based treatments for primary care (World Health Organisation 2013). Even in the modern era, the discovery of various pharmacological substances, such as aspirin (*Salix alba*), quinine (*Cinchona officinalis*), and artemisinin (*Artemisia annua*), demonstrates medicinal plants' continued importance in drug research (Fabricant and Farnsworth 2001).

Medicinal plants are the foundation of both traditional and modern healthcare systems, providing the primary source of medicinal chemicals and bioactive metabolites. The genetic variety of these plants is critical to their adaptability, survival, and therapeutic potential (Fig. 15.1). Genetic variety is the variation in genes within and between populations of a species that directly affects its ability to evolve under environmental constraints (Hamrick and Godt 1996; Frankham et al. 2010). This diversity allows medicinal plants to tolerate pests, infections, and environmental

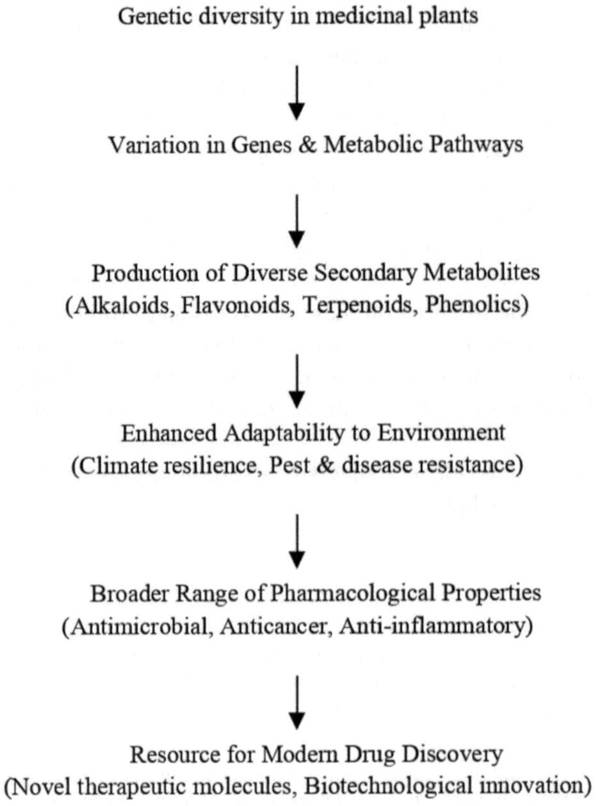

Fig. 15.1 Importance of genetic diversity in medicinal plants

stress while preserving or improving the manufacture of secondary metabolites, such as alkaloids, flavonoids, terpenoids, and phenolics (Hamilton 2004).

From an ecological standpoint, genetically varied plant populations are more resilient to environmental changes, ensuring ecosystem stability. From a pharmaceutical standpoint, such diversity acts as a reservoir for the discovery of novel therapeutic molecules, as variations in gene expression and metabolic pathways frequently result in the production of unique chemical compounds with potential medical value (Fabricant and Farnsworth 2001). Conserving the genetic variety of medicinal plants not only preserves ecological integrity but also sustains the raw biological innovation on which current medication development is based.

Even in the modern era, medicinal plants continue to play an important part in pharmaceutical development. Aside from traditional plant-derived pharmaceuticals like aspirin from *Salix alba*, quinine from *Cinchona officinalis*, and artemisinin from *Artemisia annua*, some recent advances highlight the long-term scientific and clinical usefulness of botanicals. A major accomplishment is the semi-synthetic manufacture of artemisinin, which was accomplished by synthetic biology and metabolic engineering of yeast (*Saccharomyces cerevisiae*), resulting in a consistent, large-scale supply of antimalarial chemicals (Kopetzki et al. 2013; Paddon et al. 2013).

Research on *Catharanthus roseus* reveals considerable chemotypic differences between populations, affecting the output of alkaloids such as vincristine and vinblastine—important anticancer drugs (Van der Heijden et al. 2004). Genetic erosion diminishes chemical diversity, reducing prospects for drug discovery. Furthermore, recent studies emphasise the use of AI-assisted metabolomics and high-throughput screening to identify novel plant secondary metabolites with anticancer, antibacterial, and anti-inflammatory properties (Atanasov et al. 2021). Collectively, these discoveries confirm medicinal plants' importance in modern drug discovery and future pharmacological advancements.

However, this critical diversity is under increasing threat. Climate change, habitat loss, unsustainable harvesting, and land-use conversion have led to declines in plant population size, fragmentation of habitats, and reduced genetic variation over the last decade (Atanasov et al. 2021; Mykhailenko et al. 2025). *Clivia miniata*, a medicinal species in South Africa, has lost almost 40% of its wild populations over the past 90 years due to land-cover change, harvesting, and climate fluctuations (Groner et al. 2022). Modelling of species distributions for high-altitude medicinal plants in the Himalayas indicated that 106 species' appropriate habitats are declining and 94 species' ranges are shifting under future warming scenarios (Mykhailenko et al. 2025). Thus, protecting genetic variety is more than just a conservation issue; it is critical to the future of medicinal plant science.

This chapter presents a comprehensive summary of the key risks to medicinal plant genetic diversity, focusing on both natural and anthropogenic factors. It investigates how climate change, habitat loss, deforestation, overharvesting, environmental pollution, urbanisation, invasive species, and the introduction of genetically engineered plants all contribute to the reduction of genetic variety in medicinal plant populations.

15.2 Natural Threats to Genetic Diversity

15.2.1 Climate Change

Climate change is one of the most serious dangers to the genetic diversity of medicinal plants worldwide. Temperature, precipitation, and atmospheric CO_2 levels affect the physiology, distribution, and reproductive success of these species (Thuiller et al. 2005; Parmesan 2006).

15.2.2 Drought Stress

Drought is one of the most important climate-related stresses that affects the genetic structure of medicinal plants. Extended dry periods lower water availability, photosynthetic activity, and growth, resulting in smaller and more dispersed plants. A study of 22 basil (*Ocimum basilicum*) accessions under different soil-moisture regimes found substantial differences in drought tolerance and essential oil output (Rahimi et al. 2023). Also, under controlled drought, *A. annua* demonstrated decreased biomass and altered artemisinin levels, as well as alterations in proline-metabolism gene expression, suggesting that drought diminishes population vigour and affects secondary-metabolite characteristics through genetic mechanisms (Soni and Abdin 2017). These results show that drought stress not only reduces plant performance but also puts selective pressure on genotypes, resulting in genetic erosion in medicinal species.

15.2.3 Elevation of Temperature

Elevated temperatures put severe selective pressures on medicinal plants, affecting growth, development, and secondary-metabolite production, resulting in genetic composition changes. Recent reviews of heat-stress responses across medicinal taxa summarise these findings and emphasise that repeated heat events favour heat-tolerant alleles, reduce allelic richness, and drive population differentiation, all of which contribute to the long-term genetic erosion of medicinal plant resources (Zhu et al. 2024). Another study examines how several temperature regimes (28 °C, 34 °C, and 40 °C) impact volatile oil concentration in a medicinal plant, *Atractylodes lancea (Thunb.),* relating temperature stress to chemotype change and, consequently, potential genetic consequences (Zhu et al. 2025).

15.2.4 Phenological Shifts

Phenological shifts in medicinal plants are becoming well-documented, with important implications for their genetics and therapeutic properties. For example, in American ginseng (*Panax quinquefolius*), developmental phases (leafing, blooming, and seed maturity) were found to be predominantly influenced by temperature and photoperiod, with altered timing associated with the quality of bioactive compounds (Chen et al. 2022). Similarly, the traditionally utilised Indian olive (*Olea ferruginea*) in the Himalayas has changed its blossoming date by 15–21 days over the previous century in response to increased spring minimum temperatures (Khan et al. 2023). These temporal changes affect gene flow, synchronisation with pollinators, and seed distribution, raising the danger of population isolation and reduced genetic diversity.

15.2.5 Natural Disasters

Natural outbreaks like wildfires, floods, and hurricanes all result in rapid habitat damage and population reduction. While these events are normal parts of Earth's dynamics, their frequency and intensity have grown in recent decades as a result of climate change and human-caused environmental deterioration. Natural disasters endanger the survival, population stability, and genetic diversity of medicinal plants. Wildfires in Mediterranean environments have diminished *Thymus vulgaris* populations and other aromatic medicinal plants, resulting in considerable genetic erosion (Raposo et al. 2020). Similarly, cyclones in coastal Asia have degraded mangrove habitats that support medicinally essential species such as *Avicennia officinalis* (Bhowmik et al. 2022).

15.2.6 Pest and Disease

Pests and diseases pose serious biological challenges to the genetic diversity of medicinal plants. Herbivore insects, nematodes, fungi, bacteria, viruses, and parasitic plants can affect plant growth, reproduction, and survival either directly or indirectly. Insect infestations and disease outbreaks can exert persistent and long-term stresses on the plant population. These outbreaks can decimate species with little genetic variation. Fungal diseases have been found on various *Taxus* species, and they can produce needle spots, stem cankers, and root/collar rot, all of which impact individual tree health. Pathogenic Pestalotioid fungi (such as Neopestalotiopsis and Pestalotiopsis) have been isolated from Chinese yew and found to cause needle-spot disease in inoculation trials (Wang et al. 2024).

15.2.7 Invasive Species

Invasive plants frequently outcompete native medicinal plants for nutrients, water, and space. These plants cause a significant threat to medicinal plants because they disrupt the ecosystem, which causes long-term stress on biodiversity (Table 15.1). *Lantana camara*, an aggressive intruder in tropical Asia, inhibits the regeneration of indigenous medicinal species by changing soil chemistry and shading seedlings (Sharma et al. 2005). These incursions lower the population numbers of local medicinal flora and limit genetic diversity through competitive exclusion. *Chromolaena* invades open forests and disturbed lands, suppressing regeneration of native trees and shrubs via shading, rapid biomass accumulation, and allelopathic leachates. A 2024 study shows *Chromolaena*'s negative effects on the growth of *Aegle marmelos* (a culturally important medicinal tree) under water-stress conditions—illustrating how invasion can threaten specific medicinal trees and their recruitment (Poudel et al. 2024). Such replacement reduces the local genetic pools of medicinal trees and associated understory herbs.

Prosopis juliflora grows densely in dry lands, outcompeting native shrubs and trees, altering hydrology and soil chemistry, and diminishing native species abundance (including those utilised in traditional medicine). Recent regional studies (2020–2023) show significant decreases in native plant variety and local livelihoods in areas where Prosopis prevails; community surveys also show a loss of access to traditional medicinal herbs. This replacement mechanism diminishes the effective population numbers of native medicinal taxa, causing genetic degradation (Howari et al. 2022).

15.2.8 Genetic Drift and Isolation

Genetic drift refers to changes in small, scattered populations which occur when alleles are randomly lost, reducing genetic variability (Ellstrand and Elam 1993). Unlike natural selection, genetic drift happens solely by chance, particularly in small populations. This pattern may significantly decrease genetic variety in medicinal plants, making populations less adaptable to diseases, pests, and environmental changes. Geographic isolation further limits gene flow, resulting in inbreeding depression and lower adaptation.

Genetic drift and population isolation are serious yet underappreciated risks to medicinal plants. *Melissa officinalis* populations are highly organised with restricted gene flow, indicating that drift and isolation play a key role in divergence (Koohdar and Sheidai 2022). Population genomics in high-altitude medicinal plants demonstrate that small, fragmented populations have a smaller effective size and evolve through drift (Liu et al. 2025). *Eucommia ulmoides*, an endangered medicinal tree, was investigated and found to have low within-population diversity and substantial population differentiation, indicating long-term isolation and drift (Yu and Zhang 2015). These findings illustrate how fragmentation, restricted dispersal, and reduced gene flow can lead to genetic degradation and drift, compromising both adaptation and therapeutic value.

Table 15.1 Impacts of invasive species on biodiversity

Sl. No.	Invasive species	Region(s)	Mechanism/ impacts on habitat	Implication for medicinal plants (how genetic diversity is threatened)	References
1	*Lantana camara* L.	India, South Asia, Africa, Australia	Forms dense thickets; shade, allelopathy; affects soil nitrogen cycling	Displaces understory plants and specialist medicinal species, reducing habitat and population levels and increasing the risk of genetic erosion.	Negi et al. (2019)
2	*Mikania micrantha* Kunth	South and Southeast Asia, Pacific regions	Rapid vine growth, smothering shrubs and seedlings; allelopathy; fast seed production	Replaces native medicinal understorey and forest edges, reducing recruitment and population connection, resulting in loss of alleles.	Khan (2023)
3	*Chromolaena odorata* (L.) R.M. King & H. Rob.	Africa, Asia	Dense growth, allelopathic leachates, prevent tree regeneration	Reduces recruitment of medicinal trees (e.g., *Aegle marmelos*) as well as mature and seedling recruitment, leading to genetic deterioration.	Poudel et al. (2024)
4	*Acacia mangium* Willd.	Southeast Asia, Pacific region, Africa	Changes soil chemistry through nitrogen fixation and litter; boosts own recruitment	Changes in understory composition can harm low-nutrient medicinal plants and lead to local extinction, resulting in a diminished genetic pool.	Koutika and Richardson (2019)
5	*Prosopis juliflora*	Drylands in Africa, South Asia, Middle East	Dense stands; reduce diversity; cause hydrological and soil changes, and compete for resources	Displaces natural trees and shrubs used in traditional medicine, reducing availability and population levels, and increasing the risk of genetic erosion	Howari et al. (2022)
6	*Reynoutria japonica*/ *Fallopia japonica*	Europe, North America	Dense riparian stands modify soil and trash, shading out the understory	Displaces riparian herbaceous flora, fragments habitats and decreases local population sizes, resulting in loss of genetic diversity	Liu et al. (2022)

15.3 Anthropogenic (Human-Caused) Threats

15.3.1 Deforestation and Habitat Destruction

Large-scale deforestation for agriculture, logging, and infrastructural development is one of the major causes of genetic erosion. Medicinal plants which frequently inhabit tropical forests, marshes, and wetlands are under threat of habitat loss, which directly results in genetic erosion, limited adaptive ability, and even species extinction. Approximately half of the world's biodiversity hotspots are under severe threat from deforestation, with the majority of them containing medicinally important plants. Tropical forests, home to a large number of medicinal plant species, are disappearing at an alarming rate (Myers et al. 2000). For example, Amazonian deforestation poses a threat to species such as *Uncaria tomentosa*, which is used in traditional medicine to treat inflammation and immunological diseases (Honório et al. 2017).

Regionally, the Amazon Basin has garnered special attention due to its extraordinarily high biodiversity and concentration of medicinal plant taxa. Remote sensing and national monitoring systems exhibit significant interannual variability: Brazilian monitoring (INPE/PRODES) and international syntheses reported increased Amazon losses in the late 2010s and early 2020s, with a notable increase in 2021 but decreases in 2023 following increased enforcement (European Commission—Joint Research Centre 2024; Reuters reporting on INPE data 2024). Nonetheless, the conversion of Amazonian forests to agriculture, forestry, mining, and infrastructure continues to deplete habitat at a rate that endangers many forest-dependent species.

Deforestation impacts medicinal plant genetic diversity through multiple mechanisms.

1. Reduced population size and local extinction: Direct clearing destroys individuals and entire populations of forest-dependent medicinal species, resulting in the abrupt loss of alleles and a reduction in overall allelic richness (FAO 2020).
2. Fragmentation and isolation: When forests are fragmented, remnant populations shrink and become more isolated; reduced gene flow between fragments promotes genetic drift and inbreeding, hastening the loss of rare alleles and decreasing heterozygosity (Chen et al. 2016).
3. Loss of specific habitats and mutualists: Many medicinal species rely on specialised pollinators, mycorrhizal partners, and nursing trees. Deforestation disturbs these ecological networks, resulting in fewer successful reproductive events and smaller effective population sizes over time, reducing genetic diversity (Chen et al. 2016; FAO 2020).

15.3.2 Overharvesting and Unsustainable Use

Overharvesting for high-value chemicals is one of the primary causes of population decrease and genetic erosion in medicinal plants. Medicinal plants are frequently collected in inappropriate ways, especially those whose roots, bark, or whole plants are consumed. Overharvesting lowers population density, genetic variety, and regeneration potential (Schippmann et al. 2002).

15.3.3 Agricultural Expansion and Monoculture

The conversion of forests to monoculture plantations affects landscape variability, decreasing habitats for medicinal flora (Tscharntke et al. 2005). Monocultures also promote insect outbreaks, forcing chemical treatments that kill non-target therapeutic plants. The expansion of agricultural systems, as well as the conversion of natural habitats into large-scale monocultures, has far-reaching implications for medicinal plant genetic diversity. As natural habitats are replaced, wild medicinal plant populations are decimated, fragmented, or driven into uniform cultivation. These processes results in:

(i) The loss of wild gene pools and unusual alleles
(ii) The formation of genetically homogeneous cultivated populations with limited adaptive capacity
(iii) Altered ecological contexts that further decrease genetic exchange and gene flow across remnant populations.

A recent study in China compared wild and cultivated populations of *Polygonatum odoratum, Dioscorea nipponica,* and *Acanthopanax sessiliflorus and* discovered that cultivated populations, grown under more uniform agricultural conditions, exhibit significant genetic differentiation from wild populations as well as altered secondary-metabolite profiles. Although genetic diversity remained reasonably high, gene flow (Nm) was less than one in some situations, and cultivated populations showed differentiation (Li et al. 2025). This shows that agricultural expansion of cultivation areas can result in genetic structuring, a lack of gene exchange with wild populations, and eventually the extinction of wild gene pools.

In another study of two medicinal vines, *Tripterygium wilfordii* and *Tripterygium hypoglaucum*, the authors discovered that cultivated populations (typically produced in monoculture or homogenous agro-settings) had much less genetic variation than wild populations. The decrease was linked to tiny founder populations, vegetative propagation, and simplified agricultural environments (Liu et al. 2024).

15.3.4 Environmental Pollution

Environmental pollution causes ecosystem degradation and detrimental impacts on living species. Medicinal plants which rely heavily on secondary metabolites for therapeutic benefit can affect metabolic pathways, impair growth, and lower phytochemical concentrations. Heavy metals, pesticide residues, and industrial effluents all have an impact on soil and water quality. For example, chromium contamination from tanneries has been demonstrated to impair medicinal plant development and secondary metabolite production (Shanker et al. 2005). Pollution not only threatens survival but also diminishes plant therapeutic efficiency by changing phytochemical pathways. Furthermore, long-term exposure to pollutants depletes soil fertility, contaminates water supplies, weakens ecological networks, and endangers the survival of both wild and cultivated plants.

In a study of aromatic and medicinal plants in Portugal, "Monitoring Genetic Erosion of Aromatic and Medicinal Plant Populations" (Póvoa et al. 2023), heavy-metal and other environmental pollution risk zones were found to correlate with lower genetic diversity metrics in rosemary (*Salvia rosmarinus*), thyme (*Thymus vulgaris*), and other wild medicinal species. The study examined DNA marker variety in communities around industrial zones, mining-impacted sites, and agricultural heavy-chemical use areas. It was discovered that populations in more polluted places had significantly lower allelic richness, greater inbreeding coefficients, and smaller effective population sizes than populations in less polluted sites. The scientists found that environmental contamination is "an additional driver of genetic erosion" for aromatic and medicinal vegetation. In another study, it is revealed that pollutants can cause mutations or epigenetic alterations, potentially compromising germline integrity or producing non-adaptive alleles that affect gene pool structure (Asiminicesei et al. 2024).

15.3.5 Urbanisation and Industrialisation

Urbanisation and industrialisation are defining features of modern human civilisation, yet they are also two of the most potent processes causing biodiversity loss. Rapid urbanisation, infrastructural development, and industrial complexes have put unprecedented pressure on natural ecosystems. Unlike natural processes, these activities may split landscapes, pollute the surroundings, and permanently alter ecosystems. Expanding towns and industrial zones cause the destruction of habitats, separate populations, and expose plants to new toxins (Millennium Ecosystem Assessment 2005). Medicinal species in peri-urban areas frequently confront two pressures: land conversion and commercial exploitation. These impacts of urban growth and industrial development lower the amount of natural habitats and also damage their quality, which leads to serious threat to medicinal plant diversity.

Table 15.2 Natural and human-induced threats to medicinal plants

Category	Threat	Examples	Impact on genetic diversity
Natural factors	Climate change	Rising temperature, altered rainfall	Range shifts, reduced adaptability
	Natural disasters	Wildfires, floods, earthquakes	Habitat destruction, population loss
	Pests and diseases	Pathogen outbreaks Herbivore pressure	Reduced resistance, genetic bottleneck
	Genetic drift	Random loss of alleles in a small population	Reduced genetic variation
	Natural selection and isolation	Geographical barriers	Local adaptation, but overall reduced gene flow
Human-induced factors	Overharvesting	Excessive wild collection for trade	Population decline, genetic erosion
	Deforestation and habitat loss	Logging, land use change	Fragmentation, species extinction
	Agriculture and monoculture	Replacement of wild species with crops	Loss of native diversity
	Industrialization and pollution	Heavy metals, Chemicals, Plastics	Mutations, sterility, reduced fertility
	GM plant production	Hybridization with wild relatives	Genetic contamination
	Urbanization	Infrastructure expansion	Habitat shrinkage, fragmentation

The natural and human-induced threats to medicinal plants and their impacts on genetic diversity are illustrated in Table 15.2

15.4 Conclusion

The genetic diversity of medicinal plants serves as the foundation for both traditional and modern global healthcare systems. However, it is under serious jeopardy from a combination of natural and anthropogenic stressors. Climate change, habitat destruction, overexploitation, pollution, and genetic drift all contribute to the genetic erosion of medicinal species, jeopardising their survival, adaptation, and therapeutic potential. Genetic degradation in medicinal plants has far-reaching implications such as loss of adaptive capacity, decline in phytochemical diversity, threats to traditional medicine, pharmaceutical limitations, and ecological instability. In genetic diversity conservation, in-situ conservation and ex situ conservation play a crucial role, and different approaches to the conservation of genetic diversity in medicinal plants are illustrated in Fig. 15.2. In-situ conservation protects medicinal plants in their native environments, preserving ecological interactions and promoting evolutionary adaptation. Protected areas such as national parks and biosphere reserves,

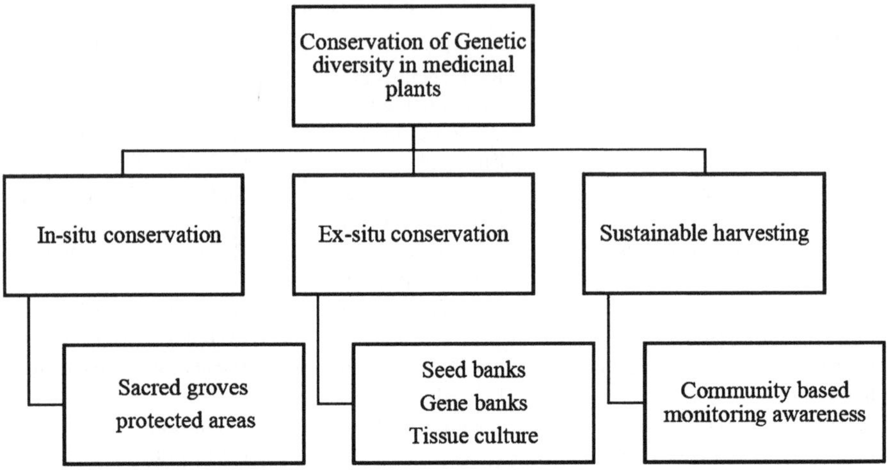

Fig. 15.2 Conservation approaches to counter genetic diversity loss in medicinal plants

notably Sri Lanka's Sinharaja Forest, play an important role in preserving indigenous medicinal plant variety.

Ex-situ conservation involves preserving plant genetic resources outside of their native habitats under controlled settings, such as seed banks, botanical gardens, field gene banks, tissue culture facilities, and cryopreservation. This strategy helps to save rare and endangered medicinal species from extinction, promotes regulated propagation and breeding, and encourages scientific study and biotechnology applications, ensuring the long-term preservation of precious plant germplasm.

References

Asiminicesei DM, Toma S, Cioroiu RI (2024) Impact of heavy metal pollution in the environment on the genetic and metabolic diversity of medicinal plants. Plants 13(6):913

Atanasov AG, Zotchev SB, Dirsch VM, Supuran CT (2021) Natural products in drug discovery: advances and opportunities. Nat Rev Drug Discov 20(3):200–216

Bhowmik AK, Cabral P, Navarro DK, Rahman M (2022) Global mangrove deforestation and its interacting social–ecological drivers. Sustainability 14(8):4433

Chen S, Yu H, Luo H, Wu Q, Li C, Steinmetz A (2016) Conservation and sustainable use of medicinal plants: problems, progress, and prospects. Chin Med 11:37

Chen T, Wang L, Wang H, Jiang S, Zhou S (2022) Photoperiod and temperature as dominant environmental drivers triggering plant phenological development of American ginseng along with its quality formation. Front Earth Sci 10:894251

Ellstrand NC, Elam DR (1993) Population genetic consequences of small population size: implications for plant conservation. Annu Rev Ecol Syst 24:217–242

European Commission — Joint Research Centre (2024) The Amazon region in 2022 and 2023: deforestation, forest degradation and risk growing soy production. https://joint-research-centre.ec.europa.eu/jrc-news-and-updates/amazon-region-2022-and-2023-deforestation-forest-degradation-and-risk-growing-soy-production-2024-02-28_en

Fabricant DS, Farnsworth NR (2001) The value of plants used in traditional medicine for drug discovery. Environ Health Perspect 109(Suppl 1):69–75

Food and Agriculture Organization of the United Nations (2020) Global forest resources assessment 2020: main report. FAO. https://www.fao.org/3/ca9825en/online/ca9825en.html

Frankham R, Ballou JD, Briscoe DA (2010) Introduction to conservation genetics, 2nd edn. Cambridge University Press, Cambridge

Groner VP, Nicholas O, Mabhaudhi T, Slotow R, Akçakaya HR, Mace GM, Pearson RG (2022) Climate change, land cover change, and overharvesting threaten a widely used medicinal plant in South Africa. Ecol Appl 32(4):e2545

Hamilton AC (2004) Medicinal plants, conservation and livelihoods. Biodivers Conserv 13:1477–1517

Hamrick JL, Godt MJW (1996) Effects of life history traits on genetic diversity in plant species. Philos Trans R Soc Lond B 351:1291–1298

Honório IC, Bertoni BW, Telles MP, Braga RD, Franca SD, Coppede JD, Correa VS, Diniz Filho JA, Pereira AM (2017) Genetic and chemical diversity of *Uncaria tomentosa* (Willd. ex. Schult.) DC. in the Brazilian Amazon. PLoS One 12(5):e0177103

Howari FM, Sharma M, Nazzal Y, El-Keblawy A, Mir S, Xavier CM, Salem IB, Al-Taani AA, Alaydaroos F (2022) Changes in the invasion rate of *Prosopis juliflora* and its impact on depletion of groundwater in the northern part of the United Arab Emirates. Plants 11(5):682

Khan MA (2023) *Mikania micrantha* Kunth: an ethnopharmacological treasure trove of therapeutic potential. Chem Biodivers 20(5):e202300392

Khan S, Gaira KS, Asgher M, Verma S, Pant S, Kesawat MS (2023) Temperature induced flowering phenology of *Olea ferruginea* Royle: a climate change effect. Sustainability 15(8):6936

Koohdar F, Sheidai M (2022) Genetic diversity and population structure in medicinal plant *Melissa officinalis* L. (Lamiaceae). Genet Resour Crop Evol 69(4):1013–1025

Kopetzki D, Lévesque F, Seeberger PH (2013) A continuous-flow process for the synthesis of artemisinin. Chem Eur J 19(17):5450–5456

Koutika LS, Richardson DM (2019) *Acacia mangium* Willd: benefits and threats associated with its increasing use around the world. For Ecosyst 6(1):2

Li Y, Zhao W, Zuo X, You J, Xing M, Du Y, Chen X, Liang C (2025) Habitat changes due to cultivation alter the genetic diversity and secondary metabolic changes of medicinal plants. Sci Rep 15(1):36185

Liu S et al (2022) *Reynoutria japonica* (Houtt.) as a threat to biodiversity: invasive species review. Plants 11(3):3393

Liu C et al (2024) Genetic diversities in wild and cultivated populations of the traditional medicinal semi-woody vines *Tripterygium wilfordii* and *T. hypoglaucum*. BMC Plant Biol 24:Article 487

Liu Y, Chen Z, Li Y (2025) Population genomics provides new insights into the genetic consequences of isolation, drift, and adaptation in a high-altitude medicinal plant. BMC Plant Biol 25:Article 345

Millennium Ecosystem Assessment (2005) Ecosystems and human well-being: biodiversity synthesis. Island Press, Washington, DC

Myers N, Mittermeier RA, Mittermeier CG, da Fonseca GAB, Kent J (2000) Biodiversity hotspots for conservation priorities. Nature 403:853–858

Mykhailenko O, Jalil B, McGaw LJ, Echeverría J, Takubessi M, Heinrich M (2025) Climate change and the sustainable use of medicinal plants: a call for "new" research strategies. Front Pharmacol 15:1496792

Negi GCS, Sharma S, Vishvakarma SCR, Samant SS, Maikhuri RK, Prasad RC (2019) Ecology and use of *Lantana camara* in India. Bot Rev 85(2):109–130

Paddon CJ, Westfall PJ, Pitera DJ, Benjamin K, Fisher K, McPhee D, Leavell MD, Tai A, Main A, Eng D, Polichuk DR (2013) High-level semi-synthetic production of the potent antimalarial artemisinin. Nature 496(7446):528–532

Parmesan C (2006) Ecological and evolutionary responses to recent climate change. Annu Rev Ecol Evol Syst 37:637–669

Poudel S, Pant RR, Thapa LB, Chettri MK (2024) Impacts of invasive weed *Chromolaena odorata* on growth and development of threatened native tree *Aegle marmelos* under water stress conditions. J Ecol Environ 48:42

Póvoa O, Duarte M, Felizardo M, Bento A (2023) Monitoring genetic erosion of aromatic and medicinal plant populations in Portugal: the role of environmental contamination in diversity loss. Plants 12(5):1120

Rahimi M, Mortazavi M, Mianabadi A, Debnath S (2023) Evaluation of basil (*Ocimum basilicum*) accessions under different drought conditions based on yield and physio-biochemical traits. BMC Plant Biol 23(1):523

Raposo MAM, Pérez IA, García LV, Marañón T, Villar R (2020) Selective shrub management to preserve Mediterranean shrub diversity. Fire 3(4):65

Schippmann U, Leaman DJ, Cunningham AB (2002) Impact of cultivation and gathering of medicinal plants on biodiversity: global trends and issues. In: Biodiversity and the ecosystem approach in agriculture, forestry and fisheries. FAO, Rome, pp 142–167

Shanker AK, Cervantes C, Loza-Tavera H, Avudainayagam S (2005) Chromium toxicity in plants. Environ Int 31:739–753

Sharma GP, Raghubanshi AS, Singh JS (2005) *Lantana* invasion: an overview. Weed Biol Manag 5:157–165

Soni P, Abdin MZ (2017) Water deficit-induced oxidative stress affects artemisinin content and expression of proline metabolic genes in *Artemisia annua* L. FEBS Open Bio 7(3):367–381

Thuiller W, Lavorel S, Araújo MB, Sykes MT, Prentice IC (2005) Climate change threats to plant diversity in Europe. Proc Natl Acad Sci U S A 102(23):8245–8250

Tscharntke T, Klein AM, Kruess A, Steffan-Dewenter I, Thies C (2005) Landscape perspectives on agricultural intensification and biodiversity: ecosystem service management. Ecol Lett 8:857–874

Van der Heijden R, Jacobs DI, Snoeijer W, Hallard D, Verpoorte R (2004) The *Catharanthus* alkaloids: pharmacognosy and biotechnology. Curr Med Chem 11:607–628

Wang Y, Tsui KM, Chen S, You C (2024) Diversity, pathogenicity and two new species of pestalotioid fungi (Amphisphaeriales) associated with Chinese Yew in Guangxi, China. MycoKeys 102:201

World Health Organization (2013) WHO traditional medicine strategy: 2014–2023. World Health Organization, Geneva

Yu J, Zhang L (2015) Genetic diversity and population structure of *Eucommia ulmoides* Oliver – an endangered medicinal plant. Genet Mol Res 14(4):13141–13149

Zhu Z, Bao Y, Yang Y, Zhao Q, Li R (2024) Research progress on heat stress response mechanism and control measures in medicinal plants. Int J Mol Sci 25(16):8600

Zhu J, Cheng Q, Alami MM, Zhong S, Li M, Mao M, Shu S (2025) The temperature-dependent changes of volatile oil content in *Atractylodes lancea* (Thunb.) DC detected by full-length transcriptomic and physiological approaches. Plant Growth Regul 31:1–4

Part III
Application of Genetic Diversity

Chapter 16
Harnessing and Preserving the Genetic Diversity of Medicinal Plants: A Comprehensive Approach to Conservation, Climate Adaptation, and Drug Discovery

Ali Devlet and Mesut Işık

Abstract Medicinal and aromatic plants (MAPs) are indispensable to global health and biodiversity, yet their genetic diversity faces escalating threats from habitat loss, overexploitation, climate change, and domestication. This study synthesizes a global perspective on the genetic and phytochemical diversity of medicinal plants, employing advanced molecular tools such as simple sequence repeat (SSR) markers, chloroplast DNA (cpDNA) sequencing, next-generation sequencing (NGS), and liquid chromatography-mass spectrometry (LC-MS) to assess diversity, identify threats, and explore conservation strategies. We investigate the determinants of diversity, including environmental, ecological, and anthropogenic factors, and evaluate the impacts of cultivation, domestication, and climate change on genetic and phytochemical profiles. Case studies across biodiversity hotspots (e.g., the Himalayas, Atlantic Forest, and the Lamiaceae family) reveal region-specific threats and conservation needs. Additionally, we explore the role of genetic diversity in climate adaptation, breeding, and transgenic development, highlighting its potential to enhance stress tolerance and phytochemical yield. By integrating omics technologies, geographic information system (GIS) mapping, and innovative conservation strategies (e.g., seed banking, cryopreservation, community engagement), this research provides a roadmap for preserving medicinal plant diversity. Expected outcomes include validated assessment tools, genetic–phytochemical correlations, and actionable conservation frameworks. This study underscores the critical importance of medicinal plant genetic diversity for drug discovery, ecological resilience, and

A. Devlet
Department of Agronomy, Faculty of Agriculture and Natural Science, Bilecik Şeyh Edebali University, Bilecik, Turkey

M. Işık (✉)
Department of Bioengineering, Faculty of Engineering, Bilecik Şeyh Edebali University, Bilecik, Turkey
e-mail: mesut.isik@bilecik.edu.tr

S. Jha et al. (eds.), *Genetic Diversity in Medicinal Plants*, Sustainable Development and Biodiversity 44, https://doi.org/10.1007/978-981-95-7281-6_16

sustainable health systems, offering solutions to ensure their survival and utility in a rapidly changing world.

Keywords Medicinal plants · Genetic diversity · Climate change · Conservation · Drug discovery · Phytochemicals

16.1 Introduction

Medicinal plants are vital to global healthcare, underpinning over 80% of traditional medicine systems and serving as sources for 25% of modern pharmaceuticals (Salmerón-Manzano et al. 2020). Their genetic diversity drives ecological resilience and variability in bioactive compounds critical for therapeutic applications. However, habitat loss, overexploitation, and climate change have accelerated genetic erosion, with a global meta-analysis estimating a 15–20% decline in within-population genetic variation over the past century due to anthropogenic pressures (Sharma et al. 2022). This loss threatens phytochemical yields and adaptation potential, necessitating robust assessment tools and conservation strategies. Globally, medicinal plant diversity is concentrated in biodiverse regions like the tropics, which face heightened risks; the Global Strategy for Plant Conservation 2020–2030 calls for aligning genetic monitoring with threat assessments to protect over 30,000 medicinal species (Alum 2024).

The determinants of genetic and phytochemical diversity in medicinal plants are complex, involving evolutionary processes like gene duplication, polyploidy, and environmental interactions. Recent studies reveal how tandem duplications in biosynthetic gene clusters enhance secondary metabolite diversity, as observed in the terpene synthase families of tea (*Camellia sinensis*), which improve aroma and medicinal profiles (Alami et al. 2022). Phytochemical variation, often linked to genetic markers, is modulated by abiotic stressors; for instance, ultraviolet-B radiation (UV-B) exposure in *Artemisia annua* upregulates transcription factors for artemisinin biosynthesis through epigenetic changes like deoxyribonucleic acid (DNA) methylation (Vashisth et al. 2018). These findings highlight the need for integrated genotypic–phenotypic approaches to predict resilience under climate change.

Assessment techniques for medicinal plant genetic diversity have evolved significantly, transitioning from morphological evaluations to advanced molecular methods. NGS and long-read technologies have enabled telomere-to-telomere genome assemblies for over 400 medicinal species by 2025 (Yu et al. 2024). Tools like restriction-site-associated DNA sequencing (RAD-seq) and RNA-seq provide precise transcriptome profiling, uncovering regulatory networks in secondary metabolism (Barrera-Redondo et al. 2020). Future prospects include pan-genomics and clustered regularly interspaced short palindromic repeats (CRISPR)-based editing for enhanced bioactive yields, though polyploid genome assembly remains challenging (Schaart et al. 2021). Molecular markers like inter-simple sequence repeats

(ISSR) and start codon targeted (SCoT) polymorphism offer cost-effective insights into population structure (Alotaibi and Abd-Elgawad 2022).

From a chloroplast perspective, cpDNA provides phylogeographic resolution due to its uniparental inheritance and slower mutation rates. Recent super-barcoding of *Fritillaria* cp genomes achieves 95% resolution in distinguishing cryptic species, surpassing standard rbcL/matK markers (Chen et al. 2022). In threatened orchids, cpDNA assessments reveal low haplotype diversity, signaling inbreeding risks; SCoT markers in *Dendrobium* species indicate 72% polymorphism, highlighting fragmented Southeast Asian populations (Bhattacharyya et al. 2013). Chloroplast phylogenomics in Paeonia elucidates reticulate evolution, guiding conservation of medicinal variants (Sang et al. 1997).

Regionally, medicinal flora face varying threats, with Asia and Africa as critical hotspots. In India and Asia, over 7500 medicinal species experience 20–30% genetic erosion from urbanization, but SCoT-ISSR hybrids have authenticated more than 20 taxa for ex situ conservation (Rajasekharan and Wani 2020). China's 1000 Medicinal Plant Genomes Project has sequenced more than 200 species since 2018, revealing high heterozygosity in *Panax ginseng* for adaptive potential (Alami et al. 2022). In Africa, RAD-seq identifies drought-tolerant alleles in *Moringa oleifera*, supporting pharmacogenomic applications across 54 countries (El Bilali et al. 2024). Climate change is projected to exacerbate threats to local biodiversity by facilitating the expansion of the invasive species *Calotropis procera* into adjacent arid and semiarid biomes (Gomes et al. 2025) Brazil is experiencing a growing prevalence in the use of medicinal plants, particularly *M. officinalis* and *P. boldus*; however, it is crucial to safeguard the genetic diversity of these species in the face of global and regional threats (da Silva et al. 2023). In Indonesia and Australia, SCoT assessments in Borneo orchids show 65% intraspecific variation, aiding reintroduction efforts (Akram 2015). Conservation strategies emphasize community-led seed banks and genomic-informed translocations, aligning with the Kunming–Montreal Framework (Shaw et al. 2024).

Recent research trends integrate ethnopharmacology with multi-omics, focusing on families like Lamiaceae and Asteraceae. High-throughput sequencing dominates, with 60% of studies using NGS for metabolite-gene correlations, as in *Saussurea involucrata* where SNP diversity informs high-altitude adaptation (Liu et al. 2025). CRISPR applications for metabolite enhancement have grown 25% (Ansori et al. 2023). While DNA barcoding has become critically important for the identification of herbal medicinal plant species and the monitoring of their quality, it also opens avenues for enhancing the accuracy and reliability of herbal medicine (Chen et al. 2023b). Species-specific studies, such as SCoT-based assessments in *Trigonella foenum-graecum*, uncover polymorphisms for fenugreek yield improvement (Naaz et al. 2024).

These findings underscore the critical imperative to mitigate genetic erosion through integrative conservation frameworks that synergize in situ preservation of native ecosystems with advanced biotechnological interventions. Ex situ strategies, including cryopreservation in germplasm banks and clonal propagation via tissue culture, are vital for safeguarding allelic diversity. Concurrently, biotechnological tools such as marker-assisted selection and CRISPR-based genomic editing present transformative potential for enhancing the adaptive resilience and bioactive compound yield of medicinal plant genetic resources (MPGRs). The sustainable management of this genetic wealth is paramount, not only for pioneering novel pharmacotherapeutic leads but also for bolstering ecological stability and socioeconomic resilience in the face of climate change and demographic pressures. By prioritizing the conservation and innovative utilization of MPGRs, we can secure the indispensable biological foundation for future advancements in both healthcare and sustainable agriculture.

16.2 Genetic Diversity Evaluation of Medicinal Plants

The evaluation of genetic diversity in medicinal plants has emerged as a critical research domain, driven by escalating threats from habitat loss, overexploitation, and climate change. Recent studies underscore a paradigm shift from traditional morphological and biochemical assessments to high-throughput genomic tools, including NGS, single-nucleotide polymorphisms (SNPs), and metabolomics-integrated approaches. This evolution facilitates precise quantification of intraspecific variation, population structure, and adaptive potential, essential for conservation and sustainable utilization. Globally, over 400 medicinal plant genomes have been sequenced by early 2025, yet challenges persist in assembly quality and functional annotation of secondary metabolite pathways (Zhao et al. 2024, Cheng et al. 2025; Mykhailenko et al. 2025). This review synthesizes trends in specific regions (e.g., Asia, Africa), families (e.g., Lamiaceae, Papaveraceae), and species (e.g., *Artemisia annua*, *Coptis* spp.), highlighting implications for biodiversity preservation.

In Asia, a biodiversity hotspot encompassing the Himalayas and Qinling Mountains, research emphasizes phylogeographic analyses to trace refugia amid anthropogenic pressures. For instance, in China, SNP-based assessments of *Acorus tatarinowii* populations revealed the genetic diversity and population structure of the species and established the core germplasm and DNA fingerprints for *A. tatarinowii* (Qin et al. 2025b). Similarly, *Coptis* spp. (Huang-lian) in the Tianshan forests exhibited structured diversity via plastid DNA, with domestication reducing variation by 15% in cultivated lines (Wang et al. 2025). In India, ISSR and SCoT markers evaluated *Podophyllum hexandrum* across altitudinal gradients, detecting low haplotype diversity ($H_d = 0.112$) in high-elevation populations, linked to fragmentation (Naik et al. 2010).

Africa's tropical regions, including East African savannas, are overharvesting orchids such as *Eulophia* spp., for which RAD-seq identified low gene flow ($N_m = 0.36$) and 68% interspecific variation loss, advocating community-led propagation (Phillips et al. 2020). In Indonesia, species distribution models integrated with SSR markers prioritized 80% native medicinal taxa for sustainable extraction, projecting 50% range contraction by 2050 (Cahyaningsih et al. 2021). DArT-based genomic screening of medicinal *Eucalyptus albens* and *E. sideroxylon* reveals high intraspecific diversity, minimal interspecific divergence, ongoing gene flow, and predominant isolation-by-distance with minor climate-driven isolation by environment (IBE), indicating substantial adaptive potential. These markers facilitate identification of resilient chemotypes for large-scale restoration and sustainable medicinal eucalypt utilization (Murray et al. 2019). High-quality genome assemblies of *Engelhardia fenzelii* and *E. roxburghiana* reveal whole-genome duplications, extensive TE-driven expansions, and adaptive enlargement of the terpene synthase (TPS) gene family, particularly in *E. fenzelii.* Phylogenetic reconstruction and demographic inference indicate that species divergence and historical population declines coincide with East Asian summer monsoon intensification. These genomic adaptations underpin ecological resilience in evergreen broadleaf forests and highlight climate-driven evolutionary processes in the region (Li et al. 2025).

Lamiaceae species exhibit remarkable chemodiversity in essential oils, strongly influenced by cultivation system, climate, and geographic origin, resulting in variable antioxidant, antifungal, antibacterial, and anti-inflammatory properties with high medicinal potential. Certain members also demonstrate insect-repellent activity, phytoremediation capacity, and suitability for thermal-regulating green roofs. This review underscores the family's pharmacological and ecological versatility, advocating intensified research to fully exploit its promising natural compounds for therapeutic and environmental applications (Ramos da Silva et al. 2021). Herbgenomics harnesses multi-omics and CRISPR technologies to elucidate biosynthetic pathways, discover novel bioactive compounds, and enable targeted metabolic engineering in medicinal plants, particularly alkaloid-rich Papaveraceae. Despite current transformation challenges, declining sequencing costs position it as a pivotal approach for developing superior cultivars and next-generation plant-derived biopharmaceuticals (Kielich et al. 2024). DArT genotyping of *Artemisia annua* germplasm reveals high polymorphism in GA/UV-B-responsive regulatory regions, with elite accessions showing enhanced NAC/MYB expression and superior artemisinin yield under combined induction, confirming DArT's utility for rapid selection of high-producing chemotypes (Ma et al. 2020). DArT-enabled genetic diversity analysis of medicinal *Acorus* species reveals high polymorphism and subgenome-specific marker partitioning in allotetraploid *A. calamus*, reflecting B-subgenome dominance and adaptive gene family restructuring (Ma et al. 2023). These transferable markers facilitate rapid chemotype identification, population structuring, and conservation prioritization of threatened wetland *Acorus* germplasm. DArT- and SSR-based genotyping of threatened medicinal Ranunculaceae species, exemplified by *Aconitum reclinatum* (69 alleles across 12 polymorphic loci), reveals high intrapopulation diversity essential for fine-scale genetic structure analysis. These high-resolution markers enable rapid identification of priority conservation units and support evidence-based ex situ strategies for rare and endangered medicinal plant germplasm (Zhou et al. 2018). DArT genotyping reveals high polymorphism in glucosinolates (GLS) biosynthetic genes across Brassicaceae, sharply distinguishing GLS-rich wild *Diplotaxis* from *Brassica* crops and accelerating discovery of elite nutraceutical and stress-tolerant chemotypes for medicinal plant breeding (Essoh et al. 2020).

DArT-based genetic diversity evaluation of medicinal plants under climate change reveals accelerated allelic erosion in stress-sensitive populations, while identifying resilient chemotypes with adaptive variation in secondary metabolite pathways, enabling targeted conservation and pre-breeding for climate-resilient medicinal cultivars (Sengar and Sengar 2014). DArT/SSR genotyping of the endangered medicinal *Paeonia decomposita* establishes baseline genetic diversity and population structure, identifying three distinct management units essential for effective conservation and sustainable exploitation of this high-value pharmaceutical species (Wang 2020). SCoT-based genotyping of wild *Dendrobium nobile* (medicinal orchid) from Northeast India reveals high genetic polymorphism (96.21%), moderate-to-high population differentiation (56.63% among-population variation), and low gene flow (N_m = 0.37), confirming substantial interpopulation diversity. These high-resolution SCoT markers provide an effective tool for rapid genetic diversity evaluation, conservation prioritization, and future breeding of elite chemotypes in this endangered medicinal orchid (Bhattacharyya et al. 2013). DArT/RAPD-ISSR genotyping of medicinal *Schisandra chinensis* germplasm reveals high polymorphism (59–79%) and diagnostic markers distinguishing rare white-fruited chemotypes from standard red-fruited populations. These high-resolution marker systems enable rapid genetic diversity assessment, chemotype authentication, and targeted conservation of elite pharmaceutical accessions (Li et al. 2014).

Table 16.1 summarizes key studies on selected species from the past 5 years on evaluating genetic diversity in medicinal plants, focusing on threats like overharvesting, domestication, and climate impacts, while highlighting tools and insights for conservation and breeding. The timeframe emphasizes emerging trends like

Table 16.1 Recent trends in genetic diversity evaluation: examples by region/family/species

Category	Example	Method/key metric	Trend/insight	References
Family (Sapindaceae)	*Acer truncatum* (China)	Chromosome-scale genome (633 Mb); 28 KCS genes	KCS expansion drives seed nervonic acid synthesis for medicinal use	Ma et al. (2019)
Family (Orchidaceae)	*Eulophia edwardii (South Africa)*	Field survey + IUCN Red List assessment	Extremely narrow endemic (single population of ~200 individuals), critically endangered due to habitat loss and transformation	Bytebier and Grieve (2025)
Family (Lamiaceae)	*Scutellaria spp.*	Comparative genomics	Tandem duplications in flavonoids	Xu et al. (2020)
Family (Papaveraceae)	*Papaver somniferum*	Herbgenomics (BGCs)	Alkaloid pathway mapping	Kielich et al. (2024)
Family (Crassulaceae)	*Rhodiola rosea*	ML-GIS + SNPs	Harvesting reduction	Gyorgy et al. (2014)
Family (Moringaceae)	*Moringa oleifera*	ISSR/SCoT (H_d = 0.21)	Diversity for breeding	Popoola et al. (2020)

high-throughput sequencing over traditional markers, reflecting advances in omics for phytochemical-linked diversity.

Recent trends in genetic diversity evaluation of medicinal plants reflect a transition to integrative omics and AI-driven tools, enhancing resolution for region-specific, family-level, and species-focused studies. This facilitates targeted conservation amid annual erosion rates, with future emphasis on pan-genomics for resilient cultivars (Dawson et al. 2019). This introductory section establishes the methodological framework for assessing genetic variability, including molecular markers, population genetics analyses, and genomic tools. It sets the stage by reviewing current techniques (e.g., NGS-based sequencing and SNP genotyping) and their relevance to medicinal plant species, providing baseline data for subsequent discussions on threats and conservation. This facilitates targeted conservation amid 10–20% annual erosion rates, with future emphasis on pan-genomics for resilient cultivars (Abdul Aziz and Masmoudi 2025). Pan-genome assemblies of over 400 medicinal species reveal pervasive whole-genome duplications and tandem expansions in alkaloid/terpenoid clusters, driving up to 2.5-fold metabolite enrichment in polyploid chemotypes (*Catharanthus roseus, Artemisia annua*). Multi-omics integration highlights stress-induced epigenetic reprogramming as a key modulator of phytochemical plasticity. Future perspectives mandate AI-augmented, climate-resilient pan-genomic surveillance coupled with CRISPR-directed allele mining to secure bioactive yields and adaptive potential across 30,000+ taxa amid escalating anthropogenic pressures (Chen et al. 2023a, Chen et al. 2025). Overall, recent advances in genomic, transcriptomic, and metabolomic technologies have substantially deepened our understanding of genetic diversity in medicinal plants, revealing both the magnitude of biodiversity loss and the potential for targeted conservation. High-resolution SNP genotyping, RAD-seq, pan-genomics, and AI-integrated spatial models now enable precise assessment of population structure, adaptive variation, and phytochemical pathways across regions, families, and species. Despite progress, persistent threats from overharvesting, habitat degradation, and climate change continue to erode genetic resources at alarming rates. Therefore, future conservation strategies must integrate multi-omics, climate-resilient pan-genomic surveillance, and genomics-assisted breeding to safeguard the adaptive capacity and medicinal value of plant taxa in the face of accelerating environmental pressures.

16.3 Threats and Conservation Challenges for Medicinal Plant Genetic Diversity

Medicinal plant genetic diversity, encompassing variations in DNA sequences, epigenetic profiles, and phytochemical composition, is critical for their adaptability and therapeutic value (Rajpal et al. 2022). Over 30,000 medicinal species support 80% of traditional medicine systems and 25% of modern pharmaceuticals globally

(Davis and Choisy 2024). However, habitat loss, overharvesting, and climate change have driven a 15–20% decline in within-population genetic variation over the past century, threatening phytochemical yields and ecological resilience (Mykhailenko et al. 2025). The Global Strategy for Plant Conservation 2020–2030 emphasizes integrating genetic monitoring with threat assessments, particularly in biodiversity hotspots like the tropics (Giakoumi et al. 2025).

Genetic diversity in medicinal plants, driven by recombination, mutations, gene flow, and genetic drift, is crucial for their adaptability, phytochemical variability, and therapeutic potential. Supporting 70–80% of traditional medicine and 25% of modern pharmaceuticals, these plants face a 15–20% genetic erosion due to habitat loss, overharvesting, and climate change, threatening food security and healthcare (Salgotra and Chauhan 2023). Advanced tools like NGS and microsatellite markers reveal diversity trends, while conservation strategies, including in situ preservation, ex situ gene banks, and biotechnological approaches like CRISPR, are vital for sustainable management (Engels and Ebert 2021).

Genetic diversity is fundamental to crop improvement, enabling the development of cultivars with enhanced yield, quality, and resilience to biotic and abiotic stresses. Diverse plant genetic resources (PGRs), including mutant lines, wild species, and breeding stocks, support breeding for climate-resilient varieties (Salgotra and Chauhan 2023). Genetic variability, driven by DNA sequence differences, mutations, and meiotic recombination, enhances adaptability across species, populations, and individuals, reducing inbreeding depression and strengthening ecosystem functionality (Kardos et al. 2016). Germplasm characterization, using phenotypic, biochemical, and molecular methods, enables identification of elite genotypes for breeding. Sustainable management of PGRs is essential for agricultural resilience and food security (Gepts 2023).

Davis and Choisy (2024) proposed an interdisciplinary framework integrating evolutionary ecology, molecular biology, and ethnopharmacology to enhance MAP research, using tools like DNA barcoding to combat genetic erosion and support conservation. These approaches are vital for preserving genetic diversity and securing pharmaceutical resources. For the assessment of genetic diversity in MAPs, Nazar et al. (2022) discussed the use of molecular techniques such as the application of NGS and DNA barcoding to analyze wild MAP populations, as discussed in the context of *Trachyspermum ammi*.

This study investigates the multi-mycotoxin contamination in 36 medicinal plant samples from South Africa. For instance, the presence of toxins such as aflatoxins (AF), ochratoxin A (OTA), and zearalenone (ZEA) has been revealed in plants like *Hippobromus pauciflorus* and *Warburgia salutaris* (Areo et al. 2020), Brazil's Atlantic Forest species (e.g., *Saraca asoca*). In India, *Saraca asoca* (Roxb.) W.J. de Wilde is rapidly contracting in its natural distribution range and is critically threatened due to excessive medicinal use and habitat loss (Nair et al. 2024).and a significant habitat loss by 2070 across all scenarios (RCP 2.6 – RCP 8.5), while also projecting a gain in suitable habitats of 52,275.17–95,994.62 km^2; this gain is directed toward the colder regions of Meghalaya, Mizoram, Assam, Uttarakhand, and Himachal Pradesh (Jena et al. 2023). These patterns, emphasized by molecular insights from NGS, RAD seq, and cpDNA barcoding, not only highlight general genetic erosion but also plant-specific abnormalities such as resilient adaptive SNPs in populations, prompting a paradigm shift toward interdisciplinary frameworks and providing a basic framework for explaining and protecting biodiversity (Theissinger et al. 2023). Our research findings contribute to understanding the microbial relationships between wild *Dendrobium* species, protecting endangered *Dendrobium* species, and developing therapeutic properties for *Dendrobium* species in industrial production (Zhang et al. 2022). Future conservation efforts can mitigate risks, preserve phytochemical efficacy for drug discovery, and bolster climate resilience, aligning with the Global Strategy for Plant Conservation to safeguard more species for sustainable health systems and biodiversity hotspots (Neergheen-Bhujun et al. 2017) and intended to protect 75% of endangered plant species and prepare for 20% of their restoration and reconstruction plans (Dhyani and Abeli 2022). Building on the evaluation, this section examines global and regional vulnerabilities, incorporating case studies on population structure, genetic diversity, threats, and conservation efforts. Examples from diverse ecosystems (e.g., Brazil, Indonesia, Australia, China, and Africa) are synthesized to illustrate general patterns and plant-specific anomalies, highlighting overexploitation, habitat fragmentation, and invasive species as overarching issues. Medicinal plant genetic diversity faces severe threats from habitat fragmentation, overharvesting, invasive species, and climate-driven range shifts. Across priority taxa, population declines and marked reductions in allelic diversity are widely documented. Genomic vulnerability assessments further highlight the progressive loss of adaptive structural variants within biosynthetic gene clusters, undermining phytochemical stability and long-term therapeutic efficacy. Current ex situ collections preserve only a limited portion of overall intraspecific variation, while in situ reserves remain vulnerable to stochastic extirpation and ongoing habitat degradation. Addressing these challenges requires integrated conservation strategies that combine dynamic seed and cryo-banking systems with pangenome-informed monitoring, allele-specific reintroduction protocols, and CRISPR-based restoration of degraded metabolic pathways. Such approaches are essential to maintain evolutionary potential and support the sustained bioprospecting value of medicinal plant resources.

16.3.1 Threats to Genetic Diversity in Medicinal Plants

Medicinal plants, encompassing over 50,000 species globally, face multifaceted threats that erode genetic diversity, thereby compromising their evolutionary potential and therapeutic utility. Primary drivers include habitat fragmentation, overexploitation, invasive species, and climate-induced shifts, collectively reducing effective population sizes and allele frequencies (Póvoa et al. 2023). Overexploitation, particularly in biodiversity hotspots, has led to a 10–20% annual decline in wild populations, with species like *Podophyllum hexandrum* in fragmented Himalayan stands (Nag et al. 2014). Habitat loss via deforestation and urbanization further exacerbates genetic bottlenecks, as evidenced by a meta-analysis of many terrestrial species showing two-thirds impacted by anthropogenic threats, resulting in within-population diversity loss over three decades (González et al. 2020). Invasive species and pollution further diminish gene flow, while global trade amplifies risks through selective harvesting, underscoring the need for integrated threat mitigation (Smith et al. 2020).

Aconitum austrokoreense in South Korea has experienced a 12% loss in genetic diversity due to fragmented habitats, as revealed by SSR and cpDNA analyses, prompting its inclusion on the IUCN Red List (Choi et al. 2016). Similarly, *Ginkgo biloba*, a widely used medicinal species, faces reduced genetic variation in fragmented forests, with studies showing a 15% decline in allelic diversity in Chinese populations (Tang et al. 2012). Altered temperature and precipitation patterns disrupt reproductive cycles and seed viability, accelerating genetic erosion. Environmental stressors and low reproductive success severely threaten the genetic diversity of medicinal plants. Invasive species and pollution (heavy metals, pesticides) reduce genetic diversity through resource competition and population decline; for example, *Centella asiatica* has experienced an 18% loss in genetic variation in polluted regions (Lee et al. 2024). Similarly, sterile and clonally reproducing species such as *Crocus sativus* exhibit near-zero genetic diversity, while *Panax quinquefolius* suffers a genetic loss due to low seed germination and clonal propagation (Ali et al. 2013). These factors particularly hinder conservation efforts for unclassified medicinal plants. Lee et al. (2018) highlighted threats like habitat fragmentation and limited gene flow in an endangered medicinal species, *Aconitum austrokoreense*, valued for its anti-inflammatory alkaloids.

16.3.2 Conservation Strategies: Gaps and Future Directions

Medicinal plants, vital for global healthcare and biodiversity, face escalating threats from overharvesting, habitat loss, and climate change, with 15,000 species at risk and around 20% annual declines in key hotspots (Chen et al. 2016). Current gaps include inadequate integration of multi-omics data for threat assessment, limited policy enforcement in underrepresented regions, and insufficient ex situ gene banking, which covers only 7% of threatened taxa (Napolitano et al. 2024). Community-driven exploitation continues to exacerbate population losses in endemic species such as *Swertia chirayita*. Future strategies focus on harmonized approaches, including (1) pan-genomic surveillance and species distribution models (SDMs) for prioritizing refugia, (2) implementing good agricultural practices (GAP) and CRISPR/Cas9 technology to develop resilient cultivars, potentially reducing wild pressure by 30%, and (3) establishing transnational networks under COP15 that integrate ethnobotanical knowledge with AI-driven monitoring to achieve protected land (Vajana et al. 2023; Thapa et al. 2024).

Protected areas and sustainable harvesting regulations play a crucial role in maintaining wild populations. For instance, the cultivation of *Rhodiola rosea* significantly reduces pressure on wild stocks in Canada and the United States (Brinckmann et al. 2021). Additionally, seed banks and DNA storage are vital for preserving genetic diversity. DNA libraries are particularly recommended for conserving rare alleles in species such as *Warburgia salutaris*. To address challenges like low seed viability, techniques such as tissue culture and shoot-tip cuttings have proven effective, especially for species like *Warburgia salutaris* (Mykhailenko et al. 2025). Moreover, biotechnological approaches, including the engineering of biosynthetic pathways in fungi, enable the production of medicinal compounds without the need to harvest wild plants (Karalija et al. 2025). Engaging local communities is another critical aspect, as their involvement enhances sustainable management practices, as evidenced by GIS-based studies (Maurya and Kumar 2024). The Kunming–Montreal Global Biodiversity Framework emphasizes the importance of conserving genetic diversity; however, its implementation often lags due to limited genetic data (Hoban et al. 2024). To further support sustainable practices, the proposal of a global medicinal plant GIS (GMPGIS) aims to map cultivation zones for sustainable use (Roopashree et al. 2024). Additionally, leveraging local biodiversity can provide affordable cancer treatments in low- and middle-income countries, supported by omics and AI techniques (Hong et al. 2024). Finally, strategies such as translocations and assisted gene flow are essential for enhancing genetic diversity. However, these methods require careful management to avoid outbreeding depression. In conclusion, combining in situ and ex situ strategies with biotechnological and community-driven approaches is essential for effective conservation. Nevertheless, global policy enforcement and the integration of genetic data remain critical challenges (Reid et al. 2025). Table 16.2 outlines practical strategies for conserving medicinal plant genetics: cultivation to ease wild harvesting pressure (e.g., *Rhodiola rosea*), fungal biotechnological production, and GIS-based community engagement, alongside ex situ methods like DNA libraries for preserving rare alleles. It highlights gaps in scaling genetic rescue due to outbreeding risks and poor policy–data integration under the Kunming–Montreal Framework, recommending AI-omics integration and GMPGIS for equitable, resilient solutions despite low-viability constraints.

India, a biodiversity hotspot, harbors a rich array of medicinal plants integral to traditional healing practices such as Ayurveda and Unani. These plants, enriched with bioactive compounds like alkaloids and flavonoids, have been employed across civilizations to treat diverse ailments, complementing modern medicine (Pandey et al. 2013). The renewed interest in herbal remedies stems from concerns over synthetic drug side effects and a shift toward sustainable healthcare (Ekor 2014). Despite their importance, escalating depletion necessitates urgent conservation to sustain ecological balance and cultural heritage. India's medicinal plant diversity, critical to traditional and pharmaceutical systems, faces endangerment from habitat loss and overexploitation (Mehta et al. 2020). Conservation efforts, including protected areas and sustainable practices, are essential to mitigate these threats and ensure resource availability (Adams et al. 2023).

Conservation of medicinal plants necessitates a multifaceted approach, beginning with the identification and protection of biodiversity-rich zones to support their

Table 16.2 Medicinal plant genetic diversity conservation strategies: recent research findings

Conservation strategy	Key application/ benefit	Target species/ context	Key research finding/ evidence	References
Cultivation and sustainable harvesting	Reduces pressure on wild populations	*Rhodiola rosea* (Canada, United States)	Cultivation significantly reduces pressure on wild stocks.	Brinckmann et al. (2021)
Advanced biotechnological production	Enables production without wild harvesting	General medicinal compounds	Engineering biosynthetic pathways in fungi allows for sustainable compound production.	Tiwari and Park (2024)
Advanced propagation techniques	Overcomes low seed viability	*Warburgia salutaris*	Tissue culture and shoot-tip cuttings are effective for propagation.	Hannweg et al. (2016)
Genetic storage (ex-situ)	Preserves rare genetic material	*Warburgia salutaris*	DNA libraries are highly recommended for conserving rare alleles.	Glennon et al. (2023)
Community engagement and technology	Enhances sustainable management	General medicinal plants	GIS-based studies prove that local community involvement improves practices.	Foo (2016)
Genomic and AI-driven discovery	Provides affordable treatments	Cancer treatments in LMICs	Leveraging local biodiversity with omics and AI supports affordable drug discovery.	Albani et al. (2025)
Genetic rescue strategies	Enhances genetic diversity	Populations with low diversity	Translocations and assisted gene flow are essential but require management to avoid outbreeding depression.	Hoffmann et al. (2020)
Policy & Data Integration	Framework for global conservation	Kunming-Montreal GBF	Implementation of global biodiversity targets is hindered by a lack of genetic data.	Li et al. (2023)
Spatial Planning Technology	Maps zones for sustainable use	Global medicinal plants	A proposed GMPGIS can map cultivation zones.	Roopashree et al. (2024)

growth, alongside reforestation and sustainable land-use practices to mitigate deforestation and overgrazing (Chen et al. 2016). In situ conservation is enhanced through the establishment of national parks and reserves, with active involvement of local communities in habitat management (Shukla 2023). Ex situ conservation efforts include the maintenance of botanical gardens and seed banks for propagating endangered species (Awazi et al. 2025). Integrating traditional knowledge through documentation and fostering community-based sustainable enterprises strengthens conservation outcomes (Ruiz-Mallén and Corbera 2013), while public education campaigns and workshops promote sustainable harvesting and ecosystem health

(Borawska 2017). Research and monitoring of population dynamics and threat assessment, supported by collaborations with institutions and nongovernmental organizations (NGOs), are critical for informed conservation strategies (Stem et al. 2005).

Molecular tools enhance conservation by assessing genetic diversity and preserving rare species through micropropagation for rapid clonal propagation, DNA banks, and gene libraries for long-term genetic storage, DNA fingerprinting with restriction fragment length polymorphism (RFLP) and polymerase chain reaction (PCR) markers for clonal identity, DNA barcoding for accurate species identification, genetic diversity analysis using SSR and SNP markers to prioritize conservation areas, genetic mapping to identify trait-linked genes for breeding, and cryopreservation to preserve genetic material at ultralow temperatures (Kress and Erickson 2008; Mondini et al. 2009). Ethnobotanical surveys document traditional knowledge to conserve underutilized medicinal plants. Sustainable cultivation reduces wild pressure, while community-based management ensures effective harvesting and local benefits. Legal frameworks support habitat protection and species recovery, forming the basis for national conservation strategies (Wanjohi et al. 2020).

Table 16.3 illustrates key international milestones shaping global awareness of MAP conservation, highlighting themes such as conservation guidelines, sustainable use, local knowledge, participatory approaches, and complementary conservation methods.

Table 16.4 highlights key needs reported in MAP gene banks worldwide, including increased funding, stronger international collaboration, better documentation of local knowledge, and expansion of research on conservation biology and complementary conservation strategies. Both tables together demonstrate that despite growing global interest in MAP conservation, significant gaps in research capacity, funding, and coordinated policy support remain.

Recent conservation genomic studies reveal persistent deficits in the protection of MAPs, including fragmented population structures, reduced adaptive variation, and unprotected functional diversity within key biosynthetic pathways. These findings underscore the need for more evolutionarily informed and systematically coordinated conservation efforts (Hendricks et al. 2018). However, medicinal plant conservation lags, with gene bank accessions for MAPs and most species represented, indicating poor genetic coverage. Improvement opportunities include combining in situ and ex situ conservation, using GIS and molecular tools for targeted collection, and involving local communities. International collaboration, digital indigenous knowledge databases, and equitable benefit-sharing initiatives support progress. Long-term sustainability requires resolving conservation-use conflicts through community-driven agriculture, better research, policy, and global partnerships (Zamani et al. 2025).

Effective conservation hinges not only on understanding these threats but also on implementing multifaceted strategies and addressing critical research gaps. Conservation efforts require an integrated approach combining in situ (on-site) and ex situ (off-site) methods. The establishment of protected areas and the enforcement of sustainable harvesting regulations are crucial for maintaining wild populations

Table 16.3 Milestones in awareness raising/consolidation of MAP conservation work

Conference	Venue	Year	Main issues
WHO-IUCN-WWF International Consultation on the Conservation of Medicinal Plants https://portals.iucn.org/library/node/5909	Chiang Mai, Thailand	1988	Conservation guidelines Greater partnership Role of IK
United Nations Conference on Environment and Development (UNCED) https://www.un.org/en/conferences/environment/rio1992	Rio de Janeiro, Brazil	1992	Conservation Sustainable use Sharing of benefits
I World Congress on Medicinal and Aromatic Plants for Human Welfare https://agris.fao.org/search/en/providers/122535/records/65dde0324c5aef494fd6484b	Maastricht, The Netherlands	1992	Documentation Sustainable harvesting practices Legal frameworks for use Commercialization guidelines
IV International Technical Conference of FAO on Plant Genetic Resources https://enb.iisd.org/events/4th-international-technical-conference-plant-genetic-resources-itcpgr-4	Leipzig, Germany	1996	Promotion of underutilized species Plan of Action for PGRFA
II World Congress on Medicinal and Aromatic Plants for Human Welfare http://www.icmap.org/icmap/icmap.php?page=congress&i=1	Buenos Aires, Argentina	1997	IK documentation Better use practices Conservation and partnership Inventory of MAP
Medicinal Plants for Survival: International Conference on Medicinal Plants https://lib.icimod.org/records/5xvp9-4yv49	Bangalore, India	1998	Priority setting Complementary conservation Participatory approaches Use promotion
International Conference of the Global Forum of Agricultural Research (GFAR) https://gfair.network/documents/gfar-2000-conference-highlights-and-follow-action-dresden-germany-21-23-may-2000	Dresden, Germany	2000	Role of underutilized species Chain production system
International Conference on Medicinal Plants, Traditional Medicine & Local Communities in Africa: Challenges and Opportunities of the Millennium https://erepository.uonbi.ac.ke/handle/11295/51999	Nairobi, Kenya	2000	Inventory Use-enhancing methods Community-based approaches Documentation/dissemination of information Networking

(continued)

Table 16.3 (continued)

Conference	Venue	Year	Main issues
Conference of the Parties (COP) of the CBD https://enb.iisd.org/biodiv/cop5/#:~:text=The%20fifth%20Conference%20of%20the,indigenous%20and%20local%20community%20organizations	Nairobi, Kenya	2000	Indigenous knowledge
SAT21–International Conference on Science and Technology for Managing Plant Genetic Diversity in the 21st Century (SAT21) https://www.nature.com/articles/6800105.pdf	Kuala Lumpur, Malaysia	2000	Complementary conservation Methodologies and techniques Priority setting

and facilitating gene flow. For instance, the cultivation of *Rhodiola rosea* has significantly reduced pressure on wild stocks in Canada and the United States (Mykhailenko et al. 2025). Ex situ strategies, such as seed banks and DNA storage, play a vital role in preserving genetic diversity. DNA libraries are highly recommended for conserving rare alleles in critically endangered species like *Warburgia salutaris* (Glennon et al. 2023). To overcome challenges like low seed viability, advanced propagation techniques, including tissue culture and shoot-tip cuttings, have proven effective for many species (León-Lobos et al. 2012). Biotechnological innovations offer promising solutions to circumvent the need for wild harvesting. Engineering biosynthetic pathways in microbes, such as fungi, enables the sustainable production of valuable medicinal compounds (Nwankwegu et al. 2025).

Furthermore, local community engagement is crucial, enhancing sustainable management as supported by GIS studies (Xu 2025). The Kunming–Montreal Global Biodiversity Framework emphasizes genetic diversity conservation but struggles with limited data. A GMPGIS maps optimal cultivation zones for sustainable use. Genetic rescue through translocations and assisted gene flow can restore diversity but carries risks like outbreeding depression, requiring careful management. Gaps include few studies on threatened medicinal plants, regional biases (e.g., China), and a lack of data on salt stress impacts. Future directions involve expanding genomic research in underrepresented regions (Africa, Latin America), developing global genetic diversity indicators, and blending traditional knowledge with modern tools. Salt-tolerant crop methods, such as transcriptomics and field trials with sorghum, quinoa, and millets, can aid medicinal plant conservation in saline environments (Stephens 2023). The mechanisms against heavy metal stress in MAPS are poorly understood, with limited information on secondary metabolites and gene expression. The phytoremediation potential of perennial MAPs, especially postharvest biomass, needs validation. WHO/FAO guidelines set toxin limits (As, Cd, Pb) but lack thresholds for Ni and Zn, requiring interdisciplinary research on toxicity levels. Future work should focus on nonedible MAPs for soil cleanup, stress-induced essential oil boosts, and plant–metal interactions for bio-fortification, fostering cost-effective phytoremediation (Mohamed et al. 2025). Cross-disciplinary collaboration is vital to develop safe, effective products while tackling conservation

Table 16.4 Needs as indicated by MAP Gene Banks around the world

Organization	Country	No. of species	No. of accessions	Needs
Gatersleben Gene Bank (IPK) https://www.ipk-gatersleben.de/en/research/genebank	Germany	931	17,757	More research on conservation biology of MAP Evaluation/research for breeding purposes
Research Institute of Medicinal Plants https://ec.europa.eu/info/funding-tenders/opportunities/portal/screen/how-to-participate/org-details/985518449	Poland	7	33	Call for greater international collaboration for ex situ conservation
Centro Nacional de Conservavion de recursos Fitogenetics	Venezuela	30	60	Funds More research on conservation biology of MAP
Biodiversity Institute https://ebi.gov.et/resources/	Ethiopia	35 (app.)	1282	Documentation of orally transmitted IK on MAP Systematizing information gathered so far
Agricultural Research Council (ARC) https://www.gov.za/about-government/contact-directory/soe/agricultural-research-council-arc	Republic of South Africa	60	560 (app.)	Funds Clarify access policy to MAP More research on conservation biology of MAP More research on complementary conservation approaches
Zhejing Research Institute of Chinese Materia Medica https://www.zcmu.edu.cn/english/info/10532/1314.htm	China	200	50,000	Greater international cooperation More research on conservation methods
National Bureau of Plant Genetic Resources (NBPGR) https://nbpgr.org.in/nbpgr2023/	India	600	849	More research on conservation biology of MAP Research on in vitro conservation methods Surveys and collecting Documentation Research on agro-techniques for use promotion

(continued)

Table 16.4 (continued)

Organization	Country	No. of species	No. of accessions	Needs
Agricultural Research Organization, Newe-Ya'ar Research Center https://www.agri.gov.il/en/research-centers/newe-yaar/	Israel	12 (app.)	350	Funds Greater research/ conservation on endemic species
Indian Institute of Horticultural Research (IIHR) https://www.fao.org/agris/data-provider/indian-institute-horticultural-research	India	11	227	Funds Research on impact of ex situ conservation on MAP Characterization of traits of economic interest

challenges—preserving ethnobotanical knowledge and biodiversity to prevent genetic loss and ecological imbalance. Sharing global expertise on cultivation and therapeutic use across generations is essential to maximize their contribution to human health (Mbelebele et al. 2024).

Despite substantial advances in molecular characterization, spatial planning, and biotechnological production systems, conservation of MAPs remains constrained by fragmented genetic coverage, limited multi-omics integration, weak policy implementation, and unequal regional research intensity. Current strategies—ranging from in situ reserve design and ex situ gene banking to micropropagation and engineered biosynthetic pathways—mitigate some pressures but do not sufficiently address the erosion of adaptive variation or functional diversity in biosynthetic gene clusters. Critical gaps persist in threatened-species genomic datasets, standardized indicators for genetic diversity under global frameworks, and risk assessments for genetic rescue, particularly regarding outbreeding depression. Future progress requires coordinated global action integrating pan-genomic surveillance, AI-driven monitoring, GMPGIS-based spatial optimization, and community-embedded sustainable agriculture. Equally important are advances in stress-resilience studies, including salinity and heavy metal responses, to guide conservation in degraded landscapes. Overall, safeguarding MAP evolutionary potential and therapeutic value demands a unified, data-rich, transdisciplinary framework that bridges traditional knowledge, modern genomics, and international policy commitments.

Table 16.5 brings together species-level evidence from recent studies to highlight the major threats to the decline of medicinal plant genetic diversity. Habitat loss and fragmentation reduce gene flow and genetic variation in *Aconitum austrokoreense*, as demonstrated by cpDNA analyses. Overharvesting similarly leads to measurable decreases in heterozygosity observed in *Rhodiola rosea*. Climate change further threatens adaptive potential in *Boswellia sacra*, which exhibits low germination due to environmental stress. Invasive species and pollution exacerbate genetic erosion in taxa such as *Centella asiatica* through competitive pressure and contamination. Furthermore, the inherently low reproductive success in *Panax quinquefolius*,

Table 16.5 Threats to medicinal plant genetic diversity

Threat	Key impact	Species	Evidence	References
Habitat loss/ fragmentation	Diversity loss, reduced gene flow	*Aconitum austrokoreense*	cpDNA analysis	Lee et al. (2018)
Overharvesting	Heterozygosity loss	*Rhodiola rosea*	Genetic depletion	Terletskaya et al. (2023)
Climate change	Loss in adaptive alleles	*Boswellia sacra*	Low germination	Al-Harrasi et al. (2019)
Invasive species/ pollution	Diversity decline	*Centella asiatica*	Competition and contamination effects	Zhang et al. (2012)
Low reproductive success	Diversity reduction	*Panax quinquefolius*	Clonal propagation, low germination	Lee et al. (2023)

demonstrated by clonal reproduction and low germination, contributes to further diversity loss. Collectively, these findings highlight that both ecological disturbances and biological constraints play critical roles in shaping genetic vulnerability in medicinal plant species.

Effective conservation of medicinal plants combines in situ and ex situ strategies. In situ approaches, including protected areas and sustainable harvesting, alleviate pressure on wild populations, as demonstrated by *Rhodiola rosea* cultivation in Canada and the United States. Ex situ methods—seed banks, DNA libraries, and advanced propagation like tissue culture and shoot-tip cuttings—preserve rare alleles and address low seed viability, especially in species like *Warburgia salutaris*. Biotechnological advances, such as fungal pathway engineering, enable compound production without wild harvesting. Community engagement, enhanced by GIS-based mapping (e.g., Global Medicinal Plant GIS), supports sustainable management. Genetic rescue through translocations and assisted gene flow increases diversity but requires careful oversight to prevent outbreeding depression (Chen et al. 2016).

16.4 The Role of Medicinal Plant Genetic Diversity in Climate Change Adaptation

Medicinal plant genetic diversity plays a pivotal role in adapting to climate change by enhancing species resilience against shifting environmental conditions. Projections indicate that a substantial proportion of medicinal plant species could face climate-driven range contractions in the coming decades, yet genetically diverse populations retain adaptive alleles essential for stress tolerance. In the Indian Himalayas, *Rheum australe* populations with elevated genetic variation exhibit upslope migration and higher podophyllotoxin content (Pandith et al. 2018). *Boswellia sacra* demonstrates improved germination success in drought-tolerant ecotypes, linked to sesquiterpene diversity (Khan et al. 2022). SNP-based studies in

Picea abies–P. obovata identified drought-adaptive loci, guiding the development of resilient cultivars (Vasileva et al. 2024). Phenotypic variation, supported by allelic richness, may provide potential for short-term environmental buffering, but prolonged extremes likely exceed this capacity; in such cases, conservation strategies emphasize maintaining or enhancing genetic diversity to preserve evolutionary potential (Reed and Frankham 2003). Future adaptation of small-island coastal ecosystems depends on preserving ecosystem diversity and restoring natural habitats to buffer erosion and sea-level rise pressures (Hernández-Delgado 2024). These findings emphasize the critical role of genetic variation in enhancing species resilience to climate change and supporting sustainable conservation strategies (Jangpangi et al. 2025).

Species highlight how medicinal plants native to arid and semiarid regions remain vulnerable to a combination of threats, including habitat degradation, unsustainable harvesting practices, and climatic stress—factors that may compromise long-term viability unless conservation strategies integrate population monitoring, habitat protection, and sustainable use frameworks. (Kougioumoutzis et al. 2024). The qualitative vulnerability assessment framework highlights the complex interplay of natural and anthropogenic factors, emphasizing that species-specific responses necessitate tailored interventions. While quantification of vulnerability parameters is a potential future step, the intricate network of causal factors cautions against oversimplification, advocating instead for nuanced, integrative approaches (Turner II et al. 2003).

The preservation of MAP genetic diversity is pivotal for climate change adaptation, offering both ecological and socioeconomic benefits. In situ conservation, supported by sustainable harvesting and cultivation practices, alongside ex situ methods like seed banks and DNA libraries, can safeguard critical genetic resources (Salgotra and Chauhan 2023). Biotechnological innovations, such as CRISPR/Cas9-enhanced hairy root systems, enhance resilience and secondary metabolite production, reducing pressure on wild populations (Elbouzidi et al. 2025). Moreover, the economic value of MAPs provides financial incentives for conservation, as seen in secondary forests serving as carbon sinks while supporting local livelihoods (Verma et al. 2024). Medicinal plant genetic diversity provides essential stress-tolerance alleles rapidly deployable via genomics and CRISPR; its conservation and characterization are critical for pharmaceutical security and climate-resilient agriculture (Ghorbanzadeh et al. 2025).

The genetic diversity of MAPs harbors essential climate-adaptive alleles; its conservation and genomic exploitation via pan-genomics and CRISPR are critical for sustaining pharmaceutical resources under climate change (Smith-Hall et al. 2025). By fostering collective stewardship among stakeholders, implementing innovative biotechnologies, and prioritizing conservation in underrepresented regions, we can ensure the resilience of genetic diversity of MAPs. This not only secures the life-support systems underpinning human wellbeing but also reinforces the critical role of MAPs in harmonizing ecological and societal needs in a rapidly changing world. Complementing the prior subsection, this section explores adaptive mechanisms, such as resilient genotypes and trait-based selection, demonstrating how intraspecific variation enables range shifts and stress tolerance, with implications for breeding climate-resilient varieties. Medicinal plant genetic diversity constitutes a pivotal adaptive reservoir against the escalating exigencies of climate change, underpinning phenotypic plasticity, stress-responsive metabolic reconfiguration, and evolutionary resilience in the face of thermal, hydric, and edaphic perturbations. Empirical evidence underscores that intraspecific allelic variation modulates acclimatory thresholds—facilitating, for instance, the retention of pharmacologically critical secondary metabolites amid abiotic stressors—whereas its erosion precipitates maladaptive bottlenecks and diminished phytochemical fidelity. Harnessing this genomic heterogeneity through targeted conservation genomics, assisted migration, and precision breeding emerges as an imperative for sustaining therapeutic efficacy and ecosystem functionality. Ultimately, safeguarding medicinal plant genetic diversity transcends mere biodiversity preservation; it represents a strategic linchpin for pharmacopoeial stability, biocultural continuity, and adaptive global health security in a climatically volatile Anthropocene.

16.5 Phytochemical Diversity of Medicinal Plants: A Resource for Drug Discovery

MAPs are increasingly imperiled by a confluence of biotic and anthropogenic perturbations that erode genomic heterogeneity and biosynthetic fidelity, thereby vitiating both evolutionary resilience and pharmacotherapeutic consistency. Genetic diversity—encompassing DNA sequence variation, gene duplications, and polyploidy events—directly influences the biosynthesis of secondary metabolites such as alkaloids, terpenoids, and phenolics that confer pharmacological properties. Phytochemical diversity, in turn, emerges from the interplay between these genetic factors and environmental cues, including abiotic stressors like UV radiation and drought. Recent genomic advances, particularly long-read sequencing and pangenome assemblies, have elucidated these drivers of variation; by early 2025, over 400 medicinal plant genomes had been sequenced, revealing patterns of whole-genome duplication (WGD) and tandem gene expansions in biosynthetic clusters (Chen et al. 2025; Gupta et al. 2024).

Polyploidy, a prevalent evolutionary mechanism in plants, involves WGD leading to multiple chromosome sets, which amplifies genetic redundancy and promotes neofunctionalization of duplicated genes. This process enhances phytochemical diversity by expanding gene families involved in secondary metabolism. A 2025 theoretical model elucidates polyploidy's dual role: it stabilizes populations in predictable environments via genomic buffering, allowing retention of deleterious mutations, while accelerating innovation in variable conditions through increased allelic diversity (Heslop-Harrison et al. 2023). In medicinal plants, induced polyploidy has been shown to boost alkaloid production; for instance, in *Catharanthus roseus* (a source of the anticancer compounds vinblastine and vincristine), colchicine-induced tetraploidy doubled chromosome numbers from 18 to 36, elevating indole alkaloid yields by upregulating biosynthetic genes such as strictosidine synthase (*STR*) and tryptophan decarboxylase (*TDC*) (Nourozi et al. 2025).

Gene duplication, often via tandem repeats or WGD, further diversifies metabolic pathways. Pan-genome analyses of polyploid species in 2025 revealed that structural variants (SVs) arising from duplications generate adaptive alleles, with polyploids accumulating more SVs than diploids, thereby contributing to local adaptation and novel trait evolution (Hämälä et al. 2024). In the Lamiaceae family, tandem duplications in terpene synthase genes have expanded volatile compound profiles in species like *Salvia*, enhancing aroma and antimicrobial properties. A March 2025 integrated genomics and metabolomics analysis of *Saussurea obvallata* (a Tibetan medicinal herb) demonstrated that high-altitude polyploidy correlates with elevated flavonoid diversity, driven by duplicated genes under hypoxic stress (Zhang et al. 2025). While genetics provides the blueprint, environmental factors act as modulators, inducing phenotypic plasticity in phytochemical profiles. Abiotic stressors like UV-B and drought trigger signaling cascades that upregulate duplicated genes, amplifying metabolite production for defense. A February 2025 study on *Quercus* species demonstrated that UV exposure enhances phenolic acid accumulation via stress-responsive transcription factors, with genetic diversity (measured by SNPs) predicting variation across populations. In arid regions, crop wild relatives (CWRs) of medicinal plants in the UAE exhibit 20–30% higher terpenoid diversity due to drought-induced polyploidy, bolstering resilience as per a March 2025 meta-analysis (Gupta et al. 2024; Shakuri et al. 2025).

Multi-omics integration has been pivotal in 2024–2025 research. For example, chemotaxonomic profiling of Asteraceae medicinals revealed that soil salinity variations correlate with alkaloid shifts, linked to polyploid-specific gene expansions. In *Carthamus tinctorius* (safflower), a May 2024 review tied genetic heterozygosity to enhanced carthamin yields under heat stress, emphasizing the role of environmental-gene interactions in therapeutic compound optimization (Qin et al. 2025a). Table 16.6 synthesizes how specific genetic mechanisms influence secondary metabolite production in medicinal plants, often enhancing therapeutic potential under stress. It underscores the role of genetic variation in driving phytochemical adaptation, with implications for breeding resilient varieties. Each row links a plant species, its genetic determinant, the observed phytochemical effect, and the reference.

Table 16.6 Key genetic determinants and their impact on phytochemical diversity in select MPAs

Plant species	Determinant	Phytochemical effect	References
Catharanthus roseus	Induced polyploidy ($2n = 36$)	↑ Indole alkaloids (vinblastine +25%)	Almagro et al. (2015)
Saussurea obvallata	Tandem gene duplication	↑ Flavonoids under hypoxia	Zhang et al. (2025)
Quercus spp.	SNP variation + UV stress	↑ Phenolic acids	Shakuri et al. (2025)
Carthamus tinctorius	WGD heterozygosity	↑ Fatty acids and carotenoid under heat	Rahnama et al. (2024)

Table 16.7 Recent multi-omics studies on environmental modulation

Study focus	Omics approach	Key finding	Citation
Chemotaxonomy in *Asteraceae*	Metabolomics + genomics	Salinity-induced alkaloid shifts	Amin and Park (2025)
CWR Resilience in UAE	Population genomics	Terpenoid diversity ↑	Gkanogiannis et al. (2025)
Rare Plant Chemo-Diversity	Systematic review	Conservation for rare plant species	Kricsfalusy and Trevisan (2014)

Recent Multi-omics studies on environmental modulation summarizes cutting-edge research that employs multi-omics approaches to investigate how environmental factors influence phytochemical diversity in medicinal plants. The table highlights three key studies, each focusing on a distinct aspect of environmental modulation, the omics methodologies applied, their primary findings, and the corresponding citations. Below is a detailed explanation of each row, followed by the reference list for the citations included (Table 16.7).

Phytochemical variability in forest trees is crucial for ecosystem resilience and offers significant research opportunities. Understanding these dynamics, as shown in a global synthesis on genetic and chemical diversity (Ogwu et al. 2025), addresses research gaps and informs future methodological approaches, particularly in *Quercus* species (Sampaio et al. 2025; Shakuri et al. 2025). Phytochemical variability in forest trees poses significant research challenges while offering substantial opportunities. The study of phytochemical variability in forest trees is critical for understanding ecosystem resilience, yet significant gaps remain in linking this diversity to environmental and anthropogenic pressures. Methodological limitations currently hinder a comprehensive understanding of these complex dynamics. Future research must therefore leverage advanced analytical techniques and interdisciplinary models to predict ecological responses and inform effective conservation strategies (Ogwu et al. 2025).

As global climate patterns shift, understanding how these changes influence phytochemical traits in forest ecosystems, including medicinal plant populations, becomes paramount (Tiwari et al. 2023). Ogwu et al. (2025) highlight that environmental factors, such as climate and soil type, significantly impact phytochemical composition, a finding that aligns with the complex determinants of phytochemical diversity in medicinal plants, where abiotic stressors like UV radiation and drought modulate secondary metabolite production. Future studies should focus on forecasting these responses to better inform conservation strategies and ecosystem management practices, especially for the medicinal species that support global healthcare.

By identifying and conserving areas with rich phytochemical diversity, we can enhance the resilience of forest ecosystems and their capacity to provide essential ecosystem services, including the sustainable supply of medicinal compounds (Nizam et al. 2021; Ogwu et al. 2025).

Furthermore, understanding the evolutionary dynamics of phytochemical diversity is crucial, especially for medicinal plants where evolutionary processes such as WGD and tandem gene expansions drive chemical variability. Studies have shown that chemical diversity in plants evolves under selective pressures imposed by herbivore communities (Salazar et al. 2018; Gupta et al. 2024), a principle that applies to medicinal species like *Camellia sinensis*, where terpene synthase duplications improve medicinal profiles. Investigating these evolutionary processes can provide insights into the maintenance of phytochemical diversity and its implications for forest health, guiding integrative conservation frameworks that synergize in situ preservation with biotechnological interventions (Hao and Xiao 2015).

The remarkable chemical adaptability and biosynthetic complexity of these metabolites arise from evolutionary pressures that drive plants to synthesize defense molecules against pathogens, herbivores, and environmental stressors. This same adaptive chemistry underpins their therapeutic relevance in humans. For instance, phenolic compounds act as natural antioxidants that mitigate oxidative damage in biological systems, while terpenoids and alkaloids contribute both to plant survival and to drug leads in modern pharmacology, such as the antimalarial artemisinin and the anticancer taxol (Newman and Cragg 2012; Dar et al. 2023; Ndhlala et al. 2024). Beyond their therapeutic efficacy, bioactive substances from medicinal plants are increasingly valued for their natural origin and ecological sustainability. They are often perceived as safer and more environmentally compatible than synthetic pharmaceuticals, aligning with global movements toward green chemistry and sustainable bioprospecting. Furthermore, the ethnobotanical knowledge maintained by indigenous cultures continues to guide the identification of novel phytochemicals, reinforcing the indispensable role of traditional wisdom in modern drug discovery (Heinrich 2013; Dar et al. 2023; Banerjee 2024).

The phytochemical diversity of MAPs serves as a vital resource for drug discovery, underpinning their therapeutic potential and ecological adaptability. Secondary metabolites, including alkaloids, flavonoids, terpenoids, phenolics, and saponins, exhibit a wide range of pharmacological activities—antioxidant, antimicrobial, anticancer, antidiabetic, and neuroprotective—driven by evolutionary responses to biotic and abiotic stressors (Newman and Cragg 2012; Saxena et al. 2013). By synergizing traditional knowledge with cutting-edge biotechnologies and sustainable bioprospecting, interdisciplinary efforts can ensure MAPs remain a strategic biological asset, addressing unmet medical needs and bolstering ecosystem resilience amid global environmental change.

16.6 Conclusion

The accelerating erosion of genetic diversity in medicinal and aromatic plants represents a profound threat to phytochemical stability, evolutionary potential, and long-term pharmacological innovation. Genomic and multi-omics studies consistently demonstrate that allelic variation, gene-family expansion, and epigenomic plasticity constitute the molecular basis of both adaptive capacity and chemodiversity in key MAP lineages. Yet, climate-driven habitat shifts, fragmentation, and unsustainable harvesting continue to erode these foundational genetic resources. A genomics-informed conservation paradigm—integrating dynamic in situ protection of genetic refugia, synchronized ex situ genome/cryobanking, sustainable community-based cultivation, and targeted restoration of biosynthetic pathways—offers the most effective strategy to stabilize threatened taxa. Safeguarding MAP genomic diversity is therefore indispensable not only for ecosystem resilience, but also for securing future drug discovery pipelines and ensuring sustainable access to bioactive metabolites in a rapidly changing global environment.

References

Abdul Aziz M, Masmoudi K (2025) Molecular breakthroughs in modern plant breeding techniques. Hortic Plant J 11(1):15–41. https://doi.org/10.1016/j.hpj.2024.01.004

Adams VM, Chauvenet ALM, Stoudmann N, Gurney GG, Brockington D, Kuempel CD (2023) Multiple-use protected areas are critical to equitable and effective conservation. One Earth 6(9):1173–1189. https://doi.org/10.1016/j.oneear.2023.08.011

Akram M (2015) The Society for Indonesian Biodiversity. https://doi.org/10.13057/BIODIV/D150201

Alami MM, Ouyang Z, Zhang Y, Shu S, Yang G, Mei Z, Wang X (2022) The current developments in medicinal plant genomics enabled the diversification of secondary metabolites' biosynthesis. Int J Mol Sci 23(24):15932. https://doi.org/10.3390/ijms232415932

Albani FG, Alghamdi SS, Almutairi MM, Alqahtani T (2025) Artificial intelligence-driven innovations in oncology drug discovery: transforming traditional pipelines and enhancing drug design. Drug Des Devel Ther 19:5685–5707. https://doi.org/10.2147/DDDT.S509769

Al-Harrasi A, Khan AL, Asaf S, Al-Rawahi AK (2019) Propagation and conservation of *Boswellia sacra*. In: Biology of Genus *Boswellia*, pp 57–72. https://doi.org/10.1007/978-3-030-16725-7_5

Ali G, Iqbal AM, Nehvi FA, Samad SS, Nagoo S, Naseer S, Dar NA (2013) Prospects of clonal selection for enhancing productivity in saffron (Crocus sativus L.). Afr J Agric Res 8(5):460–467

Almagro L, Fernández-Pérez F, Pedreño MA (2015) Indole alkaloids from *Catharanthus roseus*: bioproduction and their effect on human health. Molecules (Basel, Switzerland) 20(2):2973–3000. https://doi.org/10.3390/molecules20022973

Alotaibi MO, Abd-Elgawad ME (2022) ISSR and SCoT for evaluation of hereditary differences of 29 wild plants in Al Jubail Saudi Arabian. Saudi J Biol Sci 29(5):3223–3231. https://doi.org/10.1016/j.sjbs.2022.01.053

Alum EU (2024) Climate change and its impact on the bioactive compound profile of medicinal plants: implications for global health. Plant Signal Behav 19(1):2419683. https://doi.org/10.1080/15592324.2024.2419683

Amin A, Park S (2025) Harnessing molecular phylogeny and chemometrics for taxonomic validation of Korean aromatic plants: integrating genomics with practical applications. Plants (Basel, Switzerland) 14(15):2364. https://doi.org/10.3390/plants14152364

Ansori AN, Antonius Y, Susilo RJ, Hayaza S, Kharisma VD, Parikesit AA, Zainul R, Jakhmola V, Saklani T, Rebezov M, Ullah ME, Maksimiuk N, Derkho M, Burkov P (2023) Application of CRISPR-Cas9 genome editing technology in various fields: a review. Narra J 3(2):e184. https://doi.org/10.52225/narra.v3i2.184

Areo OM, Phoku JZ, Gbashi S, Njobeh PB (2020) A preliminary study of multi-mycotoxins contamination in some selected South Africa medicinal plants. Emir J Food Agric 10. https://doi.org/10.9755/ejfa.2020.v32.i6.2113

Awazi NP, Ambebe TF, Njamnjubo NA (2025) Ex-situ conservation of threatened flora and fauna species in Cameroon through botanic and zoological gardens: governance and policy paradigms. Discov Conserv 2:26. https://doi.org/10.1007/s44353-025-00049-9

Banerjee S (2024) Introduction to ethnobotany and traditional medicine. In: Traditional resources and tools for modern drug discovery, 1st edn., Chapter 1. Springer Nature. https://doi.org/10.1007/978-981-97-4600-2_1

Barrera-Redondo J, Piñero D, Eguiarte LE (2020) Genomic, transcriptomic and epigenomic tools to study the domestication of plants and animals: a field guide for beginners. Front Genet 11:742. https://doi.org/10.3389/fgene.2020.00742

Bhattacharyya P, Kumaria S, Kumar S, Tandon P (2013) Start Codon Targeted (SCoT) marker reveals genetic diversity of *Dendrobium nobile* Lindl., an endangered medicinal orchid species. Gene 529(1):21–26. https://doi.org/10.1016/j.gene.2013.07.096

Borawska A (2017) The role of public awareness campaigns in sustainable development. Econ Environ Stud 17(44):865–877. https://doi.org/10.25167/ees.2017.44.14

Brinckmann JA, Cunningham AB, Harter DEV (2021) Running out of time to smell the rose-roots: reviewing threats and trade in wild *Rhodiola rosea* L. J Ethnopharmacol 269(3):113710. https://doi.org/10.1016/j.jep.2020.113710

Bytebier B, Grieve KW (2025) 1135. Eulophia edwardii Bytebier. Curtis's Bot Mag 42(1):53–60. Available from: https://doi.org/10.1111/curt.12624

Cahyaningsih R, Joana MB, Nigel M (2021) Setting the priority medicinal plants for conservation in Indonesia. Genet Resour Crop Evol 68:2019–2050. https://doi.org/10.1007/s10722-021-01115-6

Chen SL, Yu H, Luo HM, Wu Q, Li CF, Steinmetz A (2016) Conservation and sustainable use of medicinal plants: problems, progress, and prospects. Chin Med 11:37. https://doi.org/10.1186/s13020-016-0108-7

Chen Q, Hu H, Zhang D (2022) DNA barcoding and phylogenomic analysis of the genus *Fritillaria* in China based on complete chloroplast genomes. Front Plant Sci 13:764255. https://doi.org/10.3389/fpls.2022.764255

Chen H, Guo M, Dong S, Wu X, Zhang G, He L, Jiao Y, Chen S, Li L, Luo H (2023a) A chromosome-scale genome assembly of *Artemisia argyi* reveals unbiased subgenome evolution and key contributions of gene duplication to volatile terpenoid diversity. Plant Commun 4(3):100516. https://doi.org/10.1016/j.xplc.2023.100516

Chen S, Yin X, Han J, Sun W, Yao H, Song J, Li X (2023b) DNA barcoding in herbal medicine: retrospective and prospective. J Pharm Anal 13(5):431–441. https://doi.org/10.1016/j.jpha.2023.03.008

Chen J, Zhang Y, Li Y, Liu Y, Li Q, Lv Z et al (2025) The last decade of medicinal plant genomics: advances and challenges. Engineering. https://doi.org/10.1016/j.eng.2025.09.024

Cheng LT, Wang ZL, Zhu QH, Ye M, Ye CY (2025) A long road ahead to reliable and complete medicinal plant genomes. Nat Commun 16(1):2150. https://doi.org/10.1038/s41467-025-57448-8

Choi JE, Kim GB, Lim CE, Yu HJ, Mun JH (2016) The complete chloroplast genome of *Aconitum austrokoreense* Koidz. (Ranunculaceae), an endangered endemic species in Korea. Mitochondrial DNA B Resour 1(1):688–689. https://doi.org/10.1080/23802359.2016.1219644

da Silva AM, Horsth AL, Timóteo ÉD et al (2023) Use of medicinal plants during COVID-19 pandemic in Brazil. Sci Rep 13:16558. https://doi.org/10.1038/s41598-023-43673-y

Dar RA, Shahnawaz M, Ahanger MA, Majid IU (2023) Exploring the diverse bioactive compounds from medicinal plants: a review. J Phytopharm 12(3):189–195. https://doi.org/10.31254/phyto.2023.12307

Davis CC, Choisy P (2024) Medicinal plants meet modern biodiversity science. Curr Biol 34(4):R158–R173. https://doi.org/10.1016/j.cub.2023.12.038

Dawson IK, Powell W, Hendre P, Jamnadass R, Hoad S, Akinwumi A, Wainaina J, Tenkouano A, Tohme J (2019) The role of genetics in mainstreaming the production of new and orphan crops to diversify food systems and support human nutrition. New Phytol 224(1):20–25. https://doi.org/10.1111/nph.15895

Dhyani A, Abeli T (2022) Plant translocation for threatened species conservation. Proceedings 80:Article 1. https://doi.org/10.3390/proceedings2022080001

Ekor M (2014) The growing use of herbal medicines: issues relating to adverse reactions and challenges in monitoring safety. Front Pharmacol 4:177. https://doi.org/10.3389/fphar.2013.00177

El Bilali H, Dan Guimbo I, Nanema RK, Falalou H, Kiebre Z, Rokka V-M, Tietiambou SRF, Nanema J, Dambo L, Grazioli F, Naino Jika AK, Gonnella M, Acasto F (2024) Research on Moringa (*Moringa oleifera* Lam.) in Africa. Plants 13(12):1613. https://doi.org/10.3390/plants13121613

Elbouzidi A, Taibi M, Haddou M, Baraich A, Sadougui I, Bellaouchi R, Asehraou A, Chaabane K, El Guerrouj B, Addi M (2025) Hairy roots as a biotechnological tool for medicinal plant secondary metabolites: a systematic review. J Biotechnol 409:117–136. https://doi.org/10.1016/j.jbiotec.2025.10.010. Advance online publication

Engels JMM, Ebert AW (2021) A critical review of the current global ex situ conservation system for plant agrobiodiversity. II. Strengths and weaknesses of the current system and recommendations for its improvement. Plants (Basel, Switzerland) 10(9):1904. https://doi.org/10.3390/plants10091904

Essoh AP, Monteiro F, Pena AR, Pais MS, Moura M, Romeiras MM (2020) Exploring glucosinolates diversity in Brassicaceae: a genomic and chemical assessment for deciphering abiotic stress tolerance. Plant Physiol Biochem 150:151–161. https://doi.org/10.1016/j.plaphy.2020.02.032

Foo J (2016) Community participation in governance for sustainable management of medicinal plants market in urban area. Jurnal Antarabangsa Alam dan Tamadun Melayu 4(1):73. https://doi.org/10.17576/IMAN-2016-04SI1-08

Gepts P (2023) Biocultural diversity and crop improvement. Emerg Top Life Sci 7(2):151–196. https://doi.org/10.1042/ETLS20230067

Ghorbanzadeh Z, Panahi B, Purhang L, Hossein Panahi Z, Zeinalabedini M, Mardi M, Hamid R, Ghaffari MR (2025) Integrative genomics and precision breeding for stress-resilient cotton: recent advances and prospects. Agronomy 15(10):2393. https://doi.org/10.3390/agronomy15102393

Giakoumi S, Richardson AJ, Doxa A, Moro S, Andrello M, Hanson JO, Hermoso V, Mazor T, McGowan J, Kujala H, Law E, Álvarez-Romero JG, Magris RA, Gissi E, Arafeh-Dalmau N, Metaxas A, Virtanen EA, Ban NC, Runya RM, Dunn DC, Fraschetti S, Galparsoro I, Smith RJ, Bastardie F, Stelzenmüller V, Possingham HP, Katsanevakis S (2025) Advances in systematic conservation planning to meet global biodiversity goals. Trends Ecol Evol 40(4):395–410. https://doi.org/10.1016/j.tree.2024.12.002

Gkanogiannis A, Almansoori SR, Kabshawi M, Shahid M, Almansoori S, Rahman H, Becerra Lopez-Lavalle A (2025) Population genomics and genetic diversity of *Prosopis cineraria* in the United Arab Emirates: insights for conservation in arid ecosystems. Plants (Basel, Switzerland) 14(19):2970. https://doi.org/10.3390/plants14192970

Glennon KL, Le Roux JJ, Thompson DI (2023) Genetic insights into pepper-bark tree (*Warburgia salutaris*) reproduction in South Africa. Conserv Genet 24(6):883-891. https://doi.org/10.1007/s10592-023-01531-4

Gomes VHF, Andrino CO, Barbosa-Silva RG, Acosta AL, González-Chaves AD, Slätis J, Giannini TC (2025) Food insecurity under global human-induced changes: plants of the future in the Amazonian biome. Biol Conserv 311:111398. https://doi.org/10.1016/j.biocon.2025.111398

González AV, Gómez-Silva V, Ramírez MJ, Fontúrbel FE (2020) Meta-analysis of the differential effects of habitat fragmentation and degradation on plant genetic diversity. Conserv Biol 34(3):711–720. https://doi.org/10.1111/cobi.13422

Gupta N, Bhattacharya S, Dutta A, Cusimamani EF, Milella L, Leuner O (2024) In vitro synthetic polyploidization in medicinal and aromatic plants for enhanced phytochemical efficacy—a mini-review. Agronomy 14(8):1830. https://doi.org/10.3390/agronomy14081830

Gyorgy L, Ladnyi M, Redryc A (2014) Genetic survey of *Rhodiola rosea* L. populations from the Swiss Alps based on SSR markers. Biochem Syst Ecol 54:137–143. https://doi.org/10.1016/j.bse.2014.01.012

Hämälä T, Moore C, Cowan L, Carlile M, Gopaulchan D, Brandrud MK, Birkeland S, Loose M, Kolář F, Koch MA, Yant L (2024) Impact of whole-genome duplications on structural variant evolution in *Cochlearia*. Nat Commun 15(1):5377. https://doi.org/10.1038/s41467-024-49679-y

Hannweg K, Sippel A, Hofmeyr M, Froneman W (2016) Strategies for the conservation of *Warburgia salutaris* (family: Canellaceae), a red data list species – development of propagation methods. Acta Hortic. https://doi.org/10.17660/ActaHortic.2016.1125.4

Hao DC, Xiao PG (2015) Genomics and evolution in traditional medicinal plants: road to a healthier life. Evol Bioinformatics Online 11:197–212. https://doi.org/10.4137/EBO.S31326

Heinrich M (2013) Ethnopharmacology and drug discovery. In: Reedijk J (ed) Elsevier reference module in chemistry, molecular sciences and chemical engineering. Elsevier, Waltham. https://doi.org/10.1016/B978-0-12-409547-2.02773-6

Hendricks S, Anderson EC, Antao T, Bernatchez L, Forester BR, Garner B, Hand BK, Hohenlohe PA, Kardos M, Koop B, Sethuraman A, Waples RS, Luikart G (2018) Recent advances in conservation and population genomics data analysis. Evol Appl 11(8):1197–1211. https://doi.org/10.1111/eva.12659

Hernández-Delgado EA (2024) Coastal restoration challenges and strategies for small island developing states in the face of sea level rise and climate change. Coasts 4(2):235–286. https://doi.org/10.3390/coasts4020014

Heslop-Harrison JSP, Schwarzacher T, Liu Q (2023) Polyploidy: its consequences and enabling role in plant diversification and evolution. Ann Bot 131(1):1–10. https://doi.org/10.1093/aob/mcac132

Hoban S, da Silva JM, Hughes A, Hunter ME, Kalamujić Stroil B, Laikre L, Mastretta-Yanes A, Millette K, Paz-Vinas I, Ruiz Bustos L (2024) Too simple, too complex, or just right? Advantages, challenges, and guidance for indicators of genetic diversity. Bioscience 74(4):269–280. https://doi.org/10.1093/biosci/biae006

Hoffmann AA, Miller AD, Weeks AR (2020) Genetic mixing for population management: from genetic rescue to provenancing. Evol Appl 14(3):634–652. https://doi.org/10.1111/eva.13154

Hong JH, Lim AH, Kaewnarin K, Chan JY, Ng CCY, Teh BT (2024) Biodiversity medicine: new horizon and new opportunity for cancer. Cancer Discov 14(3):392–395. https://doi.org/10.1158/2159-8290.CD-23-1585

Jangpangi D, Patni B, Chandola V, Chandra S (2025) Medicinal plants in a changing climate: understanding the links between environmental stress and secondary metabolite synthesis. Front Plant Sci 16:1587337. https://doi.org/10.3389/fpls.2025.1587337

Jena M, Mohanta MR, Charles B, Aravind NA, Ravikanth G, Sahu SC (2023) Predicting potential distribution, range change and niche dynamics for *Saraca asoca* (Roxb.) De Wilde: a threatened medicinal plant under climatic change. Curr Sci 125(9):989–998. https://doi.org/10.18520/cs/v125/i9/989-998

Karalija E, Macanović A, Ibragić S (2025) Revisiting traditional medicinal plants: integrating multiomics, in vitro culture, and elicitation to unlock bioactive potential. Plants (Basel, Switzerland) 14(13):2029. https://doi.org/10.3390/plants14132029

Kardos M, Taylor HR, Ellegren H, Luikart G, Allendorf FW (2016) Genomics advances the study of inbreeding depression in the wild. Evol Appl 9(10):1205–1218. https://doi.org/10.1111/eva.12414

Khan AL, Al-Harrasi A, Wang J-P, Wang X-Y (2022) Genome structure and evolutionary history of frankincense producing *Boswellia sacra*. iScience 25(7):104574. https://doi.org/10.1016/j.isci.2022.104574

Kielich N, Mazur O, Musidlak O, Gracz-Bernaciak J, Nawrot R (2024) Herbgenomics meets Papaveraceae: a promising -omics perspective on medicinal plant research. Brief Funct Genomics 23(5):579–594. https://doi.org/10.1093/bfgp/elad050

Kougioumoutzis K, Tsakiri M, Kokkoris IP, Trigas P, Iatrou G, Lamari FN, Tzanoudakis D, Koumoutsou E, Dimopoulos P, Strid A, Panitsa M (2024) Assessing the vulnerability of medicinal and aromatic plants to climate and land-use changes in a Mediterranean biodiversity hotspot. Land 13(2):133. https://doi.org/10.3390/land13020133

Kress WJ, Erickson DL (2008) DNA barcodes: genes, genomics, and bioinformatics. Proc Natl Acad Sci USA 105(8):2761–2762. https://doi.org/10.1073/pnas.0800476105

Kricsfalusy V, Trevisan N (2014) Prioritizing regionally rare plant species for conservation using herbarium data. Biodivers Conserv 23(1):39–61. https://doi.org/10.1007/s10531-013-0583-4

Lee SR, Choi JE, Lee BY, Yu JN, Lim CE (2018) Genetic diversity and structure of an endangered medicinal herb: implications for conservation. AoB Plants 10(2):ply021. https://doi.org/10.1093/aobpla/ply021

Li XK, Wang B, Zheng YC, Song X, Wu YN, & Chen L (2014) Zhong yao cai = Zhongyaocai = J Chin Med Mater 37(4):568–572

Lee J-W, Kwon N, Kim J-U, Bang K-H, Jung SM, Lee SW, Kim D-H, Kim Y-C, Jo I-H, Park Y-D (2023) In vitro micropropagation of commercial ginseng cultivars (*Panax ginseng* Meyer) via somatic embryogenesis compared to traditional seed production. Horticulturae 9(4):435. https://doi.org/10.3390/horticulturae9040435

Lee SA, Seo H, Han S, Lee KJ, Kim JH, Park SY, Choi YJ, Lee JY (2024) Exploring genetic diversity of *Centella asiatica* (L.) Urb. across six Korean island regions: implications for conservation and utilization. Plant Genet Resour 22(1):1–3. https://doi.org/10.1017/S1479262123001090

León-Lobos P, Dávila P, Lima M (2012) The role of ex situ seed banks in the conservation of plant diversity and in ecological restoration in Latin America. Plant Ecol Divers 5(2):229–237. https://doi.org/10.1080/17550874.2012.713402

Li Q, Ge Y, Sayer JA (2023) Challenges to implementing the Kunming-Montreal Global Biodiversity Framework. Land 12(12):2166. https://doi.org/10.3390/land12122166

Li M, Wu JJ, Su RP, Fang OY, Cai X, Huang PH, Gao XY, Fu XX, Ma XH, He LY, Song YG, Hu GX, Zhou SS, Tan YH, Van de Peer Y, Li J, Wu SD, Meng HH (2025) Genome analyses provide insights into Engelhardia's adaptation to East Asia summer monsoon. Plant Divers 47(5):718–732. https://doi.org/10.1016/j.pld.2025.07.003

Liu D, Luo C, Dai R, Huang X, Chen X, He L, Mao H, Li J, Zhang L, Yang QY, Mei Z (2025) AMIR: a multi-omics data platform for Asteraceae plants genetics and breeding research. Nucleic Acids Res 53(D1):D1563–D1575. https://doi.org/10.1093/nar/gkae833

Ma Q, Wang Y, Zhu L, Bi C, Li S, Li S, Wen J, Yan K, Li Q (2019) Characterization of the complete chloroplast genome of *Acer truncatum* Bunge (Sapindales: Aceraceae): a new woody oil tree species producing nervonic acid. Biomed Res Int 2019:Article 7417239. https://doi.org/10.1155/2019/7417239

Ma T, Gao H, Zhang D, Shi Y, Zhang T, Shen X, Wu L, Xiang L, Chen S (2020) Transcriptome analyses revealed the ultraviolet B irradiation and phytohormone gibberellins coordinately promoted the accumulation of artemisinin in *Artemisia annua* L. Chin Med 15:67. https://doi.org/10.1186/s13020-020-00344-8

Ma L, Liu KW, Li Z, Zhang T, Zhang Y, Chen X (2023) Diploid and tetraploid genomes of *Acorus* and the evolution of monocots. Nat Commun 14:3661. https://doi.org/10.1038/s41467-023-38829-3

Maurya AKK, Kumar A (2024) The role of GIS in the study of sustainable development and environmental management. Int J Multidiscip Res 6(6):1–13. https://doi.org/10.36948/ijfmr.2024.v06i06.32879

Mbelebele Z, Mdoda L, Ntlanga SS, Nontu Y, Gidi LS (2024) Harmonizing Traditional Knowledge with Environmental Preservation: Sustainable Strategies for the Conservation of Indigenous Medicinal Plants (IMPs) and Their Implications for Economic Well Being. Sustain 16(14):5841. https://doi.org/10.3390/su16145841

Mehta P, Bisht K, Chandra Sekar K (2020) Diversity of threatened medicinal plants of Indian Himalayan region. Plant Biosyst 155(1). https://doi.org/10.1080/11263504.2020.1837278

Mohamed HI, Ullah I, Toor MD, Tanveer NA, Din MMU, Basit A, Sultan Y, Muhammad M, Rehman MU (2025) Heavy metals toxicity in plants: understanding mechanisms and developing coping strategies for remediation: a review. Bioresour Bioprocess 12(1):95. https://doi.org/10.1186/s40643-025-00930-4

Mondini L, Noorani A, Pagnotta MA (2009) Assessing plant genetic diversity by molecular tools. Diversity 1(1):19–35. https://doi.org/10.3390/d1010019

Murray KD, Janes JK, Jones A, Bothwell HM, Andrew RL, Borevitz JO (2019) Landscape drivers of genomic diversity and divergence in woodland *Eucalyptus*. Mol Ecol 28(24):5232–5247. https://doi.org/10.1111/mec.15287

Mykhailenko O, Jalil B, McGaw LJ, Echeverría J, Takubessi M, Heinrich M (2025) Climate change and the sustainable use of medicinal plants: a call for "new" research strategies. Front Pharmacol 15:1496792. https://doi.org/10.3389/fphar.2024.1496792

Naaz N, Choudhary S, Hasan N, Sharma N, Al Aboud NM, Shehata WF (2024) Biochemical and molecular profiling of induced high yielding M3 mutant lines of two *Trigonella* species: insights into improved yield potential. PLoS One 19(7):e0305691. https://doi.org/10.1371/journal.pone.0305691

Nag A, Ahuja PS, & Sharma RK (2014) Genetic diversity of high-elevation populations of an endangered medicinal plant. AoB PLANTS, 7:plu076. https://doi.org/10.1093/aobpla/plu076

Naik PK, Alam MA, Singh H, Goyal V, Parida S, Kalia S, Mohapatra T (2010) Assessment of genetic diversity through RAPD, ISSR and AFLP markers in *Podophyllum hexandrum*: a medicinal herb from the Northwestern Himalayan region. Physiol Mol Biol Plants 16(2):135–148. https://doi.org/10.1007/s12298-010-0015-9

Nair RR, Ravikanth G, Udayan PS (2024) Conservation management of Saraca asoca (Roxb.) W. J. de Wilde (Fabaceae) using ecological niche modeling. Trop Ecol 65:239–257. https://doi.org/10.1007/s42965-024-00329-w

Napolitano C, Clavijo C, Rojas-Bonzi V, Miño CI, González-Maya JF, Bou N, Giraldo A, Martino A, Miyaki CY, Aguirre LF, Cosacov A, Milián-García Y, Prosdocimi L, Ramírez-Bravo OE, Tovar LA, Velez-Zuazo X, Barrios M, Herrera-Fernández B, Montiel-Villalobos MG, Oliveira-Miranda MA et al (2024) Understanding the conservation-genetics gap in Latin America: challenges and opportunities to integrate genetics into conservation practices. Front Genet 15:1425531. https://doi.org/10.3389/fgene.2024.1425531

Nazar N, Howard C, Slater A, Sgamma T (2022) Challenges in medicinal and aromatic plants DNA barcoding—lessons from the Lamiaceae. Plants 11(1):137. https://doi.org/10.3390/plants11010137

Ndhlala AR, Işık M, Kavaz Yüksel A, Dikici E (2024) Phenolic content analysis of two species belonging to the Lamiaceae family: antioxidant, anticholinergic, and antibacterial activities. Molecules 29(2):480. https://doi.org/10.3390/molecules29020480

Neergheen-Bhujun V, Awan AT, Baran Y, Bunnefeld N, Chan K, Dela Cruz TE, Egamberdieva D, Elsässer S, Johnson MV, Komai S, Konevega AL, Malone JH, Mason P, Nguon R, Piper R, Shrestha UB, Pešić M, Kagansky A (2017) Biodiversity, drug discovery, and the future of global health: introducing the biodiversity to biomedicine consortium, a call to action. J Glob Health 7(2):020304. https://doi.org/10.7189/jogh.07.020304

Newman DJ, Cragg GM (2012) Natural products as sources of new drugs over the 30 years from 1981 to 2010. J Nat Prod 75(3):311–335. https://doi.org/10.1021/np200906s

Nizam A, Meera SP, Kumar A (2021) Genetic and molecular mechanisms underlying mangrove adaptations to intertidal environments. iScience 25(1):103547. https://doi.org/10.1016/j.isci.2021.103547

Nourozi E, Hedayati A, Madani H, Hosseini B, Hemmaty S (2025) In vitro synthetic polyploidization and enhancement of anticancer compounds in *Catharanthus Roseus* (L.) G. Don important cultivars. Sci Rep 15(1):6563. https://doi.org/10.1038/s41598-025-91103-y

Nwankwegu AS, Hongsanan S, Nwankwegu UP, Xie N (2025) Exploring the critical environmental optima and biotechnological prospects of fungal fruiting bodies. Microb Biotechnol 18(8):e70210. https://doi.org/10.1111/1751-7915.70210

Ogwu MC, Izah SC, Joshua MT (2025) Ecological and environmental determinants of phytochemical variability in forest trees. Phytochem Rev. https://doi.org/10.1007/s11101-025-10066-0

Pandey MM, Rastogi S, Rawat AK (2013) Indian traditional ayurvedic system of medicine and nutritional supplementation. Evid Based Complement Alternat Med 2013:376327. https://doi.org/10.1155/2013/376327

Pandith SA, Dar RA, Lattoo SK, Shah MA, Reshi ZA (2018) *Rheum australe*, an endangered high-value medicinal herb of North Western Himalayas: a review of its botany, ethnomedical uses, phytochemistry and pharmacology. Phytochem Rev 17(3):573–609. https://doi.org/10.1007/s11101-018-9551-7

Phillips RD, Reiter N, Peakall R (2020) *Orchid conservation*: from theory to practice. Ann Bot 126(3):345–362. https://doi.org/10.1093/aob/mcaa093

Popoola JO, Igwe DO, Iwu V, Adegbite AE, Omonhinmin CA (2020) Genetic diversity, clustering pattern and breeding implications of 40 *Moringa oleifera* accessions using AFLP markers. J Agric Res Dev Trop Subtrop 12:659–670. https://doi.org/10.5373/JARDCS/V12SP8/20202568

Póvoa O, Lopes V, Barata AM, Farinha N (2023) Monitoring genetic erosion of aromatic and medicinal plant species in Alentejo (South Portugal). Plants (Basel, Switzerland) 12(14):2588. https://doi.org/10.3390/plants12142588

Qin Y, Fan K, Yimamu A, Zhan P, Lv L, Li G, Liu J, Hu Z, Yan X, Hu X, Liu H, Qin R (2025a) Integrated genetic diversity and multi-omics analysis of colour formation in safflower. Int J Mol Sci 26(2):647. https://doi.org/10.3390/ijms26020647

Qin Q, Dong Y, He J, Chen J, Wu D, Zhang S (2025b) Assessment of genetic diversity and construction of core germplasm in populations of *Acorus tatarinowii* based on SNP markers. J Appl Res Med Aromat Plants 44:100605

Rahnama A, Salehi F, Meskarbashee M, Mehdi Khanlou K, Ghorbanpour M, Harrison MT (2024) High temperature perturbs physicochemical parameters and fatty acids composition of safflower (*Carthamus tinctorius* L.). BMC Plant Biol 24(1):1080. https://doi.org/10.1186/s12870-024-05781-3

Rajpal VR, Rathore P, Mehta S, Wadhwa N, Yadav P, Berry E, Goel S, Bhat V & Raina SN (2022) Epigenetic variation: A major player in facilitating plant fitness under changing environmental conditions. Front Cell Dev Biol 10:1020958. https://doi.org/10.3389/fcell.2022.1020958

Rajasekharan PE, Wani SH (eds) (2020) Conservation and utilization of threatened medicinal plants. Springer Nature Switzerland. https://doi.org/10.1007/978-3-030-39793-7

Ramos da Silva LR, Ferreira OO, Cruz JN, de Jesus Pereira Franco C, Oliveira Dos Anjos T, Cascaes MM, Almeida da Costa W, Helena de Aguiar Andrade E, Santana de Oliveira M (2021) Lamiaceae essential oils, phytochemical profile, antioxidant, and biological activities. Evid Based Complement Alternat Med 2021:6748052. https://doi.org/10.1155/2021/6748052

Reed DH, Frankham R (2003) Correlation between fitness and genetic diversity. Conserv Biol 17(1). https://doi.org/10.1046/j.1523-1739.2003.01236.x

Reid BN, Hofmeier J, Crockett H, Fitzpatrick R, Waters R, Fitzpatrick SW (2025) Balancing inbreeding and outbreeding risks to inform translocations throughout the range of an imperiled darter. Evol Appl 18(3):e70088. https://doi.org/10.1111/eva.70088

Roopashree S, Anitha J, Challa S, Mahesh TR, Venkatesan VK, Guluwadi S (2024) Mapping of soil suitability for medicinal plants using machine learning methods. Sci Rep 14:Article 3741. https://doi.org/10.1038/s41598-024-3741

Ruiz-Mallén I, Corbera E (2013) Community-based conservation and traditional ecological knowledge: implications for social-ecological resilience. Ecol Soc 18(4):Article 19. https://www.jstor.org/stable/26269393

Salazar D, Lokvam J, Mesones I, Vásquez Pilco M, Ayarza Zuñiga JM, de Valpine P, Fine PV (2018) Origin and maintenance of chemical diversity in a species-rich tropical tree lineage. Nat Ecol Evol 2(6):983–990. https://doi.org/10.1038/s41559-018-0552-0

Salgotra RK, Chauhan BS (2023) Genetic diversity, conservation, and utilization of plant genetic resources. Genes 14(1):174. https://doi.org/10.3390/genes14010174

Salmerón-Manzano E, Garrido-Cardenas JA, Manzano-Agugliaro F (2020) Worldwide research trends on medicinal plants. Int J Environ Res Public Health 17(10):3376. https://doi.org/10.3390/ijerph17103376

Sampaio ACF, Hoffmann PM, Gurski EM, Bizarro OR, Velazco SJE, Carlucci MB, Borgo M, Ferreira MM, Angelo AC, Blum CT (2025) Filling knowledge gaps: a methodological approach for the rapid investigation of species with restricted populations. Perspect Ecol Conserv 23(2):130–140. https://doi.org/10.1016/j.pecon.2025.04.003

Sang T, Crawford D, Stuessy T (1997) Chloroplast DNA phylogeny, reticulate evolution, and biogeography of *Paeonia* (Paeoniaceae). Am J Bot 84(8):1120

Saxena M, Saxena J, Nema R, Singh D, Gupta A (2013) Phytochemistry of medicinal plants. J Pharmacogn Phytochem 1(6):168–182

Schaart JG, van de Wiel CCM, Smulders MJM (2021) Genome editing of polyploid crops: prospects, achievements and bottlenecks. Transgenic Res 30(4):337–351. https://doi.org/10.1007/s11248-021-00251-0

Sengar RS, Sengar K (eds) (2014) Climate change effect on crop productivity, 1st edn. CRC Press. https://doi.org/10.1201/b17684

Shakuri F, Eghlima G, Behboudi H, Babashpour-Asl M (2025) Phytochemical variation, phenolic compounds and antioxidant activity of wild populations of Iranian oak. Sci Rep 15(1):6534. https://doi.org/10.1038/s41598-025-90991-4

Sharma S, Chhabra M, Singh SK, Parmar R, Kapila RK (2022) Genetic diversity and population structure of critically endangered *Dactylorhiza hatagirea* (D. Don) Soo from North-Western Himalayas and implications for conservation. Sci Rep 12(1):11699. https://doi.org/10.1038/s41598-022-15742-1

Shaw RE, Brockett B, Pierson JC, Sarre SD, Doyle P, Cliff HB, Eldridge MDB, Miller KA, Ottewell K, Parrott ML, Rossetto M, Sunnucks P, Banks SC, Byrne M, Campbell BL, Chong C, Ens EJ, Ferraro PA, Grueber CE, Harrison PA, Hogg C, Hutton M, Roycroft E, Sowersby W, Rangers K, Rangers K, Rangers N, Rangers N, Rangers N, Neaves LE (2024) Building meaningful collaboration in conservation genetics and genomics. Conserv Genet 25(5):1127–1145. https://doi.org/10.1007/s10592-024-01636-4

Shukla S (2023) National parks of India and their role in in-situ conservation. Int J Bot Stud 8(6):49–54

Smith AL, Hodkinson TR, Villellas J, Catford JA, Csergő AM, Blomberg SP, Crone EE, Ehrlén J, Garcia MB, Laine AL, Roach DA, Salguero-Gómez R, Wardle GM, Childs DZ, Elderd BD, Finn A, Munné-Bosch S, Baudraz MEA, Bódis J, Brearley FQ et al (2020) Global gene flow releases invasive plants from environmental constraints on genetic diversity. Proc Natl Acad Sci USA 117(8):4218–4227. https://doi.org/10.1073/pnas.1915848117

Smith-Hall C, Pyakurel D, Treue T, Pouliot M, Ghimire S, Timoshyna A, Meilby H (2025) A roadmap to sustainable management of commercial medicinal and aromatic plants, fungi, and lichens in Nepal. Conserv Biol 39(4):e14442. https://doi.org/10.1111/cobi.14442

Stem C, Margoluis R, Salafsky N, Brown M (2005) Monitoring and evaluation in conservation: a review of trends and approaches. Conserv Biol 19(2):295–309. https://doi.org/10.1111/j.1523-1739.2005.00594.x

Stephens T (2023) The Kunming–Montreal Global Biodiversity Framework. Int Leg Mater 62(5):1–20. https://doi.org/10.1017/ilm.2023.16

Tang CQ, Yang Y, Ohsawa M, Yi SR, Momohara A, Su WH, Wang HC, Zhang ZY, Peng MC, Wu ZL (2012) Evidence for the persistence of wild *Ginkgo biloba* (Ginkgoaceae) populations in the Dalou Mountains, southwestern China. Am J Bot 99(8):1408–1414. https://doi.org/10.3732/ajb.1200168

Terletskaya NV, Turzhanova AS, Khapilina ON, Zhumagul MZ, Meduntseva ND, Kudrina NO, Korbozova NK, Kubentayev SA, Kalendar R (2023) Genetic diversity in natural populations of *Rhodiola* species of different adaptation strategies. Genes 14(4):794. https://doi.org/10.3390/genes14040794

Thapa B, Upadhyay K, Chhetri B, Gurung N, Meena NK (2024) From wild to cultivated: unraveling the potential of Swertia chirayita (Roxb.) H.Karst. for crop improvement and conservation. In: Mishra AP, Kumar A, Chandra N, Singh G, Pande CB (eds) Threatened medicinal plants in the Indian Himalayan region. World sustainability series. Springer, Cham. https://doi.org/10.1007/978-3-031-73687-2_10

Theissinger K, Fernandes C, Formenti G, Amorim IR (2023) How genomics can help biodiversity conservation. Trends Genet 39(7):Article 001. https://doi.org/10.1016/j.tig.2023.01.005

Tiwari P, Park KI (2024) Advanced fungal biotechnologies in accomplishing sustainable development goals (SDGs): what do we know and what comes next? J Fungi (Basel, Switzerland) 10(7):506. https://doi.org/10.3390/jof10070506

Tiwari D, Kewlani P, Gaira KS, Bhatt ID, Sundriyal RC, Pande V (2023) Predicting phytochemical diversity of medicinal and aromatic plants (MAPs) across eco-climatic zones and elevation in Uttarakhand using Generalized Additive Model. Sci Rep 13:10888. https://doi.org/10.1038/s41598-023-37495-1

Turner BL II, Kasperson RE, Matson PA, Schiller A, the other authors. (2003) A framework for vulnerability analysis in sustainability science. Proc Natl Acad Sci 100(14):8074–8079. https://doi.org/10.1073/pnas.1231335100

Vajana E, Bozzano M, Marchi M, Piotti A (2023) On the inclusion of adaptive potential in species distribution models: towards a genomic-informed approach to forest management and conservation. Environments 10(1):3. https://doi.org/10.3390/environments10010003

Vashisth D, Kumar R, Rastogi S, Patel VK, Kalra A, Gupta MM, Gupta AK, Shasany AK (2018) Transcriptome changes induced by abiotic stresses in *Artemisia annua*. Sci Rep 8(1):3423. https://doi.org/10.1038/s41598-018-21598-1

Vasileva Y, Zhulanov A, Chertov N, Sboeva Y, Boronnikova S, Pechenkina V, Nechaeva Y, Kalendar R (2024) Identification of SNPs associated with drought resistance in hybrid populations of *Picea abies* (L.) H. Karst.–*P. obovata* (Ledeb.). Genes 15(11):1440. https://doi.org/10.3390/genes15111440

Verma P, Siddiqui AR, Mourya NK, Devi AR (2024) Forest carbon sequestration mapping and economic quantification infusing MLPnn-Markov chain and InVEST carbon model in Askot Wildlife Sanctuary, Western Himalaya. Ecol Inform 79:102428. https://doi.org/10.1016/j.ecoinf.2023.102428

Wang SQ (2020) Genetic diversity and population structure of the endangered species Paeonia decomposita endemic to China and implications for its conservation. BMC Plant Biol 20(1):510. https://doi.org/10.1186/s12870-020-02682-z

Wang S, Zhang D, Jiao X, Wu L, Zhu QH, Lv H, Wang H, Han Z, Li S, He P, Chen J, Feng S, Olsson U, Alström P, Lei F (2025) Conservation genomics of two endangered buntings reveal genetic diversity before and after severe population declines. BMC Biol 23(1):287. https://doi.org/10.1186/s12915-025-02393-7

Wanjohi BK, Sudoi V, Njenga EW, Kipkore WK (2020) An ethnobotanical study of traditional knowledge and uses of medicinal wild plants among the Marakwet Community in Kenya. Evid Based Complement Alternat Med 2020:3208634. https://doi.org/10.1155/2020/3208634

Xu M (2025) Enhancing GIS models for sustainable development in human settlements using intelligent IoT infrastructure and human-machine interaction. Results Eng 27:Article 106713. https://doi.org/10.1016/j.rineng.2025.106713

Xu Z, Gao R, Pu X, Xu R, Wang J, Zheng S, Zeng Y, Chen J, He C, Song J (2020) Comparative genome analysis of *Scutellaria baicalensis* and *Scutellaria barbata* reveals the evolution of active flavonoid biosynthesis. Genom Proteom Bioinform 18(3):230–240. https://doi.org/10.1016/j.gpb.2020.06.002

Yu HJ, Byun YH, Park CK (2024) Techniques for assessing telomere length: a methodological review. Comput Struct Biotechnol J 23:1489–1498. https://doi.org/10.1016/j.csbj.2024.04.011

Zamani S, Fathi M, Ebadi M-T, Máthé Á (2025) Global trade of medicinal and aromatic plants: a review. J Agric Food Res 21(5):101910. https://doi.org/10.1016/j.jafr.2025.101910

Zhang XG, Han T, He ZG, Li XM (2012) Genetic diversity of *Centella asiatica* in China analyzed by inter-simple sequence repeat (ISSR) markers: combination analysis with chemical diversity. J Nat Med 66:241–247. https://doi.org/10.1007/s11418-011-0572-4

Zhang L, Rammitsu K, Tetsuka K, Yukawa T, Ogura-Tsujita Y (2022) Dominant *Dendrobium officinale* mycorrhizal partners vary among habitats and strongly induce seed germination in vitro. Front Ecol Evol 10:Article 994641. https://doi.org/10.3389/fevo.2022.994641

Zhang S, Wang S, Wang S, Su H, De J (2025) Genetic diversity and metabolic profile of Tibetan medicinal plant *Saussurea obvallata*. Genes 16(5):593. https://doi.org/10.3390/genes16050593

Zhao R, Huang N, Zhang Z, Luo W, Xiang J, Xu Y et al (2024) Genetic diversity analysis and prediction of potential suitable areas for the rare and endangered wild plant *Henckelia longisepala*. Plants 13(15):2093. https://doi.org/10.3390/plants13152093

Zhou J, Liu W, Kong H, Gong W (2018) Identification and characterization of microsatellites in *Aconitum reclinatum* (Ranunculaceae), a rare species endemic to North America. Appl Plant Sci 6(6):e01161. https://doi.org/10.1002/aps3.1161

Chapter 17
Applications and Future Potential of Genetic Diversity in Medicinal Plant Improvement Through Traditional Breeding and Transgenic Techniques

Mihir Halder

Abstract Medicinal plants have significant genetic diversity, which helps them to survive in different environments and produce a wide range of bioactive phytochemicals. Humanity has depended on this evolving diversity of genetic blueprints for sustainable healthcare and innovation for medicinal purposes. The understanding and characterization of this genetic diversity are essential for targeted manipulation of medicinal plant genomes through traditional breeding, biotechnology, and molecular biology, which hold promising implications for the rapid development of new and improved medicinal plant varieties that minimize the gap between availability and the growing global demand for plant-derived medicine. Future research should focus on developing new breeding programs that conserve and use the genetic diversity of medicinal plants to improve yield, quality, and disease resistance. This will help the global health sector by providing a more reliable and uninterrupted supply chain for plant-based medicines. Such improvements reduce pressure on vulnerable wild populations and also help protect the global biodiversity of medicinal plant species. Despite the focus on a few high-performing genotypes, characterizing unexploited genetic pools is essential for future use and for protection against unexpected disease and climate change. This chapter outlines methods for improving medicinal plants through a comprehensive understanding of genetics and reviews recent advancements in the production of secondary metabolites via breeding and biotechnological approaches, highlighting the role of genetic diversity in enhancing medicinal plant biotechnology. Using genetic diversity to make desired plants also comes with responsibilities of balancing with a commitment to sustainable sourcing, conservation of wild populations, and equitable benefit sharing with indigenous communities. Ethical considerations and intellectual

M. Halder (✉)
Post-Graduate Department of Botany, Barasat Government College,
Kolkata, West Bengal, India
e-mail: mihir.halder@bgc.ac.in

S. Jha et al. (eds.), *Genetic Diversity in Medicinal Plants*, Sustainable Development and Biodiversity 44, https://doi.org/10.1007/978-981-95-7281-6_17

property rights related to the use of genetic resources from indigenous communities also require careful attention.

Keywords Traditional breeding · Transgenic techniques · Genetic diversity · Medicinal plant improvement

17.1 Introduction

Medicinal plant species harbor a wide variety of genetic diversity that is responsible for morphological, physiological, anatomical, and biochemical differences. This genetic variability within and between species is essential for their ecological resilience, adaptability, and phytochemical richness (Jadaun et al. 2023; Malakar et al. 2025). From a pharmacological perspective, these plants are rich living repositories of valuable, unique genes and metabolic pathways that determine the range and variability of distinct bioactive phytochemical profiles in terms of both quantity and quality. Variability in phytochemical profiles among different species and even among different accessions within the same species is common in medicinal plants (Tiwari et al. 2024). This variability has evolved over millions of years through mutation, gene flow, and natural selection. This genetic heterogeneity forms the essential foundational chemical arsenal for traditional healing systems and provides a broad spectrum of chemical scaffolds for the discovery of new life-saving drugs and therapeutic products. Humanity has relied on this vast and evolving diversity of genetic blueprints for sustainable healthcare and innovation for medicinal purposes (Nasim et al. 2022; Monib 2024).

In the modern era, effective understanding and characterization of genetic diversity are essential for targeted manipulation of medicinal plant genomes or genes through traditional plant breeding, modern biotechnology, and molecular biology to create the desired plant (Niazian 2019; Chaturvedi et al. 2022). By combining several DNA-based molecular marker systems with advanced techniques like next-generation sequencing, genome-wide association studies, marker-assisted selection, transcriptomics, and metabolomics, researchers can assess and compare complex inter- and intra-specific genetic information (genes, regulatory elements, and variations in gene copy number) and their expression patterns at transcriptional, translational, and biosynthesis pathway levels (Pandita et al. 2021; Sharma et al. 2023). This provides a detailed view of the genetic landscape of medicinal plants and establishes a direct link between this genetic diversity and functional innovation.

Identification of the “elite genotypes” and their unique alleles or genes from high-value chemovars or germplasms responsible for rare higher yields, novel compounds, and better survival under environmental stress, or superior agronomic traits is very important for cultivation and genetic manipulation (Mayer et al. 2017; Saran et al. 2022; Wang et al. 2023). Once these genes are identified, they can be introgressed by crossing or cloned into cultivated varieties to ensure more consistent medicinal properties and reduce dependence on wild harvesting (Chaturvedi et al.

2022). This large pool of genetic variation provides a large library of molecular blueprints that can lead to compounds with therapeutic or commercial relevance. Rapid growth, plant propagation, as well as the production of similar or enhanced secondary metabolites compared to those produced by various medicinal plants grown in the wild or in field cultivation, have also been achieved in various medicinal plants using different biotechnology approaches (Marthe 2018; Mahendran and Vimolmangkang 2024; Sozoniuk et al. 2025). The genotype of the explant material is a critical determinant of morphogenetic and biochemical responses in plant tissue culture systems. If the starting explant material does not have the genetic capacity to produce a high-level target metabolite, the impacts of other modifications, such as media optimization or bioreactor fine-tuning, are unable to compensate for this deficiency. This would restrict its usefulness as a viable commercial platform for bioactive compound production.

As biotechnology advances, the link between genetic diversity and plant improvement will become even more dynamic. The success of these innovations will rely not only on cutting-edge laboratory techniques but also on our understanding and utilization of the natural variability stored in plant DNA in natural plant populations. This diversity can be used as a reservoir of raw materials for the strategic selection of the best alleles or genes. Such approach supports the conservation, breeding, and improvement of plants through genetic engineering efforts to meet growing industrial demands (Chaudhary et al. 2025). Genetic engineering allows the enhancement of valuable compounds production, the introduction of novel traits, or even the production of complex pharmaceuticals in heterologous systems (Sharma et al. 2018; Kulagina et al. 2021; Zhang et al. 2022; Zeng et al. 2022; Zhao et al. 2022).

Several studies have demonstrated that climate change significantly influences the phytochemical profile in different medicinal plants as a protective response (Pellegrini et al. 2018; Rastogi et al. 2019; Zhang et al. 2019; Cahyaningsih et al. 2021). In an era of climate uncertainty and emerging new diseases, phytochemical diversity helps plants in their adaptation and survival against biotic and abiotic stress (Jangpangi et al. 2025). Populations with a rich genetic pool are more likely to survive these changes because genetic diversity acts as a source of phytochemical diversity. The use of genetic diversity in conservation strategies and plant improvement is crucial for the enhancement of the sustainability and resilience of medicinal plant cultivation, as many wild species face critical threats from habitat loss, overharvesting, and climate change (Zaman et al. 2025). Maintaining and using diverse genetic material from wild relatives will help breeders create medicinal crops that can adapt to changing climates without compromising their pharmacological potency. Plant improvement programs can use this genetic variety to increase both the quantity and quality of medicinal compounds (Venkatesha et al. 2020; Dhawan et al. 2022; Shibano et al. 2024). In traditional breeding, scientists select plants with high levels of beneficial phytochemicals or superior disease resistance from diverse populations, and develop new plants with superior traits by cross-breeding (Gupta et al. 2016; Jhang et al. 2018; Dhawan et al. 2019; Dhawan et al. 2022; Shibano et al. 2024). By identifying and conserving these diverse genetic resources, researchers can develop cultivars that are more resistant to pests and diseases, more tolerant to adverse environmental conditions, and thus more suitable for cultivation.

This chapter discusses the goals and methods for improving medicinal plants, along with the importance of understanding genes, alleles, and regulatory elements for effective improvement of medicinal plants by plant breeding, genetic engineering, and metabolic engineering. The chapter also reviews recent advances in secondary metabolite production in medicinal plants achieved through breeding and biotechnological interventions, where genetic diversity serves as the foundation for innovative applications in genetic engineering in medicinal plant biotechnology.

17.2 Methods of Medicinal Plant Improvement

Targeted improvement of medicinal plants has been done using traditional plant breeding methods and modern biotechnological approaches, including transgenic techniques and genome editing (Dhawan et al. 2022; Zhao et al. 2022; Zeng et al. 2022; Shibano et al. 2024). Each method has its own advantages and disadvantages. Traditional plant breeding includes approaches like selective breeding, inter- and intra-specific hybridization, polyploidy breeding, and mutational breeding (Hiyama et al. 2017; Jhang et al. 2018; Kolakar et al. 2018; Madani et al. 2021; Shibano et al. 2024). All these methods use the existing natural genetic diversity of medicinal plants as raw material to achieve desired goals. In selective breeding, systematic evaluation is carried out to identify and propagate individuals with high levels of bioactive phytochemicals from diverse heterogeneous populations. Mass selection, pure line selection (for self-pollinated medicinal plants), and clonal selection (for vegetatively propagated plants) can be done to stabilize desirable quantitative traits over many generations by using the selected plants as parents for the next generation. In contrast, hybridization creates new genetic combinations in progenies through the crossing of two genetically different species within the same genus (interspecific hybridization) or different varieties of the same species (intraspecific hybridization), followed by intensive, repeated cycles of selfing and selection across multiple generations. Hybridization combines and rearranges genes that prevail among wild populations, landraces, and cultivated varieties of medicinal plants to produce desired results (De 2017; Johnson and Franz 2022). Polyploidy breeding utilizes or induces the presence of more than two complete sets of chromosomes, which can lead to increased biomass or secondary metabolite concentrations (Madani et al. 2021). When the existing natural diversity has a narrow genetic base and fails to provide the desired genetic composition, breeders use mutational breeding to induce new genetic variation by using mild physical radiation (e.g., gamma rays) or chemical mutagens (Kolakar et al. 2018). These mutagens are used to accelerate the natural mutation rate at random sites and develop new variations that do not exist in current populations. Among the new variations the desired variations are selected. In general, traditional plant breeding programs are time-consuming and require rigorous progeny testing, an indispensable step for validating and refining desirable traits.

Transgenic technology, or genetic engineering and genome editing, is now a very common biotechnological technique applied in medicinal plant improvement (Sharma et al. 2018; Zeng et al. 2022; Mishra et al. 2023; Das et al. 2024; Qamar et al. 2024). Transgenic technology involves the transfer and integration of specific foreign gene(s) into the genome of the host plant to develop new traits or improve existing ones. The development of transgenic plants and transgenic (unorganized and organized) cultures of medicinal plants is the most commonly exploited transgenic systems to achieve various targeted goals (Halder et al. 2021; Balasubramani et al. 2023). Stable integration of specific foreign DNA (transgene) derived from any source (plant, animal, or microorganism) into the plant genome across sexually incompatible barriers has been achieved through biolistic- and/or *Agrobacterium*-mediated transformation in numerous medicinal plants (Halder et al. 2021; Pawar et al. 2025). Once integrated, the transgene is expressed under the control of specific promoters and terminators, leading to the expression of the desired protein or RNA molecules that develop new traits or modify existing traits. This shift fundamentally accelerates the development of a superior, stable, high-yielding transgenic system by improving yields and compound profiles of medicinal plants (Bai et al. 2019).

Metabolic engineering is a specific application of genetic engineering that aims to improve the production of desired compounds by manipulating and optimizing the biosynthesis pathways of desired secondary metabolites. This may include increasing the production of useful compounds by manipulating genes encoding enzymes within specific metabolic pathways to increase the flux toward target compounds, divert resources from unwanted pathways, or introduce a non-native pathway for synthesis (Sharma et al. 2018; Zeng et al. 2022; Das et al. 2024). Unlike the random genetic variation introduced by mutation breeding or classical hybridization, transgenic technology and metabolic engineering allow researchers to manipulate linear or branched metabolic pathways at different key regulatory checkpoints. This can be performed by the overexpression of rate-limiting enzyme(s) to increase metabolic flux toward a desired compound (Cui et al. 2015; Jiang et al. 2016), by silencing competing pathways with the help of RNA interference (RNAi) to redirect precursor molecules (Pal et al. 2019), and by manipulating transcription factor(s) that control entire biosynthetic pathways (Alfieri et al. 2018). The introduction of novel single or multiple genes from other organisms (plants, microbes) or the construction of entirely novel biosynthetic pathways within the medicinal plant facilitates the creation of new chemical profiles. However, the application of both technologies often faces public concerns and regulatory challenges.

Genome editing is another molecular technique of genetic improvement that can be performed using meganucleases, ZFNs (zinc finger nucleases), TALENs (transcription activator-like effector nucleases), and clustered regularly interspaced short palindromic repeats-Cas 9 (CRISPR/Cas9) for the improvement of medicinal plants (Mishra et al. 2023). Among these, CRISPR-Cas9 is preferred due to its high precision and direct genetic modification or editing ability of a plant's indigenous genome in a cost-effective manner (Guo et al. 2022; Das et al. 2024). It has great potential for targeted knockouts of inhibitory genes, subtle base edits, or the insertion of beneficial alleles. This method allows for targeted improvement of metabolic pathways in medicinal plants along with genome functional studies, synthetic biology, and germplasm innovation (Guo et al. 2022). It can lead to quicker regulatory approval since it does not involve the integration of foreign DNA.

Regardless of the approach used, genetic diversity plays an important role in each intervention to achieve specific goals. Although these goals include enhancing the yield of bioactive compounds, optimizing phytochemical profiles, developing resistance against biotic and abiotic factors, improving nutritional quality through biofortification, overcoming domestication bottlenecks, ensuring long-term sustainability through effective conservation, and improving agronomic traits, this chapter emphasizes the improvement of bioactive compounds in medicinal plants. Medicinal plants often exhibit significant chemotypic variation due to environmental factors, geographical origin, and genetic heterogeneity. The primary objective is to develop stable high-yielding cultivars with optimized phytochemical profiles by targeting the biosynthesis. It ensures optimum accumulation of desired active constituents. Ensuring consistency in bioactive secondary metabolite profiles and minimizing the production of undesirable or toxic compounds through precise modification of metabolic pathways in medicinal plants is essential for quality control, safety, and regulatory approval of pharmaceutical products. Furthermore, these improvements are vital for developing standardized extracts and formulations with reproducible therapeutic outcomes. Batch-to-batch uniformity in medicinal compound profiles is essential for their wider clinical applications and integration into modern healthcare systems, achieving pharmaceutical-grade quality and patient trust. Breeders also focus on developing improved varieties with enhanced agronomic traits, such as higher biomass yield, increased resistance to pests and diseases, and enhanced tolerance to various abiotic stresses through targeted breeding efforts to promote sustainable and efficient cultivation. Such advancements reduce cultivation costs and resource consumption, thereby improving the economic viability of growers. Additionally, they contribute significantly to ecological sustainability by minimizing the dependence on synthetic pesticides and fertilizers and reducing pressure on wild plant populations. Another important goal is to improve safety by reducing harmful substances or anti-nutritional factors in plants, thereby meeting strict safety and purity standards for medicines and healthcare products (Zakaria et al. 2021). Overall, these improvement efforts are essential for meeting future demands for effective, safe, and sustainably produced phytopharmaceuticals, supporting sustainable farming, improving economic outcomes for farmers, and conserving plant diversity.

17.3 Application of Genetic Diversity of Medicinal Plants in Plant Breeding

Classical plant breeding is an applied genetic discipline dedicated to systematic improvement of plant species and development of new plant varieties through different methods of intensive selection, hybridization, and inducing new variations (via mutation breeding) to achieve specific goals. The success of the green revolution in crop plants shows the transformative ability of directed plant breeding in global food security (Eliazer Nelson et al. 2019). Despite significant economic and social benefits associated with medicinal plant breeding, this field has not received as much attention or support as crop plant breeding (Wang et al. 2020). Breeding and cultivation of medicinal plants are often limited to specific regions in comparison to crop plants. The dependency on wild populations for commercial production of plant-derived herbal remedies and modern medicines significantly impacts their global biodiversity, future availability, and the sustainable use of these resources (Wang et al. 2020; Turgut et al. 2023). Medicinal plant breeding and cultivation deserve similar scientific attention as food crops, especially given the increasing demand for plant-derived phytochemicals (Astutik et al. 2019) and the continuous emergence of new diseases (Halder et al. 2023). To ensure global health security, protect medicinal plant biodiversity, and promote sustainable use, a phytochemical revolution in the pharmaceutical sector is urgently needed in the current scenario.

Medicinal plant breeding approaches have a great scope for improvement in pharmaceutical output (Ramawat and Arora 2021). The availability of rich genetic variations at the allele, gene, and regulatory element levels within medicinal plant species is a fundamental prerequisite for successful and sustainable plant breeding. The extent and nature of preexisting, naturally evolved genetic diversity directly influence and guide traditional breeding strategies, as this diversity is used as the foundational source from which suitable parent plants are selected. Selection of appropriate parent plants is a key step in the breeding program, since the genetic makeup of the offspring depends entirely on the genetic characteristics of the parent plants. Various plant breeding approaches have made some important progress in transforming valuable wild medicinal plants into stable, high-yielding, and environmentally robust varieties by overcoming the existing genetic bottleneck of low productivity in their wild populations (Table 17.1). The combination of traditional knowledge with modern science has improved both the biomass and phytochemical yield in some medicinal plant species. These quantitative improvements are essential for meeting the rapidly escalating global demand for plant-based medicines using fewer resources and ensuring quality for medical use.

Methodologically, the field has undergone a profound transformation that includes the shift from purely conventional breeding based on phenotypic selection and pedigree tracking to integrated molecular breeding strategies, such as marker-assisted selection and genomic selection (Song and Li 2021; Kumar et al. 2025). The development of genomic and molecular screening techniques makes the preselection of superior genotypes faster and more efficient by allowing precise tracking and identifying desirable genes within a diverse population. It bypasses the time-consuming field-based phenotyping and accelerates breeding cycles. This approach helps in more efficient utilization of genetic resources and accelerates the introgression of desirable traits while minimizing the transmission of unwanted alleles in newly developed varieties compared to traditional methods.

Additionally, current breeding efforts focus on creating improved resistance in medicinal plants against diseases, pests, and environmental stresses (Bahl et al. 2013; Xiao et al. 2016; Dhawan et al. 2019). Due to the rapid evolutionary capacity

Table 17.1 Application of genetic diversity to improvement of medicinal plants through various plant breeding methods

Medicinal plant	Plant breeding approaches used	Name of the released variety	Achievement/ improvement	References
Artemisia annua	Crossing between Jeevan Raksha and CIM-Arogya followed by population enrichment with desirable genes	CIM-Sanjeevani	High artemisinin yielding	Gupta et al. (2016)
Asparagus adscendens	Half sib family selection	CIM Sunahari	Higher root yield	Singh et al. (2016)
Bacopa monnieri	Selected breeding	CIM-Jagriti	High herbage and bacoside—A yielding	Darokar et al. (2007)
Catharanthus roseus	EMS-induced mutation breeding	CIM-Sushil	High vindoline yielding	Gupta et al. (2020)
Catharanthus roseus	Chemical mutation breeding	Dhawal	0.89–1.40% of total alkaloids in leaves, 1.60–2.22% of total alkaloids in roots	Kulkarni et al. (2003)
Ephedra sinica	Selective breeding using stolon propagation	–	Stabilized production of total alkaloids	Hiyama et al. (2017)
Glycyrrhiza spp	Inter-specific hybridization between *Glycyrrhiza uralensis* strain OMP-28 and *Glycyrrhiza glabra* strain OMP-10	Strain C-18	Vigorous growth with high glycyrrhizin and production of two unique phytochemicals	Shibano et al. (2024)
Glycyrrhiza uralensis	Crossbreeding between strain A-19 and G-6 of *Glycyrrhiza uralensis*	Strain C-2	Vigorous growth with high glycyrrhizin and total flavonoid contents	Ozaki and Shibano (2014)
Mentha arvensis	Half-sib progeny selection of the variety-Gomti	CIM-Kranti	High-yielding menthol (79–81%) and cold tolerant	Bahl et al. (2013)
Mentha arvensis	Half-sib progeny selection of the cultivar-Shivalik	"CIM-Vishisht"	High pulegone yielding	Bahl et al. (2016)
Mentha arvensis	Intensive breeding efforts	CIMUnnati	High essential oil yielding	Venkatesha et al. (2020)
Mentha piperita	Mutation breeding of variety Kukrail using different doses of gamma irradiation	CIM-Suras	High menthol (68–78%) yielding	Birendra et al. (2023)
Ocimum	Interspecific hybridization between *O. basilicum* (Ob-1) × *O. kilimandscharicum* (OK)	CIM-SUVAAS	Chavibetol rich and high essential oil yielding	Dhawan et al. (2022)

(continued)

Table 17.1 (continued)

Medicinal plant	Plant breeding approaches used	Name of the released variety	Achievement/ improvement	References
Ocimum	Interspecific hybridization *O. basilicum* (Ob-1) × *O. kilimandscharicum* (OK) f1× O. basilicum Ob-2 and selection process	CIM Shishir	Lodging resistant, cold tolerant, high essential oil yielding, linalool rich inter specific hybrid	Dhawan et al. (2019)
Ocimum africanum	Intensive breeding efforts	CIM Jyoti	High citral-rich essential oil (68–76%) yielding	Lal et al. (2017)
Ocimum basilicum	Intensive breeding efforts	CIM-Sharada	High oil and methyl chavicol (85–89%) yielding	Lal et al. (2015)
Pseudostellaria heterophylla	Mass selection	Shitai No. 1	Disease and lodging resistance, root yield, polysaccharide content	Xiao et al. (2016)
Withania somnifera	Crossing between Israel and Indian chemotypes of *W. somnifera*	–	New phytochemical production	Nittala and Lavie (1981)
Withania somnifera	Intraspecific hybridization between Indian ecotypes 'Nagouri' and 'Kashmiri' of *W. somnifera*	CIM-Pushti	Leaf blight tolerant and high Withanolide-A yielding	Jhang et al. (2018)

of pathogens, breeding strategies have shifted from relying on single, major R-genes (vertical resistance) to combining multiple resistance genes from wild plants, through a process known as gene pyramiding (Chaturvedi et al. 2022). This approach aims to establish broad-spectrum, horizontal resistance that confers greater longevity and stability in commercial varieties of mainly crop plants while reducing the application of chemical control methods that enhance ecosystem health (Dormatey et al. 2020). This steady change, induced by increased genetic variance available for selection, fixes desired traits and develops stable, improved varieties of medicinal plants that can be used as raw material for industrial production, securing a reliable supply chain. The development of medicinal plants with resilience and adaptability to climatic changes, although limited, will allow breeders to grow medicinal plants quickly and produce high-quality ingredients, even in changing weather conditions and diverse geographical regions.

17.4 Potential Role of Genetic Diversity in Transgenic-Mediated Improvement

Unlike conventional breeding, which takes a decade or more because it relies on repeated cycles of pedigree selection and crossing, transgenic approach is much faster. Conventional breeding also often links desirable quantitative trait loci (QTLs) with undesirable traits, a problem known as linkage drag. In contrast, transgenesis precisely targets improvements by directly harnessing genetic diversity at the gene level, instead of chromosome or genome level. This method transforms the slow, labor-intensive, and generation-dependent process of traditional plant breeding into a rapid, targeted engineering technique. It bypasses fundamental barriers to sexual reproduction within and between species, or even within diverse germplasm, and addresses the challenges of secondary metabolite pathways. To fully realize the advantages of this innovative approach, it is critical to consider how it integrates with the inherent genetic resource base available in plant breeding.

The success and efficiency of transgenic-mediated targeted trait enhancement critically depend on the availability, thorough understanding, and utilization of the entire collection of genetic traits. This includes variations from single-nucleotide polymorphisms (SNPs) to full gene rearrangements (genetic diversity) in medicinal plant species (Do et al. 2019; Du et al. 2022; Wang et al. 2024). This rich genetic resource of medicinal plants has been shaped by millions of years of evolution, involving chemical trial and error, serving as vital molecular raw materials for biotechnological plant improvement. This genetic diversity manifests as variability in chemical compositions, morphotypes, and physiological reactions to environmental stimuli. Incorporating this inherent genetic diversity is essential for unlocking the full potential of transgenic technologies and holds the key to novel drug development, improved yields, and sustainable production. Such integration occurs at several critical stages, further illustrating the importance of a well-understood genetic foundation for the successful application of transgenic approaches.

17.4.1 Identification of Superior Genotypes and Chemotypes

The first step is to carefully characterize the genetic diversity of a medicinal plant species. This can be done using phenotypic screening, chemotyping, or molecular marker analysis (Rai et al. 2023; Sharma et al. 2023; Mahgoub et al. 2025). These methods help researchers understand the genetic reasons behind differences in plant traits and chemical types.

17.4.2 Discovery of Gene and Bioactive Pathways Elucidation

After identifying the superior genotypes or unique chemotypes, the next step focuses on discovering the specific genes responsible for these traits (Liu et al. 2017b; Shilpashree et al. 2022). Genetic and phytochemical differences among plants help in this "gene mining." an integrative approach of multi-omics techniques is very effective in identifying candidate genes or transcription factors by comparing genomes, transcriptomes, proteomes, and metabolites between high- and low-producing medicinal plant species, genotypes, or chemotypes (Renu et al. 2014; Gupta et al. 2015; Kaur et al. 2019). Sequencing the genomes of multiple diverse accessions can identify unique genes, gene duplications, or allelic variations that lead to specific chemotypes (Upadhyay et al. 2015; Liu et al. 2017a). Priyadharshini et al. (2025) highlighted the potential application of pan-genomics along with other omics techniques in unlocking deeper information related to biogenesis and regulation of bioactive compounds in medicinal plants and driving innovation in natural product drug discovery. Isolation and functional characterization of targeted enzymes can reveal novel enzymatic steps or highly efficient variants crucial for pathway engineering (Huang et al. 2012; Zhao et al. 2018; Schlesinger et al. 2019; Li et al. 2023; Pu et al. 2024).

17.4.3 Selection of Donor Plant for Transgene Source and Recipient Plant

Genetic diversity acts as a repository for selection of ideal "donor genes." If a particular medicinal plant species shows a variant that naturally produces a higher concentration of a desired compound due to a more efficient enzyme, the gene encoding that superior enzyme variant can be cloned from that specific donor plant and can be integrated in ow-yielding recipient plant. Similarly, selection of an appropriate recipient plant is equally important because not all genotypes within a species have identical morphogenic responses in vitro tissue culture media and genetic transformation efficiency. Several studies have clearly demonstrated that regeneration capacity and transformation efficiency significantly differ with the medicinal plant genetic diversity. Therefore, identification of highly regenerative and responsive genotypes to transformation protocols is very crucial for the transfer of inherently superior genetic elements into recipient plants. Once a transgenic line is established, it can then be used in traditional breeding programs, if applicable, to cross into other desirable genetic backgrounds, further leveraging existing diversity.

17.4.4 Targeted Gene Transfer for Secondary Metabolite Enhancement

With identified genes and elucidated pathways, transgenic or metabolic approaches can be applied for the improvement of secondary metabolites by transfer or modification of targeted gene(s). Apart from these two approaches, genome editing is also evolving as a tool to achieve the same goal, utilizing the genetic diversity of medicinal plants. Some of the selected transgenic approaches and their effects are illustrated in Table 17.2.

Table 17.2 Application of genetic diversity to medicinal plant improvement through various biotechnological approaches, including transgenesis, metabolic engineering and genome editing

Plant	System	Strategy applied	Effects	References
Artemisia annua	T2 transgenic plants	Co-overexpression of six pathway genes of biosynthetic enzymes (IDI, FPS, ADS, CYP71AV1, AACPR, and DBR2)	Enhance (232%) artemisinin production	Qamar et al. (2024)
	Transgenic plant	RNAi mediated gene silencing of C4H gene of *A. annua*	Increase in salicylic acid and artemisinin production, and reduction of p-coumaric acid, total phenolics, anthocyanin production	Kumar et al. (2016)
Atropa belladonna	Transgenic root cultures	Co-overexpression of PYKS and CYM82M3 genes	Increase tropinone and tropine yield	Zeng et al. (2022)
	Transgenic plant	CRISPR/Cas-based targeted mutagenesis of H6H of *A. belladonna*	High-yield hyoscyamine and without its derivatives (anisodamine and scopolamine)	Zeng et al. (2021)
	T1 transgenic plant	Overexpression of PMT from *Nicotiana tabacum* and h6h genes from *Hyoscyamus niger*	High scopolamine content	Xia et al. (2016)
	Transgenic plant	Co-overexpression ODC from *A. belladonna* and H6H from *Hyoscyamus niger*	High scopolamine yield and high tolerance of cold stress	Zhao et al. (2024)
Catharanthus roseus	Transgenic plant	Overexpression of two upstream TIA pathway genes (TDC and STR) derived from *C. roseus*	Increase (two-fold) total alkaloid production with maximum ninefold increase in vindoline and catharanthine, and fivefold increase in vinblastine production	Sharma et al. (2018)
Dioscorea zingiberensis	Transgenic plant	CRISPR/Cas-based gene silencing of farnesyl pyrophosphate synthase gene of *D. zingiberensis*	Decrease squalene content in isolated mutants compared to in wild-type plants	Feng et al. (2018)
Salvia miltiorrhiza	Transgenic hairy root culture	CRISPR/Cas-based gene silencing of bZIP2 gene	Improve phenolic acid biosynthesis	Shi et al. (2021)
Scopolia lurida	Transgenic hairy root cultures	Overexpression of SlH6H or HnH6H genes	Increase scopolamine production, enhancement of scopolamine biosynthesis higher in HnH6H than SlH6H	Lan et al. (2018)

(continued)

Table 17.2 (continued)

Plant	System	Strategy applied	Effects	References
Withania somnifera	T1 transgenic plants	Overexpression of SQS gene derived from *W. somnifera*	Increase (4–4.5 fold) Withaferin A and Withanolide production in leaf tissue	Patel et al. (2015)

IDI isopentenyl diphosphate isomerase, *FPS* farnesyl pyrophosphate synthase, *ADS* amorpha 4,11-diene synthase, *CYP71AV1* cytochrome P450 monooxygenase, *AACPR* cytochrome P450 oxidoreductase, *DBR2* artemisinic aldehyde D11 reductase, *SQS* squalene synthase, *TDC* tryptophan decarboxylase, *STR* strictosidine synthase, *PYKS* type III polyketide synthase, CYP82M3 tropinone synthase, *H6H* hyoscyamine 6β-hydroxylase, *bZIP2* basic leucine zipper2, *fps* farnesyl pyrophosphate synthase gene, *PMT* putrescine N-methyltransferase, *AbODC* ornithine decarboxylase

17.5 Future Potential of Medicinal Plant Genetic Diversity

The "vast gene pool" (genetic diversity) is the foundational blueprint responsible for an immense spectrum of biochemical capabilities harbored across diverse wild populations, landraces, and even uncatalogued genetic accessions of medicinal plants. This genetic and phytochemical variability observed in medicinal plants is often a direct result of millennia-old divergent selective pressures imposed by their specific microclimates, predator populations, or soil compositions. Apart from environmental influences, the genetic diversity of a medicinal plant species governs the intricate enzymatic pathways, the expression levels of specific genes, and the overall metabolic profile, ultimately resulting in the synthesis of potent secondary metabolites across various chemotypes found within a single species. Each genotype may possess distinct genetic pathways, including unique enzyme variants, regulatory elements, or gene cascades, that lead to the production of unique chemical profiles within the same species. Some genotypes within this vast gene pool may produce a known compound at higher concentrations, while others may synthesize entirely new derivatives by activating dormant biosynthetic pathways. The novel molecules with unique scaffolds and biological activities previously unseen in closely related individuals may have developed in response to specific ecological pressures in their native habitats. The role of genetic diversity in facilitating the discovery of novel secondary metabolites represents a critical frontier for pharmaceutical innovation and therapeutic development.

The future of medicinal plant improvement is exceptionally bright due to the rich genetic diversity available. The synergistic application of traditional breeding methods with modern molecular techniques holds great promise for the rapid development of new and improved medicinal plant varieties. These plants may play an important role in minimizing the gap between production and the growing demands of the pharmaceutical industry. In comparison to crop plants, breeding and transgenic-mediated improvement efforts in medicinal plants are limited to a very small number of species. Broader application of multi-omics technologies will provide deeper insights into genetic diversity, variants of key biosynthetic genes, regulatory networks, and epigenetic mechanisms. Advancing medicinal plant improvement is essential for sustainable healthcare systems and requires a concerted effort comparable to the green revolution to ensure a consistent and ethical global supply of medicinal compounds. Future research should focus on the development of new breeding programs that conserve and utilize the genetic diversity of medicinal plants, on the application of advanced molecular techniques to improve the yield, quality, and disease resistance of medicinal plants for global health benefits, and on ensuring a more reliable supply chain for plant-based medicines. Such improvement reduces pressure on vulnerable wild populations of medicinal plants and contributes directly to global biodiversity conservation.

Moreover, harnessing genetic diversity for plant improvement carries responsibilities. Conservation and characterization of wild genetic pools of medicinal plant genetic resources are essential for their future use and for protection against unexpected diseases or climate change. Narrowing the focus to just a few high-performing genotypes risks creating a genetic bottleneck and increasing plant vulnerability. The effort to use the diversity of medicinal plants must be balanced with a commitment to sustainable sourcing, conservation of wild populations, and equitable benefit sharing with indigenous communities. Conservation of wild genetic pools and their ecosystems from threats like overharvesting and habitat loss remains essential for both as a reservoir for future improvements and as a safeguard against unforeseen environmental shifts, and can be done by ex situ conservation methods (seed banks and botanical gardens) and in situ methods. Sustainable practices in the collection, cultivation, and establishment of genomic repositories, as well as the traditional knowledge associated with their use, are crucial for keeping this valuable resource for both local communities and global innovation that scientists depend upon. Ethical considerations and intellectual property rights related to the use of genetic resources from indigenous communities also require careful attention.

Acknowledgements MH gratefully acknowledges the Principal of Barasat Government College, Kolkata, for the continuous support and encouragement in research activities.

References

Alfieri M, Vaccaro MC, Cappetta E, Ambrosone A, De Tommasi N, Leone A (2018) Coactivation of MEP-biosynthetic genes and accumulation of abietane diterpenes in *Salvia sclarea* by heterologous expression of WRKY and MYC2 transcription factors. Sci Rep 8(1):11009

Astutik S, Pretzsch J, Ndzifon Kimengsi J (2019) Asian medicinal plants' production and utilization potentials: a review. Sustainability 11(19):5483

Bahl JR, Bansal RP, Dhawan OP, Dhawan SS, Upadhyay RK, Singh JP, Lal RK, Shasany AK, Chanotiya CS, Patra DD, Kalra A (2013) Registration of a high yielding cold tolerant menthol mint, variety CIM-Kranti of *Mentha arvensis*. J Med Aromat Plant Sci 35(3–4):184–189

Bahl JR, Verma RS, Dhawan SS, Shasany AK, Chanotiya CS, Yadav AK, Upadhyay RK, Tandon S, Misra A, Akhtar MQ, Singh JP (2016) Registration of a pulegone rich variety CIM-Vishisht of *Mentha arvensis*. J Med Aromat Plant Sci 8:71–77

Bai F, Li S, Yang C, Zhao T, Zhang T, Lan X, Chen M, Liao Z (2019) Overexpression of the AbSAUR1 gene enhanced biomass production and alkaloid yield in *Atropa belladonna*. Ind Crops Prod 140:111705

Balasubramani S, Chen Q, Zhou Z, Moola AK, Duraisamy SM, Prakash P, Gayathiri E, Satish L, Swamy MK (2023) Transgenic medicinal plants for improved plant metabolites production. In: Swamy MK, Kumar A (eds) Phytochemical genomics: plant metabolomics and medicinal plant genomics. Springer Nature, Singapore, pp 403–415

Birendra K, Prasad P, Gupta A, Kishor R, Kumar R, Kumar N, Verma RS, Tandon S, Chanotiya CS, Samad A, Verma RK, Kumar D, Lal RK, Singh V, Kushwaha HK, Aftab N, Jhang T, Yadav A, Singh S, Singh V, Chaudhary N, Venkatesha KT, Dwivedi A, Lakhan R (2023) CIM-Suras: a menthol rich, erect growth habit, sucker producing and high yielding peppermint (*Mentha piperita* L.) variety. J Med Aromat Plant Sci 45(1–2):52–61

Cahyaningsih R, Phillips J, Brehm JM, Gaisberger H, Maxted N (2021) Climate change impact on medicinal plants in Indonesia. Glob Ecol Conserv 30:e01752

Chaturvedi T, Gupta AK, Lal RK, Tiwari G (2022) March of molecular breeding techniques in the genetic enhancement of herbal medicinal plants: present and future prospects. Nucleus 65(3):413–436

Chaudhary N, Chaudary D, Kumar R, Aggarwal K, Sharma H, Roy J, Gupta RC, Sharma V (2025) SSR based genetic diversity and population structure of important Herb *Tribulus terrestris* L. from North India. J Appl Res Med Aromat Plants 25:100675

Cui L, Ni X, Ji Q, Teng X, Yang Y, Wu C, Zekria D, Zhang D, Kai G (2015) Co-overexpression of geraniol-10-hydroxylase and strictosidine synthase improves anti-cancer drug camptothecin accumulation in *Ophiorrhiza pumila*. Sci Rep 5(1):1–9

Darokar MP, Khanuja SP, Gupta AK, Sushil Kumar SK, Gupta MM, Singh AK, Shasany AK, Raju BV, Mishra PK, Laiq-ur-Rehman LU, Kalra A (2007) 'CIM-Jagriti' a variety of Brahmi (*Bacopa monnieri*) with high herbage and bacoside yield potential. J Med Aromat Plant Sci 29:90–93

Das S, Kwon M, Kim JY (2024) Enhancement of specialized metabolites using CRISPR/Cas gene editing technology in medicinal plants. Front Plant Sci 15:1279738

De LC (2017) Breeding of medicinal and aromatic plants-an overview. Int J Bot Res 7(2):25–34

Dhawan SS, Lal RK, Gupta P, Chanotiya CS, Kalra A, Mishra A, Singh SK, Yadav A, Srivastava S, Singh S, Singh VR, Tandon S, Singh R, Verma RK, Singh S, Maurya R (2019) Registration of a new variety of basil- CIM Shishir: a multicut, lodging resistant, cold tolerant, high essential oil yielding, linalool rich inter specific hybrid of *Ocimum*. J Med Aromat Plant Sci 41:86–91

Dhawan SS, Lal RK, Gupta P, Chanotiya CS, Kalra A, Singh VR, Singh S, Mishra A, Yadav A, Srivastava S, Maurya R (2022) CIM-SUVAAS: a new source of chavibetol, multi-cut, highessential oil yielding inter-specific hybrid of *O. basilicum* and *O. kilimandscharicum*. J Med Aromat Plant Sci 44:120–126

Do HD, Jung J, Hyun J, Yoon SJ, Lim C, Park K, Kim JH (2019) The newly developed single nucleotide polymorphism (SNP) markers for a potentially medicinal plant, *Crepidiastrum denticulatum* (Asteraceae), inferred from complete chloroplast genome data. Mol Biol Rep 46(3):3287–3297

Dormatey R, Sun C, Ali K, Coulter JA, Bi Z, Bai J (2020) Gene pyramiding for sustainable crop improvement against biotic and abiotic stresses. Agronomy 10(9):1255

Du Q, Li J, Wang L, Chen H, Jiang M, Chen Z, Jiang C, Gao H, Wang B, Liu C (2022) Complete chloroplast genomes of two medicinal *Swertia* species: the comparative evolutionary analysis of *Swertia* genus in the Gentianaceae family. Planta 256(4):73

Eliazer Nelson AR, Ravichandran K, Antony U (2019) The impact of the Green Revolution on indigenous crops of India. J Ethnic Foods 6(1):1–0

Feng S, Song W, Fu R, Zhang H, Xu A, Li J (2018) Application of the CRISPR/Cas9 system in *Dioscorea zingiberensis*. Plant Cell Tissue Org Cult 135(1):133–141

Guo M, Chen H, Dong S, Zhang Z, Luo H (2022) CRISPR-Cas gene editing technology and its application prospect in medicinal plants. Chin Med 17(1):33

Gupta P, Goel R, Agarwal AV, Asif MH, Sangwan NS, Sangwan RS, Trivedi PK (2015) Comparative transcriptome analysis of different chemotypes elucidates withanolide biosynthesis pathway from medicinal plant *Withania somnifera*. Sci Rep 5(1):18611

Gupta AK, Gupta MM, Srivastava AB, Bansal RP, Lal RK, Shasany AK, Saikia DH, Dhawan OP, Tandon SU, Mishra RI, Maurya PR (2016) CIM-Sanjeevani: a high artemisinin yielding population of Artemisia (*Artemisia annua*). J Med Aromat Plant Sci 38:78–83

Gupta AK, Shukla AK, Singh PO, Mall M, Yadav SE, Shasany AK, Shanker K, Baskaran K, Sundaresan V, Talha M, Srivastava A (2020) CIM-Sushil: a high vindoline yielding variety of *Catharanthus roseus* (L.) G. Don. J Med Aromat Plant Sci 42(1–2):51–63

Halder M, Majumder A, Ray S, Jha S (2021) Medicinal plant research at crossroads: biotechnological approaches for conservation, production and stability in tissue cultures and regenerated plants. In: Ekiert HM, Ramawat KG, Arora J (eds) Medicinal plants: domestication, biotechnology and regional importance. Springer, Cham, pp 459–544

Halder M, Bose R, Jha S (2023) The potential role of medicinal plants, traditional herbal medicines, and formulations to overcome SARS-CoV-2 induced health crisis. In: Jha S, Halder M (eds) Medicinal plants: biodiversity, biotechnology and conservation. Springer Nature, Singapore, pp 465–524

Hiyama H, Ozawa A, Kumazawa H, Takeda O (2017) Stabilization of ephedrine alkaloid content in Ephedra sinica by selective breeding and stolon propagation. Biol Pharm Bull 40(1):43–48

Huang W, Sun W, Wang Y (2012) Isolation and molecular characterisation of flavonoid 3′-hydroxylase and flavonoid 3′, 5′-hydroxylase genes from a traditional Chinese medicinal plant, *Epimedium sagittatum*. Gene 497(1):125–130

Jadaun JS, Yadav R, Yadav N, Bansal S, Sangwan NS (2023) Influence of genetics on the secondary metabolites of plants. In: Natural secondary metabolites: from nature, through science, to industry. Springer International Publishing, Cham, pp 403–433

Jangpangi D, Patni B, Chandola V, Chandra S (2025) Medicinal plants in a changing climate: understanding the links between environmental stress and secondary metabolite synthesis. Front Plant Sci 16:1587337

Jhang T, Singh P, Ahmed N, Shukla S, Maurya P, Joshi P, Gupta M, Shastry K, Kumar S, Tomar V, Samad A (2018) Registration of 'CIM-Pushti': Withanolide-a rich, leaf blight tolerant high yielding variety of ashwagandha (*Withania somnifera*) with good root textural quality. J Med Aromat Plant Sci 40(1–4):58–66

Jiang W, Fu X, Pan Q, Tang Y, Shen Q, Lv Z, Yan T, Shi P, Li L, Zhang L, Wang G (2016) Overexpression of AaWRKY1 leads to an enhanced content of artemisinin in *Artemisia annua*. Biomed Res Int 2016(1):7314971

Johnson CB, Franz C (2022) Breeding research on aromatic and medicinal plants. 1st ed., The Haword Press, Inc., New York, USA, pp 1–437

Kaur P, Pandey DK, Gupta RC, Dey A (2019) Assessment of genetic diversity among different population of five *Swertia* species by using molecular and phytochemical markers. Ind Crops Prod 138:111569

Kolakar SS, Nadukeri S, Jakkeral SA, Lakshmana D, Hanumanthappa M, Gangaprasad S (2018) Role of mutation breeding in improvement of medicinal and aromatic crops. J Pharmacogn Phytochem 7(3S):425–429

Kulagina N, Guirimand G, Melin C, Lemos-Cruz P, Carqueijeiro I, De Craene JO, Oudin A, Heredia V, Koudounas K, Unlubayir M, Lanoue A (2021) Enhanced bioproduction of anticancer precursor vindoline by yeast cell factories. Microb Biotechnol 14(6):2693–2699

Kulkarni RN, Baskaran K, Chandrashekara RSR, Khanuja SPS, Darokar MP, Shasany AK, Uniyal GC, Gupta MM, Kumar S (2003) 'Dhawal', a high alkaloid producing periwinkle plant. U. S. Patent No. 6,548,746 B1

Kumar R, Vashisth D, Misra A, Akhtar MQ, Jalil SU, Shanker K, Gupta MM, Rout PK, Gupta AK, Shasany AK (2016) RNAi down-regulation of cinnamate-4-hydroxylase increases artemisinin biosynthesis in *Artemisia annua*. Sci Rep 6(1):26458

Kumar N, Sindhu S, Gawaria B, Ansari A, Dhariwal A, Sharma P, Kumar R (2025) GWAS approaches in medicinal plants for enhanced secondary metabolite production. In: Biotechnology, multiple omics, and precision breeding in medicinal plants. CRC Press, Boca Raton, pp 119–134

Lal RK, Shaseny AK, Chandra R, Singh S, Sarkar S, Yadav R, Sattar A, Dhawan OP, Gupta AK, Chanotiya CS, Dhawan SS, Gupta P, Yadav A, Kumar B, Zaim M, Singh VR, Pandey R, Dubey BK (2015) Registration of variety CIM – Sharda: an early maturing, methyl chavicol rich and high essential oil yielding of basil (*Ocimum basilicum*). J Med Aromat Plant Sci 37:61–64

Lal RK, Shasany AK, Singh VR, Gupta AK, Singh S, Sarkar S, Chanotiya CS, Dhawan OP, Kumar B, Dhawan SS, Gupta P, Zaim M (2017) Registration of high citral rich essential oil yielding variety-CIM Jyoti of *Ocimum africanum*. J Med Aromat Plant Sci 39:135–138

Lan X, Zeng J, Liu K, Zhang F, Bai G, Chen M, Liao Z, Huang L (2018) Comparison of two hyoscyamine 6β-hydroxylases in engineering scopolamine biosynthesis in root cultures of *Scopolia lurida*. Biochem Biophys Res Commun 497(1):25–31

Li DS, Shi LL, Guo K, Luo SH, Liu YC, Chen YG, Liu Y, Li SH (2023) A new sesquiterpene synthase catalyzing the formation of (R)-β-bisabolene from medicinal plant *Colquhounia coccinea* var. *mollis* and its anti-adipogenic and antibacterial activities. Phytochemistry 211:113681

Liu X, Liu Y, Huang P, Ma Y, Qing Z, Tang Q, Cao H, Cheng P, Zheng Y, Yuan Z, Zhou Y (2017a) The genome of medicinal plant *Macleaya cordata* provides new insights into benzylisoquinoline alkaloids metabolism. Mol Plant 10(7):975–989

Liu Y, Wang Y, Guo F, Zhan L, Mohr T, Cheng P, Huo N, Gu R, Pei D, Sun J, Tang L (2017b) Deep sequencing and transcriptome analyses to identify genes involved in secoiridoid biosynthesis in the Tibetan medicinal plant *Swertia mussotii*. Sci Rep 7(1):43108

Madani H, Escrich A, Hosseini B, Sanchez-Muñoz R, Khojasteh A, Palazon J (2021) Effect of polyploidy induction on natural metabolite production in medicinal plants. Biomolecules 11(6):899

Marthe F (2018) Tissue culture approaches in relation to medicinal plant improvement. In: Gosal S, Wani S (eds) Biotechnologies of crop improvement. Volume 1, Springer, Cham, pp 487–497 https://doi.org/10.1007/978-3-319-78283-6_15

Mahendran G, Vimolmangkang S (2024) Elicitation boosts withaferin A and withanolide A production in *Withania somnifera* (L.) Dunal hairy root cultures. Plant Cell Tissue Organ Cult 158(2):29

Mahgoub YA, Mahmoud HA, El-Sebakhy NA, Abdallah II (2025) Unravelling *Artemisia* species genetic variation via DNA barcoding, ISSR and RAPD with the development of eco-specific SCAR markers. BMC Plant Biol 25(1):1034

Malakar A, Dutta S, Shukla N, Dasgupta S, Singh RP, Dwivedi S, Maheshwari S, Sharma PD, Pattanaik A, Gorain S, Sinha A (2025) Genetic variability and ecosystem services of Neem (*Azadirachta indica* A. Juss): a two-decadal systematic study for harnessing global sustainable goals. Trees Forests People 14:101014

Mayer M, Unterseer S, Bauer E, de Leon N, Ordas B, Schön CC (2017) Is there an optimum level of diversity in utilization of genetic resources? Theor Appl Genet 130(11):2283–2295

Mishra R, Agarwal P, Mohanty A (2023) Applications of genome editing techniques for the improvement of medicinal plants. In: Swamy MK, Kumar A (eds) Phytochemical genomics: plant metabolomics and medicinal plant genomics. Springer Nature, Singapore, pp 545–569

Monib PN (2024) The role of plants in traditional and modern medicine. J Pharmacogn Phytochem 13(2):643–647

Nasim N, Sandeep IS, Mohanty S (2022) Plant-derived natural products for drug discovery: current approaches and prospects. Nucleus 65(3):399–411

Niazian M (2019) Application of genetics and biotechnology for improving medicinal plants. Planta 249(4):953–73

Nittala SS, Lavie D (1981) Chemistry and genetics of withanolides in *Withania somnifera* hybrids. Phytochemistry 20(12):2741–2748

Ozaki K, Shibano M (2014) Aim for production of *Glycyrrhizae* Radix in Japan (3): development of a new licorice cultivar. J Nat Med 68(2):358–362

Pal S, Rastogi S, Nagegowda DA, Gupta MM, Shasany AK, Chanotiya CS (2019) RNAi of sterol methyl transferase1 reveals its direct role in diverting intermediates towards withanolide/phytosterol biosynthesis in *Withania somnifera*. Plant Cell Physiol 60(3):672–686

Pandita D, Pandita A, Wani SH, Abdelmohsen SA, Alyousef HA, Abdelbacki AM, Al-Yafrasi MA, Al-Mana FA, Elansary HO (2021) Crosstalk of multi-omics platforms with plants of therapeutic importance. Cells 10(6):1296

Patel N, Patel P, Kendurkar SV, Thulasiram HV, Khan BM (2015) Overexpression of squalene synthase in *Withania somnifera* leads to enhanced withanolide biosynthesis. Plant Cell Tissue Organ Cult 122(2):409–420

Pawar AT, Tagalpallewar AA, Mishra M, Swami AG, Baheti AM (2025) Medicinal plant biotechnology. In: Odoh UE, Gurav SS, Chukwuma MO (eds) Pharmacognosy and phytochemistry: principles, techniques, and clinical applications, vol 13. Wiley, Hoboken, pp 347–370

Pellegrini E, Campanella A, Cotrozzi L, Tonelli M, Nali C, Lorenzini G (2018) Ozone primes changes in phytochemical parameters in the medicinal herb *Hypericum perforatum* (St. John's wort). Ind Crops Prod 126:119–128

Priyadharshini S, Samuel P, Gnanaraj M, Sundar MB, Rajadurai M, Rajasudhakar D, Jebastin T (2025) Pangenomes and pangenomics in medicinal plants. In: Biotechnology, multiple omics, and precision breeding in medicinal plants. CRC Press, Boca Raton, pp 70–78

Pu X, Zhang J, He J, Ai Z, He X, Zhou X, Tong S, Dai X, Wu Q, Hu J, He J (2024) Discovery of a novel flavonol O-methyltransferase possessing sequential 4′-and 7-O-methyltransferase activity from *Camptotheca acuminata* Decne. Int J Biol Macromol 266:131381

Qamar F, Mishra A, Ashrafi K, Saifi M, Dash PK, Kumar S, Abdin MZ (2024) Increased artemisinin production in *Artemisia annua* L. by co-overexpression of six key biosynthetic enzymes. Int J Biol Macromol 281:136291

Rai AK, Khan S, Kumar A, Dubey BK, Lal RK, Tiwari A, Trivedi PK, Elliott CT, Ch R (2023) Comprehensive metabolomic fingerprinting combined with chemometrics identifies species- and variety-specific variation of medicinal herbs: an *Ocimum* study. Metabolites 13(1):122

Ramawat KG, Arora J (2021) Medicinal plants domestication, cultivation, improvement, and alternative technologies for the production of high value therapeutics: an overview. In: Ekiert HM, Ramawat KG, Arora J (eds) Medicinal plants: domestication, biotechnology and regional importance. Springer, Cham, pp 1–29

Rastogi S, Shah S, Kumar R, Vashisth D, Akhtar MQ, Kumar A, Dwivedi UN, Shasany AK (2019) *Ocimum* metabolomics in response to abiotic stresses: cold, flood, drought and salinity. PLoS One 14(2):e0210903

Renu IK, Haque I, Kumar M, Poddar R, Bandopadhyay R, Rai A, Mukhopadhyay K (2014) Characterization and functional analysis of eugenol O-methyltransferase gene reveal metabolite shifts, chemotype specific differential expression and developmental regulation in *Ocimum tenuiflorum* L. Mol Biol Rep 41(3):1857–1870

Saran PL, Damor HI, Lodaya DH, Suthar MK, Kalariya KA, Roy S (2022) Identification of potential accessions of *Bacopa monnieri* L. for herbage yield and bacosides A content. Ind Crops Prod 176:114348

Schlesinger D, Rikanati RD, Volis S, Faigenboim A, Vendramin V, Cattonaro F, Hooper M, Oren E, Taylor M, Sitrit Y, Inbar M (2019) Alkaloid chemodiversity in *Mandragora* spp. is associated with loss-of-functionality of MoH6H, a hyoscyamine 6β-hydroxylase gene. Plant Sci 283:301–310

Sharma A, Verma P, Mathur A, Mathur AK (2018) Overexpression of tryptophan decarboxylase and strictosidine synthase enhanced terpenoid indole alkaloid pathway activity and antineoplastic vinblastine biosynthesis in *Catharanthus roseus*. Protoplasma 255(5):1281–1294

Sharma RS, Vaidya N, Maloo SR, Kumar A, Sharma S, Ramkrishnan RS, Kumari V (2023) Application of molecular markers in assessment of genetic diversity of medicinal plants. In: Kumar N (ed) Molecular marker techniques: a potential approach of crop improvement. Springer, Singapore, pp 103–116

Shi M, Du Z, Hua Q, Kai G (2021) CRISPR/Cas9-mediated targeted mutagenesis of bZIP2 in *Salvia miltiorrhiza* leads to promoted phenolic acid biosynthesis. Ind Crops Prod 167:113560

Shibano M, Matsuda K, Ozaki K (2024) Chemical quality evaluation and characteristics of interspecific hybrid (*G. uralensis*× *G. glabra*) C-18 strain as a new Licorice cultivar. J Nat Med 78(4):876–886

Shilpashree HB, Sudharshan SJ, Shasany AK, Nagegowda DA (2022) Molecular characterization of three CYP450 genes reveals their role in withanolides formation and defense in *Withania somnifera*, the Indian Ginseng. Sci Rep 12(1):1602

Singh VR, Lal RK, Shanker K, Dhawan OP, Singh S, Dhawan SS, Gupta P, Kumar B, Gaur NP, Singh HP, Singh AK, Chauhan A, Sarkar S, Chandra R (2016) CIM-Sunahari: high-yielding, yellow (pili) satawar variety of (*Asparagus adscendens* Roxb.). J Med Aromat Plant Sci 38:84–88

Song Z, Li X (2021) Recent advances in molecular marker-assisted breeding for quality improvement of traditional Chinese medicine. Curr Pharm Biotechnol 22(6):867–875

Sozoniuk M, Petrova M, Mishev K, Miladinova-Georgieva K, Geneva M (2025) Identification and validation of reference genes with stable expression under elicitor treatments of the medicinal plant *Arnica montana* L. BMC Plant Biol 25(1):546

Tiwari S, Shukla PK, Chauhan D, Khatoon S (2024) Variability study of active metabolites and identification of elite chemotype of *Withania somnifera* L. Dunal collected from Central India. J Planar Chromatogr Mod TLC 37(2):189–198

Turgut K, Baydar H, Telci İ (2023) Cultivation and breeding of medicinal and aromatic plants in Turkey. In: Medicinal and aromatic plants of Turkey. Springer International Publishing, Cham, pp 131–167

Upadhyay AK, Chacko AR, Gandhimathi A, Ghosh P, Harini K, Joseph AP, Joshi AG, Karpe SD, Kaushik S, Kuravadi N, Lingu CS (2015) Genome sequencing of herb Tulsi (*Ocimum tenuiflorum*) unravels key genes behind its strong medicinal properties. BMC Plant Biol 15(1):212

Venkatesha KT, Singh VR, Padalia RC, Upadhyay RK, Chauhan A, Tiwari AK, Singh S, Bhatt G, Divya, Pandey P, Verma RS, Mall M, Shukla AK, Chanotiya CS, Kumar S, Kumar B, Gupta AK, Verma RK, Samad A, Kumar P, Rawat MC (2020) CIM-Unnati: an essential oil rich variety of menthol mint (*Mentha arvensis* L). J Med Aromat Plant Sci 42:73–79

Wang W, Xu J, Fang H, Li Z, Li M (2020) Advances and challenges in medicinal plant breeding. Plant Sci 298:110573

Wang Q, Chen B, Chen X, Mao X, Fu X (2023) Squalene epoxidase (SE) gene related to triterpenoid biosynthesis assists to select elite genotypes in medicinal plant: *Cyclocarya paliurus* (Batal.) Iljinskaja. Plant Physiol Biochem 199:107726

Wang X, Guo L, Ding L, Medina L, Wang R, Li P (2024) Comparative plastome analyses and evolutionary relationships of 25 East Asian species within the medicinal plant genus *Scrophularia* (Scrophulariaceae). Front Plant Sci 15:1439206

Xia K, Liu X, Zhang Q, Qiang W, Guo J, Lan X, Chen M, Liao Z (2016) Promoting scopolamine biosynthesis in transgenic *Atropa belladonna* plants with pmt and h6h overexpression under field conditions. Plant Physiol Biochem 106:46–53

Xiao CH, Jiang WK, Zhou T, Liao MW, Yang CG, Zhang E (2016) Breeding and extension of *Pseudostellaria heterophylla* new variety "Shitai No.1" in Guizhou province. Zhongguo Zhong Yao Za Zhi 41(13):2381–2385

Zakaria MM, Schemmerling B, Ober D (2021) CRISPR/Cas9-mediated genome editing in comfrey (*Symphytum officinale*) hairy roots results in the complete eradication of pyrrolizidine alkaloids. Molecules 26(6):1498

Zaman W, Ayaz A, Park S (2025) Climate change and medicinal plant biodiversity: conservation strategies for sustainable use and genetic resource preservation. Genet Resour Crop Evol 21:1–34

Zeng L, Zhang Q, Jiang C, Zheng Y, Zuo Y, Qin JZ et al (2021) Development of *Atropa belladonna* L. plants with high-yield hyoscyamine and without its derivatives using the CRISPR/Cas9 system. Int J Mol Sci 22:1731

Zeng J, Liu M, Zeng L, Yang C, He P, Lin M, Liao Z, Qiu F (2022) Engineering the production of medicinal tropane alkaloids through enhancement of tropinone and littorine biosynthesis in root cultures of *Atropa belladonna*. Ind Crops Prod 189:115778

Zhang C, Yang D, Liang Z, Liu J, Yan K, Zhu Y, Yang S (2019) Climatic factors control the geospatial distribution of active ingredients in *Salvia miltiorrhiza* Bunge in China. Sci Rep 9(1):904

Zhang M, Gao Q, Liu Y, Fang Z, Gong Z, Zhao ZK, Yang X (2022) Metabolic engineering of *Rhodotorula toruloides* for resveratrol production. Microb Cell Factories 21(1):270

Zhao Q, Cui MY, Levsh O, Yang D, Liu J, Li J, Hill L, Yang L, Hu Y, Weng JK, Chen XY (2018) Two CYP82D enzymes function as flavone hydroxylases in the biosynthesis of root-specific 4′-deoxyflavones in *Scutellaria baicalensis*. Mol Plant 11(1):135–148

Zhao L, Zhu Y, Jia H, Han Y, Zheng X, Wang M, Feng W (2022) From plant to yeast—advances in biosynthesis of artemisinin. Molecules 27(20):6888

Zhao T, Yang M, Htun ZL, Zhou J, Zeng J, Qiu F, Zhang H, Lan X, Chen M, Liao Z (2024) Engineering the production of scopolamine and cold tolerance through overexpressing ODC and H6H in *Atropa belladonna*. Ind Crops Prod 208:117886

MIX
Papier aus verantwortungsvollen Quellen
Paper from responsible sources
FSC® C105338

If you have any concerns about our products,
you can contact us on
ProductSafety@springernature.com

In case Publisher is established outside the EU,
the EU authorized representative is:
Springer Nature Customer Service Center GmbH
Europaplatz 3, 69115 Heidelberg, Germany

Printed by Libri Plureos GmbH
in Hamburg, Germany